parat

Index of Acronyms and Abbreviations in Electrical and Electronic Engineering

© VCH Verlagsgesellschaft mbH, D-6940 Weinheim (Federal Republic of Germany), 1989

Distribution:

VCH Verlagsgesellschaft, P.O. Box 10 11 61, D-6940 Weinheim (Federal Republic of Germany)

Switzerland: VCH Verlags-AG, P.O. Box, CH-4020 Basel (Switzerland)

Great Britain and Ireland: VCH Publishers (UK) Ltd., 8 Wellington Court, Wellington Street, Cambridge CB1 1HW (Great Britain)

USA and Canada: VCH Publishers, Suite 909, 220 East 23rd Street, New York, NY 10010-4606 (USA)

ISBN 3-527-26842-1 (VCH Verlagsgesellschaft)　　　　　　　　　　　　　　　　ISSN 0930-6862
ISBN 0-89573-812-0 (VCH Publishers)

parat

Index of Acronyms and Abbreviations in Electrical and Electronic Engineering

Compiled by
Büro Scientia, Berlin

Büro Scientia
Fasanenstraße 43
D-1000 Berlin 15
Federal Republic of Germany

> This book was carefully produced. Nevertheless, authors, editors and publishers do not warrant the information contained therein to be free of errors. Readers are advised to keep in mind that statements, data, illustrations, procedural details or other items may inadvertently be inaccurate.

Published jointly by
VCH Verlagsgesellschaft, Weinheim (Federal Republic of Germany)
VCH Publishers, New York, NY (USA)

Editoral Director: Dr. H.-D. Junge
Production Manager: Elke Littmann
Text processing and composition: Fa. U. Hellinger, D-6901 Heiligkreuzsteinach
Printing: betz-druck, D-6100 Darmstadt 12
Bookbinding: Verlagsbuchbinderei Kränkl, D-6148 Heppenheim
Cover design: TWI, Herbert J. Weisbrod, D-6943 Birkenau

Library of Congress Card No. applied for

British Library Cataloguing-in-Publication Data:
Index of acronyms and abbreviations in
electrical and electronic engineering.
1. Electrical engineering 2. Electronic
engineering
I. Series
621.3
ISBN 3-527-26842-1

Deutsche Bibliothek Cataloguing-in-Publication Data:
Index of acronyms and abbreviations in electrical and electronic
engineering / comp. by Büro Scientia, Berlin. - Weinheim ;
Basel (Switzerland) ; Cambridge ; New York, NY : VCH, 1989
(Parat)
ISBN 3-527-26842-1 (Weinheim ...) Gb.
ISBN 0-89573-812-0 (New York) Gb.
NE: Büro Scientia <Berlin, West>

© VCH Verlagsgesellschaft mbH, D-6940 Weinheim (Federal Republic of Germany), 1989
All rights reserved (including those of translation into other languages). No part of this book may be reproduced in any form - by photoprinting, microfilm, or any other means - nor transmitted or translated into a machine language without written permission from the publishers. Registered names, trademarks, etc. used in this book, even when not specifically marked as such, are not to be considered unprotected by law.
Printed in the Federal Republic of Germany

Introduction

This file contains approximately 45,000 entries of acronyms and abbreviations used in electrical and electronic engineering and related fields. Many of them are standardized all over the world (e.g. by IEC, IEEE or other standardizing organizations), others originate from public or private organizations or institutions, firms etc. or from scientific or military usage.

The entries are alphabetically listed beginning with the acronym or abbreviation and followed by the corresponding full text. Due to the varying usage of capitals and lower-case letters, this book employs capitals for acronyms and abbreviations and lower-case letters for the full text. It is believed that the usage of unified notation simplifies finding the requested entry.

A

A
  abbreviation
  absolute
  accelerating contactor
  accelerating relay
  acceleration
  accumulator
  acoustic
  acre
  adjustment
  aerial
  aeronautics
  airborne
  aircraft
  airman
  amateur
  ampere
  amperemeter
  amplification
  amplifier
  amplitude
  analog analyzer
  analysis
  angstrom
  angstrom associate
  anode
  answer
  antenna
  area
  arithmetic
  arm
  armature
  asbestos wire
  aspect ratio
  assembly
  atomic association
  atomic weight
  attenuation
  atto
  austenite
  automatic
  auxiliary
  availability
  aviation
  azimuth
AA
  abampere
  absolute altitude
  access authorization
  aerotronic associates
  air to air
  airborne alert
  aluminum association
  always afloat
  american airlines
  anti-aircraft
  approach aid
  approximate absolute
  arithmetic average
  armature accelerator
  atomic age
  audible alarm
  automobile association
AAA
  active acquisition aid
  all american aviation
  american accounting association
  anti-aircraft artillery
AAAI
  american association for artificial intelligence
AAAIS
  advanced army airborne indicating system
  anti aircraft artillery information service
AAARDR
  anti-aircraft artillery radar
AAARDRCRM
  anti-aircraft artillery radar crewman
AAAS
  american association for the advancement of science
AAB
  aircraft accident board
AABCP
  advanced airborne command post
AABM
  association of american battery manufacture
AABP
  aptitude assessment battery program
AAC
  arithmetic and controls
  automatic amplitude control
  automatic approach control
AACB
  aeronautics and astronautics coordination board
AACC
  american association for contamination control
  american association of cereal chemists
  american automatic control council
  application of automatic control committee
  automatic approach control complex
AACE
  american association of coste engineers
AACOM
  army area communciations

AACOMS
army area communications system
AACP
advanced airborne command post
AACR
all america cables and radio
AACS
airborne astrographic camera system
airways and air communciations service
army airways communications service
AACSCEDR
associate and advisory committee to the special committee on electronic data retrieval
AACSR
airways and air communications service regulations
AACSRON
airways and air communications service squadron
AACSWG
airways and air communications service wing
AAD
address adder
assembly and disassembly
AADC
advanced avionics digital computer
all application digital computer
AADIS
automatic air defense information system
AADS
area air defense system
AAE
american association of engineers
armament and electronics
assembly and equipment
assembly and erection
automatic answering equipment
azimuth and elevation
AAEC
australian atomic energy commission
AAEE
aeronautical and aircraft experimental establishment
american academy of environmental engineers
american association of electrical engineers
AAES
american association of engineering societies
AAEW
atlantic airborne early warning
AAF
allied air forces
army air force
austenite and ferrite

AAFB
andrews air force base
auxiliary air force base
AAFC
anti-aircraft fire control
AAFPS
army and air force postal service
AAFSS
advanced aerial fire support system
AAG
aeronautical assignment group
AAGMC
anti-aircraft artillery and guided missile center
AAGMS
anti-aircraft guided missile school
AAGW
air to air guided weapon
AAI
air aid to intercept
air to air identification
american association of immunologists
american astrinics incorporated
american audio institute
AAIEE
associate of the american institute of electrical engineers
AAIM
american association of industrial management
AAL
absolute assembly language
ames aeronautical laboratory
arctic aeromedical laboratory
AALA
american association for laboratory accreditation
AALC
advance airborne launch centers
AAM
air to air missile
amplitude and angel modulation
anti-aircraft missile
assembly and maintenance
auxiliary aiming mark
AAMA
architectural aluminum manufacturers' association
AAMG
anti-aircraft machine gun
AAMI
association for the advancement of medical instrumentation
AAML
arctic aeromedical laboratory

AAMM
  anti anti missile missile
AAMS
  american air mail society
AANDE
  armament and electronics
AANDEMAINTSQ
  armament and electronics maintenance squadron
AAO
  anti-air output
  assembly and checkout
AAOR
  anti-aircraft artillery operations room
AAP
  apollo applications program
AAPG
  american association of petroleum geologists
AAPM
  american association of physicists in medicine
AAPT
  american asocciation of physics teachers
  association of asphalt paving technologists
AAPTC
  assistant airport traffic controller
AAR
  against all risks
  assembly and repair
  association of american railroads
AARL
  army aircraft radio laboratory
AARP
  atomic air raid precautions
AARR
  argonne advanced research reactor
AARS
  army amateur radio system
AAS
  advanced administrative system
  advanced aerial system
  advanced antenna system
  american astronautical society
  american astronomical society
  apollo abort system
  arithmetic assignment statement
  automated accounting system
  azimuth alignment system
AASC
  army area signal center
AASE
  association for applied solar energy
AASHO
  american association of state highway officials
AASR
  airport and airways surveillance radar
  airport and surveillance radar
AASS
  army area signal system
AAT
  anti-aircraft tank
  assembly and test
  automatic answer trunk
AATC
  anti-aircraft training center
  automatic air traffic control
AATCAN
  army air traffic control and navigation system
AATCC
  american association of textile chemists and colorists
AATP
  assembly and test pit
AATRI
  army air traffic regulations and identification
AAUP
  american association of university professors
AAVD
  automatic alternate voice data
AAVS
  aerospace audiovisual service
AAWTC
  assistant airway traffic controller
AB
  able bodied
  accumulator and buffer
  adapter booster
  aeronautical board
  afterburner
  air base
  air blast
  air break
  airborne
  allen bradley
  anchor bolt
  arithmetic bus
  automatic blow down
ABA
  american bankers association
  american bar association
ABACUS
  air battle analysis center utility system
  architecture and building aids computer unit of strathclyde university
  autonetics business and control united systems
ABAMP
  absolute ampere

ABAR
  advanced battery acquisition radar
ABB
  automatic back bias
ABBE
  advisory board on the built environment
ABC
  air blast cooled
  american broadcasting company
  american broadcasting corporation
  approach by concept
  atanasoff berry computer
  australian broadcasting commission
  automatic background control
  automatic bandwidth control
  automatic bass compensation
  automatic bass control
  automatic beam control
  automatic bias control
  automatic binary computer
  automatic branch control
  automatic brightness control
ABCB
  air blast circuit breaker
  australian broadcasting control board
ABCC
  atomic bomb casualty commission
ABCCTC
  advanced base combat communication training center
ABCN
  airodrome beacon
ABCOULOMB
  absolute coulomb
ABCS
  automatic broadcast control system
ABCSI
  american bowling computer service incorporated
ABCU
  association of burroughs computer users
ABD
  alloy bulk diffusion
ABDL
  automatic binary data link
ABER
  aberration
ABES
  aerospace business environment simulator
ABET
  accreditation board for engineering and technology
ABETS
  airborne beacon electronic test set
ABF
  average branching factor
ABFARAD
  absolute farad
ABFS
  auxiliary building filter system
ABG
  aural bearing generator
ABGTS
  auxiliary building gas treatment system
ABHENRY
  absolute henry
ABIS
  associate of the british interplanetary society
ABL
  atlas basic language
ABLE
  activity balanced line evaluation
ABM
  anti ballistic missile
  automated batch mixing
ABMA
  army ballistic missile agency
ABMDA
  advanced ballistic missile defense agency
ABMEWS
  anti-ballistic missile early warning system
ABMHO
  absolute mho
ABN
  aerodrome beacon
  airborne
ABNDIV
  airborne division
ABNED
  abnormal end
ABNRADARARADIOINSTLTM
  airborne radar and radio installation team
ABNRADIOMAINTTM
  airborne radio maintenance team
ABNSIGBN
  airborne signal battalion
ABO
  astable blocking oscillator
ABOHM
  absolute ohm
ABP
  active band pass
  american broadcasting paramount theatres
ABPC
  associated british picture corporation

ABPLM
   asynchronous bipolar pulse length modulation
ABPW
   agency broadcast producers workshop
ABR
   abbreviation
   automatic backup and recovery
   automatic band rate
ABRC
   advisory board for the research councils
ABRES
   advanced ballistic reentry system
ABS
   absolute
   acrylonitrile butadiene styrene
   air break switch
   allied business system
   american broadcasting system
   american bureau of shipping
   anti blocking system
   automatic beam current stabilizing
   auxiliary building sump
ABSAP
   airborne search and attack plotter
ABSE
   absolute error
ABSIE
   american broadcasting station in europe
ABSORP
   absorption
ABSS
   advanced beach signal station
ABSV
   absolute value
ABSVM
   absolute voltmeter
ABSVOLT
   absolute volt
ABSW
   air break switch
ABT
   air blast transformer
ABTF
   airborne task force
ABU
   asian broadcasting union
AC
   abcoulomb
   absolute ceiling
   access cycle
   accounting computer
   accumulator
   acetic acid
   acid
   adaptive control
   add carry
   address carry
   adjacent channel
   advisory committee
   aerial current
   aerodynamic center
   air conditioning
   air cooled
   aircraft
   allocation counter
   alternating current
   analog computer
   antenna current
   anti-clutter
   application control
   application controller
   armored cable
   audio center
   autodecoder
   automatic check
   automatic checkout
   automatic computer
   automatic control
   automation center
   auxiliary console
ACA
   adjacent channel attenuation
   american chain association
   american communications association
   american crystallographic association
   asynchronous communication adapter
   automatic circuit analyzer
   automatic communication association
ACAD
   air containment atmosphere dilution
ACAN
   army command and administrative network
ACAP
   automatic circuit analysis program
ACAR
   aluminium cable aluminium reinforced
   american cable and radio
ACARD
   advisory council on applied research and development
ACARS
   arinc communication addressing reporting system
ACAS
   aircraft collision avoidance system
   automatic component assembly system
ACAU
   automatic calling and answering unit

ACAW
  air communications and weather
ACB
  access method control block
  air circuit breaker
ACBB
  american council for better broadcasts
ACBS
  accrediting commission for business school
ACC
  acceleration
  acceptance
  accessories
  according
  accumulate
  accumulator
  accumulator shift circuit
  acknowledge control
  adaptive control with constraint
  additive card code
  advanced communications course
  aft cargo carrier
  air control center
  area control center
  automatic carrier control
  automatic chroma control
  automatic chrominance control
  automatic color control
  automatic contrast control
ACCA
  asynchronous communications control
  attachment
ACCAP
  autocoder to cobol conversion aid program
ACCCE
  association of consulting chemists and chemical engineers
ACCEL
  accelerator
  automated circuit card etching layout
ACCESS
  air canada cargo enquiry and service system
  aircraft communication control and electronic signaling system
  automatic computer controlled electronic scanning system
ACCM
  advisory council on calibration and measurement
ACCOM
  aircraft communicator
ACCS
  aerospace command and control system
ACCT
  account

ACCUM
  accumulate
ACCW
  alternating current continuous wave
ACD
  addressed cable delivery
  aerial control display
  automatic call distributor
  automatic control distribution
ACDA
  arms control and disarmament agency
ACDIAL
  alternating current dialing
ACDIS
  association for development of computer based instructional systems
ACDS
  accept data state
ACE
  absorption of conversion electrons
  accelerated cathode excitation
  acceptance checkout equipment
  acknowledge enable
  advanced control experiment
  altimeter control equipment
  american council of education
  amplifier controlled euphonic
  animated computer education
  area control error
  association for cooperation in engineering
  association of computer educators
  association of conservation engineers
  association of consulting engineers
  attitude control electronics
  automated computing engine
  automated cost estimating
  automatic calling equipment
  automatic checkout equipment
  automatic circuit exchange
  automatic computer evaluation
  automatic computing engine
  automatic computing equipment
  aviation construction engineers
ACEAA
  advisory committee on electrical appliances and accessories
ACEC
  american consulting engineers' council
  army communications and electronic command
ACEE
  aircraft energy efficiency
ACELSCO
  associated civil engineers and land surveyors of santa clara county

ACEM
  actinium emanation
ACERP
  advanced communications electronics requirements plan
ACES
  automated code evaluation system
  automatic checkout and evaluation system
  automatic control evaluation simulator
ACESA
  arizona council of engineering and scientific associations
  australian commonwealth engineering standards association
ACESC
  acceptance checkout equipment spacecraft
ACESS
  accessory
ACET
  advisory committee for electronics and telecommunications
ACEW
  association of communications workers
ACF
  alternate communications facility
  access control facility
  advanced communication function
  alternate communications facility
  area computing facilities
  autocorrelation function
ACFF
  alternating current flip flop
ACFG
  automatic continuous function generation
ACFM
  actual cubic feet per minute
ACG
  adjacent charging group
ACGIH
  american conference of government industrial hygienists
ACH
  acknowledge hold
ACHP
  advisory council on historic preservation
ACI
  acoustic comfort index
  air combat intelligence
  american concrete institute
  attitude control indicator
  automatic card identification
  automation center international
ACIA
  asynchronous communication interface adapter

ACIC
  aeronautical chart and information center
ACID
  automatic classification and interpretation of data
ACIL
  american council of independent laboratories
ACIS
  aircraft crew interphone system
  army central intelligence service
  association for computing and information sciences
ACK
  acknowledge character
  acknowledged
  acknowledgement
  automatic course keeping
ACKI
  acknowledge input
ACKO
  acknowledge output
ACL
  aoeronautical computer laboratory
  application control language
  association for computational linguistics
  astro communication laboratory
  atlas commercial language
ACLI
  american council of life insurance
ACLMRS
  advisory committee for land mobile radio services
ACLP
  above core load pad
ACLS
  all weather carrier landing system
  american council of learned societies
  automatic carrier landing system
ACM
  active countermeasures
  amplitude comparison monopulse
  asbestos covered metal
  association for computing machinery
  audio center module
ACME
  advanced computer for medical research
  association of consulting management engineers
  attitude control and maneuvering electronics
ACMI
  air combat maneuvering instrumentation
ACMO
  afloat communications management office
ACMR
  air combat maneuvering range

ACMSC
  acm standards committee
ACN
  automatic celestial navigation
ACNA
  analog[ue] computer for net adjustment
ACNI
  automatic called number identification
ACNMR
  alternating current normal mode rejection
ACNOCOMM
  assistant chief of naval operations for communications and cryptology
ACO
  adaptive control with optimization
  air communication officer
  air control officer
ACOE
  automatic checkout equipment
ACOM
  automatic coding machine
  automatic coding system
  awards committee
ACOPP
  abbreviated cobol preprocessor
ACORN
  associative content retrieval network
  automatic checkout and recording equipment
ACOS
  advisory committee on safety
ACOUS
  accoustics
ACP
  action current, potential
  advanced computational processor
  aerospace computer program
  airline control program
  allied communications publication
  auxiliary control panel
  azimuth change pulse
ACPA
  adaptive controlled phased array
  association of computer programmers and analysts
ACPDP
  alternating current plasmadisplay panel
ACPDS
  advisory committee on personal dosimetry services
ACPI
  automatic cable pair identification
ACPM
  aerospace computer program model
  attitude control propulsion motor

ACPR
  advanced core performance reactor
  annular core pulse reactor
ACQ
  acquisition
ACQADJ
  acquisition adjustment
ACQN
  acquisition
ACR
  abandon call and retry
  accumulator register
  administrative communications requirements
  advanced capabilities radar
  aerodrome control radar
  air control room
  air field control radar
  aircraft control room
  airfield control radar
  antenna coupling regulator
  approach control radar
  automatic call recording
  automatic carriage return
  automatic compression regulator
ACRC
  audio center receiver
ACRE
  action committee for rural electrification
  automatic checkout and readiness
  automatic checkout and readiness equipment
ACREL
  alternating current relay
ACRES
  airborne communication relay station
ACRI
  air conditioning and refrigeration institute
ACRL
  association of college and research libraries
ACRM
  aviation chief radioman
ACRS
  advisory committee on reactor standards
ACRT
  aviation chief radio technician
ACS
  accumulator switch
  adaptive control system
  administrative control system
  admiralty computing service
  advanced communications service
  advanced communications system
  advanced computer services
  advanced computer system
  advice called subscriber
  aircraft control and surveillance

aircraft control system
alternating current synchronous
analog computer system
application customizer service
applied computer sciences
army communications service
assembly control system
attitude control system
audio conducted susceptibility
automated communications set
automatic checkout system
automatic coding system
automatic control system
auxiliary cooling system
auxiliary core storage
ACSB
  apollo crew systems branch
ACSC
  association of casualty and surety companies
ACSCE
  army chief on staff for communications electronics
ACSD
  army communication service division
ACSF
  attack carrier striking force
ACSI
  affiliated computer system incorporated
ACSIG
  alignment countdown set inertial guidance
ACSIL
  admiralty center for scientific information and liaison
ACSL
  advanced continuous simulation language
ACSM
  american congress on surveying and mapping
  apollo command and service module
ACSOC
  acoustical society
ACSP
  advisory council on scientific policy
  alternating current spark plug
ACSPNL
  access panel
ACSR
  aluminium cable steel reinforced
  aluminium conductor steel reinforced
ACSS
  air combat and surveillance system
  analog computer subsystem
ACST
  access time
ACSWS
  admiralty civilian shore wireless service

ACT
  acting
  active
  actuating
  advanced communications technology
  advanced computer techniques
  air control team
  air cooled triode
  algebraic compiler and translator
  anti comet tail
  apparatus carrier telephone
  applied computer techniques
  automated continuity tester
  automatic capacitance testing
  automatic code translation
  automatic code translator
  automatic component tester
ACTE
  automatic checkout test equipment
ACTER
  active filter
ACTH
  adrenocorticotropic hormone
ACTM
  audio center transmitter
ACTO
  automatic computing transfer oscillator
ACTOR
  askania cine theodolite optical tracking range
ACTP
  advanced computer techniques project
  advanced computer technology project
  allied electronics publications
ACTR
  actuator
ACTRAC
  accurate tracking
ACTRAN
  autocoder to cobol translating service
ACTRUS
  automatically controlled turbine run up system
ACTS
  acoustic control and telemetry system
  all channel television society
  arc current time simulator
  automatic computer telex services
ACTSA
  advanced computer tape standards association
ACTT
  association of cinematograph and television technicians
ACTU
  automatic control of training unit

ACTVTR
  activator
ACU
  acknowledgement signal unit
  address control unit
  address control unit
  arithmetic and control unit
  automatic calling unit
  automatic control unit
  availability control unit
ACUI
  automatic calling unit interface
ACUTE
  accountants computer users technical exchange
ACV
  access control verification
  air cushion vehicle
  alarm check value
  alarm check valve
ACVTVM
  alternating current vacuum tube voltmeter
ACW
  air control and warning
  airborne collision warning
  aircraft control and warning
  aircraft control warning
  alternating continuous wave
ACWS
  air control and warning stations
  aircraft control and warning system
ACWSS
  aircraft control and warning system station
AD
  adapted
  adapter
  address
  advanced design
  air dried
  air dry
  alloy diffused
  alloy diffusion
  ampere demand meter
  amplifier detector
  analog devices
  analog to demand meter
  analog to digital
  apollo development
  applied dynamics
  armored
  automatic data acquisition
  automatic detection
  average depth
  average deviation
ADA
  action data automation
  address adder
  airborne data automation
  american dental association
  analog digital analog
  automatic data acquisition
ADABAS
  adaptable data base system
ADAC
  analog to digital to analog converter
  applications in design automation committee
  astrodata data acquisition and control
  automated direct analog computer
  automatic data aquisition and comuter complex
  automatic direct analog computer
ADACC
  automatic data acquisition and computer complex
ADACS
  automatic data acquisition and control system
ADALINE
  adaptive linear
ADAM
  advanced data access method
  advanced direct landing apollo mission
  associometrics data management system
  automatic distance and angle measurement
ADAML
  advise by airmail
ADAPS
  automatic display and plotting system
ADAPSO
  association of data processing service organizations
ADAPT
  adaption of automatically programmed tools
  adoption of automatically programmed tools
ADAPTICOM
  adaptive communication
ADAPTS
  air delivered antipollution transfer system
  analogue digital analogue process and test system
ADAR
  advanced design array radar
ADAS
  action data automation small
  automatic data acquisition system
  auxiliary data annotation set
ADAT
  automatic data accumulator and transfer
ADC
  aerodrome control

air data computer
airborne digital computer
ampex disk controller
analog to digital conversion
analogue to digital converter
antenna dish control
applied data communications
audio data communications
automatic data collection
automatic data computing
automatic digital calculator
automatic digital computer
automatic drift control
automatic drive control
ADCAD
airways data collection and distribution
ADCC
air defense command computer
air defense control center
ADCCOMNET
air defense command communications network
ADCCP
advanced data communications control procedure
ADCLAMPLIFIER
automatic direct current level amplifier
ADCOM
association of data center owners and managers
ADCON
address constant
ADCONVERSIONOFVIDEO
analogue to digital conversion of video
ADCP
advanced data communictons control procedure
ADCSP
advanced defense communication satellite project
advanced defense communications satellite program
ADD
address decoding unit
address defined
address definition
aerospace digital development
after diversity demand
analog digital designer
automatic digital depth
ADDAC
analog data distributor and computer
ADDAR
automatic digital data acquisition and recording

ADDAS
automatic digital data assembly system
ADDC
air defense direction center
ADDDS
automatic direct distance dialing system
ADDON
add on memory
ADDR
adder
ADDRESOR
analog to digital data reduction system for oceanographic research
ADDS
advanced data display system
applied digital data system
automatic data digitizing system
ADE
address enable
audible doppler enhancer
automated design engineering
automatic drafting equipment
ADEBUG
assembly language symbolic debug
ADEC
address decoding
ADEE
air defense electronic environment
ADEM
automatically data equalized modem
ADEMS
automated data entry measurement system
ADEPO
automatic dynamic evaluation by programmed organizations
ADEPT
automatic data extractor and plotting table
automatic dynamic evaluation by programmed testing
ADES
automated data entry system
automatic digital encoding system
ADF
air direction finder
airborne direction finder
automatic direction finder
automatic direction finding
ADFAP
automatic direction finder approach
ADG
automatic degaussing
ADGE
air defense ground environment
ADH
a digit hunter

ADHA
  analog data handling assembly
ADI
  acceptable daily intake
  alterate digit inversion
  alternating direction implicit
  altitude direction indicator
  american documentation institute
  attitude direction indicator
  attitude director indicator
ADIE
  automatic dialing and indicating equipment
ADINTELCEN
  advanced intelligence center
ADIOS
  automatic digital input output system
ADIS
  air defense integrated system
  association for the development of instructional systems
  automatic data interchange system
ADIT
  analog digital integration translator
  analog to digital integrating translator
ADIZ
  air defense identification zone
ADJ
  adjacent
  adjust
  adjustable
  adjusting
  adjustment
ADJSPD
  adjustable speed
ADK
  automatic depth keeping
ADL
  artifical delay line
  automatic data link
ADLS
  air dispatch letter service
ADM
  activity data method
  adaptive delta modulation
  add magnitude
  application data management
  automated data management
ADMA
  automatic drafting machine
ADMIR
  automatic diagnostic maintenance information retrieval
ADMS
  asynchronous data multiplexer synchronizer
  automatic digital message switching

ADMSC
  automatic digital message switching center
ADMSG
  advise by message
ADO
  address out
  audio decade oscillator
  avalanche diode oscillator
ADOC
  air defense operation center
ADONIS
  automatic digital on line instrumentation system
ADP
  acid dip
  air defense position
  airborne data processor
  airport development program
  ammonium dihydrogen phosphate
  automatic data processing
  automatic data processor
  automatic destruct program
ADPACS
  automated data processing and communications service
ADPC
  automatic data processing center
ADPCM
  adaptive differential pulse code modulation
  association for data processing and computer management
ADPE
  auxiliary data processing equipment
ADPESO
  automatic data processing equipment selection office
ADPP
  automatic data processing program
ADPS
  active delayed phase shift
  automatic data processing system
ADPSO
  association of data processing service organizations
ADPT
  adapter
ADPU
  automatic data processing unit
ADQ
  almost differential quasi ternay code
ADR
  add register
  addar
  address register
  advisory route

aircraft direction room
applied data research
audit discrepancy report
austin data recorder
automatic distortion reduction
ADRA
automatic dynamic response analyzer
ADRAC
automatic digital recording and control
ADRG
automatic data routing group
ADRMP
automated dialer and recorded message player
ADRSA
assistant data recording system analyst
ADRT
analog data recorder transcriber
ADS
accurately defined system
activated data sheet
activity data sheet
address data strobe
address display subsystem
address display system
adsorption
air defense sector
air development service
anker data system
applied digital data systems
automated data system
automatic depressurization system
automatic door seal
ADSAF
automatic data system for the army in the field
automatic data system within the army in the field
ADSAS
air derived separation assurance system
ADSC
automatic data service center
automatic digital switching center
ADSCOM
advanced shipboard communication
ADSD
addressed
ADSE
addressee
ADSEC
air defense systems engineering committee
ADSG
atomic defense and space group
ADSID
air defense systems integration division

ADSIGS
advanced signals
assitant director of signals
ADSP
automatic dispatching stick repeater
ADSTP
administrative data system teleprocessing
ADSUP
automated data systems uniform practices
ADT
accelerated development test
american district telegraph
application dedicated terminal
automatic data translator
average daily traffic
ADTECH
advanced decoy technology
ADTELP
advise be teletype
ADTRANSISTOR
alloyed diffused transistor
ADU
accumulation and distribution unit
ammonium diuranate
angular dialing unit
angular display unit
automatic dialing unit
automatic dialling unit
ADV
advance
arc drop voltage
ADVG
advancing
ADVON
advanced echelon
ADW
air defense warning
analogue to digital conversion
ADWKP
air defense warning key point
ADX
add index
automatic data exchange
automatic data exchange add index
AE
absolute error
acoustic emission
address effective
aerial
aeroelectronic
aeronautical engineering
air escape
airborne electronics
angle of elevation

apollo engineering
application engineer
architect engineer
arithmetic element
arithmetic expression
attenuation equalizer
automatic electric
automation device
aviation electrician's mate
aviation engineer
aviation engineering
order reception

AEA
agriculture engineers association
aircraft electronics association
american electronics association
american engineering association
association of engineers and associates
atomic energy authority
automotive electric association

AEACII
atomic energy advisory committee on industrial information

AEAI
association of engineers and architects in israel

AEB
aera electricity board
analyses economiques budget
apollo engineering bulletin
auxiliary equipment building

AEBA
automatic emergency broadcast alerting

AEBRD
electrically activated bank release device

AEC
american engineering council
army electronics command
atomic energy commission

AECB
atomic energy control board

AECC
aerospace energy conversion committee

AECL
atomic energy of canada ltd

AECM
atomic energy commission manual

AECNASA
atomic energy commission national aeronautics and space administration

AECOM
army electronics command

AECPR
aec procurement regulations

AECS
american electro chemical society
apollo environmental control system

AECT
association for educational communications and technology

AED
aerospace electrical division
algol extended for design
association of electronic distributors
automated engineering design
average and excess demand

AEDNET
automated engineering design of networks

AEDP
association for educational data processing

AEDS
association for educational data systems
association for electronic data system
atomic energy detection system

AEE
airborne evaluation equipment
association of energy engineers
atomic energy establishment

AEEC
airlines electronic engineering committee
arinc's electronic engineering committee

AEEGS
american electroencephalographic society

AEEL
aeronautical electronic and electrical laboratory

AEENG
aeronautical engineer

AEEP
association of environmental engineering professors

AEETF
army's electronic environmental test facility

AEF
architectural engineering firm
aviation engineer force

AEFC
alkine electrolyte fuel cell

AEFLIC
allied expeditionary force long lines control

AEG
active element group
association of engineering geologists

AEGIS
astro electronic gimballess inertial system

AEI
associated electrical industries
association of electrical industries

average efficiency index
azimuth error indicator
AEIC
association of edison illuminating companies
AEIMS
administrative engineering information management system
AEL
aeronautical engine laboratory
american electronic laboratories
average effectiveness level
AELECTECH
associate in electrical technology
AELECTRTECHN
associate in electronics technology
AELP
allied electrical publications
AEM
association of electronic manufacturers
aviation electrician's mate
AEMED
association of electronic manufacturers, eastern division
AEMS
american engineering model society
AEMT
association of electrical machinery trades
AEMU
absolute electromagnetic unit
AEN
attenuation equivalent nettiness
AENGELECT
associate in engineering electronis
AEO
acoustoelectric oscillators
air electronics officer
air engineer officer
airborne electronics operator
AEOG
air ejection of gas
AEP
american electric power
atomic electrical project
automatic electronic production
AEPEM
association of electronic parts and equipment manufacturers
AEPG
army electronic proving ground
AEPS
american electroplaters' society
AEPSC
atomic energy plant saftey committee
AER
address extension register

aerial
aerodrome
aerodynamics
aeronautics
aeroplane
association for education by radio
AERA
american educational research association
american electric railway association
AERATIO
absorptivity emissivity ratio
AERCOM
aeronautical communications
AERDL
army electronics research and development laboratory
AERE
atomic energy research establishment
AEREA
american electric railway engineering association
AERIS
airborne electronic ranging instrumentation system
airways environment radar information system
AERNO
aeronautical equipment reference number
AERO
association of electronic reserve officers
AEROE
aeronautical engineering
AEROF
aerological officer
AEROSAT
aeronautical satellite system
AES
abrasive engineering society
activ electromagnetic system
active electromagnetic system
aerospace and electronic systems
aerospace electrical society
aerospace electronics systems
aircraft electrical society
airways engineering society
american electrochemical society
american electroencephalographic society
apollo extension system
artificial earth satellite
audio engineering society
auger electron spectroscopy
AESOP
an evolutionary system for on-line processing
artificial earth satellite observation program

AESS
aerospace and electronics systems
AESU
absolute electrostatic unit
AET
actual exposure time
associate in electrical technology
AETM
aviation electronic technician's mate
AETP
allied electronics publication
AETR
advanced engineering test reactor
AEVMOST
angle evaporated vertical channel power mosfet
AEW
airborne early warning
aircraft early warning
AEWA
airborne early warning aircraft
AEWCON
airborne early warning and control
AEWPC
army electronic warfare policy committee
AEWRADAR
airbone early warning radar
AEWTU
airborne early warning training unit
AF
audio frequency
AFP
advanced function printing
AG
air to ground
AH
air over hydraulic
ampere hour
AHSR
airborne height-surveillance radar
AJ
anti-jamming
ALADIN
advanced laser-aided defect inspection in nondestructive testing
ALT
amber light
AM
abrasive machining
ammeter
ampere minute
ampere per meter
amplitude modulated
amplitude modulation
automatic manual

AMME
automated multi media exchange
AMSWITCH
automatic manual switch
AN
alphanumeric
as needed
ANAV
air navigator
ANSI
american national standards institute
AO
answer originate
AOAMPL
and or amplifier
AP
autopilot
APA
all points adressable
APCTL
autopilot control
APFC
assembly process flow chart
APIS
air position indicating station
APL
a programming language
additional programming language
advanced product line
algorithmic procedural language
applied physics laboratory
association of programmed learning
associative programming language
authorized possession limits
automatic phase lock
average picture level
APLE
association of public lighting engineers
APLHGA
average planar heat generation rate
APLL
automatic phase-locked loop
APLM
asynchronous pulse length modulation
APM
air particulate monitor
aluminium powder material
analog panel meter
APMA
automatic phonograph manufacturers' association
APMCU
autopilot monitor and control unit

APMI
  american powder metallurgy institute
  area precipitation measurement indicator
APMON
  autopilot monitor
APMS
  advanced power management system
APNA
  american power net association
APNIC
  automatic programming national information center
APO
  advanced post office
  airforce post office
  apollo project office
  army post office
  australian post office
APOSS
  automatic point of sale system
APOTA
  automatic positioning of telemetering antenna
APP
  air parcel post
  apparatus
  appendix
  applications program system
  applied
  approved
  approximate
  associative parallel processor
  automatic plate processor
  automatic position planning
  auxiliary power plant
APPA
  american public power association
APPC
  automatic power plant checker
APPECS
  adaptive pattern-perceiving electronic computer system
APPI
  advanced planning procurement information
APPL
  applicable
  application
APPOI
  autopilot positioning indicator
APPR
  army package power reactor
APPS
  application support system
APPTU
  all-pakistan post and telegraph union

APQ
  arithmetic processor queue
APR
  advanced production release
  airborne profile recorder
  alternate path retry
  analog parameter record
  automatic passbook recording
  automatic presure relief
  automatic production recording
  automatic production recording system
APRAPS
  active passive reliable acoustic path smar
APRCAS
  all-purpose rocket for collecting atmospheric soundings
APRD
  atmosphere particulate radioactivity detector
APRIL
  automatically programmed remote indication logged
  automatically programmed remote indication logging
APRM
  automatic position reference monitor
  average power range monitor
APRS
  affirmative poll response state
  association of professional recording studios
  automatic position reference system
APS
  accessory power system
  airborne power system
  allegheny power system
  alphanumeric photocomposer system
  american physical society
  american physics society
  american physiological society
  american phytopathological society
  applications program system
  applied peripheral system
  arizona public service
  army postal service
  array processor software
  assembly programming system
  atomic power station
  automated patching system
  automated production control
  automatic patching system
  automatic pilot system
  automatically rotatable phase shifter
  auxiliary power supply
  auxiliary power system
  auxiliary program storage
  auxiliary propulsion system

APSE
  ada programming support environment
APSI
  ampere per square inch
  amperes per square inch
APSR
  axial power shaping rod
APSTECHNIQUE
  arc plasma spray technique
APSTLGRU
  air postal group
APSTRON
  air postal squadron
APT
  actual parameter area
  airport tower
  analog program tape
  armour-piercing tracer
  augmented programming training
  automatic parts testing
  automatic picture taking
  automatic picture transmission
  automatic position telemetering
  automatic position telemetry
  automatic programmed tools
  automatic programming for tools
  automatic programming of tools
  automatically programmed tool language
  automatically programmed tools
  automation planning and technology
APTA
  american public transit association
APTC
  airport traffic controller
APTR
  advanced pressure tube reactor
APTS
  automatic picture transmission subsystem
  automatic picture transmission system
  automatic programming testing system
APTSTMN
  autopilot test monitor
APTSTPG
  autopilot test programmer
APTSYSTEM
  automatic picture transmission system
APTT
  apollo parts task trainer
APTU
  african postal and telecommunications union
APU
  accessory power unit
  airborne power unit
  arab postal union
  army postal unit
  audio playback unit
  automatic program support
  auxiliary power unit
APUHS
  automatic program unit, high-speed
APULS
  automatic program unit, low-speed
APV
  armoured personnel vehicle
APW
  augmented plane wave
APWA
  american public works association
APWL
  automatically processed wire list
APX
  ampex
  automatic programming system extended
AQ
  acquisition message
  air quenched
  any quality
  any quantity
  apollo qualification
AQDM
  air quality display model
AQGV
  azimuth quantized gated video
AQL
  acceptable quality level
  acceptance quality limit
  average quality of the lot
AQS
  automated quotation systems
AQT
  acceptable quality test
AR
  accounts receivable
  accumulative register
  acid resisting
  acoustic reflex
  acquisition radar
  action register
  actual range
  aeronautical radio
  aeronautical research
  air reconnaissance
  all risks
  amateur radio
  amplifier
  analytical reagent
  anti-reflective
  antireflection
  application review
  area

area-funktion
argon
arithmetic register
army receiving-valve
arrester
aspect ratio
assembly and repair
at the rate of
auxiliary routine
aviation radionavigation
avionic requirements
ARA
abbreviated registered address
actual range angle
airborne radar attachment
amateur rocket association
american radio association
auxiliary recovery antenna
ARAA
airodrome radio approach aids
ARAC
aerospace research applications center
ARAD
airborne radar and doppler
ARADCOM
army air defne command
ARAL
automatic record analysis language
ARB
aircraft reactor branch
ARBM
association of radio battery manufacturers
ARBOR
argonne boiling reactor experiment
argonne boiling water reactor
ARBS
angle rate bombing system
ARC
advanced reentry concepts
aeronautical research council
airborne radio communicating
aircraft radio corporation
ames research center
amplitude and rise time compensation
analog recursive computer
analyzer-recorder-controller
arcus
argonne reactor computation
augmentation research center of stanford research institute
automatic range control
automatic ratio control
automatic reception control
automatic relay calculator
automatic relay computer
automatic remote control
autopilot rate control
average response computer
ARCA
airconditioning and refrigeration contractors of america
ARCAN
aeronautical radio of canada
ARCAS
all-purpose rocket for collecting atmospheric soundings
automatic radar chain acquisition system
ARCAT
army radio code aptitude test
ARCCOS
arc cosine
ARCG
american research committee on grounding
ARCH
articulated computing hierarchy
ARCO
automatic reservations and communications
ARCOM
arctic communication satellite
ARCON
automatic rudder control
ARCS
adaptive reliability control system
air resupply and communication service
ARCT
army radio code aptitude test
ARCW
arc welding
ARDA
analog recording dynamic analyzer
atomic research and development authority
ARDC
air research and development command
american research and development corporation
ARDIS
army research and development information system
ARDM
asynchronous time-division multiplexing
ARDS
advanced remote display station
aviation research and development service
ARDT
automatic remote data terminal
ARE
aircraft reactor experiment
apollo reliability engineering
asymptotic relative efficiency
automated responsive environment

AREA
   amateur radio emergency association
   american railway engineering association
   army reactor area
   army reactor experimental area
AREC
   amateur radio emergency corps
ARED
   aperture relay experiment definition
AREE
   admiralty regional electrical engineer
   apollo reliability engineering electronics
AREG
   a register
   accumulator register
ARENTS
   advanced research environmental test satellite
   arpa environmental test satellite
ARES
   advanced research electromagnetic simulation
ARF
   aeoronautical research foundation
   american radio forum
   armon research foundation
ARFA
   allied radio frequency agency
ARFC
   average rectified forward current
   average rectified foward current
ARG
   argument
   atlas reliability group
ARGMA
   army rocket and guided missile agency
ARGO
   advanced research geophysical observatory
ARGONAUT
   argonne nuclear assembly for university training
ARGP
   aeronautical radionavigation glide path
ARGUS
   automatic routine generating and updating system
ARI
   air-conditioning and refrigeration institute
   airborne radio installation
   airborne radio instrument
   aircraft radio instrument
ARIA
   advanced range instrumentation aircraft
   american radio importers association
   apollo range instrumentation
   apollo range instrumentation aircraft
   apollo range instrumentation ships
   apollo range instrumented aircraft
ARIEL
   automated real time investment exchange ltd.
ARINC
   aeronautical radio, incorporated
ARIP
   automatic rocket impact predictor
ARIS
   advanced range instrumentation ship
   atomic reactor in space
   attitude and rate indicating system
ARISTOTLE
   annual review and information symposium on the technology of training, learning, and education
ARITH
   arithmetical
ARL
   acceptable reliability level
   aerospace research laboratory
   aircraft radio laboratory
   amateur radio league
   association of research libraries
   automatic record level
ARM
   aircraft radio mechanic
   amateur radio monitor
   ampex replacement memories
   anti-radiation missile
   apollo requirements manual
   application reference manual
   area radiation monitor
   armature
   asynchronous response mode
   atmosphere radiation monitor
   automated route management
   avialibility, reliability, maintainability
   aviation radioman
ARMA
   american records management association
   autoregressive moving average
ARMACCEL
   armature acceleration
ARMC
   automatic repeat request mode counter
ARMEL
   armament and electronics
ARMF
   advanced reactivity measurement facility
ARMMS
   automated reliability and maintainability measurement system
ARMS
   aerial radiological measurements and surveys

amateur radio mobile society
automatic radio meteorological measurements and survey
ARMSH
armature shunt
ARNA
association of radio-television news analysts
ARO
after receipt of order
air radio officer
airborne range only
all rods out
applied research objective
army research office
automatic range only
AROD
airborne range and orbit detemination
ARODS
airborne radar orbital determination system
AROM
alterable read-only memory
associative read-only memory
AROU
aviation repair and overhaul unit
ARP
accept response
active recording program
active rest point
advanced reentry program
advanced research projects agency
aeronautical recommmended practice
airborne radar platform
automatic receiver program
automatic recovery program
automatic relative plotter
avalanche resonance-pumped mode
memory resident program
ARPA
advanced research projects agency
ARPAT
arpa terminal
ARPC
australian radio propagation committee
ARPES
angle-resolved photoelectron spectroscopy
ARPI
absolute rod position indication
analog rod position indicator
ARPSC
amateur radio public service corps
ARQ
accept request
automatic repeat request
automatic repetition request
automatic request

automatic request for repeat
automatic request for repetition
automatic retransmission request
ARR
aeronautical radionavigation radar
airborne radio receiver
aircraft radio regulations
antenna rotation rate
anti-repeat relay
arrester
ARRC
air reserve records center
ARRE
average relative representation error
ARRL
aeronautical radio and radar laboratory
american radio relay league
ARRS
aerospace rescue and recovery service
ARS
access relay set
active repeater satellite
advanced reconnaissance satellite
advanced record system
advanced reentry system
agricultural research service
aircraft radiosight
american radium society
american recorder society
american rocket society
amplified response spectrum
analog control system
army radio school
army radio station
attitude reference system
ARSB
anchored radio sound buoy
ARSP
aeorospace research support program
ARSR
air route surveillance radar
ARSTN
airways radio station
ART
additional reference transmission
advanced reactor technology
advanced reasoning tool
advanced research and technology
airborne radiation thermometer
aircraft reactor test
antenna-receiver-transmitter
arcresistance tester
artificial
augmented reentry test
automatic range tracker

automatic range tracking
automatic reporting telephone
aviation radio technician
ARTA
  american radio telegraphists' association
ARTC
  air route traffic control
ARTCC
  air route traffic control center
ARTE
  admiralty reactor test establishment
ARTIC
  associometrics remote terminal inquiry control system
ARTNA
  association of radio-television news analysts
ARTOB
  amateur radio translation on balloon
ARTRAC
  advanced range testing, reporting and control
  advanced real-time range control
ARTRON
  artificial neutron
ARTS
  advanced radar terminal system
  advanced radar traffic control system
  advanced radar traffic system
  amateur radio-teletype society
  associated radio and television servicemen
  automated radar terminal system
ARTU
  automatic range tracking unit
ARTVS
  australian rocket test vehicle simulator
ARU
  acoustic resistance unit
  audio response unit
  automatic range unit
ARV
  alternate record voice
ARWIRE
  address read wire
ARX
  automatic retransmission exchange
AS
  add-subtract
  address syllable
  aeronautical standard
  air speed
  airseasoned
  airspace
  airspeed
  alloter switch
  alloy steel
  ammeter switch
  ampere-second
  angle of sight
  antenna assembly
  antisubmarine
  anti-submarine
  armature shunt
  arsenic
  ascent stage
  assembler
  asymmetric
  atlantic semiconductor
  australian standard
  autotech symap
  auxiliary storage
  axis select
  connection control unit
  output signal
ASA
  abort sensor assembly
  accelerated storage adapter
  acoustical society of america
  american society of agronomy
  american society of appraisers
  american standards association
  american statistical association
  anti-static agents
  army security agency
  atomic security agency
ASAC
  asian standards advisory committee
  automatic selection of any channel
ASACS
  airborne surveillance and control system
ASAE
  american society of aeronautical engineers
  american society of agricultural engineers
  american society of association executives
ASALM
  advanced strategic air launched missile
ASAP
  a scientific application programmer
  analog system assembly pack
  anti-submarine attack plotter
  antisubmarine attack plotter
  army scientific advisory panel
  as soon as possible
ASARC
  army systems acquisition review council
ASAT
  antisatellite
ASB
  air safety board
  aircraft safety beacon
  asymmetric sideband

ASBC
  advanced standard buried collector
  american society of biological chemists
  american standard building code
ASBD
  advanced sea-based deterrent
ASBE
  american society of body engineers
ASBL
  assemble
ASBM
  air-to-surface ballistic missile
ASBO
  association of school business officials
ASBPA
  american shore and beach preservation association
ASBU
  arab states broadcasting union
ASC
  adaptive signal correcting
  adaptive signal correction
  advanced scientific computer
  aeronautical systems center
  american semiconductor corporation
  american society for cybernetics
  ampere per square centimeter
  analog signal converter
  analog-to-stochastic converter
  army signal corps
  associative structure computer
  atlantic systems conference
  automated system charter
  automatic selectivity control
  automatic sensitivity control
  automatic system control
  automatic system controller
  auxiliary switch closed
  auxiliary switch normally closed
ASCA
  automatic science citation alerting
ASCAP
  american society of composers, authors and publishers
ASCAT
  analog self-checking automatic tester
ASCATS
  apollo simulation checkout and training system
ASCC
  air standardization coordinating committee
  automatic sequence control calculator
  automatic sequence controlled calculator
  automatic sequence controlled computer
  automatic sequence controller calculator
  automatic sequency-controlled calculator
  automatic sequency-controlled computer
ASCE
  american society of civil engineers
ASCENT
  assembly system for central processor
ASCET
  american society of certified engineering technicians
ASCII
  american national standard code for information interchange
  american standard code for information interchange
ASCO
  abort sensing control unit
  automatic sustainer cut-off
ASCP
  automatic system checkout program
ASCS
  agricultural stablization and conservation service
  area surveillance control system
  automatic stabilization and control system
ASCTA
  association of short circuit testing authorities
ASCU
  atomatic scanning control unit
ASD
  aeronautical systems division
  assign symbolic device
  automatic synchronized discriminator
ASDA
  atomic and space development authority
ASDD
  apollo signal definition document
ASDE
  airport surface detection equipment
  antenna slave data equipment
ASDEC
  applied system development evaluation center
ASDIC
  allied submarine detection investigation committee
  anti-submarine depth indicating control
  anti-submarine detection and identification committee
  anti-submarine detection investigation committee
ASDIRS
  army study documentation and information retrieval system
ASDJ
  american society of disk jockeys

ASDL
  automated ship data library
ASDNG
  ascending
ASDR
  airport surface detection radar
  american society of dental radiographers
ASDSRS
  automatic spectrum display and signal recognition system
ASE
  admiralty signal establishment
  aerospace support equipment
  air standard efficiency
  airborne search equipment
  airport surface-detection equipment
  albany society of engineers
  american society of engineers
  american stock exchange
  amplified spontaneous emission
  association for science education
  automatic stabilization equipment
  automatic support equipment
ASEC
  american standard elevator code
ASEE
  american society for engineering education
  american society for environmental education
  associate in science in electronics engineering
  association of supervising electrical engineers
ASEL
  aircraft single engine land
ASEM
  american society for engineering management
ASEP
  array structure experiment package
ASER
  amplification by stimulated emission of radiation
ASESA
  armed services electro-standards agency
ASET
  american society for engineering technology
ASETC
  armed services electron tube committee
ASF
  ampere per square foot
  automatic signal filtration
  auxiliary supporting feature
ASFDO
  anti-submarine fixed defenses officer
ASFE
  association of soil and foundation engineers

ASFIP
  accelerometer scale factor input panel
ASFIR
  active swept frequency interferometer radar
ASFTS
  airborne systems functional test stand
ASFX
  assembly fixture
ASG
  aeronautical standards group
ASGB
  aeronautical society of great britain
ASGE
  american society of gas engineers
ASGLS
  advanced space ground link system
ASGN
  assignment
ASGPD
  attitude set and gimbal position display
ASH
  advanced scout helicopter
  armature shunt
ASHE
  american society for hospital engineering
ASHVE
  american society of heating and ventilating engineers
ASI
  advanced scientific instruments
  advanced systems
  air speed indicator
  airsignal international
  american society of inventors
  american standards institute
  application software
  augmented spark igniter
  axial shape index
  azimuth speed indicator
ASIA
  army signal intelligence agency
ASIC
  aerospace instrumentation committee
  air service information circular
ASID
  automatic station identification device
ASIDP
  american society of information and data processing
ASII
  american science information institute
ASIM
  analog simulation

ASIS
  abort sensing and implementation system
  abort sensing and intrumentation system
  advanced scientific instruments
  american society for information science
ASISES
  american section, international solar energy society
ASIST
  advanced scientific instruments symbolic translator
ASIT
  adaptive surface interface terminal
ASK
  amplitude-shift keying
ASKA
  automatic system for kinematic analysis
ASKS
  automatic station keeping system
ASL
  above sea level
  advanced systems laboratory
  arithmetic shift left
  association for symbolic logic
  atomic safety line
ASLAP
  atomic safety and licensing appeal panel
ASLB
  atomic safety licensing board
ASLBM
  air-to-ship launched ballistic missile
ASLBP
  atomic safety and licensing board panel
ASLE
  american society of lubrication engineers
ASLEEP
  automated scanning low-energy electron probe
ASLIB
  association of special libraries and information bureau
ASLO
  american society of limnology and oceanography
ASLT
  advanced solid logic technology
ASM
  adaptive system
  advanced surface-to-air missile
  air-to-surface missile
  airfield surface movement indicator
  allocation strategy module
  american society for metals
  american society for microbiology
  apollo service module
  apollo systems manual
  assembly
  associated semiconductor manufacturers
  association for computing machinery
  association for systems management
  asynchronous state machine
  audiosonometry
ASMB
  acoustical standards management board
ASMBL
  assemble
ASME
  american society of mechanical engineers
ASMI
  airfield surface movement indicator
ASMM
  american society of machine manufacturers
ASMPE
  american society of motion picture engineers
ASMPTE
  american society of motion picture and television engineers
ASMS
  advanced surface missile system
ASMT
  air-space multiple-twin
ASN
  average sample number
ASNDT
  american society for nondestructive testing
ASNE
  american society of naval engineers
ASO
  advanced solar observatory
  aeronautical supply office
  atomic spin orbital
  auxiliary switch normally open
ASODDS
  asweps submarine oceanographic digital data system
ASOP
  automatic scheduling and operating program
ASP
  advanced support processor
  all south pole
  american society of photogrammetry
  apollo spacecraft project
  association storage processor
  association storing processor
  associative-data structure package
  atmospheric sounding projectile
  attached support processor
  automatic schedule procedure
  automatic schedule processor
  automatic scheduling procedure

automatic service panel
automatic servo plotter
automatic switching panel
automatic synthesis program
ASPC
   air space paper core
   analysis of spare parts change
ASPDE
   automatic shaft position data encoder
ASPE
   american society of plastic engineers
   american society of plumbing engineers
ASPEP
   association of scientists and professional engineering personnel
ASPER
   assembly system for the peripheral-processor
   assembly system peripheral processors
ASPI
   apollo information
ASPIL
   american standard practice for industrial lighting
ASPJ
   advanced self-protective jammer
ASPM
   automated system for production management
ASPO
   american society of planning officials
   apollo spacecraft project office
ASPP
   alloy-steel protective plating
   american society of plant physiologists
ASPR
   armed services procurement regulations
ASQC
   american society of quality control
ASR
   accumulator shift right
   address start register
   air search radar
   air surveillance radar
   airborne search radar
   airborne surveillance radar
   airport surveillance radar
   arithmetic shift right
   automatic send and receive
   automatic send and receive teletype terminal
   automatic send receive
   automatic send receive set
   automatic sending and receiving
   automatic speech recognition
   automatic step regulator
   automatic strength regulation
   available supply rate
ASRA
   american society of refrigeration engineers
   automatic stereophonic recording amplifier
ASRAL
   analog schematic translator for algebraic language
ASRDL
   army signal research and development laboratory
ASRE
   admiralty signal and radar establishment
ASROC
   anti-submarine rocket
   anti-submarine rocket computer
ASRS
   automatic seat reservations system
ASRT
   air support radar teamsupplemental procedural
ASRU
   automatic signal recognition unit
ASS
   advanced space station
   aerospace support systems
   analogue switching subsystem assembler
   army signal school
   army signal squadron
ASSA
   assembly area
ASSB
   asynchronous single sideband
ASSC
   aerospace support systems committee
   air service signal corps
ASSE
   american society of safety engineers
   american society of sanitary engineers
ASSET
   aerothermodynamic elastic structural systems environmental test
   air surveillance subsystem evaluation and training
ASSG
   assignment
ASSOTW
   airfield and seaplane stations of the world
ASSP
   acoustics, speech, and signal processing
   assembly position
ASSRS
   adaptive step-size random search
ASSS
   aerospace systems safety society

ASSU
air support signal unit
AST
activa segment table
advanced simulation technology
alaskan standard time
anti-sidetone
apollo systems test
atlantic standard time
ASTA
american statistical association
association of short-circuit testing authorities
automatic system trouble analysis
ASTC
allied signal training center
army satellite tracking center
ASTD
american society for training and development
ASTE
american society of tool engineers
ASTEC
advanced solar turbo-electric concept
advanced solar turbo-electric conversion
advanced systems technology
anti-submarine technical evaluation center
ASTIA
armed services technical information agency
ASTM
american society for testing and materials
american society for testing materials
american standard test method
ASTME
american society of tool and manufacturing engineers
ASTOR
antiship torpedo
ASTP
army specialized training program
ASTR
asynchronous synchronous transmitter receiver
ASTRA
advanced structural analyser
application of space techniques relating to aviation
automatic scheduling with time integrated resource allocation
automatic sorting, testing, recording analysis
automatic strobe tracking
ASTRAC
arizona statistical repetitive analog compute
ASTRAL
analog schematic translator to algebra language

analog schematic translator to algebraic language
ASTRO
advanced spacecraft trainer resuable orbiter
advanced spacecraft transport resuable orbiter
advanced spacecraft truck resuable orbiter
air space travel research organization
artificial satellite time and radio orbit
ASTROPHYS
astrophysics
ASTS
advanced sage tracking study
ASTU
automatic systems test unit
ASU
altitude sensing unit
auxiliary storage unit
ASV
accerleration-switching valve
air-to-surface vessel
air-to-surface vessel radar
airborne radar for detecting surface vessels
ampere-seconds per volt
angle stop value
angle stop valve
automatic self verification
automatic self-verification
ASVIP
american standard vocabulary for information processing
ASVOL
assign volume
ASVP
application system verification and transfer program
ASW
acoustic surface wave
anti-aircraft switch
anti-satellite weapon
anti-submarine warfare
antisubmarine weapons
application software
auxiliary switch
ASWCR
airborne surveillance warning and control radar
ASWEPS
anti-submarine warfare environmental prediction service
ASWG
american steel and wire gage
ASWO
air stations weekly orders
ASWS
aerospace surveillance and warning system

ASWSIGNAL
  all-seems-well signal
ASWSPO
  anti-submarine warfare systems project office
ASYM
  asymmetrical
AT
  absolute term
  acceptance tag
  action time
  adjacent tone
  advanced technology
  air temperature
  air transmit
  air transportable
  ambient temperature
  ampere turn
  analysis time
  army transmitting-valve
  astronomical time
  atlantic time
  atmosphere
  atom
  atomic time
  attenuator
  automatic telephone
  automatic ticketing
  automatic tracking
  automatic transmitter
  automatic typewriter
  autothrottle
ATA
  absolute atmosphere
  actual time of arrival
  air transport association
  air turbine alternator
  albanian telegraph agency
  associated television authority
  association of technical artists
  asynchronous terminal adapter
ATABE
  automatic target assignment and battery evaluation
ATACC
  advanced tactical airborne command and control
ATACCS
  advanced tactical airborne command and control system
ATAE
  associated telephone answering exchanges
  association of telephone answering exchanges
ATAF
  allied tactical air force
ATAMS
  advanced tactical attacks manned system
ATARS
  aircraft traffic advisory resolution system
ATAS
  three-axis stabilization
  academy of television arts and sciences
  association of telephone answering services
  automatic terrain avoidance system
ATB
  all trunk busy
  apollo test box
ATBCB
  architectural and transportation barriers compliance board
ATBE
  absolute time base error
ATBM
  advanced tactical ballistic missile
  anti-tactical ballistic missile
  average time between maintenance
ATC
  acoustic tile ceiling
  adiabatic toroidal compressor
  aerial tuning capacitor
  air temperature control
  air traffic conference
  air traffic control
  air transport committee
  airborne test conductor
  aircraft technical committee
  alert transmit console
  antenna tuning capacitor
  applied technology council
  assitant test conductor
  automatic tap-chaning
  automatic telephone call
  automatic temperature compensation
  automatic threshold control
  automatic time control
  automatic timing control
  automatic tint control
  automatic tone correction
  automatic tool changer
  automatic tool-changing device
  automatic train control
  automatic tuning control
  automation training center
ATCA
  air traffic confernce of america
  air traffic control association
ATCAC
  air traffic control advisory committee
ATCBI
  air traffic control beacon interrogator

ATCC
  air traffic control center
ATCCC
  air traffic control coordination center
ATCCE
  air traffic control chief engineer
ATCEU
  air traffic control evaluation unit
ATCF
  air traffic control facility
ATCO
  air traffic control office
  air traffic control officer
ATCOG
  air traffic communications operators' guild
ATCOM
  air traffic compiler
ATCRB
  air traffic control radar beacon
ATCRBS
  air traffic control radar beacon system
ATCS
  air traffic communications station
  air traffic communications service
ATCSS
  air traffic control signaling system
ATCW
  air traffic control and warning
ATD
  automatic tape dispenser
ATDA
  augmented target docking adapter
ATDD
  average total diametrical displacement
ATDESA
  automatic three-dimensional electronics scanning array
ATDM
  asynchronous time division multiplexing
ATDMA
  advanced time-division multiple access
ATDS
  airborne tactical data system
  automatic telemetry decommutation system
ATE
  altitude transmission equipment
  altitude transmitting equipment
  atlantic city electric
  automatic telephone and electric
  automatic test equipment
  automatic testing equipment
ATEA
  american technical education association
ATEC
  automated technical control

ATENE
  asahi television news
ATES
  air transportable earth station
ATF
  accounting tabulating form
  actuating transfer function
  alcohol, tobacco and firearms bureau
  automatic target finder
  automatic target follower
  automatic terrain following
  automatic transmission fluid
  snap acceptance test facility
ATFP
  alliance of television film producers
ATH
  automatic tape handler
ATHESA
  automatic three-dimensional electronic scanned array
ATHT
  atomic heat
ATI
  aerial tuning inductance
  american television, incorporated
  antenna tuning inductance
  average total inspection
ATIC
  air technical intelligence center
ATIP
  analog tune in progress
ATIS
  air technical intelligence service
  automatic terminal information service
ATJS
  advanced tactical jamming system
ATK
  aviation turbine kerosine
ATL
  analog threshold logic
  artificial transmission line
  automatic telling
  automatic test line
ATLAS
  abbreviated test language for all systems
  abbreviated test language for avionic systems
  anti-tank laser assisted system
  automatic tape load audit system
ATLASICBM
  atlas intercontinental ballistic missile
ATLIS
  automatic tracking laser illumination system
ATM
  address translation memory
  air turbine motor

ampere turn per meter
ampere turns per motor
antenna turning motor
apollo telescope mount
asynchronous time multiplexing
atmosphere
atmospheric
atmospherics
automated teller machine
ATMCHGS
atmospheric changes
ATMOS
adjustable threshold mos
atmospheric
atmospheric
ATMPR
atmospheric pressure
atmospheric pressure
ATMR
advanced technology medium range transport
ATMS
advanced text management system
ATN
attention
ATNO
atomic number
ATO
aeronautical telecommunications operator
antimony tin oxide
apollo test and operations
assisted take off
automatic train operation
ATOD
analog to digital
ATOLL
acceptance, test or launch language
ATOMDEF
atomic defense
ATOMS
automated technical order maintenance sequences
ATOS
assisted take-off system
ATP
acceptance test procedure
airline transport pilot certificate
alternative term plan
anode tapping point
array transform processor
assembly test program
automated test plan
automated testing procedure
ATPC
assisted for telecommunications program and control

ATPG
automated test pattern generation
automatic test pattern generation
ATPSK
adjacent tone-reference phase-shift keying
ATR
advanced test reactor
air traffic regulations
air transport radio
airborne test reactor
aircraft transmitter-receiver
all transistor
answer time recorder
antenna transmit receive
anti-transmit-receive
anti-transmitter-receiver
apollo test requirement
assembly test recording
attenuated total reflectance
ATRAN
automatic terrain recognition and navigation
ATRC
anti-tracking control
antitracking control
ATRCE
advanced test reactor critical experiment
ATRIB
average transfer rate of information bits
ATRT
anti-transmit-receive tube
ATS
accelerometer-timer switch
acquisition and tracking station
acquisition and tracking system
administrative terminal system
advanced technological satellite
advanced teleprocessing system
aided tracking system
air technical service
air traffic services
air-to-ship
american television society
ampere turns
applications technology satellite
astronomical time switch
atlantic tracking ship
automated trading system
automatic telephone set
automatic test system
automatic testing system
ATSC
assistant for telecommunications systems coorporation

ATSE
    assistant for telecommunications systems engineering
ATSIT
    automatic techniques for the selection and identification of targets
ATSR
    argonne thermal source reactor
ATSS
    acquisition and tracking subsystem
ATST
    atlantic standard time
ATT
    american telephone and telegraph
    attenuated
    attenuator
    avalanche transit time
ATTC
    american telephone and telegraph company
ATTCS
    automatic take-off thrust control system
ATTEN
    attenuation
ATTN
    attention
ATTO
    avalanche transit time oscillator
ATTR
    all thrust terminate relay
    average time to repair
ATTUN
    automatic tuning
ATU
    aerial tuning unit
    alliance of telephone unions
    antenna tuning unit
    automatic tracking unit
ATUR
    apollo test unsatisfactory report
ATV
    air-to-vessel
    air-to-vessel radar
    all terrain vehicle
    associated television
    automatic threshold variation
ATW
    aircraft tailwarning
    atomic hydrogen weld
    automatic tape winder
ATWS
    anticipated transient without scram
    automatic track while scanning
ATWT
    atomic weight

AU
    air university
    angstrom unit
    answer unit
    arbitrary unit
    arithmetic unit
    assembler unit
    astronomical unit
    audion
    aurum
    automatic
AUC
    american used computer corp.
AUD
    asynchronous unit delay
    audibility
    audible
    audio
AUDAR
    autodyne detection and ranging
AUDEQUIP
    audio equipment
AUDIT
    automatic unattended detection inspection transmitter
AUDREY
    audio reply
AUDSNL
    audio signal
AUI
    attachment unit interface
AUM
    air-to-underwater missile
AUMN
    air-to-underwater missile - nuclear
AUNT
    automatic universal translator
AURA
    automated reasoning assistant
AUSTP
    australian patent
AUT
    advanced user terminal
    association of university teachers
AUTEC
    atlantic underwater test evaluation center
AUTODIN
    automatic digital network
AUTODOC
    automated documentation
AUTOMAP
    autocoder to map translator
    automatic machining program
    automatical map

AUTOMAST
  automatic mathematical analysis and symbolic translation
AUTOMEX
  automatic message exchange services
AUTONET
  automatic network display
AUTOOVLD
  automatic overload
AUTOPASS
  autopass
AUTOPIC
  automatic personal identification code
  automatic personnel identification code
AUTOPLOT
  automatic plotter
AUTOPOL
  automated programming of lathes
  automatic programming of lathes
AUTOPROMT
  automated programming of machine tools
  automatic programming of machine tools
AUTOPSY
  automatic operating system
AUTORECL
  automatic reclosing
AUTOSATE
  automated systems analysis technique
AUTOSCAN
  automatic stereo broadcast scanner
AUTOSCRIPT
  automated system for composing, revising, illustrating and phototypesetting
AUTOSEVOCOM
  automatic secure voice communication
AUTOSPOT
  automated system for positioning of tools
  automatic system for positioning of tools
  automatic system for positioning tools
AUTOSTATIS
  automatic statewide theft inquiry system
AUTOSTRAD
  automated system for transportation data
AUTOSTRT
  automatic starter
AUTOSTRTG
  automatic starting
AUTOSYN
  automatically synchronous
AUTOTR
  autotransformer
AUTOTRACK
  automatic tracking
AUTOVON
  automatic voice network

AUTPOL
  automatic polarity indication
AUTRAN
  automatic utility translator
AUTRAX
  automatic traffic recording and analysis complex
AUTSWITCH
  automatic switching
AUTTUN
  automatic tuning
AUW
  all-up weight
  anti-underwater warfare
  automatic winder
AUX
  auxiliary
  auxiliary store
AUXS
  auxiliaries
AV
  abvolt
  acid value
  angular velocity
  anion vacancy
  atomic volume
  audio-visual
  average
  aviation
  avionics
AVA
  active van ata array
  azimuth versus amplitude
AVC
  automatic valve control
  automatic vent control
  automatic voltage control
  automatic volume control
AVCS
  advanced vidicon camera system
AVD
  alternate voice and data service
  anode voltage drop
  automatic voice data
AVDS
  automatic vacuum deposition system
AVE
  aerospace vehicle electronics
  automatic volume expander
  automatic volume expansion
AVEC
  amplitude vibration exciter control
AVEFF
  average efficiency

AVERT
    association of voluntary emergency radio teams
AVF
    azimthally varying field
AVHRR
    advanced very high resolution radiometer
AVL
    automatic vehicle location
AVLABS
    aviation laboratories
AVLB
    armored vehicle launched bridge
AVM
    advanced virtual machine
    automatic vehicular monitoring
AVMS
    annulus vacuum maintenance system
AVNET
    aviation network
AVNL
    automatic video noise levelling
    automatic video noise limiter
    automatic video noise limiting
AVO
    ampere, volt, ohm
    avoid verbal orders
AVOID
    accelerated view of input data
    airfield vehicle obstacle indicating device
AVR
    adjustable voltage rectifier
    automatic voice relay
    automatic voltage regulation
    automatic voltage regulator
    automatic volume recognition
    automatic volume regulation
AVS
    adjustable voltage screw-down
    american vacuum society
AVSEP
    audio visual superimposed electrocardiogram presentation
AVSL
    available space list
AVSS
    apollo vehicle systems section
AVT
    all volatile treatment
    applications vertical test program
AVTA
    automatic vocal transaction analysis
AVTR
    airborne videocassette tape recorder

AW
    a-wire
    above water
    acual weight
    air warning
    air-to-water
    aircraft warning
    all water
    all weather
    apparent watt
    arming wire
    atomic weight
    automatic weapons
    automatic word
    auxiliary winding
AWA
    american wire association
    aviation space writers association
AWACS
    advanced warning and control system
    airborne warning and command system
    airborne warning and control system
AWADS
    adverse weather aerial delivery system
    all weather aerial delivery system
AWAR
    area weight average resolution
AWARS
    airborne weather and reconnaissance system
AWAT
    aera weighted average t-number
AWC
    anaconda wire and cable
AWCS
    air weapons control system
AWCT
    association of wireless and telegraphists
AWE
    acme wire company
AWEASVC
    air weather service
AWG
    american wire gage
    american wire gauge
AWGN
    additive white gaussian noise
AWIS
    automatic maintenance information system
AWLS
    all-weather landing system
AWN
    automated weather network
AWO
    army wireless officer

AWOL
  absent without leave
AWR
  adaptive waveform recognition
AWRA
  american water resources association
AWRNCO
  aircraft warning company
AWRS
  airborne weather radar system
AWRTV
  american women in radio and television
AWS
  air warning system
  air weapon system
  air weather service
  aircraft warning service
  aircraft warning set
  american welding society
AWSF
  alpha waste storage facility
AWSG
  armstrong-whitworth and sperry gyroscope
AWSVC
  air warning service
AWT
  atomic weight
AWTC
  airway traffic controller
AWU
  atomic weight unit
AWWA
  american water works association
AWY
  airway
AX
  axis
AXD
  auxiliary drum
AXP
  axial pitch
AXS
  auxiliary store
AXSIGCOMM
  axis of signal communication
AXT
  address to index register, true
AY
  airway
  anyone
AZ
  arizona
AZAR
  adjustable zero, adjustable range

AZAS
  adjustable zero, adjustable span
AZEL
  azimuth and elevation
AZON
  azimuth only
AZRAN
  azimuth and range
AZS
  automatic zero set

B

B
  ballast
  bandwidth
  bar
  barn
  base
  base electrode
  base region
  battery
  baud
  beam
  beam unit
  bel
  best
  bias
  binary digit
  bit
  black
  blank
  block
  blower
  bonded
  booster
  boron
  braid
  branch
  breadth
  break contact
  brightness
  british
  british thermal unit
  broadcast
  byte
BA
  backlash allowance
  barium
  basic assembler
  battery
  bell alarm
  bending allowance

binary add
binary addition
blanking amplifier
blind approach
breathing apparatus
bridging amplifier
british association
buffer amplification
buffer amplifier
bus available

BAA
buffer address array

BAAS
british association for the advancement of science

BABI
broadcast advertising bureau

BABS
beam approach beacon system
blind approach beacon system

BAC
bell aerospace corporation
bendix aviation corporation
binary asymmetric channel
boeing aircraft company
british aircraft constructors
british aircraft corporation
british automation company

BACA
british advisory committee for aeronautics

BACAIC
boeing airplane company algebraic interpretive computing

BACE
basic automatic checkout equipment
british association of consulting engineers

BACS
banks automated clearing services
boeing applied computing service

BACUS
booz, allen and hamilton computer utilization system

BADC
binary asymmetric dependent channel

BADGE
base aera defense ground environment
base air defense ground environment

BADIC
biological analysis detection instrumentation and control

BAE
beacon antenna equipment

BAF
bioaccumulation factor
british air force

BAFCOM
basic armed forces communication

BAGR
bureau of aeronautics general representative

BAGS
bullpup all-weather guidance system

BAI
boeing airborne instrumentation

BAIC
binary asymmetric independent channel
british aviation insuranced company

BAID
boolean array identifier

BAIT
bacterial automated identification technique

BAL
balance
balanced
balancing
basic assembler language

BALL
ballast

BALLLP
ballast lamp

BALLOTS
bibliographic automation of large library operations using time sharing

BALLT
ballast tube

BALMI
ballistic missile

BALNET
balancing network

BALOG
base logistical command

BALOP
balopticon

BALRHEO
balancing rheostat

BALS
balancing set

BALTR
balancing transformer

BALUN
balanced-to-unbalanced
balancing unit

BALVOLT
balanced voltage

BAM
basic access method
block allocating map
block availability map
broadcasting amplitude modulation

BAMBI
ballistic anti-ballistic missile boost intercept

ballistic missile bombardment interceptor
ballistic missile boost intercept
ballistic missile burning intercept
BAMCL
british aircraft manufacturing company limited
BAMIRAC
ballistic missile radiation analysis center
BAMM
balloon altitude mosaic measurement
BAMO
bureau of aeronautics material officer
BAMOS
batch processing multilanguage operating system
BAMS
band archiv management service
broadcast to allied merchant ship
bulletin of the american mathematical society
BAMT
boric acid mixing tank
BANEC
balanced nuclear economy code
BANIR
bantam inertial reference
BAP
band amplitude product
basic assembly program
beacon aircraft position
branch arm piping
BAPE
branch arm piping enclosure
BAPL
base assembly parts list
BAPO
british army post office
BAPS
branch arm piping shielding
BAR
barometer
barrier gate
base address register
base register
battery acquisition radar
broadcast advertisers reports
buffer address register
bureau of aeronautics representative
BARC
british aeronautical research committee
BARITT
barrier injection transit time
BARITTDIODE
barrier injection transit time diode
BAROSWITCH
barometric switch
BARR
bureau of aeronautics resident representative
BARSTUR
barking sands tactical underwater range
BART
bay area rapid transit system
BARTD
bay area rapid transit district
BAS
beacon airborne s-band
block automation system
boolean assignment statement
british association standard
bulgarian astronautical society
BASC
berlin air safety center
BASE
bank americard authorization system
BASEEFA
british approvals service for electrical equipment in flammable atmospheres
BASEX
basic experimental language
basic-extension
BASIC
basic algebraic symbolic interpretive compiler
basic appraisal system for incoming components
basic automatic stored instruction computer
battle area surveillance and integrated communications
beginners' all-purpose symbolic instruction code
BASICPAC
basic processor and computer
battle area surveillance and integrated communications processor and computer
BASIS
bank automated service information system
burroughs advanced statistical inquiry system
burroughs applied statistical inquiry system
BASNET
basic network
BASP
biomedical analog signal processor
BAST
boric acid storage tank
BASW
bell alarm switch
BASYS
basic system
BAT
battery
blind approach training

BATC
  base air traffic control
  british amateur television club
BATCH
  block allocator transfer channel
BATCHG
  battery charger
  battery charging
BATCO
  battery cut-off
BATE
  base activation test equipment
  base assembly and test equipment
BATEX
  batch executive
BATFU
  battery fuse
BATP
  boric acid transfer pump
BATRS
  british amateur tape recording society
BATS
  basic additional teleprocessing system
BATSUP
  battery supply
BATTOPER
  battery operated
BAU
  british association unit
BAUD
  baudot
  baudot code
BAX
  beacon airborne x-band
BAY
  bayonett
BAYS
  british association of young scientists
BB
  back-to-back
  ball bearing
  bandblock
  baseband
  beacon buoy
  best best
  boole and babbage co.
  breadboard
  breaker block
  broadband
  bulk burning
  busy bit
  butane and butylene
BBC
  british broadcasting corporation
  broadband conducted
BBCRD
  british broadcasting corporation research department
BBD
  bucked brigade device
  bucket bridge device
  bucket brigade device
BBDC
  before bottom dead centre
BBES
  bang-bang erection system
BBG
  board of broadcast governors
BBIM
  buoyant ballistic inertial missile
BBL
  basic business language
  beacons and blind landing
  buried-bit-line
BBM
  break-before-make
BBO
  booster burn-out
  boster burn-out
BBR
  broadband radiated
BBRR
  brookhaven beam research reactor
BBUG
  bode and babbage users group
BC
  back-connected
  balanced current
  bare copper
  base connection
  base count
  basic control
  battery charger
  bayonet cap
  bayonet catch
  bell cord
  between centers
  binary code
  binary counter
  body-centred
  bonded single-cotton
  borroughs corporation
  branch conditional
  branch on condition
  breaking capacity
  broadcast
  broadcast control
  broadcasting
  buffer cell
  buffer cycle

bus compatible
business computer
BCA
   buffered communications adapter
BCAC
   british conference on automation and computation
BCAM
   basic communication access method
BCAMPL
   broadcast amplifier
BCB
   broadcast band
BCC
   basis-computer-center
   battery control center
   beacon control console
   binary convolution[al] code
   block character check
   block check character
   block checking character
   block coded decimal
   body-centered cubic
   broadcast control center
   broadcasting corporation of china
BCCD
   bulk ccd
   buried ccd
   buried-channel charge-coupled device
BCCS
   body-centered cubic system
BCD
   barrels per calendar day
   base circle diameter
   between comfort and discomfort
   binary coded decimal
   binary counted decimal
BCDD
   binary coded decimal digit
BCDIC
   binary coded decimal information code
   binary coded decimal interchange code
BCDP
   battery control data processor
BCE
   bachelor of civil engineering
   beam collimation error
   boolean-controlled elements
   british columbia electric
BCEC
   battle creek engineers club
BCEE
   belgian committee for electroheat and electrochemistry

BCFSK
   binary-code frequency-shift keying
BCH
   binary coded hollerith
   bits per circuit per hour
BCHPA
   british columbia hydro and power authority
BCI
   binary coded information
   broadbast interference
BCIP
   belgium centre for information processing
BCJ
   broadcasting corporation of japan
BCL
   base-coupled logik
   basic counter line
   battelle columbus laboratory
   bechtel client letter
   biological computer laboratory
   broadcast listener
   burroughs common language
   business computer ltd
BCM
   balanced crystal mixer
   basic control monitor
   battery control and monitor
BCMOS
   bipolar cmos
BCN
   beacon
BCO
   binal coded octal
   binary coded octal
   bridge cut off
BCOM
   burroughs computer output to microfilm
BCP
   basic control program
   bureau of consumer protection
BCPL
   basic combined programming language
BCR
   battery control radar
   bayes cramer-rao
   best critical region
   boro-carbon resistor
BCRT
   binary coded range time
BCRTS
   binary coded range time signal
BCRUM
   british committee on radiation units and measurement

BCS
  banking communication system
  basic control system
  bcs richland, inc., subsidiary of boeing computer services, inc.
  binary synchronous communications
  biomedical computing society
  block check sequence
  bridge control system
  british calibration service
  british computer society
  business communications sciences
  business control system
BCSTN
  broadcasting station
BCT
  branch on count
  bushing current transformer
BCU
  basic counter unit
  binary counting unit
BCW
  buffer control word
  buffing control word
BD
  band
  barrels daily
  barrels per day
  base detonating
  base diameter
  baud
  baudot-code
  bead
  bernoulli disk
  binary decode
  binary decoder
  binary divide
  binary-to-decimal
  block delete
  block diagram
  boar
  board
  bond
  bone-dry
  boundary
  bulk density
  bus driver
  bus duct
BDA
  booster-distribution amplifier
BDAM
  basic direct-access method
BDC
  binary-decimal converter
  binary-decimal counter
  bonded double-cotton
  bottom dead centre
BDD
  binary-to-decimal decoder
BDF
  base detonating fuse
  bus differential
BDG
  beacon data generation
  bridge
  bridging
BDH
  bearing, distance and heading
  binding head
BDHI
  bearing, distance and heading indicator
BDHT
  blowdown heat transfer program
BDI
  base-diffusion isolation
  bearing deviation indicator
  buffered direct injection
BDIA
  base diameter
BDITECHNIQUE
  base diffusion isolation technique
BDL
  battery data link
  business definition language
BDLS
  battery data link system
BDLT
  bow designation light
BDM
  bomber defence missile
BDMAA
  british direct mail advertising association
BDN
  blue danube network
BDO
  bus for data output
BDOS
  basic disk operating system
BDP
  bonded double paper
  bulk data processing
  business data processing
BDRY
  boundary
BDS
  base design section
  bonded double silk
BDSA
  business and defense services administration

BDSOFI
　base design section - operational facilities installations
BDSSFI
　base design section - support facilities installations
BDST
　british double summer time
BDT
　base deck to binary tape
　binary deck to binary tape
BDTR
　basic data transmission routine
BDU
　basic display unit
BDV
　breakdown voltage
BDWB
　bone-dry-weight basis
BDXAV
　bendix aviation
BE
　back end
　band elimination
　band elimination filter
　base ejection
　base injection
　beryllium
　best estimate model
　binding energy
　bird electronic
　bose-einstein
　breaker end
BEA
　british electricity authority
　british european airways
　bureau of economic analysis
　business education association
BEAB
　british electrical approvals board
BEAC
　boeing electronic analog computer
　boeing engineering analog computer
BEACON
　british european airways computer network
BEAIRA
　british electrical and allied industries research association
BEAM
　burroughs electronic accounting machine
BEAMA
　british electrical and allied manufacturers' association
BEAMOS
　beam accessible mos
　beam addressed mos
　beam-addressed metal-oxide-semiconductor
BEAR
　bearing
　biological effects of atomic radiation
BEASY
　boundary element analysis system
BEC
　beckman instruments
　beginning of the equilibrium cycle
　binary erasure channel
　british electric council
　burst-error channel
　business electronics computer
BECO
　booster engine cut off
BECTO
　british electric cable testing organisation
BED
　bridge-element delay
BEDA
　british electrical development association
BEE
　bombardment enhanced etch rate
　business efficiency exhibition
BEEC
　binary error erasure channel
BEEF
　business and engineering enriched formula translation
　business and engineering enriched fortran
BEER
　binary-element error ratio
BEF
　band-elimination filter
BEIR
　biological effects of ionizing radiation
BEJ
　base-emitter junction
BEL
　bell character
　bell
BELGP
　belgian patent
BELL
　bell telephone laboratories
　bell telephone system
　british electric lamp, limited
BEM
　boundary element method
BEMA
　business equipment manufacturers association
BEME
　brigade electrical and mechanical engineer

BEMF
  back electromotive force
BENSEC
  bentley's second phrase telegraphic code
BEP
  bit error probability
BEPA
  binding energy per atom
BEPC
  british electrical power convention
BEPO
  british experimental pile operation
BEPOC
  burrough's electrographic printer-plotter for ordinance computing
BEPP
  binding energy per particle
BER
  bit error rate
  board of engineers registration
BERCO
  british electric resistance company
BERM
  basic encyclopedic redundancy media
BERT
  basic energy reduction technology
  bit error rate test
BES
  balanced electrolyte solution
  bessel's functions
  biological engineering society
  block error status
BESA
  british engineering standards association
BESRL
  behavioral science research laboratory
BESS
  binary electromagnetic signal signature
BEST
  basic executive scheduler and timekeeper
  better electronic service technicians
  business edp system technique
  business electronic data processing systems technique
  business electronic system technique
  business equipment software technique
  business executive system for time sharing
BET
  balanced-emitter technology
  best estimate of trajectory
BETA
  battlefield exploitation and target acquisition
  british equipment trade association
  business equipment trade association

BETAP
  bell laboratories formula translation assembly program
BETSIE
  bets settlement investment engine
BEV
  billion electron volts
BEVR
  broadcast electronic video recording
BEX
  broadband exchange
  broadcast exchange
BF
  back feed
  backface
  base fuse
  beam forming
  beat frequency
  best fit algorithm
  block factor
  blocking factor
  boldface
  bottom face
  branching filter
  breaker failure
  brought forward
BFA
  broadcasting foundation of america
BFBS
  british forces broadcasting service
  british forces broadcasting station
BFBU
  british forces broadcasting unit
BFCO
  band filter cut off
BFD
  back focal distance
  beat frequency detection
  boolean function designator
BFE
  beam forming electrode
BFG
  binary frequency generator
BFID
  boolean function identifier
BFL
  back focal length
  buffered fet logic
BFN
  british forces network
BFO
  beat frequency oscillator
BFP
  battery fuse panel

BFPDDA
    binary floating point digital differential analyzer
BFPHC
    basic filter power handling capacity
BFPR
    basic fluid power research program
BFR
    buffer
BFS
    basic file system
    beam-foil spectroscopy
    bredth first search
    brute force supply
BFTC
    boeing flight test center
BFTU
    boeing field test unit
BG
    back gear
    background
    bevel gear
    birmingham gauge
    blasting gelatine
    broad gage
    bus grant
BGAM
    basic graphic access method
BGCOB
    background compiler cobol
BGE
    bull general electric
BGIRA
    british glas industry research association
BGPROGRAM
    background program
BGRR
    brookhavens' graphite research reactor
BGRV
    boost glide reentry vehicle
BGSIA
    beginning standard instrument approach
BGSTA
    beginning straight - in approach
BH
    basic heterostructure
    binary-to-hexadecimal
    brake horsepower
    brinell hardness
    buried heterostructure
    busy hour
BHA
    base helix angle
BHC
    benzene hexachloride
BHD
    bulkhead
BHN
    brinnel hardness number
BHP
    boiler horsepower
BHPHR
    brake horsepower-hour
BHRA
    british hydromechanics research association
BHS
    binding head steel
BHWR
    boiling heavy water moderated and cooled reactor
BI
    base ignition
    base injection
    battery inverter
    biot
    bismuth
    blanking input
BIA
    booster insertion and abort
    buyers information advisory
BIAPS
    battery inverter accessory power supply
BIAR
    base interrupt address register
BIAS
    battlefield illumination airborne system
BIAX
    biaxial
BIBOM
    binary input - binary output machine
BIBOMM
    binary input - binary output moore machine
BIC
    bombardment-induced conductivity
    british insulated cables
    bureau of international commerce
BICC
    british insulated callender's cables
BICEPS
    basic industrial control engineering programming system
BICS
    burroughs inventory control system
BIDEC
    binary-to-decimal
BIDOPS
    bi-doppler scoring
BIDS
    burroughs input and display terminal

BIE
british institute of engineers
BIEE
british institute of electrical engineers
BIF
basic in flow
BIFET
bipolar field effect transistor
BIFR
before encountering instrument flight rules
BIG
bevan information generator
BIGFET
bipolar insulated gate fet
bipolar insulated gate field effect transistor
bipolar-igfet
BIGIT
binary digit
BII
beckman instruments, incorporated
BIIPS
battery inverter instrument power supply
battery inverter instrumentation power supply
BIL
base isolation level
basic impulse insulation level
basic insulation level
block input length
blue indicating lamp
blue indicating light
buried injector logic
BILA
battelle institute learning automation
BILE
balanced inductor logical element
BILLI
billionth
BIM
beginning of information mark
beginning of information marker
bimonthly
british institute of management
BIMAC
bistable magnetic core
BIMAG
bistable magnetic
BIMAP
bill of material processor
BIMCAM
british industrial measuring and control apparatus manufacturers' association
BIMM
base installation-minuteman

BIMOS
bipolar metal oxide semiconductor
bipolar mos
BIN
billion instructions
binary
bomarc interception
BINAC
binary automatic computer
BINCODE
binary code
BINET
bicentennial information network
BINIT
binary digit
BINOMEXP
binomial expansion
BIO
basic input output support program package
BIOALRT
bioastronautics laboratory research tool
BIONICS
biological electronics
biology electronics
BIOR
business input-output rerun
business input output rerun compiling system
BIOS
basic input output supervisor
basic input-output system
biological investigation of space
biological satellite
BIOSIS
bioscience information service of biological abstracts
BIP
background limited infrared photoconductor
balanced in plane
binary image processor
BIPAD
binary pattern detector
BIPCO
built-in-place component
BIR
british institute of radiology
BIRDIE
battery integration and radar display equipment
BIRE
british institute of radio engineers
BIRS
basic indexing and retrieval system
basic information retrieval system
BIS
british interplanetary society

business information system
business instruction set
BISAD
business information system analysis and design
BISAM
basic index sequential access method
basic indexed sequential access method
basic indexing sequential access method
BISCUS
business information system customer service
BISEPS
bomarc integration sage evaluation program simulation
BISFA
british industrial and scientific film association
BISP
business information system program
BISS
base and installation security systems
BISU
base interface surveillance unit
BISYNC
binary synchronous
binary synchronous communication
BIT
basic indissoluble information unit
basic information unit
binary digit
boric acid injection tank
boron injection tank
built-in-test
BITE
base installation test equipment
built-in-test equipment
BITN
bilateral iterative networt
BITS
binary information transfer system
BIU
basic information unit
BIVAR
bivariant
bivariant function generator
BIX
binary information exchange
BJ
ball joint
barrage jamming
BJCEB
british joint communications and electronics board
BJF
batch job foreground

BJT
bipolar junction transistor
BK
break
break-in keying
BKCA
bank cable
BKF
blocking factor
BKGD
background
BKR
breaker
BKSP
backspace
BL
bale
barrel
base line
bell
bend line
blade
blanking
block
block label
block length
block size
blue lamp
blue light
bottom layer
bus link
BLA
blocking acknowledgement signal
BLADE
base level automation of data through electronics
basic level automation of data through electronics
BLADES
bell laboratories automatic design system
BLC
balance
boundary layer control
british lighting council
BLCS
boundary layer control system
BLD
beam lead device
BLE
block length error
BLESSED
bell little electrodata symbolic system for the electrodata
BLEU
blind landing experimental unit

BLEVE
   boiling liquid expanding vapor explosion
BLF
   blocking factor
BLG
   blooming gate
BLIP
   background limited infrared photoconductor
   block diagramm interpreter program
BLIS
   baffle liner interface seal
   bell laboratories interpretative system
   bell labs interpretive system
BLK
   black
   blank
   blink
   block
   bulk
BLKCNT
   blick count
BLL
   below lower limit
BLLE
   balanced line logic[al] element
BLM
   basic language machine
   bureau of land management
BLMUX
   block multiplex channel
BLNK
   blank
BLO
   blocking signal
BLODI
   block diagram
BLODIB
   block diagram compiler b
BLODIC
   block diagram compiler
BLP
   blind loaded and plugged
BLPC
   backward limit photocell
BLR
   baseline restorer
BLS
   bureau of labor statistics
BLSJ
   beam lead sealed junction
BLST
   ballast
BLT
   basic language translator
   blanket
   blind loaded and traced
   blind loaded with tracer
BLTC
   bottom loading transfer cask
BLTIN
   built-in
BLU
   basic link unit
   basic logic unit
   bipolar line unit
BM
   ballistic missile
   barrels per month
   beam
   bearing magnetic
   bench mark
   bending moment
   binary multiply
   bistable multivibrator
   board measure
   boundary marker
   brightness merit
   buffer mark
   buffer module
   buffer multiplexer
   byte machine
   byte multiplex mode
BMANT
   boom antenna
BMAPS
   buckbee-mears automated plotting system
BMAR
   ballistic missile acquisition radar
BMAT
   beginning of morning astronomical twilight
BMB
   ballistic missile branch
   british metrication board
   broadcast measurement bureau
BMC
   ballistic missile center
   ballistic missile committee
   basic machine cycle
   bubble memory controller
   buckbee-mears company
   bulk media conversion
   bulk molding compound
BMCO
   ballistic missile construction office
BMCP
   biomedical measurement and control panel
BMCS
   business management control system

BMCSRP
business management control system research project
BMCT
beginning of morning civil twilight
BMD
ballistic missile defence
ballistic missile division
baseband modulator-demodulator
biomedical computer programs
biomedical statistical programs
bubble memory device
BMDC
ballistic missile defense committee
BMDS
ballistic missile defense system
BME
bench maintenance equipment
birmingham electric company
BMEP
brake mean effective pressure
BMES
biomedical engineering society
BMEWS
ballistic missile early-warning system
BMI
ballistic missile interceptor
battelle memorial institute
broadcast music, incorporated
BMIGT
ballistic missile inertial guidance technician
BMIGTM
ballistic missile inertial guidance technician mechanic
BMILS
bottom-mounted impact and location system
BMIS
british medical information society
BML
bulk material length
BMNT
beginning of morning nautical twilight
BMOM
base maintenance and operational model
BMOS
back-gate mos
BMOSFET
back-gate metal-oxide semiconductor field-effect transistor
BMR
british marine radar
brookhavenß's medical reactor
BMRS
ballistic missile reentry system

BMS
ballistic missile specification
basic mapping support
bio-metrics society
bit mark sequencing
burroughs management system
business management system
BMT
basic motion time
basic motion time study
british mean time
BMW
beam width
BMWS
ballistic missile weapon system
BN
binary number
boron nitride
BNC
baby "n" connector
bayonet nut connector
bulk negative conductance
BNCH
bench
BNCIEC
british national committee of the iec
BNCNDT
british national committee for non-destructive testing
BNCS
british numerical control society
BNCSR
british national committee for space research
BND
band
bend
bond
BNDC
bulk negative differential conductivity
BNDDIS
band display
BNDG
bonding
BNEC
british nuclear energy conference
BNES
british nuclear energy society
BNF
backus naur form
backus normal form
best noise figure
bomb nose fuse
british nuclear forum
BNG
branch no group

BNGS
  bomb navigation guidance system
BNL
  brookhaven national laboratory
BNR
  burner
BNRC
  baltic and north sea radio-telephone conference
BNS
  binary number system
  bomb navigation system
BNTM
  baltic and north sea telecommunication meeting
BNU
  basic notch unit
BNWL
  battelle northwest laboratories
BO
  beat frequency oscillator
  beat oscillator
  beating oscillator
  binary-to-octal
  blackout
  blocking oscillator
  blow-out
  branch office
BOAC
  british overseas airways corporation
BOADICEA
  british overseas airways digital information computer for electronic automation
BOB
  bobbin
  bomarc b program
BOBCODE
  bipolar offset binary code
BOBHIP
  bomarc b history printout
BOC
  beginning of cycle
  best output and color
  bevitron orbit code
  block oriented computer
  blow out coil
  body on chassis
  bottom of conduit
BOCA
  building officials conference of america
BOCES
  board of cooperative educational services
BOCOL
  basic operating consumer oriented language

BOCS
  bendix optimum configuration satellite
BOD
  biochemical oxygen demand
BOE
  beginning of extent
BOF
  beginning of file
BOHP
  boiler horsepower
BOI
  branch output interrupt
BOL
  boundary light
BOLD
  bibliographic on-line display
BOLT
  beam-of-light transistor
BOM
  basic operating monitor
  basic operation memory
  beginning of message
  bill of material
BOMARC
  boeing-michigan air research conference
BOMOS
  buried-oxide mos
BOMP
  base orginazition and maintenance processor
  bill of material processor
BOMST
  bombsight
BOND
  bonding
BONUS
  banken-online-universal-system
  boiling nuclear superheat reactor
  boiling water reactor nuclear superheat project
BONUSCX
  boiling nuclear superheat critical experiment
BOOSTRKT
  booster rocket
BOP
  balance of plant
  balance of power
  basic overall polarity
  binary output program
BOPTEST
  basic overall polarity test
BOR
  beginning of record
BORAM
  block organized ram
  block oriented ram

block oriented random access
block oriented random access memory
BORAX
   boiling reactor experiment
BORE
   beryllium oxide reactor experiment
BOREQ
   broadcast request
   broadcast requested
BORSCHT
   battery power feed, overvoltage protection, ringing, supervision, coding, hybrid, and testing
BOS
   background operating system
   basic operating system
   batch operating system
   building-out section
BOSCO
   bomarc sage compatibility program
BOSPROCESS
   basic oxygen steelmaking process
BOSS
   basic operating system software
   basis of standard system
   bioastronautical orbiting space system
   biological orbiting space station
   bistable optically controlled semiconductor switch
   bmews operational simulation system
BOT
   beginning of tape
   beginning-of-tape mark
   board of trade
BOTMARKER
   beginning-of-tape marker
BOTMG
   bottoming
BOTOHM
   board-of-trade-ohm
BOUND
   boundary
BP
   band pass
   batch processing
   between perpendicular
   blueprint
   boiling point
   bonded part
   bonded single paper
   bornier programmable
   break pointer
   british patent
   broadcast pioneer

bubble pulse
by-pass
BPA
   bonneville power administration
   break-point address-register
   broadcasters' promotion association
BPAM
   basic partitioned access method
   basic program access method
BPB
   user program library
BPC
   back-pressure control
BPD
   barrel per day
   bushing potential device
BPDMS
   basic point defense missile system
BPF
   band pass filter
   blue print files
BPG
   break pulse generator
BPH
   barrels per hour
BPI
   bit per inch
   broadcasting publications, incorporated
   byte per inch
BPID
   book physical inventory difference
BPIT
   basic parameter input tape
BPKT
   basic program knowledge test
   basic programming knowledge test
BPL
   band pass limiter
   beam packing loss
   bell propulsion laboratory
   burroughs programming language
BPM
   barrels per month
   batch processing monitor
   bi-phase modulation
BPN
   breakdown pulse noise
BPO
   base post office
   bomarc prelaunch output
   british post office
BPOS
   batch processing operating system
BPP
   beacon portable packset

BPPMA
  british power press manufacturers association
BPR
  bar-pattern response
  bubble position register
  bureau of public roads
BPRA
  british pattern recognition association
  burnable poison rod assembly
BPS
  basic programming support
  basic programming system
  baud per second
  beacon portable set
  bit per second
BPSK
  binary phase-shift keyed
  binary phase-shift keying
BPSS
  basic production scheduling system
BPT
  boiling point
BPU
  basic pole unit
  basic processing unit
  bomarc pick-up
BPVC
  boiler and pressure vessel committee
BPWR
  burnable poison water reactor
BPWS
  banked position withdrawal sequence
BQL
  basic query language
BR
  band reject
  base register
  basic research
  bend radius
  boiling range
  brake relay
  branch
  breeder reactor
  bridge
  bromine
  brush
  bulk resistance
  bunker ramo
  bus request
  butadiene rubber
  butyl rubber
BRA
  british refrigeration association
  british robot association
BRAB
  building research advisory board
BRAD
  brookhaven raster display
BRAMS
  british trans-atlantic air mail service
BRANE
  bombing radar navigation equipment
BRB
  benefits review board
BRC
  boonton radio corporation
  breeder reactor corporation
  british radio communication
  bunker-ramo corporation
BRCMA
  british radio cabinet manufacturers' association
BRCT
  broad-band rectangular-to-circular transition
BRD
  braid
BRDCST
  broadcast
BRDG
  bridge
BREMA
  british radio equipment manufacturers association
BREN
  bare reactor experiment nevada
BRF
  band-rejection filter
BRG
  bearing
  budget review group
BRH
  bureau of radiological health
BRHLR
  brush holder
BRI
  base recirculation insulation
BRIG
  bmews raid input generator
BRIL
  brilliance
BRITE
  bright radar indicator tower equipment
BRITIRE
  british institute of radio engineers
BRITPAT
  british patent
BRJE
  bsc-remote job entry

BRK
　break
BRKT
　bracket
BRL
　ballistic research laboratory
　barrel
　bit rate low
　branch and link
BRLESC
　ballistic research laboratories electronic scientific computer
BRLG
　bomb, radio, longitudinal, generator-powered
BRLS
　barrier ready light system
BRLSYS
　barrier ready light system
BRM
　barometer
　binary rate multiplier
BROFICON
　broadcast fighter control
BROM
　bipolar read-only memory
BROOM
　ballistic recovery of oribiting man
BRP
　beacon ranging pulse
　british patent
BRR
　battelle research reactor
　bridge receiving room
　brookhaven research reactor
BRRL
　british radio relay league
BRS
　binary ring sequence
　boron recycle system
　break request signal
　bristol record society
　british receiving station
　british roentgen society
　business radio service
BRSL
　british record society, limited
BRST
　burst
BRSTD
　british standard
BRT
　binary run tape
　binary run tape brightness

　bright
　brightness
　british time
BRTD
　bright radar tube display
BRTG
　bomb, radio, transverse, generator-powered
BRU
　bearing repeater unit
BRUIN
　brown university interpreter
BRVMA
　british radio valve manufacturers association
BS
　backspace
　balance sheet
　band setting
　band stop
　bandsystem
　base
　base shield
　bessemer steel
　big system
　binary scale
　binary subtract
　binary subtraction
　biophysical society
　bit space
　bonded single silk
　bristol siddeley
　british size
　broadcast station
　broadcasting station
　bureau of standards
BSAM
　basic sequential access method
　binary sequential access method
BSAR
　babcock and wilcox safety analysis report
BSB
　baseband
　both sidebands
BSBG
　burst and synchronous bit generator
BSC
　backspace contact
　basic message switching center
　basic synchronous communication
　beltsville space center
　binary symmetric channel
　binary synchronous communication procedure
　binary synchronous control

british standard channel
british standard cycle
BSCA
 binary synchronous communications adapter
BSCN
 bit scan
BSD
 ballistic systems division
 bulk storage device
BSDA
 business and defense services administration
BSDC
 binary symmetric dependent channel
 british space development company
 british standard data code
BSDL
 boresight datum line
BSE
 boston edison
 boston stock exchange
BSET
 bachelor of science in engineering technology
BSF
 british standard fine
 bulk shielding facility
BSFC
 brake specific fuel consumption
BSG
 bootstrap gyro
 british standard gauge
BSH
 bushel
BSI
 booster situation indicator
 branch and store instruction
 british standard institution
 british standards institute
BSIC
 binary symmetric independent channel
BSIL
 basic switching impulse insulation level
BSIRA
 british scientific instrument research association
BSIT
 bipolar mode sit
BSL
 basic switching-surge level
 beam shape loss
BSM
 basic storage module
BSMV
 bistable multivibrator
BSNDT
 british society for non-destructive testing

BSO
 base signal officer
 beach signal office
BSP
 british standard pipe screw thread
BSR
 backspace register
 balloon-supported rocket
 better sound reproduction
 blip-scan radar
 blip-scan ratio
 board of standards review
 bulk shielding reactor
BSRA
 british sound recording association
BSRL
 boeing scientific research laboratory
BSS
 beach signal station
 bit storage and sense
 british standard specification
BSSRS
 bureau of safety and supply radio services
BST
 basic storage unit
 beam switching tube
 binary search tree
 block nesting depth
 booster
 booster amplifier
 british standard time
 british summer time
 burst
 business systems technology
BSTJ
 bell system technical journal
BSU
 basic selection unit
BSV
 boolean simple variable
BSW
 barrel switch
 british standard whithworth
BSWCC
 british short wave correspondence club
BSWG
 british standard wire gauge
BSWM
 bureau of solid waste management
BSY
 busy
BSZ
 block store zeros
BT
 basse tension

bathythermograph
beginning of tape
beginning of tape mark
bellini-tosi
bias temperature
bit
bus tie
busy tone
byte
BTAM
basic tape access method
basic telecommunication access method
BTC
batch terminal controller
bell-telephone company
block terminating character
butt-treated cedar
BTCC
basic traffic control center
BTCCODE
bipolar twos-complement code
BTD
binary to decimal
bomb testing device
BTDFLC
btt-treated douglas-fir, larch and cedar
BTDL
back transient diode logic
basic transient diode logic
BTE
battery terminal equipment
battery timing equipment
bidirectional transceiver element
boltzmann transport equation
BTEC
business and technician education council
BTF
binary transversal filter
bomb tail fuse
BTG
battery timing group
beacon trigger generator
boiler turbine generator
british technology group
BTH
british thomson houston
BTHU
british thermal unit
BTI
bank and turn indicator
basic timesharing, inc.
bridged tap isolator
BTL
backtell
balanced transformer

beginning tape label
bell telephone laboratories
BTLP
butt-treated lodgepole pine
BTM
batch time-sharing monitor
bell telephone manufacturing
buffered terminal multiplexer
BTMA
basic telecommunication
BTN
button
BTO
blocking tube oscillator
BTP
branch technical position
BTR
behind tape reader
broadcast and television receivers
BTRCONNECTION
behind-tape-reader connection
BTRY
battery
BTS
brazilian thorium sludge
british television society
BTSS
basic time-sharing system
BTST
busy-tone start-lead
BTT
beginning to tape test
BTTP
british towing tank panel
BTU
basic transmission unit
block transfer unit
board of trade unit
british thermal unit
BTUSQFTMIN
btu per square foot per minute
BU
buzzer
BUAER
bureau of aeronautics
BUEC
back-up emergency communication
BUG
bomarc unintegrated guidance
BUIC
back-up interceptor control
BUILD
base for uniform language definition
BUIS
barrier up indicator system

BUISYS
　barrier-up indicator system
BULKIO
　bulk input output
BULLRAC
　bull random access
　bull random access memory
BUMED
　bureau of medicine and surgery
BUORD
　bureau of ordonance
BUPERS
　bureau of naval personnel
BUPS
　beacon ultra portable s-band
BUPX
　beacon ultra-protable x-band
BURSTDS
　bureau of standards
BUS
　bomarc - universal automatic computer - sage
　bus line
　busbar
BUSANDA
　bureau of supply and accounts
BUSARB
　british-united states amateur rocket bureau
BUSEN
　bus enable
　bus enable signal
BUSH
　bushing
BUSHIPS
　bureau of ships
BUT
　button
BUWEPS
　bureau of naval weapons
BUZ
　buzzer
BV
　balanced voltage
　baudot-verdan
　breakdown voltage
BVA
　british radio valve manufacturers' association
BVC
　black varnish cambric
BVD
　beacon video digitizer
BVMA
　british valve manufacturers' association
BVP
　boundary value problem

BW
　backward wave
　bandwidth
　beam width
　bendix-westinghouse
　biological warfare
　biological weapon
　black and white
　braided wire
　butt-welded
　instruction word
BWA
　backward wave amplifier
　british waterworks association
BWC
　backward wave converter
BWG
　birmingham wire gauge
　british wire gauge
BWM
　backward wave magnetron
BWMS
　british wireless marine service
BWO
　backward wave oscillator
BWPA
　backward wave parametric amplifier
　backward wave power amplifier
BWR
　bandwidth ratio
　boiling water reactor
BWST
　borated water storage tank
BWT
　british winter time
BX
　box
BY
　busy
BYDA
　busy don't answer
BYMUX
　byte multiplex channel
BYP
　bypass
BYPCAP
　bypass capacitor
BYPCOND
　bypass condenser
BZ
　blank when zero
　buzzer
BZFX
　brazing fixture

C

C
  call
  calorie
  candella
  candle
  capacitance
  capacitor
  capacity
  carbon
  cathode
  cell
  cellatron
  celsius
  center[-re]
  centi
  centigrade
  centigrade degree
  centimeter
  character
  chirp
  chromiance
  chrominance
  circuit
  code
  coefficient
  coil
  collector
  compare
  compute
  computer
  computing
  concentrate
  concentration
  condenser
  conductivity
  conductor
  consolrdated computer inc.
  control
  controller
  controlling
  core
  correct
  correction value
  cotton
  coulomb
  course
  cubic
  curie
  current
  cycle
  cycles per second
  degrees centigrade
  microcomputer
  single-cotton
  single-cotton covered
CA
  cable
  california
  cancel
  cancelled
  candle
  capacitor
  cathode
  cellular automation
  circuitry adapter
  civil aviation
  clear aperture
  close annealed
  close-annealed
  coaxial
  commercial aviation
  communications adapter
  commutator assembly
  computer access
  computer and automation
  computer assisted accounting
  computer associates
  computer automation
  consonant amplification
  construction authorization
  contact ammeter
  content addressable memory
  contract administrator
  contract authorization
  control accelerometer
  control area
  controlled approach
  corrective action
  counterpoise antenna
  current address
  current address register
CAA
  central african airways corporation
  civil aeronautics administration
  computer assisted accounting
  computers and automation
CAAC
  civil aviation administration of china
CAADRP
  civil aircraft airworthiness data recording program
CAAIS
  computer assisted action information system
CAAK
  civil aviation administration of korea
CAAM
  civil aeronautics authority manual

CAAR
compressed air accumulator rocket
CAARC
commonwealth advisory aeronautical research council
CAAS
ceylon association for the advancement of science
computer assisted approach sequencing
CAATC
civil aeronautics administration-type-certificate
CAB
cabinet
canadian association of broadcasters
civil aeronautics board
computer address bus
CABMA
canadian association of british manufacturers and agencies
CABO
council of american building officials
CABOWV
captured air bubble over water vehicle
CABRA
copper and brass research association
CABT
cesium atomic beam tube
CAC
calibration and certification
calibration and checkout
california aeronautics commission
clear all channels
command and control
commonwealth aircraft corporation
communication accessories company
communication and control
complete address constant
containment atmosphere control
control and coordination
current account
CACB
compressed-air circuit breaker
CACC
civil aviation communications center
CACD
computer automated cargo documentation
CACDP
california association of county data processors
CACM
communications of the acm
communications of the association for computing machinery

CACOM
chief aircraft communicator
CACP
corrosion and cathodic protection
CACS
continental airways and communications service
controller active state
CACW
continental aircraft control and warning
core auxiliary cooling water
CACWS
core auxiliary cooling water system
CAD
calculator aided design
cartridge activated device
cartridge-activated device
command address
communications access device
communications and data
compensated avalanche diode
computer access device
computer application digest
computer-aided design
computer-aided detection
computer-aided drafting and design
computer-assisted design
containment atmosphere dilution
control and display
CADAM
computer augmented design and manufacturing
computer graphics augmented design and manufacturing system
CADAPSO
canadian association of data processing organizations
CADAR
computer aided design, analysis and reliability
CADC
cambridge automatic digital computer
central air data computer
CADCO
core and drum corrector
CADDAC
central analog data distributing and computing
CADE
computer-aided data entry
computer-assisted data entry system
CADEM
computer-aided design engineering and manufacturing

CADEP
  computer-aided design experimental translator
  computer-aided design of electronic products
CADETS
  classroom-aided dynamic educational time-sharing system
CADEX
  computer-aided design exhibition
CADF
  cathode-ray tube automatic direction finding
  commutated antenna direction finder
CADFISS
  computation and data flow integrated subsystem
CADI
  computer access device input
CADIAP
  comment and data integration and printing
CADIC
  computer-aided design of integrated circuits
CADIS
  computer-aided design interactive system
  computer-aided design of information systems
CADL
  communications and data link
CADM
  clustered airfield defeat submunition
CADNCTAPE
  computer-aided design numerical control tape
CADPIN
  customs automated data processing intelligence network
CADPO
  communications and data processing operation
CADRE
  current awareness and document retrieval for engineers
CADS
  central air data subsystem
  containment atmosphere dilution system
CADSS
  combined analog-to-digital systems simulator
  communications and data subsystem
CAE
  canadian aviation electronics
  commission for aerology
  communications and electronics
  computer-aided education
  computer-aided engineering
  computer-assisted education
  computer-assisted enrollment
  computer-augmented education

CAEI
  canadian aeronautical institute
CAEM
  commission for aeronautical meteorology
CAESQ
  communications and electronics squadron
CAF
  chemical analysis facility
CAFD
  contact analog flight display
CAFE
  computer-aided film editor
CAFS
  content addressable file store
CAFTN
  commonwealth air forces telecommunications network
CAGC
  coded automatic gain control
CAGE
  canadian airground environment
  compiler and assembler by general electric
  computerized aerospace ground equipment
CAGLI
  city and guilds of london institute
CAHE
  core auxiliary heat exchanger
CAI
  canadian aeronautical institute
  close approach indicator
  computer analog input
  computer applications, incorporated
  computer automation inc.
  computer-administered instruction
  computer-aided industry
  computer-aided information
  computer-aided instruction
  computer-assisted instruction
  control and indication
  control and instrumentation
CAIL
  computer-aided information logistics
CAIN
  calculation of inertia
CAINS
  carrier aircraft inertial navigation system
CAINT
  computer-assisted interrogation
CAIOP
  computer analog input output
CAIS
  canadian association for information science
  central abstracting and indexing service
CAL
  calculated average life

calibrate
calibrated
calibration
common assembler language
common assembly language
compressed air loudspeaker
computer animation language
conversational algebraic language
cornell aeronautic[al] laboratory
kilocalorie
CALAS
computer-aided laboratory automation system
CALASOAP
calcium lanthanum silicate oxyapatite
CALC
calculate
calculated
calculating
calculator
cargo acceptance and load control
CALCN
calculation
CALCOMP
california computer products
CALDATA
california data processors
CALL
carat assembled logical loader
CALM
collected algorithms for learning machines
computer-assisted library mechanization
CALT
cleared altitude
CAM
care and maintenance
central address memory
checkout and automatic monitoring
civil aeronautics manual
clear and add magnitude
communication access method
computer address matrix
computer applications to manufacturing
computer-aided manufacturing
computer-assisted mailing
containment atmospheric monitoring
content address memory
content addressable memory
content addressed memory
continuous air monitor
cybernetic anthropomorphous machine
CAMA
centralized automatic message accounting
control and automation manufacturers association

CAMAC
computer application for measurement and control
computer-aided measurement and control
computer-automated measurement and control
CAMAL
continuous airborne missile alert
CAMAR
common-aperture multifunction array radar
CAMEL
component and material evaluation loop
CAMERA
cooperating agency method for event reporting and analysis
CAMESA
canadian military electronics standards agency
CAMP
compiler for automatic machine programming
computer applications of military problems
computer-assisted mathematics program
computer-assisted movie production
control and monitoring processor
CAMRAS
computer-assisted mapping and records activities systems
CAMS
consumer and marketing service
CAMSAT
camera satellite
CAN
cancel
cancelled
CANARI
communications and navigation airborne radio instrumentation
CANC.
cancellation
CANDSC
candelabra screw
CANEL
connecticut advanced nuclear engineering laboratory
connecticut aircraft nuclear engine laboratory
CANGO
committee for air navigation and ground organization
CANJPHYS
canadian journal of physics
CANPAT
canadian patent
CANS
computer-assisted network scheduling system

CANSG
civil aviation navigational services group
CANTAT
canadian transatlantic telephone
CANTRAN
cancel in transmission
cancel transmission
CANUKUSJCEC
canadian-united kingdom-united states joint communications-electronics committee
CANUNET
canadian university computer network
CANUSE
canada united states eastern
CAO
collective analysis only
computer aided office
CAOC
constant axial offset control
CAOS
completely automated operational system
completely automatic operational system
CAP
california electric power
canadian association of physicists
capacitor
capacity
card assembly program
civil air patrol
communications pool assembler program
component acceptance procedure
computer aided presentation
computer analysis and programmer
computer analysis and programming
computer analysts and programmers
computer analysts and programmers of england
computer-aided planning
computer-aided programming
computer-aided publishing
computer-assisted production
console action processor
continuous audit program
cordic arithmetic processor
council of advanced programming
council to advance programming
cryotron associative processor
CAPAL
computer and photographically assisted learning
CAPCHE
component automatic program checkout equipment
CAPCO
central area power coordination
CAPCOST
capital cost of shield and vessels
CAPE
coalition of aerospace employees
communication automatic processing equipment
CAPERTSIM
computer-assisted evaluation review technique simulation
computer-assisted program evaluation review technique simulation
CAPF
capacity factor
CAPICS
computer aided processing of industrial cabling systems
CAPL
controlled assembly parts list
CAPM
computer-aided patient management
computer-aided production management
CAPO
canadian army post office
CAPOSS
capacity planning and operation sequencing system
CAPP
computer-aided partitioning program
CAPPI
constant altitude plan-position indicator
CAPRI
coded address private radio intercommunication
compact all-purpose range instrument
computerized advance personnel requirements and inventory
CAPRISTOR
capacitor-resistor
CAPS
capacitors
cell atmosphere processing system
computer assisted problem solving capacitors
computer-assisted problem solving
computerized agency processing system
courtauld's all-purpose simulator
CAPST
capacitor start
CAPSTAR
capacitor start and run
CAPT
conversational parts programming language
CAPY
capacity
CAR
canadian association of radiologists

carrier
channel address register
civil air regulation
computer assisted research
computer-aided repair
computer-aided retrieval
computer-assisted research
containment air removal fan
control-unit accumulator register
CARAC
civil aviation radio advisory committee
CARAM
content addressable random access memory
CARB
carbon
CARBINE
computer automated real-time betting information network
CARC
chicago amateur radio club
CARCC
chicago area radio club council
CARD
channel allocation and routing data
compact automatic retrieval device
compact automatic retrieval display
computer-assisted route development
CARDE
canadian armament research and development establishment
CARDIS
transportation cargo data interchange system
CARE
computer-aided reliability estimation
CARF
consumer affairs and regulatory functions, office of assistant secretary
CARIS
computerized audio report information and status
constant angle reflection interference spectroscopy
CARP
computed air release point
CARR
carriage
carried
carrier
CARS
coherent anti-stokes raman spectroscopy
community antenna relay service
computer-aided routing system
computerized automotive reporting service
CART
cartridge

central automated replenishment technique
central automatic reliability tester
centralized automatic recorder and tester
complete automatic reliability testing
computer automaticrating technique
computerized automatic rating technique
CARTA
contour analysis by random triangulation algorithm
CARTB
canadian association of radio and television broadcasters
CARVA
carolina-virginia
CAS
calculated air speed
calibrated air speed
central alarm station
chemical abstracts service
china association for standardization
circuits and system
collision avoidance system
column address select
column address strobe
communication access system
compare accumulator with storage
computer accounting system
computer aided selling
computer application summary
computerized accounting system
computers and systems
control automation system
controlled access system
controller access system
customer application summary
CASASME
computer and automated systems of sme
CASB
cost accounting standards board
CASCADE
colorado automatic single crystal analysis diffraction equipment
CASD
computer-aided system design
CASE
common access switching equipment
computer and system engineering
computer automated support equipment
computer-aided software engineering
computer-aided system evaluation
computer-assisted system evaluation
consolidated aerospace supplier evaluation
CASEE
canadian army signals engineering establishment

CASH
call accounting system for hotels
CASI
canadian aeronautics and space institute
CASM
command and service module
CASMIT
control automation system manufacturing interface tape
CASMOS
computer analysis and simulation of metaloxide semiconductor circuits
computer analysis and simulation of mos circuits
CASPAR
cambridge analog[ue] simulator for predicting atomic reactions
clarksons automatic system for passenger and agents reservations
CASPWI
collision avoidance system proximity warning indicator
CASS
command active sonobuoy system
CASSANDRA
chromatogram automatic soaking, scanning, and digital recording apparatus
CAST
clearinghouse announcements in science and technology
computerized automatic system tester
CASTE
civil aviation signals training establishment
CASTOR
college applicant status report
CAT
cado-actions-terminal
carburetor air temperature
centralized automatic testing
character assignment table
chemical addition tank
city air terminal
classical analytic technique
clear air turbulence
community antenna television
compile and test
component acceptance test
computer analysis of transistors
computer of average transients
computer-aided teaching
computer-aided test
computer-aided testing
computer-aided topographie
computer-aided translation
computer-assisted testing
conditionally accepted tag
controlled attenuator timer
controlled avalanche transistor
cooled anode transmitting value
cooled-anode transmitting tube
cooled-anode transmitting valve
cooled-anode tube
current-adjusting type
CATCH
character allocated transfer channel
CATH
cathode
CATHFOL
cathode follower
CATIS
computer-aided tactical information system
CATO
catapult-assited take-off
compiler for automatic teaching operation
CATP
computer-aided text processing
CATS
category switch
centralized automatic test system
chicago area teleprinter society
comprehensive analytical test system
computer automated test system
computer-aided teaching system
computer-aided trading system
computer-automated test system
centralized automatic test system
CATT
centralized automatic toll ticketing
controlled avalanche transit time
controlled avalanche transit time triode
conveyorized automatic tube tester
cooled-anode transmitting tube
CATTDEVICE
controlled avalanche transit time device
CATTTRIODE
controlled avalanche transit time triode
CATV
cable antenna television
cable television
cable tv
community antenna television
community antenna television system
community antenna tv
community antenna tv system
cooled-anode transmitting valve
CATVS
community antenna television system
CAU
command and arithmetic unit
command arithmetic unit

CAUML
   computers and automation universal mailing list
   computers and automation's universal mailing list
CAUSE
   college and university systems exchange
CAV
   cavity
   constant angular velocity
   constant angular velocity cavity
CAV.
   cavity
CAVU
   ceiling and visibility unlimited
CAW
   cable and wireless
   cables and wireless
   channel address word
   coal and waggon
   common antenna working
   control and warning
CAX
   community automatic exchange
CB
   capacitor bank
   center of buoyancy
   centibel
   central battery
   certification body
   check bit
   circuit breaker
   citizen's band
   citizens band
   common base
   common battery
   common bias
   contact breaker
   containment building
   continuous blowdown
   continuous breakdown
   control board
   control buffer
   control button
   cubic
   current bit
CBA
   c-band transponder antenna
   catholic broadcasters association
   central battery apparatus
   cocoa beach apollo
   community broadcasters association
   cost benefit analysis
CBAI
   chicago bridge and iron

CBAL
   counterbalance
CBAS
   central battery alarm signaling
CBASIC
   commercial basic
CBAST
   conc boric acid storage tank
CBAT
   central bureau for astronomical telegrams
CBATDS
   carrier-based airborne tactical data system
CBC
   canadian broadcasting corporation
   columbia broadcasting corporation
   computer based concentrator
   contraband control
CBD
   commerce business daily
CBDI
   control red bank demand indicator
CBE
   central battery exchange
   compression-bonded encapsulation
   computer based education
CBEMA
   computer and business equipment manufacturers' association
CBER
   citizen's bander
CBFS
   caesium beam frequency standard
CBI
   chesapeake bay institute
   compound batch identification
   computer-based instruction
CBIC
   complementary bipolar integrated circuit
CBICODE
   complementary binary code
CBIS
   computer-based instructional system
CBL
   cable
   cable electric products
   chesapeake biological laboratory
   counterbalance
CBLMN
   cableman
CBM
   constant boiling mixture
   continental ballistic missile

CBMA
  canadian business manufacturers' association
  certified ballast manufacturers' association
CBMPE
  confederation of british manufacturers of petroleum equipment
CBMS
  computer-based management system
  conference board of the mathematical sciences
CBMU
  current bit monitor unit
  current bit motor unit
CBN
  cubical boron nitride
CBO
  congressional budget office
CBPT
  clira backup plug tool
CBR
  chemical, biological, radiological
  citizens' band radio
  commercial breeder reactor
  comprehensive beacon radar
  current balance record
CBRM
  cash by return mail
CBS
  call box station
  central battery signalling
  central battery supply
  central battery system
  columbia broadcasting system
  common battery signalling
  common battery supply
  consolidated business systems
  controlled blip scan
CBSC
  common bias single control
CBSR
  coupled breeding supperheating reactor
CBT
  central battery telephone
CBU
  coefficient of beam utilization
CBW
  chemical and biological warfare
  constant bandwidth
CBX
  cam box
CC
  cable connector
  calculator
  card code
  card column

carriage control
carrier current
cast copper
center to center
center-to-center
central computer
central control
ceramic capacitor
chain command
change of course
channel command
charge coupled
chrysler corporation
circuit closing
clock control
close control
close-coupled
closed circuit
closed-circuit
closing coil
coarse control
code converter
coin collect
coincident current
color code
combined carbon
command chaining
command computer
command console
command control
common carrier
common code
common collector
common control
communication center
communication control
communication controller
communications center
communications computer
compact cassette
compass course
computer calculator
computer center
computer community
computer consulting
computer controlled
computing center
concurrent concession
condition code
conductive channel
connecting circuit
connector circuit
consolidated computer
constant current
continuous current

control center
control code
control computer
control connector
control console
coomon collector
coordinates computed
copper constantan
cotton covered
cross correlation
cross coupling
crystal control
crystal controlled
cubic centimeter
cursor control
cyclic check

CCA
canadian construction association
carrier controlled approach
central computer agency
communications channel adapter
communications control area
component checkout area
computer corporation of america
copper conductors association
cross country award
current cost accounting

CCANDS
central computer and sequencer

CCAP
communications control application program

CCAS
central computer and sequencing

CCATS
communications command and telemetry system

CCB
channel command block
channel control block
circuit concentration bay
close control bombing
combined communications board
command control block
configuration control board
console-to-computer buffer
contraband control base
convertible circuit breaker
cyclic check bit
cyclic check byte
cyclic-checking byte

CCBA
central canada broadcasters' association

CCBP
combined communications board publications

CCBS
clear channel broadcasting service

CCC
caldwell computer corporation
canadian computer conference
carrier current communication
central classification committee
channel command control
channel control check
command control center
command control console
commodity credit corporation
common control circuit
communications control console
computer circuit corporation
computer communication console
computer communications converter
computer control company
console control circuit
coordinate conversion computer
copy control character

CCCE
computer communication center europe

CCCEALS
california council of civil engineers and land surveyors

CCCEC
commission for certification of consulting engineers, colorado

CCCF
central committee on communications facilities

CCCL
complementary constant current logic

CCCR
communication and command control requirements

CCCRAM
continuously charge-coupled random access memory

CCCS
core component conditioning station
core components cleaning system
current controlled current source

CCD
charge coupled device
coarse control damper
cold cathode discharge
command control destruct
complementary binary-coded decimal
computer control division
computer controlled display
core current driver
counter-current digestion
counter-current distribution

CCDCODE
  complementary binary-coded decimal code
CCDP
  central computer development program
CCDS
  computer communications and data service
CCE
  channel command entry
  command control equipment
  communication control equipment
  complex control equipment
CCEU
  council on the continuing education unit
CCF
  central computer facility
  central control facility
  chain command flag
  cobol communications facility
  communication central facility
  cross correlation function
CCFM
  cryogenic continuous film memory
CCFT
  controlled current feedback transformer
CCG
  communications change group
  constant current generator
CCGCR
  closed cycle gas-cooled reactor
CCGCS
  containment combustion gas control system
CCH
  channel-check handler
  computerized criminal history
  connections per circuit hour
CCHX
  component cooling heat exchanger
CCI
  century computer international inc.
  charge-coupled imager
  com-share communications interface
  computer communications inc.
  computer communications interface
  connecticut consolidated industries
  consultative committee international
  convert clock input
  current-controlled inductor
CCIA
  computer and communications industry association
  console computer interface adapter
CCIC
  constant cost integer code

CCIR
  communications change initiation request
CCIS
  command and control information system
  command control information system
  common and control information system
  common channel interoffice signalling
CCITT
  consultative committee on international telegraph and telephone
CCK
  coherent carrier keying
CCKW
  counterclockwise
CCL
  clocked cmos logic
  common command language
  communications change log
  consolidated computer ltd.
  contact clock
  conversion and check limit
CCLKW
  counterclockwise
CCM
  charge coupled memory
  communications controller multichannel
  constant current modulation
  controlled carrier modulation
  counter countermeasures
CCMD
  continuous current monitoring device
CCMOS
  clocked complementary metal oxide semiconductor
CCMP
  conversion complete
CCN
  cluster controller node
  communication control number
  contract change notice
CCNR
  current controlled negative resistance
CCO
  constant control oil
  constant control oil pressure
  constant current operation
  crystal controlled oscillator
  current controlled oscillator
CCOCP
  corporate customer order control program
CCOP
  consolidated customer order processing
  constant control oil pressure
CCP
  carbonless copying paper

centrifugal charging pump
certificate computer programming
command control panel
communication control package
communications control panel
communications control program
console command processor
control command processor
core component pot
critical compression pressure
cross-connection point
cubic close-packed
CCPD
coupling capacitor potential device
CCPE
canadian council of professional engineers
CCPM
concurrent cp m
cubic centimeter per minute
CCPROCESS
close confinement process
CCPS
consultative committee for postal studies
CCR
channel command register
channel control routine
closed cycle refrigerator
coaxial cavity resonator
command control receiver
communications change request
communications control room
component catalog review
computer-controlled information retrieval
control circuit resistance
control contactor
critical compression ratio
cyclic catalytic reforming
CCRC
core component receiving container
CCROS
card capacitor read only storage
card capacity read only storage
CCS
canadian ceramic society
capital computer suites ltd.
central computing site
change control system
collective call sign
collector coupled structures
colour vision constant speed
column code suppression
command control system
commitment control system
communications control system
computer control station

computer scheduling and control system
containment cooling system
continuous color sequence
continuous commercial service
contour control system
control computer subsystem
conversational compiling system
custom computer system
CCSA
common control switching arrangement
CCSEP
cement-coated single epoxy
CCSK
cyclic shift keying
CCSL
camp coles signal laboratory
compatible current sinking logic
CCSR
copper cable steal reinforced
copper cable steel-reinforced
CCST
center for compiler sciences and technology
CCSU
computer cross-select unit
CCSW
component cooling service water
copper-clad steel wire
CCT
circuit
circuit continuity tester
clarkson college of technology
communications control team
connecting circuit
constant current transformer
correlated color temperature
crystal controlled transmitter
CCTEP
cement-coated triple epoxy
CCTEST
component check test
CCTL
casing cooling tank level
core component test loop
CCTSIM
circuit simulation
CCTV
close circuit television
closed circuit television
CCTVS
closed circuit television system
CCU
camera control unit
central control unit
channel control unit
chart comparison unit

communication control unit
communication coupling unit
computer control unit
coupling control unit
CCV
  code converter
  control configured vehicle
CCVS
  cobol compiler validation system
  current controlled voltage source
CCW
  channel command word
  circulation controlled wing
  command control word
  component cooling water
  counter clockwise
CCWS
  component cooling water system
CD
  cable duct
  cadmium
  call detector
  call dispatch
  call per day
  candela
  capacitive discharge
  capacitor diode
  card
  card deck
  carrier detected
  cathode of diode
  center distance
  chaining data
  charge discharge
  circuit description
  civil defence
  clock driver
  coast[al] defense
  cold-drawn
  collision detection
  compact disk
  conductance
  connecting devices
  continuous duty
  control rod drive
  cord
  cored
  countdown
  crystal diode
  crystal driver
  current density
  current discharge
  current driver
CDA
  carbide diamond abrasive

chain data address
chaining data address
command and data acquisition
containment depressurization activation
containment depressurization alarm
control data corporation
core disruptive accident
cradle of democracy award
CDB
  corporate data base
  current data bit
CDBN
  column digit binary network
CDC
  call directing character
  call directing code
  central data control
  characteristics distortion compensation
  code directing character
  command and data-handling console
  computer development center
  computer diode corporation
  configuration data control
  control data computer
  control data corporation
  count double count
  course and distance calculator
  course and distance computer
CDCE
  central data-conversion equipment
CDCL
  command document capability list
CDCR
  center for documentation and communication research
CDCU
  communications digital control unit
CDD
  coded
  combat data director
  command document discard
  current discontinuity device
CDDP
  console digital display programmer
CDE
  certification in data education
  command document end
  cornell-dubilier electronics
CDEBUG
  cobol symbolic debugging package
CDEVC
  computer development center
CDF
  combined distributing frame
  combined distribution frame

cummulative distribution of frequency
cumulative damage function
cumulative distribution function
cumulative distribution of frequency
CDG
  capacitor diode gate
  coding
  converter display group
CDH
  cable distribution head
CDI
  capacitor discharge ignition
  collector diffusion insulation
  collector diffusion isolation
  control data institute
  course deviation indicator
CDITECHNIQUE
  collector diffusion isolation technique
CDJM
  canadian journal of mathematics
CDL
  command definition language
  common display logic
  compiler description language
  computer description language
  computer design language
  core diode logic
  current discharge line
CDM
  central data management
  circuit directory maintenance
  code division multiplex
  code division multiplexing
  control data mathematical programming
  core division multiplexing
CDMA
  code division multiple access
CDMS
  commercial data management system
CDO
  community dial office
CDOH
  coupling display optical hand
CDP
  call routine display panel
  central data processing
  central data processor
  central distribution programmer
  centralized data processing
  certificate in data processing
  certification in data processing
  checkout data processor
  communication data processor
  compression discharge pressure
  correlated data processor

CDPA
  computer development personnel association
CDPC
  central data processing computer
  commercial data processing center
CDPIR
  crash data position indicator recorder
CDPS
  computing and data processing society
CDR
  central data recording
  central date recording
  command destruct receiver
  command document resynchronization
  compare and difference right-half-words
  conceptual design requirement
  cornell-dubilier electric
  critical design review
  current directional relay
CDRA
  committee of directors of research associations
CDRBTE
  canadian defense research board's telecommunications establishment
CDRC
  conductivity-recording controller
CDRD
  computations and data reduction division
CDROM
  compact disk read only memory
CDS
  central distribution system
  century data systems
  command document start
  common diagram system
  communications and data subsystem
  compatible duplex system
  component diassembly station
  comprehensive display system
  computer duplex system
  conceptual design study
  configuration data services inc.
  single cotton double silk
CDSC
  coupling display scanning telescope manual control
CDSE
  computer driven simulation environment
CDSL
  connect data set to line
CDST
  central daylight saving time
CDT
  central daylight time

combined double tee
command destruct transmitter
communications display terminal
control data terminal
coordinate data transmitter
countdown time

CDTI
cockpit display of traffic information

CDU
central display unit
coastal defense radar for the detection of u-boats
control and display unit
coupling data unit
coupling display unit

CDUI
command document user information

CDUO
coupling display unit optics

CDV
check digit verification

CDW
computer data word

CE
calculation experiment
card error
cellular automation cerium
chip enable
circular error
civil engineering
clear entry
coal equivalent
combustion engine
combustion engineering corporation
common emitter
communications electronics
commutator end
comparing element
compass error
compute element
computer engineer
conducted emission
consumer electronics
control equipment
controller error
corp of engineers
counducted emission
customer engineer
customer engineering

CEA
cambridge electron accelerator
canadian electrical association
central electricity authority
circular error average
commodity exchange authority
communications electronics agency
constant extinction angle
control element assembly
council of economic advisers

CEAC
committee on european airspace coordination
consulting engineers association of california
control element assembly computer
county engineers association of california

CEARC
computer education and applied research center

CEAU
continuing education achievement unit

CEB
central electricity board
communication and electronics board

CEC
canadian electric[al] code
caution exchange capacity
certification of equipment completion
charlotte engineers club
colorado engineering council
commission of the european communities
commonwealth edison company
commonwealth engineering conference
communications and electronics command
computer equipment corporation
consolidated electrodynamics corporation
consolidated electronics corporation
constant electric contact
consulting engineers council

CECC
cenelec electronic components committee
consulting engineers council of colorado

CECIL
compact electronic components inspection laboratory

CECMINN
consulting engineers council of minnesota

CECMRL
communications electronics consolidated mobilization reserve list
communications electronics material reserve list

CECNYS
consulting engineers council of new york state, inc.

CECO
commonwealth edison company
consulting engineers council of oregon
cost estimate change order

CECRCA
commonowealth and empire conference on radio for civil aviation

CECS
communications electronics coordinating section
CECTK
committee for electro-chemical thermodynamics and kinetics
CECU
consulting engineers council of utah
CED
communications and electronics division
communications electronics doctrine
CEDAC
central differential analyzer control
cooling effect detection and control
CEDAR
computer aided environmental design analysis and realization
CEDM
control element drive motor
CEDPA
california educational data processing association
certificate in edp auditing
CEDPO
communications electronics doctrinal project office
CEERI
central electronics engineering research institute
CEF
carrier elimination filter
centrifugal electrostatic focusing
complementary emitter follower
CEGB
central electricity generating board
CEI
cleveland electric illuminating
communication electronics instructions
computer exchange, incorporated
computer extended instruction
computing efficiency inc.
consolidated electronics industries
council of engineering institutions
cycle engineers' institute
cycle engineers' institute screw thread
CEIP
communications electronics implementation plan
CEIR
council for economic and industry research
CEL
carbon equilibrium loop
computer economics ltd.
crowding effect laser

CELT
coherent emitter location testbed
CEM
chief electrician's mate
circular electric mode
counter electromotive
CEMA
committee of european promoters of exhibitions of measurement and automation
communications equipment manufacturers' association
conveyor equipment manufacturers association
CEMAC
committee of european association of manufacturers of active electronic components
CEMAD
coherent echo modulation and detection
CEMAST
control of engineering material, acquisition, storage and transport
CEMCBC
chief electrician's mate, construction battalion, communications
CEMCBL
chief electrician's mate, construction battalion, line-and-station
CEMCO
continental electronics manufacturing company
CEMEC
committee of european associations of manufacturers of electronic components
CEMF
counter electromotive force
CEMINDY
cement industry
CEMON
customer engineering monitor
CEMS
central electronic management system
CEMSRG
chief electrician's mate, ship repair, general-electrician
CEMSRS
chief electrician's mate, ship repair, shop-electrician
CEN
center
central
centralization
centralize
CENCER
cen's certification body

CENEL
  committee for the coordination of engineering standards in the electrical field
CENG
  chartered engineer
CENT
  centigrade
  central
CENTIG
  centigrade
CEO
  communications electronics officer
  comprehensive electronic office
CEP
  circle of equal probability
  circular error probability
  circular error probable
  civil engineering package
  computed ephemeris position
  computer entry punch
  corporate electronic publishing
CEPA
  society for computer applications in engineering, planning and architecture, inc
CEPC
  committee of the engineering professors' conference
CEPEC
  committee of european associations of manufacturers of passive electronic components
CEPS
  command-module electrical power system
CEQ
  council on environmental quality
CER
  ceramics
  civil engineering report
  coarse element refinement
  committee electrotechnical, rumanian
  complete engineering release
  critical experiment reactor
CERA
  civil engineering research association
CERAM
  ceramic
  ceramics
CERAMETERM
  ceramic-metal terminal
CERC
  coastal engineering research center
CERCA
  commonwealth and empire radio for civil aviation

CERDIP
  ceramic dual-in-line package
CERL
  central electricity research laboratory
  computer based education research laboratory
CERMET
  ceramic and metal fuel
  ceramic metal element
  ceramic metallic
  ceramic-to-metal
CERPACK
  ceramic package
CERT
  combined environmental reliability test
CERTICO
  certification committee
CES
  candelabra edison screw
  capstone engineering society
  cleveland engineering society
  common-equipment system
  constant elasticity of substitution
  control electronics section
  corporate engineering standard
  critical experiment station
CESA
  canadian engineering standards association
CESAC
  communications electronics special accound code
CESAR
  combustion engineering safety analysis report
CESC
  communications equipment subcommittee
  consulting engineers of south carolina
CESD
  composite external symbol dictionary
  composite external symbol directory
CESI
  closed entry socket insulation
  closed entry socket insulator
  council for elementary science international
CESL
  camp evans signal laboratory
CESR
  canadian electronic sales representatives
CESSE
  council of engineering and scientific society executives
CET
  central european time
  corrected effective temperature
  critical experiment tank for htre-2
  cumulative elapsed time

CETOS
    corporate engineering transfer and obsoletion system
CETR
    consolidated edison thorium reactor
CETS
    conference on european telecommunications satellites
    control element test stand
CEV
    combat engineer vehicle
    corona extinction voltage
CEVAR
    consumable electrode vacuum arc remelt
CEVM
    consumable electrode vacuum melting
CEWA
    canadian electronic wholesalers' association
CEWR
    centimetric early warning radar
CEY
    century electric company
CE
    consolidated edison company
CF
    candle foot
    candle-foot
    capacity factor
    card feed
    carried forward
    carrier frequency
    carry flag
    cathode follower
    center frequency
    center of floatation
    central file
    centrifugal force
    change frequency
    circuit finder
    communications facility
    communications factor
    completion flag
    concept formulation
    confer
    confinement factor
    context free
    control flag
    controller function
    conversion factor
    cost and freight
    count forward
    counterfire
    crest factor
    critical fusion frequency
    crystal filter
    cubic feet
    cubic foot
    current feedback
    current force
    current forcing
CFA
    canadian forestry association
    cascade failure analysis
    cleared for approach
    code of federal regulations
    colour filter array
    core flood alarm
    council of iron foundry associations
    cross field amplifier
    crossed field amplifier
CFAR
    constant false alarm rate
CFB
    c.f. braun and company
CFBS
    canadian federation of biological societies
CFC
    capillary filtration coefficient
    central fire control
    coin and fee checking
    complex facility control
    configuration control register
    crossed field cryotron
CFCS
    cross field closing switch
CFE
    continued fraction expansion
    contractor furnished equipment
CFEC
    cape fear engineers club
CFF
    carry flip-flop
    critical flicker frequency
    critical fusion frequency
CFG
    context free grammar
CFH
    cobol file handler
CFI
    crystal frequency indicator
CFIA
    core flood isolation valve assembly
CFIS
    california fiscal information system
CFL
    calibrated focal length
    context free language
CFLG
    counter flashing

CFLIPFLOP
  carry flip-flop
CFM
  computer field maintenance
  cathode follower mixer
  computer field maintenance ltd.
  confirm
  cubic foot per minute
CFMS
  chained file management system
CFN
  canadian forces network
  cubic feet per hour
CFO
  carrier frequency oscillator
  critical flashover
  critical flashover voltage
CFOR
  conversional fortran
CFP
  computer forms printer
CFPM
  crossed field photomultiplier
CFR
  carbon film resistor
  catastrophic failure rate
  chance failure rate
  code of federal regulations
  commercial fast reactor
  contact flight rules
  cooperative fuel research committee
  coordinating fuel research
  cumulative failure rate
CFRE
  circulating fuel reactor experiment
CFRS
  customer float reporting service
CFS
  carrier frequency shift
  center frequency stabilization
  central forecasting station
  combined file search
  computer firmware systems inc.
  cubic foot per second
CFSG
  cometary feasibility study group
CFSSB
  central flight status selection board
CFSTI
  clearinghouse for federal scientific and technical information
CFT
  charge flow transistor
  core flood tank

CFTC
  commodity futures trading commission
CFUNCTION
  controller function
CFVD
  constant frequency variable dot
CG
  capacitance of grid
  cathode grid
  center of gravity
  centigram
  clutter gate
  coast guard
  code generator
  coincidence gate
  commercial ground
  conjugate gradient
  control grid
  coring
CGA
  compressed gas association
  contrast gate amplifier
CGAE
  cincinnate gas and electric
CGB
  convert gray to binary
CGE
  canadian general electric
CGEL
  cover gas evaluation loop
CGGC
  central germany gee chain
CGH
  commercial ground high
CGI
  compressed gas insulation
  computer generated imagery
  computer guided instruction
  computer managed instruction
CGK
  cathode grid capacitance
CGLI
  city and guilds of london institute
CGLS
  coast guard loran station
CGN
  computer group news
CGP
  capacitance grid plate
CGRA
  chinese government radio administration
CGRADSTA
  coast guard radio station
CGRM
  containment gaseous radiation monitor

CGS
  canadian geotechnical society
  centimeter-gram-second
  circuit group congestion signal
CGSB
  canadian government specifications board
CGSE
  centimeter-gram-second electrostatic
  coast and geodetic survey electrostatic
CGSM
  centimeter-gram-second magnetic
  coast and geodetic survey electromagnetic
CGSSYSTEM
  centimeter-gram-second system
CGT
  current gate tube
CH
  cards per hour
  case-hardened
  chain home
  chamfer
  chamfered
  channel
  chapter
  character
  check
  choke
  compass heading
CHABA
  committee on hearing and bioacoustics
CHAD
  code to handle angular data
CHAG
  compact high performance aerial gun
CHAL
  challenge
CHAN
  channel
CHARAC
  character
  characteristic
CHARM
  checking, accounting and reporting for member firms
CHARSEC
  characters per second
CHAS
  containment heat removal system
CHAT
  clira holddown assembly tool
CHB
  compagnie honeywell bull
CHCF
  component handling and cleaning facility
CHCS
  computer hardware consultants and services, inc.
CHDB
  compatible high-density bipolar
CHE
  channel end
  channel hot-electron
  chemical engineer
CHEC
  channel evaluation and call
CHESNAVFAC
  chesapeake naval division facilities
CHG
  change
  charge
CHGP
  charging pump
CHI
  computer human interaction
CHIC
  complex hybrid integrated circuit
CHICO
  coordination of hybrid and integrated circuit operations
CHIEF
  controlled handling of internal executive functions
CHIL
  current hogging injection logic
  current hogging logic
CHILD
  cognitive hybrid intelligent learning and development
  computer having intelligent learning and development
CHIO
  character orientated input output processor
CHIP
  chip hermeticity in plastic
CHIPS
  chemical engineering information processing system
  clearing house interbank payment system
CHK
  check
CHKPT
  checkpoint
CHL
  current hogging logic
CHN
  chain
CHNL
  channel

CHORI
chief of office of research and inventions
CHPAE
critical human performance and evaluation
CHPS
characters per second
CHR
candle hour
chrominance
condenser heat rejection
CHRG
charge
CHS
canadian hydrographic service
characters per second
CHSEC
characters per second
CHU
centigrade heat unit
CI
card input
carrier-to-interference
cast iron
center island
chlorine institute
circuit interrupter
cold iron soldered joint
color index
combustion institute
compression ignition
configuration item
containment integrity
control interval
course indicator
crystal impedance
curie
current interruption
cut in
cybernetics international
CIA
central intelligence agency
chemical industries association
communication interrupt analysis
computer industry association
computer interface adapter
containment isolation a
CIAC
construction industry advisory council
CIAM
computerized integrated and automated manufacturing
CIB
cobol information bulletin
containment isolation b
CIBSE
chartered institute of building services engineers
CIC
change indicator control
chemical industries council
chemical institute of canada
command interface control
communications intelligence channel
control inquiry card
customer initiated call
CICA
chicago industrial communications association
CICS
customer information control system
customer inventory control system
CICSVS
customer inventory control system virtual storage
CID
centre for information and documentation
charge injection device
commercial item description
communication identifier
compatibility initialization deck
component identification
compositional interdiffusion
core image dictionary
CIDAS
conversational interactive digital analog simulator
CIDP
computer industry development potential
CIE
coherent infrared energy
computer integrated environment
CIEUSA
chinese institute of engineers usa
CIF
central information file
central instrumentation facility
central integration facility
customer information file
CIG
computer investors group, inc.
CIL
clear indicating lamp
computer instrumentation ltd.
core image library
current injection logic
CILOP
conversion in lieu of procurement
CILRT
containment integrated leak rate test

CIM
  canadian institute of mining and metallurgy
  cincinnati milacron
  coded impulse modulation
  communications improvement memorandum
  communications interface module
  computer input microfilm
  computer input microfilming
  computer input multiplexer
  computer integrated manufacturing
  continuous image microfilm
  crystal impedance meter
  cubic inches per minute

CIMCO
  card image correction

CIMM
  canadian institute of mining and metallurgy

CIMOS
  cincinnati milacron operating system

CIMS
  computer installation management system

CIN
  carrier input
  communication identification navigation

CINDA
  computer index of neutron data

CINS
  cento institute of nuclear science
  cryogenic inertial navigating system

CIO
  central input output
  control input output

CIOCS
  communication input output control system

CIOMS
  council for international organizations of medical sciences

CIOU
  custom input output unit

CIP
  cast-iron pipe
  cincinnati milacron, process control division
  common input processor
  compatible independent peripherals
  computer information processing
  construction inspection procedure
  current injection probe

CIPASH
  committee on international programs in atmospheric sciences and hydrology

CIPM
  council for international progress in management

CIPS
  canadian information processing
  canadian information processing society

CIR
  canada-india reactor
  characteristic impedance ratio
  circuit
  computerized information research
  controlled intact reentry

CIRANT
  circular antenna

CIRC
  centralized information reference and control
  circulator
  circumference
  cross-interleaved reed-solomon code

CIRCAL
  circuit analysis

CIRCPROGRAM
  circuit design analysis program

CIRCUS
  circuit simulator

CIRES
  communication instructions for reporting enemy sightings

CIRF
  customer integrated and or reference file system

CIRGA
  critical isotope reactor, genersal atomics

CIRIA
  construction industry research and information association

CIRIN
  circular inch

CIRM
  committee international radio maritime

CIRMIL
  circular mil

CIS
  canadian institute of surveying
  central instructor school
  character instructions set
  communication information system
  computer information service
  computer informations service
  computer oriented information system
  computerized information system
  conductor insulator semiconductor
  containment isolation system
  contental information systems corp.
  convention information system
  cue indexing system
  current information selection

CISC
  complex intruction set computer
CISCO
  compass integrated system compiler
CISIR
  ceylon institute of scientific and industrial research
CISR
  center for international systems research
CIT
  california institute of technology
  call-in-time
  carnegie institute of technology
  compressor inlet temperature
  computer interface technology
CITAB
  computer instruction and training assistance for the blind
CITB
  construction industry training board
CITC
  construction industry training center
CITE
  cargo integration test equipment
  compression ignition and turbine engine
  current information tapes for engieers
CITEC
  contractor independent technical effort
CIU
  computer interface unit
CIV
  center island vessel
  containment isolation valve
CIWS
  close -in weapons systems
CJ
  cold joint
  cold junction
CJF
  connecticut joint federation
  conversational job facility
CJP
  communication jamming processor
CK
  check
  circuit check
  crystal kit
CKD
  completely knocked down
CKMTA
  cape kennedy missile test annex
CKPT
  check point
CKW
  clockwise

CL
  car load lots
  center line
  centilitre
  chlorine
  clear
  closed loop
  computational linguistics
  computer language
  confidence limit
  control language
  conversion loss
  counter logic
  current logic
  cutter location
  cylinder
CLA
  center line average
  clear and add
  communication line adapter
  communication link analyzer
  computer lessors association
  custom logic array
CLAIRA
  chalk lime and allied industries research association
CLAM
  chemical low altitude missile
CLARA
  cornell learning and recognizing automation
CLAS
  computer library applications service
CLASP
  closed line assembly for single particles
  computer language for aeronautics and space programming
CLASS
  capacity loading and scheduling system
  classification
  closed loop accounting for store sales
  computer based laboratory for automated school systems
  containerized lighter aboard ship system
CLASSIC
  classroom interactive computer
CLASSMATE
  computer language to aid and simulate scientific, mathematical, and technical education
CLAT
  communication line adapter for teletype
  communications line adapter
CLBA
  current logical byte address

CLC
  cholesteric liquid crystal
  communications link controller
  computer leasing company
  computer link corporation
  containment leakage control
  contrast light compensation
  course line computer
CLCGM
  closed loop cover gas monitor
CLCIS
  closed loop cover and instrumentation system
CLCLPE
  california legislative council of professional engineers
CLCR
  controlled letter contract reduction
CLCS
  closed loop control system
  consequence limiting control system
  current switching
CLCV
  cold leg check valve
CLD
  called line
  central library and documentation branch
  chloride leak detector
  condensed logic diagram
  current limiting device
CLDAS
  clinical laboratory data acquisition system
CLDATA
  cutter location data
CLEA
  conference on laser engineering and applications
CLEAN
  commonwealth law enforcement assistance network
CLEHA
  conference of local environmental health administrators
CLEM
  closed loop ex-vessel machine
  composite for the lunar excursion module
CLEO
  clear language for expressing orders
CLETS
  california law enforcement telecommunications system
CLF
  capacitive loss factor
CLFM
  coherent linear frequency modulated

CLG
  calling line
  cancelling
CLI
  computer linguistics, inc.
CLIC
  communication linear integrated circuit
CLIM
  cellular logic-in-memory
CLIP
  cellular logic image processor
  compiler language for information processing
  compiler language for information processing system
CLIRA
  closed loop in-reactor assembly
CLIS
  computer linked information for container shipping
CLIV
  cobol version iv
  cold leg isolation valve
CLJA
  closed loop jumper assembly
CLK
  clock
CLKW
  clockwise
CLM
  column
CLO
  cellular logic operation
CLR
  center of lateral resistance
  clean liquid radwater
  clear
  combined line and recording trunk
  combustible limit relay
  common line receiver
  computer language recorder
  computer language research
  contact load resistor
  coordinating lubricant and equipment research
  coordinating lubricants research
  council on library resources
  current limiting resistor
CLRU
  cambridge language research unit
CLS
  cask loading station
  clear and substract
  close
  closed loop system
  common language system

concept learning system
containment leakage system
control language services
CLSMDA
  closed loop system melt-down accident
CLSSEARCH
  conditional limited semantics search
CLT
  code language telegram
  communication line terminal
  communications lines terminator
  computer language translator
CLU
  central logic unit
  circuit line-up
CLUE
  computer learning under evaluation
CLUT
  computer logic unit tester
CLW
  clockwise
CM
  calibration magnification
  center of mass
  centimeter
  central memory
  circular mil
  clear memory
  command module
  common mode
  communication multiplexer
  communications multiplexer
  comparator
  computer microtechnology
  computer module
  configuration management
  connecting machine
  construction and machinery
  control mark
  controlled mine field
  core memory
  corrective maintenance
  countermeasure
  cross modulation
  cubic centimeter
  metric carat
CMA
  canadian medical association
  classified mail address
  computer monitor adapter
  contact making ammeter
CMAA
  crane manufacturers association of america
CMAC
  cerebellar model arithmetic computer

CMANY
  communications managers association of new york
CMC
  code for magnetic characters
  coherent multi-channel
  communicating magnetic card
  communications magnetic card
  communications mode control
  computer machinery corp.
  computer machinery corporation
  concurrent media conversion
  contact making clock
CMCA
  cruise missile carrier aircraft
CMCTL
  current mode complementary transistor logic
CMD
  computer memory device inc.
  core memory driver
CMDAC
  current mode digital-to-analog converter
CMDR
  command reject
CME
  computer measurement and evaluation
CMERI
  central mechanical engineering research institute
CMES
  computer and management education services
CMF
  central maintenance facility
  coherent memory filter
  command file
  common mode failure
  comprehensive management facility
CMFA
  common mode failure analysis
CMFLPD
  core maximum fraction of limiting power density
CMG
  computer management group
  computer measurement group
  control moment gyro
  control moment gyroscope
CMI
  cambridge memories inc.
  christian michelsen's institute
  common mode interface
  commonwealth mycological institute
  computer managed instruction
  core element assembly motion inhibit

CMIC
   central management information center
CMIL
   circular mil
CMIN
   cards per minute
CMIS
   common manufacturing information system
   computer-oriented management information system
CMISCIRCUIT
   complementary metal-insulator-semiconductor circuit
CML
   common machine language
   critical mass laboratory
   current mode logic
CMLCENCOM
   chemical corps engineering command
CMM
   commission for maritime meteorology
   communications multiplexer module
   coordinate measuring machine
   core mechnanical mockup
CMMF
   component maintenance and mockup facility
CMMP
   carnegie mellon multi mini processor
   commodity management master plan
CMMU
   cache memory management unit
CMN
   communication
CMOS
   complementary metal oxide semiconductor
   complementary mos
CMOSSOS
   complementary metal oxide semiconductor silicon on sapphire
CMP
   central monitoring position
   compare
   configuration management plan
   controlled materials production
   corporate manufacturing practice
CMPF
   core maximum power fraction
CMPL
   complement
CMPLX
   complex
CMPS
   centimeter per second
CMPT
   component
   computer
CMR
   committee on manpower resources
   common mode rejection
   communications monitoring report
   continuous maximum rate
   continuous maximum rating
CMRR
   common mode rejection ratio
CMS
   cambridge monitor system
   cash management system
   clay minerals society
   compiler monitor system
   computer based message system
   computer management system
   conversational monitor system
   current mode switching
CMT
   circuit master tape
   conversational mode terminal
   core measurement table
   corrected mean temperature
CMTDETECTOR
   cadmium-mercury-telluride detector
CMTOCM
   common mode-to-common mode
CMTODM
   common mode-to-differential mode
CMV
   common mode voltage
   contact making voltmeter
CMVM
   contact making voltmeter
CMVPCB
   california motor vehicle pollution control board
CMX
   cable multiplexer
CN
   carrrier-to-noise
   cetane number
   commutated network
   compass north
   compensator
   connection machine
   coordination number
   cosine
CNA
   canadian nuclear association
   common network architecture
   communictions network architecture

copper nickel alloy
cosmic noise absorption
CNAA
  council for national academic awards
CNAS
  civil navigation aids system
CNC
  communications network controller
  computer aided numerical control
  computer numerical control
  computerized numerical control
  consecutive number control
CNCE
  council for non-collegiate continuing education
CNCP
  communications network control processor
CNCT
  connect
CND
  condition
CNDP
  communication network design program
CNI
  called number identification
  communication navigation identification
  consolidated national intervenors
CNJ
  copper nickel jacket
CNL
  cancellation
  cancellation message
  circuit net loss
  control
CNLA
  council on national library associations
CNM
  communication network management
CNO
  computer not operational
CNOP
  conditional non operation
CNP
  communications network processor
CNR
  carrier- to- noise ratio
  clutter- to- noise ratio
CNRC
  canadian national research council
CNS
  communication network system
  congress of neurological surgeons
  control network system
CNSL
  console

CNT
  celestial navigation trainer
  count
  counter
CNTOR
  contractor
CNTP
  committee for a national trade policy
CNTR
  center
CO
  care of
  cash order
  cathode ray oscilloscope
  certificate of origin
  changeover
  checkout
  chief operator
  classifier overflow
  close open operation
  cobol
  colorado
  combined operations
  communciations officer
  communication
  contracting officer
  crystal oscillator
  customer order set
  cutt off
COA
  college of aeronautics
COAC
  clutter operated anti clutter
COAM
  customer owned and maintained
COARIN
  committee on arabic in informatics
COAX
  coaxial
  coaxial cable
COB
  close of business
COBESTCO
  computer based estimating technique for contractors
COBIS
  computer based instruction system
COBLIB
  cobol library
COBLOC
  codap language block oriented compiler
COBOL
  common business oriented language

COC
  coded optical character
  computer on the chip
COCAS
  customer order control automated system
COCODE
  compressed coherency detection
COCOM
  coordinating committee for exports to communist areas
  coordinating committee on east west trade
COCOS
  control communications software
COD
  carrier on board delivery
  cash on delivery
  chemical oxygen demand
  coding
  crack opening displacement
CODACOP
  cobol data communication philips
CODAN
  carrier operated device antinoise
  coded analysis
  coded weather analysis
CODAP
  control data assembly program
CODAR
  correlation display analyzing and recording
CODAS
  customer oriented data retrieval and display system
CODASYL
  conference on data system languages
CODATA
  committe on data for science and technology
CODEC
  coder decoder
CODEL
  computer developments limited automatic coding system
CODEM
  computer graphics augmented design and manufacturing
CODES
  commutating detection system
  computer design and evaluation system
CODEVER
  code verification
CODEX
  codex alimentarius commission
CODIC
  color difference computer
  computer directed communication

CODIL
  control diagram language
CODIT
  computer direct to telegraph
CODOT
  classification of occupations and directory of occupational titles
CODSIA
  council of defense and space industry associations
COE
  computer operations europe
COED
  char oil energy development
  computer operated electronic display
COEF
  coefficient
COF
  computer operations facility
  confusion signal
COFG
  center of gravity
COFIL
  core file
COFIRS
  cobol for ibm rpg systems
COG
  center of gravidity
  computer operations group
COGENT
  compiler and generalized translator
COGO
  coordinate geometry
COGS
  continuous orbital guidance sensor
  continuous orbital guidance system
COH
  coherent
COHO
  coherent oscillator
COI
  communication operation instructions
COINS
  computer and information sciences
COL
  co latitude
  collector
  column
  computer oriented language
COLASL
  compiler los alamos scientific laboratory
COLIDAR
  coherent light detection and ranging
COLINGO
  compile on line and go

COLL
  collector
COLOG
  cologarithm
COLT
  central office line tester
  communication line terminator
  computer oriented language translator
  computerized on line testing
  control language translator
COLTS
  communication on line testing system
COM
  coefficient of merit
  commercial
  common
  communication switching
  communication system disciplines
  communication technology
  communication theory
  communications
  commutator
  compiler
  complement
  computer output film
  computer output microfiche
  computer output microfilm
  computer output microfilming
  computer output microform
  computer output on microfilm
  computer output on microfilm or microfiche
  computer output to microfilm
  curve of merit
COMAC
  continuous multiple access
  continuous multiple access collator
  continuous multiple access comparator
COMAL
  common algorithmic language
COMAR
  committee on man and radiation
COMAT
  computer-assisted training
COMATS
  corporate manufacturing transfer system
COMB
  combine
COMCM
  communications countermeasures and deception
  communications countermeasures and detection
COMD
  command

COMEINDORS
  composite mechanized information and documentation retrieval system
COMET
  computer operated management evaluation technique
COMJAM
  communications jamming
COMLOGNET
  combat logistics network
  communications logistics network
COMM
  communication
  commutator
COMMEL
  communications electronics
COMMEN
  compiler oriented for multiprogramming and multiprocessing environments
COMMEND
  computer aided mechanical engineering design system
COMMSWITCH
  communications failure detecting and switching equipment
COMNET
  communications network
COMP
  comparator
  compensator
  compiler
  component
  composition
  compuer
  computer
COMPAC
  computer program for automatic control
COMPACS
  computer oriented manufacturing production and control system
COMPACT
  compatible algebraic compiler and translator
  computer oriented modular planning and control technique
  computer planning and control technique
COMPANDER
  compressor expander
COMPARE
  computerized performance and analysis response evaluator
  console for optical measurement and precise analysis of radiation from electronics
COMPASS
  compiler assembler
  complete parallel activity and security system

computer for advanced spare systems
computer-assisted
computer-assisted classification and assignment system
COMPCON
computer convention
COMPCORP
computer design corp
COMPEL
compute parallel
COMPOOL
communication pool
COMPOW
committee on professional opportunities for women
COMPRESS
computer research, systems and software
COMPROG
computer program
COMPSO
computer software and peripheral show
COMRADE
computer aided design environment
COMSAL
computer method of sequencing operations for assembly lines
COMSAT
communication satellite
COMSEC
communications security
COMSL
communication system simulation language
COMSOAL
computer method of sequencing operations for assembly lines
COMSS
compare string with string
COMSTAT
competitive statistical analysis
COMSTORE
computer storage
COMSW
compare string with word
COMSYSDISC
communication system disciplines
COMTHEORY
communication theory
COMTRAC
computer aided centralized traffic control system
computer aided traffic control
COMTRAN
commercial translator
COMZ
communications zone

CON
concentration
connection
console
constant
contactor
continued
control
controller
converter
CONAD
configuration advisor
CONC
concentration
CONCEPT
computation on line of networks chemical engineering process technology
COND
condenser
condition
conductivity
conductor
CONDOX
conditions for dx communication
CONELRAD
control of electromagnetic radiation
CONN
connecting
CONO
confirmation of number of order
CONRAD
committee on radiology
CONS
carrier operated noise suppression
CONSORT
conversational system with on line remote terminals
CONST
constant
construction
CONSTR
construction
CONSUL
control subroutine language
CONT
contact
contents
continued
continuous
control
controller
CONTD
continued
CONTHP
continental horsepower

CONTR
  control
  controller
CONTRAN
  control translator
CONTRANS
  conceptual thought, random net simulation
CONTS
  contents
CONUS
  continental united states complex
CONV
  convergence
  conversion
  converter
COOL
  checkout oriented language
  control oriented language
COOLS
  community office on line system
COP
  central operator's panel
  central ordering point
  coaxial output printer
  coefficient of performance
  communications output printer
  computer optimization package
  control optimization
  customer order processing
COPANT
  council of the pan american standards commission
COPE
  committee on political education
  communciations oriented processing equipment
  communications oriented peripheral equipment
COPI
  computer oriented programmed instruction
COPIC
  computer program information center
  computer program information system
COPICS
  communication oriented production and control system
  communication oriented production information control system
COPOLCO
  consumer policy committee
COPPS
  committe on power plant siting matopma
COPSAC
  computer order processing and sales accounting

COR
  carrier operated relay
  exclusive or
CORA
  coherent antenna array
  coherent radar array
  conditioned reflex analog
  conditioned response analog
CORAD
  correlation radar
CORAL
  computer on line real time applications language
  correlated radio lines
CORAPRAN
  cobelda radar automatic preflight analyzer
CORC
  cornell computing language
CORDIC
  coordinate rotation digital computer
CORDS
  coherent on receive doppler system
CORE
  computer oriented reporting efficiency
  computer related equipment
  core computer related equipment
CORELAP
  computerized relationship layout planning
COREP
  combined overload repair control
CORNET
  control switching arrangement network
CORR
  corrected
CORREGATE
  correctable gate
CORRESP
  corresponding
CORS
  canadian operational research society
CORTS
  convert range telemetry system
COS
  calculator on substrate
  card operating system
  centralized operating system
  commercial operating system
  communication operating system
  communication operation station
  communication oriented system
  compact operating system
  compatibility operating system
  compatible operating system
  complementary switching
  complementary symmetry

concurrent operating system
cosine
cosinus
cosmic
cosmic rays and trapped radiation committee
COSAR
compression scanning array radar
COSATI
committee on scientific and technical information
COSBA
computer services and bureaux association
computer services bureau association
COSEC
cosecant
COSI
committee on scientific information
COSINE
committee on computer science in electrical engineering education
COSIP
college science improvement program
COSMIC
computer software management and information center
COSMON
component open short monitor
COSMOS
complementary symmetric mos
complementary symmetrical metal oxide semiconductor
complementary symmetrical metal oxide semiconductor field effect transistor
complementary symmetry mos
complementary symmetric metal oxide semiconductor
computer oriented system for management order synthesis
computer oriented system for manufacturing order synthesis
COSOS
conference on self-operating systems
COSPAR
committee on space research
COSPUP
committee on science and public policy
COSRIMS
committee on research in the mathematical sciences
COSRO
conical scan on receive only
COSTING
cost account and information system guide
COSWITCH
closed open switch

COT
cotangent
COTAR
correlation tracking and ranging
cosine trajectory angle and range
COTRAN
cobol to cobol translator
COUL
coulomb
COUTS
computer operated universal test system
COV
covered
cut out valve
COWPS
council on wage and price stability
COZI
communications zone indicator
CP
calculator printing
calorific power
candlepower
card punch
card to printer
card to punch
centipoise
central processor
center of pressure
change point
change proposal
charter party
checkpoint
chemical pure
chemically pure
circuit package
circular pitch
circularly polarized
clock phase
clock pulse
code of practice
coefficient of performance
cold press
command processor
command pulse
communication processor
computer
conference paper
constant potential
constant pressure
construction permit
continuous path
control panel
control processor
control program
cosmogenic

counter poise
current paper
customized processor
CPA
    canadian pharmaceutical association
    canadian psychological association
    channel program area
    closest point of approach
    color phase alternation
    computer power australia
    critical path analysis
    cross program auditor
CPAB
    computer programmer aptitude battery
CPAL
    containment person air lock
CPB
    channel program block
CPC
    card programmed calculator
    card programmed computer
    ceramic printed circuit
    clock pulse control
    coated powder cathode
    computer process control
    computer program component
    computer programming concepts
    construction project control
    core protection calculator
    core protection computer
    current papers on computer and control
    cycle program control
    cycle program counter
    cyclic permutation code
CPCEI
    computer program contract end item
CPCODE
    cyclic permutated code
CPCOM
    capsule communications
CPCONTROL
    continuous path control
CPCYCLE
    constant pressure cycle
CPD
    call per day
    card per day
    coil predriver
    community planning and development office of assistant secretary
    compound
    computer produced drawing
    consolidated programming document
    contact potential difference
    cummulative probability distribution

CPDD
    conceptual project design description
CPDS
    computer program design specification
CPE
    central processing element
    central programmer and evaluator
    charged particle equilibrum
    circular probable error
    computer performance evaluation
    contractor performance evaluation
    counter position exit
    cross program editor
    current papers in electrical and electronics engineering
CPEA
    college physical education association
    cyprus professional engineers' association
CPEM
    conference on precision electromagnetic measurement
CPEUG
    computer performance evaluation users group
CPF
    central processing facility
    complete power failure
CPFF
    cost plus fixed fee
CPG
    certified program generator
    college publishers group
    current pulse generator
CPH
    candle power hour
    cards per hour
    closed packed hexagonal
CPHS
    containment pressure high signal
CPI
    changes per inch
    characters per inch
    clock pulse interval
    computer peripheral inc
    computer prescribed instruction
    control position indicator
CPIA
    chemical propulsion information agency
CPILS
    correlation protected instrument landing system
    correlation protected integrated landing system
    correlation protection integrated landing system

CPIOCS
   communication physical input output control system
CPIP
   computer program implementation process
CPL
   canadian program library
   characters per line
   charge pumping logic
   combined programming language compiler
   computer program library
   conversational planning language
   core performance log
   couple
   current product line
CPLG
   coupling
CPM
   calls per minute
   cards per minute
   cathode pulse method
   cathode pulse modulation
   characters per minute
   cobol performance monitor
   count per minute
   critical path method
   cycles per minute
CPMA
   computer peripheral manufacturers' association
CPMM
   characters per millimeter
CPO
   concurrent peripheral operations
CPOI
   customer order processing and invoicing
CPP
   card print processor
   card punching printer
   center for plutonium production
   computer professionals for peace
   containment pressure protection
   controllable pitch propeller
CPPS
   critical path planning and scheduling
CPR
   committee on polar research
   control program for real time
   critical power ratio
CPRG
   computer personnel research group
CPS
   capacity planning system
   card programming system
   cards per second
   cathode potential stabilized
   central processing system
   characters per second
   circuit package schematic
   clock pulse
   console programming system
   control program services
   control programs support
   conversational programming system
   conversion program system
   conversional programming system
   corporate planning system
   count per second
   critical path scheduling
   cycle per second
CPSA
   consumer product safety act
CPSC
   consumer product safety commission
CPSCI
   central personnel security clearance index
CPSE
   counterpoise
CPSK
   coherent phase shift keying
CPSS
   common programming support system
CPT
   capacitive pressure transducer
   chief programmer team
   cockpit procedures trainer
   control power transformer
   critical path technique
   customer provided terminal
CPTA
   computer programming and testing activity
CPU
   central processing unit
   collective protection unit
   communications processor unit
   computer peripheral unit
CPV
   control program five
CPX
   charged pigment xerography
CPY
   copy
CQ
   commercial quality
   congressional quarterly
   correspondence quality
CQE
   cognizant quality engineer

CQM
  class queue management
CR
  call request
  card reader
  card reproducer
  carriage return
  carry register
  catalytic reforming
  cathode ray
  chain radar
  character reader
  chloroprene rubber
  citizen radio
  code receiver
  cold rolled
  command receiver
  command register
  common return
  computing reviews
  conference report
  congressional record
  contact resistance
  containment rupture
  contract report
  control register
  control relay
  control rod
  controlled rectifier
  crate
  credit
  crystal
  crystal rectifier
  current rate
  current relay
CRA
  carry ripple adder
  catalog recovery area
  control rod assembly
  cosmic ray altimeter
CRAC
  careers research and advisory center
CRAD
  committee for research into apparatus for the disabled
CRAF
  civillian reserve air fleet
CRAFT
  changing radio automatic frequency transmission
  computerized relative allocation of facilities technique
CRAM
  card ram
  card random access memory
  computerized reliability analysis method
  conditional relaxation analysis method
CRAMM
  coupon reading and marking machine
CRB
  computer resale broker
  computer resale brokers international ltd
CRBE
  conversational remote batch entry
  conversional remote batch entry
CRBR
  clinch river breeder reactor
  controlled recirculation boiling water reactor
CRBRP
  clinch river breeder reactor project
CRC
  carriage return character
  carriage return contact
  character recognition circuit
  civil rights commision
  coordinating research council
  copy research council
  critical reactor component
  cumulative results criterion
  cyclic redundancy check
  cyclic redundancy checking
  cyclic redundant code
CRCC
  cyclic redundancy check character
  cyclic redundancy check code
CRCIES
  cedar rapids chapter iowa engineering society
CRCN
  counterreconnaissance
CRCPD
  conference of radiation control program directors
CRCTA
  composite reactor components test activity
CRD
  capacitor resistor diode
  control rod drive
CRDA
  control rod drive assembly
CRDF
  canadian radio direction finding or finder
  cathode ray direction finder
  cathode ray direction finding
CRDL
  chemical research and development laboratories
CRDM
  control rod drive mechanism
  control rod drive motor

CRDME
  committee for research into dental materials and equipment
CRDSD
  current research and development in scientific documentation
CRE
  controlled residual element
  corrosion resistant
CREAM
  computer realtime access method
CREDO
  central reliability data organization
CRES
  corrosion resistant
  corrosion resistant steel
CRESS
  central regulatory electronic stenographic system
  combined reentry effort for small systems
  computerized reader enquiry service system
CRESSAU
  center for research in social systems of the american university
CREST
  committee on reactor safety and technology
CRESTS
  courtauld's rapid extract, sort and tabulate system
CREVS
  control room emergency ventilation system
CRF
  capital recovery factor
  carrier frequency telephone repeater
  central retransmission facility
  compressor research facility
  control relay forward
  correspondence routing form
  cryptographic repair facility
CRI
  color rendering index
  committee for reciprocity information
CRIG
  capacitor rate integrating gyroscope
CRIME
  censorship records and information, middle east
CRIMP
  customer report on importance
CRIS
  command retrieval information system
  command retrieval system
  current research information system

CRIT
  critical
  critical mass
CRJE
  conversational remote job entry
CRLB
  cosmic ray logic box
CRM
  computer resource management
  containment radiation monitor
  control and reproducibility monitor
  core restraint mechanism
  count rate meter
  counter radar measures
  counter radar missile
CRN
  charge routing network
  common random numbers
CRNL
  chalk river nuclear laboratories
CRO
  cathode ray oscillograph
  cathode ray oscilloscope
  control room operator
CROM
  capacitive read only memory
  capacitive rom
  control read only memory
  control rom
CROS
  capacitor read only store
  card capacitor read only store
CRP
  card reader punch
  common reference point
  computer room people
  controlled reliability program
CRPL
  central radio propagation laboratory
CRPM
  communication registered publication memoranda
  crankshaft revolutions per minute
CRR
  conversion result register
CRREL
  cold regions research and engineering laboratory
CRS
  chain radar system
  citizen radio service
  cold rolled steel
  congressional research service
  containment rupture signal

CRSA
  control rod scram accumulator
CRST
  cold rolled steel
CRT
  cathode ray tube
  channel response time
  circuit requirement table
  control relay translator
CRTC
  cathode ray tube controller
CRTF
  create test file
CRTPB
  canadian radio technical planning board
CRTS
  controllable radar target simulator
CRTU
  combined receiving and transmitting unit
CRU
  combined rotating unit
  communication register unit
  computer resource unit
  customer replaceable unit
CRUD
  chalk river unidentified deposit
CRV
  cone resistance value
  constant reflector voltage
CRW
  community radio watch
CRWO
  coding room watch officer
CRYOSAR
  cryo switching by avalanche recombination
  cryogenic switching by avalanche and recombination
CRYPTO
  cryptographic
CRYS
  crystal
  crystallography
CRYST
  crystalline
CS
  cable ship
  call signal
  card socket
  cast steel
  catalysis society
  centistoke
  channel status
  check sorter
  check surface
  chief of staff
  chipselect
  coding specification
  colliery screened
  column split
  commercial standard
  communication system
  communications
  community service
  computer science
  conducted susceptibility
  containment spray
  context sensitive
  continuous commercial service
  continuous scan
  control section
  control signal
  control station
  control store
  control switch
  control system
  core sharing
  cotton silk
  countersink
  countersunk
  cross section
  current strength
  current switch
  current switching
  cycle shift
  cycle stealing
  cycles per second
  hydrocarbons
CSA
  canadian standard association
  canadian standards association
  carry save adder
  certificate in systems analysis
  command session abort
  common system area
  community service activities
  community services administration
  computer sciences association
  computer sciences of australia
  computer systems association
  computer systems of america
  cyrogenic society of america
CSAE
  canadian society of agricultural engineering
CSAP
  control system analysis program
CSAR
  communications satellite advanced research
  control, security, auditing, recovery
CSAS
  cove standy actuation signal

CSB
  carrier and sidebands
  channel status byte
  consumer sounding board
CSBCODE
  complementary straight binary code
CSC
  care store control
  circuit switching center
  civil service commission
  command scheduling chain
  common signaling channel
  communcation simulator console
  communication systems center
  computer sciences corp
  computer sciences corporation
  computer society of canada
  construction specifications canada
  core standby cooling
  cosecant
  course and speed computer
CSCC
  command session change control
CSCE
  connecticut society of civil engineers
CSCHE
  canadian society for chemical engineering
CSCI
  canadian society for computational studies of intelligence
CSCS
  core standby cooling system
CSD
  cold shutdown
  communication system development
  computer system department
  computer systems design
  constant speed drive
  controlled slip differentials
  cross spectral density
CSDD
  conceptual system design description
CSDF
  core segment development facility
CSDL
  current switching diode logic
CSDN
  circuit switched digital network
CSE
  circuit switched exchange
  colorado society of engineers
  command session end
  containment steam explosion
  containment systems experiment
  control and switching equipment
  control systems engineering
  core storage element
CSECT
  control section
CSEE
  canadian society for electrical engineers
CSEF
  current switch emitter follower
CSEPA
  central station electrical protective association
CSF
  carrier suppression filter
  communication serviceability facilities
  containment support fixture
  coulter steel and forge
CSG
  casing
  constructive solid geometry
  context sensitive grammar
CSI
  computer sciences international
  computer systems international
  construction specifications institute
CSIC
  computer system interface circuits
CSINK
  countersink
CSINPUT
  chip select input
CSIR
  council of scientific and industrial research
CSIRAC
  commonwealth scientific and industrial research automatic computer
CSIRO
  commonwealth scientific and industrial research organization
CSIS
  core spray injection system
CSJFET
  charge storage junction field effect transistor
CSK
  countersink
CSL
  code selection language
  component source list
  computer sensitive language
  console
  context sensitive language
  control and simulation language
  controlled saturation logic
  current sink logic

current sinking logic
current sourcing logic
current switch logic
CSM
  command and service module
  continuous sheet memory
  crosspoint switching matrix
  current switching mode
CSMA
  carrier sense multiple access
  chemical specialities manufacturers association
  communications systems management association
CSMACD
  carrier sense multiple access collision detection
CSME
  canadian society for mechanical engineering
CSMOL
  control station manual operating level
CSMP
  continuous system modeling program
CSMPS
  computerized scientific management planning system
CSMT
  clock and simulation tape maintenance message type
CSN
  computer service network
CSO
  chained sequential operation
  chief signal officer
  commanding signal officer
  computer service office
CSP
  coder sequential pulse
  commercial subroutine package
  communications satellite program
  communications security publication
  communications symbiont processor
  continuous sampling plan
  control and switching point
  control setting panel
  control switching point
  council for scientific policy
CSPG
  code sequential pulse generator
CSPM
  communications security publication memorandum
CSPO
  communications satellite project office

CSR
  circulating shift register
  clamped speed register
  clamped speed regulator
  console send receive
  constant stress rate
  control and status register
  control shift register
  controlled silicon rectifier
  counter, shift register
CSRA
  chicago suburban radio association
CSRS
  cooperative state research service
CSS
  cask support structure
  clock subsystem
  colorado scientific society
  com share southern
  command session start
  communications subsystem
  computer sales and services
  computer scheduling system
  computer system simulator
  containment spray system
  conversational software system
  core support structure
  corporate shareholder system
  cryogenic storage system
CSSA
  computer society of south africa, ltd
  crop science society of america
CSSB
  compatible single sideband
CSSE
  conference of state sanitary engineers
CSSG
  computer software and services group
CSSL
  continuous system simulation language
CSSS
  canadian soil science society
CSST
  computer system science training
CST
  centistoke
  central standard time
  channel status message
  channel status table
  code segment table
  combined station tower
  commercial subsurface transformer
  count down sequence timer
CSTC
  canadian signals training center

CSTI
   committee on scientific and technical information
CSTS
   combined system test stand
   computer sciences teleprocessing system
   computer sciences time sharing system
CSU
   central switching unit
   channel service unit
   circuit switching unit
   clear and subtract
   constant speed unit
   customer service unit
   cycle stealing unit
CSUI
   command session user information
CSV
   corona start voltage
   corona starting voltage
CSW
   channel status word
   continuous seismic wave
   control power switch
CT
   cable test
   cable transfer
   carrier telegraph channel
   carrier telephone
   carrier telephony
   center tap
   center tapped
   central time
   checkout tape
   chronometer time
   cipher telegram
   circuit
   circuit technology
   circuit theory
   clem transporter
   coastal telegraph
   coastal telegrapher
   code table
   code telegram
   command transmitter
   commercial translator
   communications technician
   computer technology
   computer terminal
   computer tomography
   conductivity transmitter
   conduit
   connecticut
   console typewriter
   control transformer
   correct time
   count
   counter
   counter timer
   counter tube
   current
   current transformer
   total capacitance
CTA
   call time adjuster
   compatibility test area
   concrete technicians association of hawaii
   control area
   controlled airspace
CTAP
   circuit transient analysis program
   contact approach
CTB
   commonwealth telecommunications board
   computer time brokers
CTC
   cam timing contact
   cards to tape converter
   cassette tape controller
   central tracking center
   central train control
   central transformer corporation
   centralized traffic control
   channel traffic control
   coaxial thermal converter
   color television committee
   computer time sharing corporation
   computer transmission corporation
   contact
   counter timer circuit
CTCA
   channel and traffic control agency
CTCCODE
   complementary twos complement code
CTD
   charge transfer device
   charged tape detection
CTDH
   command and telemetry data handling
CTDS
   code translation data system
   count, time data system
CTE
   cable termination equipment
   central timing equipment
   channel translating equipment
   coefficient of thermal expansion
   computer telex exchange

CTEL
   chief of telecommunications
CTF
   cask tilting fixture
   certificate
   change to tower frequency
   common test facility
   core test facility
CTFM
   continuous transmission frequency modulated
   continuous transmission frequency modulation
CTG
   cartridge
   coating
   cotangent
CTI
   cartridge television inc
   charge transfer inefficiency
   color television, incorporated
   communications technician intercept radioman
   communications technology, incorporated
CTL
   cage test language
   central
   checkout test language
   complementary transistor logic
   component test laboratory
   computer technology ltd
   constructive total loss
   control
   control line
   core transistor logic
CTLD
   controlled
CTM
   communication terminal module
CTMC
   communication terminal module controller
CTMT
   containment
CTN
   cable termination network
   cotangent
CTO
   central telegraph office
   charge transforming operator
   cut through operate
CTOC
   center-to-center
   communications technical operations center
CTOL
   conventional take off and landing
CTOS
   cassette operations system
CTP
   central transfer point
   charge transformating parameter
   charge transforming parameter
   chemical treatment pond
   command translator and programmer
   confidence test program
   construction test procedure
   controlled temperature profile
CTR
   cathode ray tube
   center
   certified test record
   certified test results
   complementary transistor register
   contour
   control zone
   controlled thermonuclear reaction
   controlled thermonuclear reactor
   core transistor register
   counter
   current transfer ratio
   cutter
CTRI
   catholic tape recorders, international
CTRL
   complementary transistor restistor logic
   control
CTRS
   centers
CTS
   cable with cabtyre-sheathing
   cabtyre sheathing
   carriage tape simulator
   carrier test switch
   chicago telephone supply
   circuits to specifications
   clear to send
   command telemetry system
   communication and tracking subsystem
   communication terminal synchronous
   communications technician special devices operator
   communications technology satellite
   component test set
   component test system
   computer telewriter system
   controlled thermal severity
   conversational terminal system
   conversational time sharing
   courier transfer station
CTSR
   continuously fed stirred tank reactor

CTSS
  compatible time shared system
  compatible time sharing system
CTT
  card- to- tape tape
  central trunk terminal
  color trace tube
CTTL
  complementary transistor transistor logic
CTU
  centigrade thermal unit
  central terminal unit
  central timing unit
  commercial telegraphers' union
  components test unit
CTUNA
  commercial telegraphers' union of north america
CTV
  color television
  command test vehicle
  commercial television
CTW
  console typewriter
  control typewriter
CTWT
  counterweight
CTZ
  control zone
CU
  central unit
  computing unit
  consumer union
  control unit
  copper
  crosstalk unit
  crystal unit
  cube
  cubic
  cumulative
  customer premise
  customer use
CUA
  computer aided instruction
  computer users association
CUAS
  computer utilization accounting system
CUB
  central unit buffer
CUBE
  cooperating users of burroughs equipment
CUBOL
  computer usage business oriented language
CUC
  computer usage company

CUCB
  control unit control block
CUCM
  cubic centimeter
CUCOSS
  california universities council on space sciences
CUD
  craft union department
CUDM
  cubic decimeter
CUDN
  common user data network
CUDOS
  continuously updated dynamic optimizing system
CUE
  command uplink electronics
  communications unit executor
  computer update equipment
  computer updating equipment
  computer utilization efficiency
  configuration utilization evaluator
  cooperating users exchange
CUERL
  columbia university electronic research laboratory
CUES
  computer utilities for education systems
  computer utility educational system
CUFT
  cubic feet
  cubic foot
CUI
  control unit interface
CUIN
  cubic inch
CUJT
  complementary unijunction transistor
CULP
  computer usage list processor
CUM
  computer utilization monitor
  cumulative
CUMM
  council of underground machinery manufacturers
  cubic millimeter
CUMMFU
  complete utter monumental military foul up
CUMN
  cubic micron
CUP
  communication user program

CUR
  currency
  current
CURES
  computer utilization reporting system
CURLIMITING
  current limiting
CURR
  current
CURRT
  current
CURT
  cubic root
  current
CURTS
  common user radio transmission system
CURV
  cable controlled underwater research vehicle
CUS
  clean up system
  control, utility and support
CUSC
  channel unit signal controller
CUSEC
  cubic feet per second
CUSIP
  committee on uniform security identification procedures
CUT
  circuit under test
  control unit tester
  coordinated universal time
CUTS
  computer utilized turning system
CUW
  committee on undersea warfare
CUYD
  cubic yard
CV
  calorific value
  capacitance as a function of voltage
  capacitance voltage
  constant velocity
  constant voltage
  constant volume
  continuous vulcanization
  continuously variable
  convair
  converter
  convertible
  coulomb volt
  counter voltage
  single cotton varnish
CVAC
  consolidated vultee aircraft

CVB
  convert binary
  convert to binary
CVBEM
  complex variable boundary element method
CVC
  conserved vector current
  consolidated vacuum corporation
  current voltage converter
CVCC
  constant voltage constant current
CVCL
  constant voltage current limiting
CVCP
  committee of vice chancellors and principals
CVCS
  chemical and volume control system
CVD
  chemical vapor deposit
  chemical vapor deposition
  chemical vapor deposition technique
  chemically vapor deposited
  convert decimal
  convert to decimal
  coupled vibration dissociation
  current voltage diagram
CVDV
  coupled vibration dissociation vibration
CVI
  certified vendor information
  communication, navigation and identification
CVIC
  conditional variable incremental computer
CVP
  containment vacuum pump
CVPDS
  command video prelaunch distribution system
CVPS
  central vermont public service
CVR
  carrier vessel reactor
  cockpit voice recorder
  constant voltage reference
  continuous video recorder
  controlled visual rules
  current voltage regulator
CVS
  computer controlled vehicle system
CVSD
  continuously variable slope delta modulation
  continuously variable slope delta modulator
CVSG
  channel verification signal generator

CVT
  communication vector table
  constant voltage transformer
  convertible
CVTR
  carolinas virginia tube reactor
CVU
  constant voltage unit
CW
  call waiting
  canada wire and cable
  carrier wave
  cement water ratio
  channel word
  chemical warfare
  clean water
  clockwise
  code word
  cold worked
  composite wave
  computerworld
  connected with
  continuous wave
  control word
  coursewriter
  curtiss wright
CWA
  communications workers of america
CWAR
  continuous wave acquisition radar
CWAS
  contractor weighted average share
CWC
  clear write condition
CWD
  continuous wave detector
CWE
  commonwealth edison company
  current working estimate
CWF
  course write facility
CWFFEMM
  continuous wave fixed frequency electromechanically modulated
CWFM
  continuous wave frequency modulated
CWG
  communications working group
  corrugated wire glass
CWIC
  compiler for writing and implementing compilers
CWIF
  continuous wave intermediate frequency

CWIP
  construction work in progress
CWIR
  continuous wave illuminator radar
CWL
  cable and wireless, limited
CWMTU
  cold weather material test unit
CWO
  capital work order
  carrier wave oscillator
  continuous wave oscillator
CWP
  coal worker's pneumoconiosis
  communication word processor
  computer word processing
  contractor work plan
CWPM
  correct words per minute
CWS
  caution and warning system
  center wireless station
  central wireless station
  cooling water system
CWSD
  continuous wave space duplexed
CWSIGGEN
  continuous wave signal generator
CWT
  carrier wave telegraphic
  carrier wave telegraphy
  carrier wave transmission
  central winter time
  command word trap
  computer world trade
  continuous wave tunable
CWV
  continuous wave video
  crest working voltage
CX
  central exchange
  convex
CXHLS
  cargo experimental heavy logistics system
CY
  calender year
  capacity
  copy
  cubic yard
  cybernetics

cycle
cycling
CYCLO
  cyclotron
CYL
  cylinder
  cylindrial
CZ
  canal zone
  czochralski
CZT
  chirp z transform

D

D
  data
  data telephone
  data type
  deameter
  debye
  deci
  declination
  deflection
  degree
  delay
  demand meter
  density
  department
  depth
  derivative
  destrorotatory
  deterministic
  deuterium
  dextro
  dextrorotary
  diameter
  dielectric
  differential
  differential coeffizient
  differentiation
  digit
  digital
  dimensional
  diode
  diopter
  dip
  directivity
  discharged
  displacement
  display
  dissipation
  dissipation factor
  distance

  distortion
  division
  double
  drain
  drum
  duty cycle
  dynamotor
  dyne
DA
  data acquisition
  data administrator
  data analyzer
  data available
  data output
  deca
  decimal add
  decimal to analog
  design automation
  detector amplifier
  device adapter
  differential analyzer
  diffused base alloy
  digital to analog
  direct access
  discrete address
  dissolved acetylene
  distribution amplifier
  double amplitude
  double armoured
  double displacement amplitude
  dummy antenna
DAA
  data access arrangement
  direct access arrangement
DAANDP
  data analysis and processing
DAARECORDENGR
  data recording engineer
DAB
  digital audio broadcasting
  display attention bit
DABS
  discrete address beacon system
  discretely addressed beacon system
DAC
  data acquisition and control
  data acquisition and control system
  data acquisition chassis
  data analysis and control
  data analysis center
  data analysis control
  data automation corporation
  design augmented by computer
  digital analog converter
  digital arithmetic center

digital associates corp
digital to analog circuit
digital to analog control
digital to analog converter
digital to analogue conversion
digital to analogue converter
direct access computing
disassemble communications pool
display analysis console
douglas aircraft company
DACA
digital to analog control apparatus
DACAC
digital to analog converter, alternating current
DACAD
strategic air command addressee designator
DACAPS
data collection and processing system
DACC
direct access communications channel
direct access computing corporation
DACCC
defense area communications control center
DACCCEUR
defense area communications control center, europe
DACCCPAC
defense area communications control center, pacific
DACDIC
digital to analog converter, direct current
DACE
data acquisition and control executive
DACI
direct adjacent channel interference
DACK
dma acknowledge
DACL
dynamic analysis and control laboratory
DACO
douglas aircraft company
douglas aircraft company overseas
DACOM
datascope computer output microfilmer
DACON
digital to analog converter
DACONVERSION
digital to analog conversion
DACONVERTER
digital to analog converter
DACOR
data correction
DACPO
data count print out

DACQ
data acquisition
DACS
data acquisition control system
data and analysis center for software
dataset control system
DACTION
derivative action
DACU
data acquisition and control unit
DAD
data dictionary assembler
drum and display
DADC
digital air data computer
direct access data channel
DADEE
dynamic analog differential equation equalizer
DADIC
data dictionary
DADIT
daystrom analog to digital integrating translator
DADSM
direct access device space management
DAE
data acquisition equipment
data connector
DAEMON
data adaptive evaluator and monitor
DAER
department of aeronautical and engineering research
DAF
delay amplification factor
destination address field
discard at failure
dry, ash free
dual access feature
DAFC
digital automatic frequency control
DAFCS
digital automatic flight control system
DAFM
discard at failure maintenance
DAFT
digital analog function table
digital to analog function table
DAG
design advisory group
DAGC
delayed automatic gain control

DAGMAR
   drift and ground speed measuring airborne radar
DAI
   drift angel indicator
DAIR
   direct attitude and identity readout
   driver air
   driver air, information, and routing
DAIS
   defense automatic integrated switch
   defense automatic integrated switching system
   digital avionics information system
DAISY
   data acquisition and interpretation system
   data analysis of the interpreter system
   double precision automatic interpretive system
DAJS
   distributed area jamming system
DAL
   digital access line
DALC
   divided access line circuit
DALS
   digital approach and landing system
DAM
   data address memory
   data addressed memory
   data association message
   decameter
   descriptor attribute matrix
   digital to analog multiplier
   direct access memory
   direct access method
   down range anti missile
DAMA
   demand assigned, multiple access
DAMIT
   data analysis computer program by massachusetts institute of technology
DAMP
   down range anti missile measurement program
DAMPS
   data acquisition multiprogramming system
DAN
   document accession number
DANAC
   data analysis and classification
DANDA
   detail and assembly
DANDP
   development and printing
DANDS
   demand and supply
   display and storage
DANDT
   double and twist
DANP
   danish patent
DAOB
   digital and analog output basic
   digital and analog output basis
DAP
   acquisition and processing
   data access protocol
   data acquisition and processing
   data acquisition package
   deformation of aligned phases
   detail assembly panel
   diffused alloy power
   digital assembly program
   distributed array processor
   double amplitude peak value
DAPR
   digital automatic pattern recognition
DAPS
   direct access performance software
   direct access programming system
DAR
   damage assessment routine
   damage assessment routines
   data article requirements
   data output register
   defense acquisition radar
   delayed automatic reclose
   differential absorption ratio
DARAC
   damped aerodynamic righting attitude control
DARACS
   damped aerodynamic righting attitude control system
DARB
   distressed airman recovery beacon
DARC
   data acquisition and reduction center
DARE
   data automatic reduction equipment
   data automation research and experimentation
   delay asymptotic relative efficiency
   document abstract retrieval equipment
   document automated reduction equipment
   documentation automated retrieval equipment
   doppler automatic reduction equipment
   doppler velocity and position automatic reduction equipment

DARES
  data analysis and reduction system
DARLI
  digital angular readout by laser interferometry
DARPA
  defense advanced research projects agency
  defense applied research projects agency
DARS
  digital adaptive recording system
  digital attitude and rate system
DART
  daily automatic rescheduling technique
  data analysis recording tape
  data reduction translator
  design automation routing tool
  development advanced rate technique
  digital automatic read out tracker
  diode automatic reliability tester
  director and response tester
  dual axis rate transducer
  dynamic acoustic response trigger
DARTS
  design aid for real time systems
  digital azimuth range tracking system
DAS
  data acquisition station
  data acquisition system
  data automation system
  data output control
  datatron assembly system
  digital analog simulator
  digital analog system
  digital attenuator system
  digital output for static signals
  digital to analog simulation
  digital to analog simulator
  direct access store
  disturbance analysis system
  documentation aid system
DASCOTAR
  data acquisition system correlation tracking and ranging
DASD
  direct access storage device
DASDI
  direct access storage device initialization program
DASH
  drone anti submarine helicopter
  dual access storage handling
  dynamic algol string handling
DASI
  data security
DASPAN
  data spanning
DASS
  data access security system
  demand assignment signaling and switching
  diesel air start system
DASSS
  demand assigned signaling and switching subsystem
DAST
  data selection control
  data transmission control
  division for advanced systems technology
DASTARD
  destroyer anti submarine transportable array detector
DASTRA
  data storage and retrieval system
DAT
  data file
  data set
  designation acquisition track
  digital audio tape
  director of advanced technology
  disconnect actuating tools
  duration adjusting type
  dynamic address table
  dynamic address translation
  dynamic address translator
  dynamic allocation translator
DATABUS
  datapoint business language
DATAC
  data analog computer
  digital automatic tester and classifier
DATACOL
  data collection system
DATACOM
  data communication
  data communication systems
DATAGEN
  data file generator
DATAMATION
  data automation
DATAN
  data analysis
DATAR
  digital automatic tracking and ranging
  digital automatic transducer and recorder
  digital autotransducer and recorder
DATASTOR
  data storage
DATATELEX
  data processing telecommunication exchange

DATCO
  data coordinating committee
DATCOL
  data collection
DATDC
  data analysis and technique development center
DATE
  data exchange service
DATEL
  data telecommunication
DATEPLAN
  data tabulation and editing programming language
DATEX
  data exchange
DATICO
  digital automatic tape intelligence checkout
DATIN
  data inserter
DATRAN
  data transmission corporation
DATS
  despun antenna test satellite
  digital avionics transmission system
  dynamic accuracy test system
DAU
  data acquisition unit
  digital to analog conversion
  digital to analog converter
DAV
  data above voice
  data valid
  delayed automatic volume
DAVC
  delayed automatic volume control
DAVI
  department of audio visual instruction
  dynamic antiresonant vibration isolator
DAWID
  device for automatic word identification and discrimination
DAZD
  double anode zenerdiode
DB
  data base
  dead band
  debit
  decibel
  decimal to binary
  diffused base
  digital block
  display buffer
  distribution box
  double bayonet base
  double biased
  double bottom
  double bounce
  double braid
  double braided
  double break
  dry bulb
  dynamic braking
  dynamic braking contactor to relay
DBA
  data base administration
  data base administrator
  daytime broadcasters' association
  decibel adjusted
  decibel related to amperage
  design basis accident
DBAO
  digital block and or
  digital block and or gate
DBB
  detector back bias
  detector balance bias
  detector balanced bias
DBC
  data base computer
  decibel relative to the carrier
  decimal to binary conversion
  decomposed block code
  diameter bolt circle
  digital binary converter
  digital to binary converter
DBCO
  digital block clock oscillator
DBD
  data base description
  data base descriptor
  data base directory
  double base diode
DBDA
  design basis depressurization accident
DBDL
  data base definition language
DBE
  design basis earthquake
  design basis event
DBF
  demodulator band filter
  design basis fault
  digital block flop
DBFF
  digital block flip flop
DBI
  differential bearing indicator
  double byte interleaved

DBIA
   digital block inverting amplifier
DBIL
   data base input language
DBK
   decibel referred to one kilowatt
   decibels above one kilowatt
DBL
   data base load
   detail billing number required
   double
DBLACT
   double acting
DBLCNT
   double contact
DBLR
   doubler
DBM
   data buffer mode
   data buffer module
   decibel above one milliwatt
   decibel based on milliwatt
   decibel below one milliwatt
   decibel meter
   decibel referred to one milliwatt
   double balanced mixer
DBMI
   data base management, inc
DBMS
   data base management software
   data base management system
DBMV
   digital block multivibrator
DBMW
   decibel referred to one milliwatt
DBNA
   digital block noninverting amplifier
DBOA
   delayed breeder or alternative
DBOMP
   data base organization and maintenance processor
   data bomp
DBOS
   data base operating system
   disk based operating system
DBP
   data base processor
DBPW
   decibel referred to one picowatt
DBR
   descriptor base register
   distributed bragg reflector
   power difference in db between any point and a reference point
DBRAP
   decibel above reference acoustical power
DBRN
   decibel above reference noise
   decibel referred to the reference noise
DBRNC
   db above reference noise, c message weighted
   decibel above reference noise, c message
   decibels above reference noise, c message weighted
DBS
   data base management system
   data base software
   data base system
DBSC
   digital block slave clock
DBSM
   decibel above one square meter
DBSP
   double based solid propellant
DBST
   digital block schmitt trigger
   double british standard time
   double british summer time
DBT
   depleted base transistor
   design basis tornado
DBTG
   data base task group
DBUT
   data base update time
DBV
   db above 1 microvolt
   decibel referred to one volt
   decibel related to voltage
DBVMHZ
   db above 1 microvolt per megahertz
DBVMMHZ
   db above 1 microvolt per meter per megahertz
DBW
   decibel referred to one watt
   decibel related to wattage
   design bandwidth
DBWP
   double braid weather proof
   double braided weather proof
DBX
   decibel above reference coupling
   decibels above the reference coupling
DC
   data central
   data channel
   data check
   data classifier

data collection
data communication
data communications
data concentrator
data control
data conversion
data counter
decade counter
decimal classification
define constant
definition of a constant
deposited carbon
design change
design contractor
desk checking
development center
device control
differential crosstalk
digit control
digital code
digital comparator
digital computer
digital control
diode cathode
direct control
direct coupled
direct coupling
direct current
direct cycle
directional coupler
disc controller
disc to card
discrete
disk controller
dispersion coefficient
display console
display control
distance
distorted communications
district of columbia
dot cycle
double cap
double column
double concentric
double conductor
double contact
double cotton
double current
drift correction
driver cell

DCA
   data corporation of america
   decade counting assembly
   defense communications advisory committee
   defense communications agency
   device control area
   digital command assembly
   digital computer association
   distributed communications architecture
   doppler count accumulator
   double conversion adapter
   drift correction angle

DCAA
   defense contract audit agency

DCAC
   direct current alternating current

DCAI
   defense communications agency instructions

DCAM
   data communication access method

DCAO
   digital card and or

DCAOC
   defense communications agency operations center

DCAR
   discrepancy and corrective action report

DCAS
   data collection and analysis system
   deputy commander aerospace system

DCB
   data control block
   defense communications board
   device control block
   distance controlled boat

DCBRL
   defense chemical, biological, and radiation laboratories

DCC
   data communication channel
   data condition code
   device control character
   digital communications console
   direct computer control
   direct current clamp
   discrimination and control computer
   display channel complex
   district communications center
   double call club
   double cotton covered

DCCC
   data communication control character
   defense communications control center
   defense communications control complex
   double current cable code

DCCEAS
   district of columbia council of engineering and architectural societies, inc

DCCL
digital charge coupled device
digital charge coupled logic
DCCO
digital card clock oscillator
DCCS
defense communications control system
digital camera control system
digital command communications channel
digital command communications system
DCCT
direct current current transformer
DCCU
data communication control unit
data correlation control unit
DCD
data carrier detect
data carrier detector
data collecting device
data correlation and documentation system
defense communications department
diode capacitor diode
direct current dialing
double channel duplex
dynamic computer display
DCDCCONVERTER
dc to dc converter
DCDM
digitally controlled delta modulation
DCDMA
diamond core drill manufacturers association
DCDR
direct cycle diphenyl reactor
DCDS
digital control design system
double cotton double silk
double cotton double silk covered
DCDT
direct current displacement transducer
DCE
data circuit terminating equipment
data circuit termination equipment
data communication equipment
data control equipment
DCEO
defense communications engineering office
DCES
delaware council of engineering societies
DCF
data communications formatter
digital flight controller
discounted cash flow
disk control field
DCFEM
dynamic crossed field electron multiplication
dynamic crossed field electron multiplier
dynamic crossed field electron multiplication
dynamic crossed fields, electric and magnetic
DCFF
digital card flip flop
direct current flip flop
DCFL
direct coupled fet logic
DCG
data coordiantion group
diode capacitor gate
doppler controlled gain
DCI
decision concepts, inc
differential current integrator
direct channel interface option
direct coupled inverter
DCIA
digital card inverting amplifier
DCIB
data communication input buffer
DCIP
data correction indicator panel
DCIS
downrange computer input system
DCKP
direct current key pulsing
DCL
data control list
design choice logic
designer's choice logic
diebold computer leasing inc
digital computer laboratory
direct coupled logic
DCLCS
data conversion and limit check submodule
DCM
data communication methods
data communication multiplexer
data communication noise margin
digital circuit module
direct current mains
direct current noise margin
directory control module
DCMA
direct current milliamps
DCMS
data capture and management system
DCMV
digital card multivibrator
DCN
databases in computer networks

DCNA
  data communication network architecture
  digital card non inverting amplifier
DCO
  data control office
  data control officer
  detailed checkout
  directional coupler oscillator
  district communications officer
  dynamic checkout
DCONTROL
  derivative action control
DCONTROLLER
  derivative action controller
DCOP
  detailed checkout procedure
DCOS
  data communication output selector
  downrange computer output system
DCOT
  distant central office transceiver
DCP
  data change proposal
  data collection platform
  data communication processor
  data control processor
  design change package
  design criteria plan
  development concept paper
  diagnostic control program
  differential computing potentiometer
  digital computer processor
  digital computer programming
  direct current panel
  display control panel
  distributed communication processor
DCPA
  defense civil preparedness agency
DCPBH
  double channel planar buried heterostructure
DCPF
  displaced cosine pulse function
DCPG
  defense communications planning group
  direction center programming group
DCPS
  data control panel submodule
  digitally controlled power source
DCPSK
  differentially coherent phase shift keying
DCPSP
  direct current power supply panel

DCPV
  direct current peak voltage
DCR
  data conversion receiver
  data coordinator and retriever
  decrease
  design change request
  development council for research
  digital conversion receiver
  direct conversion reactor
  direct conversion reactor study
  direct current resistance
  direct current restorer
DCRN
  dashpot cup retention nut
DCRS
  data collection and reduction system
  document control remote station
DCS
  data classification system
  data collection system
  data communication subsystem
  data communication system
  data conditioning system
  data control service
  data control system
  data conversion system
  defense communication system
  defense communications service
  design control specifications
  digital command system
  digital communication system
  direct couple operating system
  direct coupled system
  distributed commercial system
  distributed computer system
  document control station
  double channel simplex
  double cotton single silk
DCSC
  digital card slave clock
DCSP
  defense communications satellite program
  direct current straight polarity
DCST
  digital card schmitt trigger
DCT
  data communication terminal
  data conversion transmitter
  decoding part
  destination control table
  digital computer trainer
  digital curve tracer
  direct carbon transfer

direct coupled transistor
divide check test
DCTL
    diode capacitor transistor logic
    direct coupled transistor logic
DCU
    data command unit
    data control unit
    decade counting unit
    decimal counting unit
    device control unit
    digital control unit
    digital counting unit
    disk control unit
    display and control unit
    display control unit
    dynamic checkout unit
DCUTL
    direct coupled unipolar transistor logic
DCV
    digitally coded voice
    direct current voltage
    direct current, volts
    double cotton varnish
    double cotton varnished
DCVGLA
    digitally controlled variable gain linear amplifier
DCW
    data control word
    define constant with wordmark
    digital display control and warning light
DCWV
    direct current working voltage
    direct current working volts
DCX
    direct current experiment
DD
    data definition
    data demand
    data dictionary
    data directory
    data file definition
    decimal divide
    delay driver
    deuterium deuterium reaction
    digital data
    digital differential analyzer
    digital display
    direct drive
    disconnecting device
    disk to disk
    dividend
    dot and dash

    double density
    double diffused
    double diode
    double drift
    drum demand
    duplex drive
DDA
    demand deposit accounting
    digital differential analysis
    digital differential analyzer
    digital display alarm
    direct digital analysis
    direct disk attachment
    dynamic differential analyzer
DDAS
    digital data acquisition system
DDB
    two dimensional burnup reactor computer code
DDBMS
    distributed data base management system
DDC
    data description committee
    data device corporation
    data display central
    data distribution center
    deck decompression chamber
    defense documentation center
    dewey decimal classification
    digital data converter
    digital display converter
    digital to digital conversion
    digital to digital converter
    direct data channel
    direct digital computer
    direct digital control
    direct distance dialing
    director digital control
    display data controller
    dual dielectric charge storage cell
    dual diversity comparator
DDCE
    digital data conversion equipment
DDCMP
    digital data communication message protocol
DDCONVERTER
    digital to digital converter
DDCS
    digital data calibration system
DDD
    digital display detection
    direct distance dialling
    display decoder driver

DDDA
    decimal to digital differential analyser
DDE
    decentralized data entry
    differential difference equation
    director design engineering
    double diffused epitaxial process
DDG
    digital data generator
    digital data group
    digital display generator
DDGE
    digital display generator element
DDH
    digital data handling
DDHA
    digital data handling assembly
DDHADS
    digital data handling and display system
DDI
    data dimensions, inc
    depth deviation indicator
    direct digital interface
DDL
    data definition language
    data description language
    digital data line
    digital design language
    digital system design language
    dispersive delay line
    document description language
DDLC
    data description language committee
    data description language computer
DDM
    data demand module
    decision direct measurement
    derived delta modulation
    difference in depth of modulation
    digital display make up
    double diffused mesa
    dynamic depletion mode
DDN
    data file definition code
    digital data network
DDOCE
    digital data output conversion element
    digital data output conversion equipment
DDP
    data distribution panel
    differential dynamic programming
    digital data processor
    distributed data processing
    distributed data processing system
    double diode pentode

DDPE
    digital data processing equipment
DDPS
    digital data processing system
    discrimination data processing system
DDPU
    digital data processing unit
DDR
    decouy discrimination radar
    digital data receiver
    digital demand recorder
    double drift region
    drawing data requirement
    dual discrimination ratio
    dynamic device reconfiguration
DDRAE
    directorate for defense research and engineering
DDRR
    directional discontinuity ring radiator
DDRS
    digital data recording system
DDS
    data dialog system
    data display subsystem
    data display system
    data distribution system
    data telephone digital service
    dataphone digital service
    deployable defense system
    deviation dependent sensitivity
    digital data service
    digital data servo
    digital data system
    digital display scope
    digital dynamics simulator
    diversified data systems, inc
    doppler detection station
    doppler detection system
DDSS
    diversified data services and sciences
DDT
    data description table
    deduct
    deflagration to detonation transition
    design data transmittal
    design development test
    dichlorodiphenyltrichloroethane
    digital data transceiver
    digital data transmission
    digital data transmitter
    digital debugging tape
    digital debugging technique

doppler data translator
double diode triode
dual deflection tube
dynamic debugging tape
dynamic debugging technique
DDTE
digital data terminal equipment
DDTESS
digital data terminal equipment service submodule
DDTL
diode diode transistor logic
double diffused transistor logic
DDTS
digital data transmission system
direct dialing telephone system
DDU
digital distributing unit
disk data unit
dual diversity unit
DDV
direct data processing
DE
data entry
data entry system
decision element
deemphasis
deflection error
degree of elasticity
delaware
design engineering
detroit edison
dictating equipment
diesel electric
differential equation
digital element
digital encoder
digital equipment
display electronics
display equipment
double end
double ended
double entry
DEA
data encryption algorithm
display electronic assembly
draft environmental statement
drug enforcement administration
DEACON
direct english access and control
DEAL
decision evaluation and logic
DEAP
diffused eutectic aluminum process

DEB
data extend block
digital european backbone
DEBUG
debugging
DEC
danville engineers club
data exchange control
decade
decimal
decimeter
decision
declination
decrease
detector
digital equipment corporation
direct energy conversion
distant electric control
document effected code
document evaluation center
duluth engineers club
dynamic equilibrium cycling
DECA
display electronic control assembly
DECAL
dec's cai author language
desk calculator
detection and classification of an acoustic lens
digital equipment corporation's adaption to algol
DECB
data event control block
DECCO
defense commercial communications office
DECEO
defense communication engineering office
DECFA
distributed emission crossed field amplifier
DECIS
decision
DECLAB
dec laboratory system
DECM
defense electronic countermeasure
DECMIRROR
diffusion controlled electro chromic mirror
DECN
decision
DECO
detroit edison company
direct energy conversion operation
DECOM
telemetry decommutator

DECOMP
  decomposition mathematical programming
DECOR
  digital electronic continuous ranging
DECR
  decrease
  decrement
DECUS
  digital equipment computer users society
DED
  data element descriptor
  defense electronics division
  design engineering directorate
  diesel engine driven
DEDD
  diesel electric direct drive
DEDGETRIGGEREDFLIPFLOP
  delay edge triggered flipflop
DEDS
  data entry and display system
DEDUCOM
  deductive communicator
DEE
  digital evaluation equipment
  digital events evaluator
DEEPDET
  double exposure end point detection technique
DEER
  directional explosive echo ranging
DEF
  defined
  definite
  definition
  delay equalizer, fixed set
  dielectric foil
DEFEC
  defective
DEFL
  deflect
  deflection
DEFS
  definitions
DEFT
  dynamic error free transmission
DEG
  degree
DEGADEP
  degaussing and deperming
DEGCENT
  degree centigrade
DEI
  defense electronics, incorporated
  development engineering inspection
  digital electronics incorporated

DEIMOS
  development investigations in military orbiting systems
DEION
  deionized
DEL
  delay
  delete
DELPHO
  deliver by phone
DELRAC
  decca long range area coverage
DELT
  deck edge light
DELTA
  detailed labor and time analysis
  differential electronically locking test accessory
DELTAM
  delta modulation
DELTIC
  delay line time compression
DEM
  demodulator
DEMA
  data entry management association
  diesel engine manufacturers association
DEMATRON
  distributed emission magnetron amplifier
DEMEDRIVE
  decoding memory drive
  decoding memory driving
DEMOD
  demodulation
  demodulator
  depletion etch method
DEMON
  digital electric monitor
DEMUX
  demultiplexer
DENI
  damage equivalent of normally incident
DENS
  density
DEP
  deflection error probable
  dense electronic population
  design external pressure
DEPA
  defense electric power administration
DEPI
  differential equation pseudo-code interpreter
DEPSK
  differential encoded phase shift keying

DEPT
  department
  departure
DEQ
  dose equivalent
DER
  destroyer escort radar
  diesel engine, reduction drive
DERD
  diesel electric reduction drive
DERE
  dounreay experimental reactor establishment
DERM
  delayed echo radar marker
DES
  data encryption standard
  data encryption system
  department of education and science
  department of emergency services
  designate
  designation
  diesel electric ship
  differential equation solver
  digital expansion system
  division engineering standards
DESC
  defense electronics supply center
  digital equation solving computer
DESS
  dual environment safety switch
DET
  detail
  detector
  determine
  device error tabulation
  diesel electric trawler
DETAB
  decision table
DETABGT
  decision table general translator
DETABX
  decision tables, experimental
DEU
  data encryption and decryption unit
  data exchange unit
DEUA
  diesel engines and useres association
  digitronics equipment users association
DEUCE
  digital electronic universal calculating engine
  digital electronic universal computing engine
DEV
  delay equalizer, variable
  develop
  development
  deviation
  device
DEVCO
  development committee
DEVIL
  direct evaluation of indexed language
DEVR
  distortion eliminating voltage regulator
DEW
  directed energy weapons
  distant early warning
DEWITZ
  distant early warning identification zone
DEWIZ
  distant early warning identification zone
DEWLINE
  distant early warning line
DEWPO
  distant early warning project office
DEXAN
  digital experimental airborne navigator
DF
  day frequency
  dead freight
  decimal fraction
  decontamination factor
  deflection factor
  degree of freedom
  depreciation factor
  derivation of frequency with respect to time
  describing function
  deuterium fluoride
  device function
  difference frequency
  direct flow
  direction finder
  direction finding
  disk file
  dissipation factor
  distortion factor
  distribution factor
  distribution frame
  diversity factor
  dose factor
  double feeder
  duty factor
DFA
  digital fault analysis
DFB
  distributed feed back
  distributed feedback semiconductor
  distribution fuse board
DFC
  data flow control

DFCS

design field change
diagnostic flow charts
disc file controller
document flow component
double frequency change
double frequency changing
dual feed carriage
dual feed channel

DFCS
direction finding control station

DFE
decision feedback equalizer

DFET
drift fet

DFF
delay flip flop

DFFME
direction finding frequency measuring equipment

DFFR
dynamic forcing function report

DFG
digital function generator
diode function generator
discrete frequency generator

DFGA
distributed floating gate amplifier

DFGS
digital flight guidance system

DFI
deep foundations institute
developmental flight instrumentation

DFING
direction finding

DFL
display formatting language

DFLIPFLOP
delay flip flop

DFM
distortion factor meter

DFMSR
directorate of flight and missile safety research

DFO
decade frequency oscillator
director flight operations

DFOP
direction finder operator

DFP
diesel fire pump
dipole flat plate antenna

DFPT
disk file protection table

DFQ
day frequency

DFR
decreasing failure rate
degradation failure rate
dounreay fast reactor

DFRL
differential relay

DFS
depth first search
distance finding station
dividends from space
dynamic flight simulator

DFSK
double frequency shift keying

DFSM
dispersion flattened single mode

DFSR
directorate of flight safety research

DFSTN
direction finding station

DFT
diagnostic function test
discrete fourier transform
discrete fourier transformation

DFTI
distance form threshold indicator

DFTM
direction finder team

DFU
data file utility

DG
data general corp
decigram
degaussing
degree
diesel generator
differential gain
differential generator
diode gate
directional grid
directional gyro
directional gyroscope
double gear
double glass
double groove

DGB
disk gap bond

DGBC
digital geoballistic computer

DGC
data general corp

DGDGE
distributor to group display generator electronics

DGDP
   double groove double petticoat
DGNL
   diagonal
DGP
   data generation program
   directional gyroscope position
DGS
   data gathering system
   data ground station
   degaussing system
   double green silk
DGZ
   desired ground zero
DGZPRO
   desired ground zero program
DH
   decay heat
   decimal to hexadecimal
   decision height
   directly heated
   display hold
   document handling
   double heterostructure
DHC
   data handling center
DHCC
   decay heat closed cooling
DHD
   double heat sink diode
DHE
   data handling equipment
DHFA
   dominion high fidelity association
   double conductor, heat and flame resistant, armored
DHI
   directional horizon indicator
DHP
   delivered horsepower to propeller
   designed horsepower
DHR
   department of human resources
DHRS
   decay heat removal system
   direct heat removal system
DHT
   discrete hilbert transform
DHX
   dump heat exchanger
DHXCS
   dump heat exchanger control system
DI
   dark ignition
   data input
   decilitre
   demand indicator
   deviation generator
   deviation indicator
   dielectric isolation
   digital input
   diodes
   direct injection
   directivity index
   double injection
DIA
   defense intelligence agency
   design and industries association
   diameter
   distributed lumped active
DIAC
   defense industry advisory council
   diode alternating current switch
   diode, alternating current
DIAD
   drum information assembler and dispatcher
DIAG
   diagram
DIAL
   disk interrogation alternation and loading
   display interactive assembly language
   drum interrogation alteration and loading system
DIALGOL
   dialect of algol
DIAM
   data independent accessing model
   diameter
   diametral
DIAN
   decca integrated airborne navigation
   decca integrated airborne navigator
   digital analog
   digital to analog
DIANE
   digital integrated attack and navigation equipment
   direct information access network for europe
DIAP
   digitally implemented analogue processing
DIAS
   dynamic inventory analysis system
DIB
   department of information and broadcasting
   digital input basic
DIBA
   dialog basic
   digital internal ballistic analyzer

DIBL
　drain induced barrier lowering
DIBOL
　digital business oriented language
　digital equipment business oriented language
DIC
　data input clerk
　data insertion converter
　data interchange code
　data item catalog
　detailed interrogation center
　deviation indicating controller
　digital input control
　digital input group contact
　digital integrated circuit
　discrete integrated circuit
　dual in line case
DICBM
　defense intercontinental ballistic missile
　detection of intercontinental ballistic missiles
DICE
　development interim control equipment
　digital integrated circuit element
　dynamic inputs to control center equipment
DICEF
　digital communications experimental facility
DICMOS
　dielectrically isolated cmos
DICON
　digital communication through orbiting needles
DICOSE
　digital communication system evaluator
DICS
　digital channel selection
DICTA
　dictaphone
　digital integrated circuit training aid
DID
　data input device
　datamation industry directory
　digital information detection
　digital information display
　direct in dialing
　direct inward dialing
　division of isotopes development
　drum information display
DIDA
　dynamic instrumentation digital analyzer
DIDAC
　digital data communications
DIDACS
　digital data communications system
DIDAD
　digital data display

DIDAP
　digital data processor
DIDAS
　digital data systems
DIDDF
　dual input discrete describing function
DIDF
　dual input describing function
DIDO
　data input data output
　device independent disk operation
　digital input digital output
DIDS
　defense integrated data system
　defense logistics service center integrated data system
DIE
　document of industrial engineering
DIEL
　dielectric
　diesel electric
DIF
　data interchange format
　difference
　different
　direction finder
　document interchange facility
DIFCE
　difference
DIFF
　difference
　different
　differential
DIFFAMP
　difference amplifier
DIFFCALC
　differential calculus
DIFFEQ
　differential equations numerical integration
DIFFFWR
　differentiator and full wave rectifier
DIFFTR
　differential time relay
DIG
　digit
　digital
　digital image generated
　distinctness of image gloss
DIGACC
　digital guidance and control computer
DIGACE
　digital guidance and control equipment
DIGCIRENGR
　digital circuits engineer

DIGCOM
  digital computer
DIGCOMP
  digital computer
DIGEST
  diebold generator for statistical tabulation
DIGICOM
  digital communication
  digital communications system
DIGILIN
  digital linear
DIGIPLOT
  digital plotting system
DIGIRALT
  digital radar altimeter
DIGITAC
  digital tactical automatic control
DIGITAR
  digital airborne computer
DIGRM
  digit record mark
DIGRMGM
  digit record mark group mark
DIGRO
  digital read out
DIGSCOM
  digital selective communication
DIGSIGPROC
  digital signal processing
DIIC
  dielectrically isolated integrated circuit
DIL
  data in line
  dilute
  dilution
  doppler inertial loran
  dual in line
DILIC
  dual in line integrated circuit
DILPACKAGE
  dual in line package
DILS
  doppler inertial loran system
DIM
  dimension
  diminutive
  display image manipulation
  dynamic impedance measurement
DIMATE
  depot installed maintenance automatic test equipment
DIME
  dual independent map encoding
DIMEN
  dimension

DIMES
  defense integrated management engineering system
DIMPLE
  deuterium moderated pile low energy reactor
DIMSYSTEM
  distributed intelligence microcomputer system
DIMUS
  digital multibeam steering
DIN
  data in line
  device initialize
  dinuclear
DINA
  direct noise amplification
  direct noise amplifier
DINADE
  diode interrogation, navigation, and detection
DINOS
  distributed interactive operating system
DINPUT
  data input
DIO
  data input output
  diode
  district intelligence officer
  doppler inertial omega
DIOB
  digital input output buffer
DIOS
  distribution information and optimizing system
DIP
  design internal pressure
  display information processor
  dual in line package
DIPS
  development information processing system
  digital program selection
  dual impact prediction system
DIR
  direct
  directed
  direction
  director
  disassembly and inspection report
DIRAM
  digital range machine
DIRCOL
  direction cosine linkage
DIRCONN
  direct connecting
DIRCOUP
  directional coupler

DIS
   data input supervisor
   defense investigative service
   dialog terminal system
   digital integration system
   disconnected
   discontinued
   discontinuity
   disposition system
   distance
   distant
   doppler inertial system
DISAC
   digital simulator and computer
DISASSY
   disassembly
DISC
   digital information storage corporation
   disconnect
   disconnect command
   disconnect switch
   disconnected
   disconnecting
   disconnector
   discontinue
   discrimination
   discriminator
   domestic international sales corporation
DISCH
   discharge
   discharger
DISCOM
   digital selective communication
DISCON
   disconnect
   disconnection
DISCR
   discriminator
DISD
   data and information systems
   division
DISINT
   discrete integrator
DISM
   delayed impact space missile
DISOSS
   distributed office support system
DISP
   dispatcher
DISS
   data input subsystem
DISSEC
   disintegrations per second
DISSIG
   distress signal

DISSYS
   distribution system
DIST
   distance
   distant
   distort
   distribute
   distribution
DISTR
   distribute
   distribution
   distributor
DISTRAM
   digital space trajectory measurement system
   digital space trajectory measurement
DISTRAN
   diagnostic fortran
DITRAN
   diagnostic fortran
DIU
   data interchange utility
   digital input unit
DIV
   desired intermediate vertex
   digital input group voltage
   divergence
   divided
   divider
   division
   divisor
DIVA
   digital input voice answerback
DIVAD
   division air defense
DIVOT
   digital to voice translator
DIY
   do it yourself
DJ
   disk jockey
DJSU
   digital junction switching unit
DK
   disk
DL
   data field width
   data laboratories ltd
   data language
   data length
   data link
   data reader
   dead load
   delay
   delay line
   description leaf

dielectric loading
difference limen
difference limit
diode logic
disjunctively linear
dynamic load
DLA
  data link address
  defense logistic agency
DLATCH
  display latch
DLAYER
  day time layer
DLC
  data link control
  direct lift control
  down left center
  duo lateral coil
  duplex line control
DLCC
  data link control chip
DLCO
  decade lc oscillator
DLD
  direction level detector
DLDR
  differential line driver receiver
DLE
  data link escape
  data link escape character
DLF
  dielectric loading factor
DLGA
  decorative lighting guild of america
DLH
  data link hardware
  data lower half byte
DLI
  defense language institute
DLIEC
  defense language institute, east coast
DLK
  data link
DLL
  delay locked loop
DLLD
  direct linear loop detector
DLM
  daily list of mails
DLN
  digital ladder network
DLO
  delayed output
  dispatch loading only

  double local oscillator
  dual loop oscillator
DLOCK
  dial lock
DLP
  data listing program
  double layer polysilicon
DLRV
  dual mode lunar roving vehicle
DLS
  data link set
  data link simulator
  data link software
  dmf based landing system
DLSC
  defense logistics service center
DLT
  daily letter telegram
  data line terminal
  data line translator
  data link terminal
  data loop transceiver
  decision logic table
  decision logic translator
  depletion layer transistor
  distributed language translation
DLTM
  data link test message
DLTS
  deep level transient spectroscopy
DLU
  data line unit
DLWL
  designed load waterline
DLY
  delay
  dolly
DM
  data management
  decimal multiply
  decimeter
  decision maker
  deflection modulation
  delta modulation
  demand meter
  demodulate modulate
  demodulator modulator
  design manual
  detecting magnetometer
  detecting mechanism
  diagnostic monitor
  differential mode
  differentiating mechanism
  diffused mesa
  digital modulator

digital multimeter
disconnected mode
disconnecting manhole
disk monitor
doppler missile
double make
drive magnet
driving and maintenance
drum module
dynamo
dynamotor
DMA
  direct memory access
  direct memory address
DMAA
  direct mail advertising association
DMAS
  distribution management accounting system
DMC
  dielelctric magnetic and capacitor
  digital microcircuit
  digital multiplex control
  direct multiplex control
DMCL
  device media control language
DMD
  digital message device
DME
  distance measuring equipment
  draftsman, electrical
  dropping mercury electrode
  dynamic mission equivalent
DMECOTAR
  distance measuring equipment correlation tracking and ranging
DMED
  digital message entrance device
  digital message entry device
DMESFET
  depletion mode metal semiconductor fet
DMET
  distance measuring equipment tactical air navigation
  distance measuring equipment, terminal
DMF
  digital matched filter
  disk management facility
DMH
  decimeter height finder
  drop manhole
DMI
  data memory, incorporated
DMIRR
  demand mode integral rocket ramjet

DMIS
  datico missile interface simulator
DML
  data management language
  data manipulation language
  direct memory line
DMLS
  doppler microwave landing system
DMM
  digital multimeter
  direct metal mastering
DMMF
  dry, mineral matter free
DMN
  differential mode noise
  dimension
  dimensional
DMNI
  device multiplexing nonsynchronized inputs
DMNSTR
  demonstrator
DMO
  data management office
DMOS
  diffused metal oxide semiconductor
  diffused mos
  double diffused mos
DMP
  direct maximum principle
  display make up
  display making parameters
  display masking parameters
DMR
  data management routine
  defective material report
  digital meter reader
  disk monitor system
  dynamic modular replacement
DMS
  data base management system
  data management service
  data management system
  data multiplex system
  defense missile systems
  delta milliohm sensor
  development management system
  deviation from mean standard
  digital multiplexing synchronizer
  digital mumps standard
  disk monitor system
  documentation of molecular spectroscopy
DMSP
  defense meteorological satellite program
DMSR
  director of missile safety research

DMSS
   data multiplex subsystem
DMST
   demonstrate
   demonstration
DMT
   digital message terminal
   dimensional motion time
   disk operating system module tester
   dispersive mechanism test
DMTC
   digital message terminal computer
DMTI
   doppler moving target indicator
DMTR
   deuterium materials testing reactor
   dounreay materials testing reactor
DMU
   data measurement unit
   digital message unit
   distance measuring unit
   dual maneuvering unit
DMUX
   demultiplexer
DMW
   decimeter wave
DMX
   data multiplex demultiplexer
DMZ
   demilitarized zone
DN
   datanet
   decanewton
   decimal number
   decineper
   delayed neutron
DNA
   data network architecture
   desoxyribonucleic acid
   digital network architecture
   distributed network architecture
   does not apply
DNB
   departure from nucleate boiling
DNBR
   departure from nucleate boiling ratio
DNC
   data network corporation
   direct numerical control
DNCCC
   defense national communications control center
DNCCCS
   defense national communications control center system
DNCTL
   down control
DNE
   department of nuclear engineering
DNET
   division of nuclear education and training
DNL
   do not load
   dynamic noise limiter
DNM
   delayed neutron monitor
DNP
   dummy nose plug
DNRC
   democritus nuclear research center
DNS
   decimal number system
   department of national savings
   doppler navigation system
DNSR
   director of nuclear safety research
DNT
   device name table
DO
   dashpot relay
   data output
   decimal to octal
   design objective
   deviating oscillator
   digital output
   diode outline
   dissolved oxygen
   doppler
   drop out
DOA
   duty orbital analyst
DOC
   data and operations center
   data optimizing computer
   decimal to octal conversion
   direct operating cost
   documentation
DOCGEN
   document generator
DOCS
   disk oriented computer system
DOCUS
   display oriented compiler usage system
   display oriented computer usage system
DOD
   date and dock

department of defense
development operations division
diameter over the dielectric
dielectric outer diameter
direct out dialing
direct outward dialing
DODAR
determination of direction and range
DODCI
department of defense computer institute
DODCODSIA
department of defense council of defense and space industry associations
DODGE
department of defense gravity experiment
DODIS
distribution of oceanographic data at isentropic levels
DODISS
department of defense index of specifications and standards
DODREE
department of defense research and engineering
DOE
device oriented electronic
DOES
distribution order entry system
DOETS
dual object electronic tracking system
DOF
degree of freedom
direction of flight
DOFIC
domain originated functional integrated circuit
DOI
descent orbit insertion
differential orbit improvement
digital operation interpreter
DOIT
digital output input translator
DOL
department of labor
director of laboratories
display oriented language
dynamic octal load
DOLARS
digital off line automatic recording system
doppler location and ranging system
DOM
digital ohmmeter
DOMAPP
domestic appliance

DOMSATS
domestic communications satellite systems
DON
distribution octane number
DOP
detailed operating procedure
dioctyl phosphate
DOPA
dynamic output printer analyser
DOPODT
doped polysilicon diffusion technology
DOPIC
documentation of program in core
DOPLOC
doppler location
doppler phase lock
DOPOS
doped polysilicon diffusion source
DOPP
doppler
DOPS
digital optical protection system
DOR
daily operation report
digital optical record
division of operating reactors
DORAN
doppler range
doppler range and navigation
DORDISK
digital optic recording disk
DORIS
direct order recording and invoicing system
DORV
deep ocean research vehicle
DOS
data organization service
decentralized operating system
digital operation system
disk operating system
distributed operating system
DOSAR
dosimetry applications research facility
DOSE
distributed office support executive
DOSES
disc operating system eser
DOSRS
disc operating system real storage
DOSTOVS
dos to os vs
DOSV
deep ocean survey vehicle
DOSVM
disc operating system, virtual memory

DOSVS
  disk operating system virtual storage
DOT
  deep ocean technology
  department of transportation
  designating optical tracker
  domain tip
  domain tip propagation
  duplex one tape
DOTAG
  doppler tactical air navigation system
DOTMEMORY
  domain tip propagation memory
DOUB
  doubler
DOUSER
  doppler unbeamed search radar
DOUT
  data out line
DOVAP
  doppler velocity and position
DOWB
  deep operating work board
DP
  dash pot
  data acquisition package
  data path
  data processing
  data pulse
  decimal place
  deep penetration
  deep pulse
  deflection plate
  degree of polymerization
  department
  desktop publishing
  dew point
  dial pulse
  dial pulsing
  diametral pinch
  diametral pitch
  diaphragm
  difference of potential
  differential phase
  differential pressure
  diffused planar
  digit present
  digital plotter
  digitally programmed
  diode plate
  dipole
  disc to printer
  disk pack
  disk to printer
  distribution point
  distribution programmer
  document publishing
  double paper
  double petticoat
  double play
  double pole
  double precision
  double punch
  drip proof
  driving power
  drum processor
  dual purpose
  dynamic programming
DPA
  data processing activities
  destructive physical analysis
  dial pulse acceptor
DPAT
  drum programmed automatic tester
DPB
  data processing branch
  defense policy board
DPBC
  double pole back connected
  double pole both connected
DPC
  damp proof course
  data path control
  data processing center
  data processing central
  direct power conversion
  direct program control
  disc pack controller
  double paper covered
  double paper single cotton
DPCA
  displaced phase center antenna
DPCELL
  differential pressure cell
DPCM
  delta pulse code modulation
  difference pulse code modulation
  differential pulse code modulation
DPCS
  difference pressure control switch
DPD
  data processing department
  data processing division
  digit plane driver
  digital phase difference
  double plug diode
DPDC
  double paper double cotton
DPDT
  double pole double throw

DPDTSW
  double pole double throw switch
DPE
  data processing equipment
  development project engineer
DPEK
  differential phase exchange keying
DPESE
  densely packaged encased standard element
DPF
  data processing facility
  depression position finder
  driving point function
DPFAG
  data processing financial and general
DPFC
  double pole front connected
DPFM
  discrete time pulse frequency modulation
DPG
  data processing group
  digital pattern generator
DPH
  diamond pyramid hardness
  disk pack handler
DPHGM
  diaphragm
DPI
  data pathing inc
  data processing installation
  digital process instrument
  digital pseudorandom inspection
  dot per inch
DPIS
  differential pressure isolation switch
DPL
  dayton power and light
  design and programming language
  diploma
  discrete phase loop
  display and panel
  document processing language
  duplex
DPLCS
  digital propellant level control system
DPLL
  digital phase locked loop
  digital pll
DPLO
  district postal liason officer
DPLX
  duplex
DPLXR
  diplexer

DPM
  data processing machine
  data processing magazine
  data processing manager
  digital panel meter
  digital power meter
  documents per minute
DPMA
  data processing management association
DPMOAP
  data processing machine operators and programmers
DPMS
  data project management system
DPN
  data processing network
  diamond pyramid hardness number
DPO
  delayed pulse oscillator
  digital processing oscilloscope
  distributing post office
  district post office
  district post officer
  district postal office
  district postal officer
  double pulse operation
DPP
  drip pan pot
DPPA
  double pumped parametric amplifier
DPR
  dial pulse repeater
  double pure rubber
  double pure rubber lapped
DPS
  data presentation system
  data process service
  data processing service
  data processing standards
  data processing station
  data processing subsystem
  data processing system
  delayed printer simulator
  descent power system
  disk processing system
  disk programming system
  distributed presentation services
  distributed processing system
  distributed programming system
  document processing system
  double pole snap
  double pole snap switch
DPSA
  data processing suppliers association
  data processing supplies association

DPSC
  double paper single cotton
DPSK
  differential phase shift keying
  differentially coherent phase shift keying
DPSPECIALIST
  data processing specialist
DPSS
  data processing subsystem
  data processing system simulator
  direct program search system
  display presentation subsystem
  double pole snap switch
DPST
  double pole single throw
  double pole single throw switch
DPSTNC
  double pole single throw normally closed
DPSTNO
  double pole single throw normally open
DPSTSW
  double pole single throw switch
DPSTSWITCH
  double pole single throw switch
DPSW
  double pole switch
DPT
  design and partitioning for restability
  diopter
DPTS
  digital programming test set
DPTT
  double pole triple throw
DPU
  data processing unit
  display processor unit
  document processing unit
DPVS
  digitally programmed voltage source
DPWR
  data process work request
DQC
  data quality control
DQCM
  data quality control monitor
DQE
  descriptor queue element
DQM
  data quality monitor
  dynamic quantity control
DQU
  duquesne light
DR
  damping ratio
  data receiver

  data recorder
  data reduction
  data register
  data reorganizer
  dead reckoning
  decontamination cleaning with a damp rag method
  demodulation remodulation
  detection radar
  differential rate
  differential relay
  digital computer
  digital resolver
  direct or reverse
  direct recording
  directional radio
  directive antenna with reflector
  discrepancy report
  discrimination radar
  dispatch rider
  distant range
  distant reading
  distant reception
  divided ringing
  division of research
  division register
  dram
  drawn
  drif
  drive
  driver
  driver gear
  drum
DRA
  dead reckoning analyzer
  digital record analyzer
  digital recorder analyzer
  doppler radar
  drum read amplifier
DRAA
  data reduction and analysis
DRAC
  distributed read address counter
DRACWG
  data reduction and computing working group
DRADS
  degradation of radar defense system
DRAI
  dead reckoning analog indicator
  dead reckoning analyzer indicator
DRAM
  display random access memory
  dynamic random access memory
DRAMA
  digital radio and multiplex acquisition

DRANS
　data reduction and analysis subsystem
DRAPF
　data reduction and processing facility
DRAT
　data reduction and analysis tape
DRAWDISK
　direct read after write disk
DRB
　decade resolver bridge
DRC
　damage risk criterion
　data reduction compiler
　data reduction computer
　distant reading compass
DRCCC
　defense regional communications control center
DRCG
　discrimination radar control group
DRCPR
　differential reactive current protection relay
DRCS
　dynamically redefinable character set
DRD
　data recording device
　director of research and development
　drum read driver
DRDTO
　detection radar data take off
DRDWTECHNIQUE
　direct read during write technique
DRE
　data reduction equipment
　dead reckoning equipment
　director of research and engineering
　disassembly reassembly equipment
DREAC
　drum experimental automatic computer
DRED
　data router and error detector
　detection radar environmental display
DREG
　d register
DREPO
　district reserve electronics program officer
DRES
　direct reading emission spectrograph
DRET
　direct reentry telecommunications
　direct reentry telemetry
DRETS
　direct reentry telemetry system
DREWS
　direct read out equatorial weather satellite

DRF
　data reporting form
　discharge ringing frequency
　dose reduction factor
DRG
　drawing
DRHLA
　double conductor, radio, high tension, lead armored
DRI
　data recording instrument co ltd
　data recording instruments
　data reduction interpreter
　dead reckoning indicator
DRICO
　data recording instrument company
DRIFT
　diversity receiving instrumentation for telemetry
　dynamic reliability instantaneous forecasting technique
DRIP
　data reduction for interception program
　data reduction input program
DRIPP
　drip proof
DRIR
　direct read out infrared
DRL
　diode resistor logic
DRM
　decimal rate multiplier
　depositional remanent magnetization
　digital radiometer
　direction of relative movement
DRML
　defense research medical laboratories
DRMO
　district records management officer
DRN
　data reference number
DRO
　destructive read only
　destructive read out
　digital read out
　distructive read out
　doubly resonant oscillator
DROD
　delayed read out detector
　digital read out device 1
DROMDI
　direct read out miss distance indicator
DRON
　data reduction

DROS
    direct read out satellite
    disk remote operating system
    disk resident operating system
DROT
    delayed range on target
DRP
    data reduction program
    dead reckoning plotter
    diebold research program
    digital recording process
    discoverer research program
DRPI
    digital rod position
DRQ
    dma request
DRR
    data redundancy reduction
    digital radar relay
    document release record
DRRL
    digital radar relay link
DRS
    data reaction system
    data recording system
    data reduction software
    data reduction system
    data relay satellite
    degrees
    digital radar simulator
    disassembly reassembly station
    disk resident system
    distributed resource system
DRSA
    data recording system analyst
DRSC
    direct radar scope camera
DRSS
    data relay satellite system
DRT
    dead reckoning tracer
    decade ratio transformer
    device rise time
    diode recovery tester
    direct reading telemetering
    discrimination radar transmitter
    distant remote transceiver
DRTC
    documentation research and training center
DRTE
    defense research telecommunications establishment
DRTL
    diode resistor transistor logic

DRTR
    dead reckoning trainer
DRV
    data recovery vehicle
    development reentry vehicle
DRVR
    driver
DRWG
    data reduction working group
DS
    data scanning
    data security
    data series
    data set
    data sheet
    data system
    day's sight
    decimal subtract
    decoder simulator
    define storage
    define symbol
    definition of a storage area
    definition of a symbol
    depth sounder
    depth sounding
    descent stage
    desk stand
    device selector
    dial system
    digit select
    digit selector
    digit switch
    diode switch
    directing station
    disconnect switch
    disconnection switch
    disk storage
    display started
    display station
    distant
    distant surveillance
    domestic service
    doppler sonar
    double screened
    double silk
    drive surface
    drum storage
    drum switch
    dummy section
    dyna soar
    dynamic speaker
    dynamic static analysis
DSA
    deep space antenna
    define symbol address

dial service assistance
dial service auxiliary
dial system a switchboard
diffusion self-aligned
diffusion self-aligning
digital signal analyzer
direct storage access
dynamic signal analyzer
dynamic storage area
DSAMOSFET
dsa metal oxide semiconductor field effect transistor
DSAMOST
diffusion self aligned mos transistor
DSAP
data systems automation program
DSAR
data sampling automatic receiver
DSARC
defense systems acquisition review council
DSASBL
disassemble
DSB
data services bureau
decade synchronic bridge
defense science board
dial system b position
dial system b switchboard
digital storage buffer
distribution switchboard
double sideband
DSBAMRC
double sideband amplitude modulation reduced carrier
DSBK
data set by key
DSBMODULATION
double sideband modulation
DSBSC
double sideband suppressed carrier
double sideband with suppressed carrier
DSBWC
double sideband with carrier
DSC
data set controller
data stream compability
data subcentral
data synchronizer channel
design safety criteria
difference signal control
differential scanning calorimetry
digital set point control
digital signal converter

disc storage controller
discone
disk storage controller
doctor of science
double silk covered
dynamic sequential control
DSCAN
d scanner
DSCAT
data set catalog
DSCB
data set control block
DSCC
deep space communications complex
double silk cotton covered
DSCG
digital sine cosine generator
DSCS
defense satellite communications system
defense system communications satellite
desk side computer system
DSCT
double secondary current transformer
DSD
data systems division
digital system design
dual speed drive
DSDT
deformographic storage display tube
discrete space and discrete time
DSE
dartmouth society of engineers
data set extension
data storage equipment
data switching exchange
data systems engineering
detroit stock exchange
digit selector emitter
distributed system environment
DSEA
data storage electronics assembly
DSEG
data systems engineering group
DSF
data scanning and formating
design safety factor
DSG
designate
designation
DSGA
double conductor, shipboard general use, armored
DSGN
design
designation

DSGS
data set groups
DSH
deactivated shutdown hours
DSI
data systems, incorporated
digital speech interpolation
DSIF
deep space instrument facility
deep space instrumentation facility
DSIR
department of scientific and industrial research
DSIS
directorate of scientific information services
DSK
delay shift keying
DSKY
display and keyboard
display system keyboard
DSL
data set label
data simulation language
data structures language
deep scattering layer
development support library
development system library
digital simulation language
direct swift link
drawing and specification listing
DSLT
deck surface light
DSM
delta sigma modulator
device strategy module
digital simulation model
disk space management
dynamic scattering mode
DSMG
designated systems management group
DSMT
dual speed magnetic transducer
DSN
data smoothing network
deep space net
deep space network
distributed systems network
DSO
data set optimizer
digital storage oscilloscope
district signal officer
divisional signal officer
DSORG
data set organization

DSOT
daily system operability test
DSOTS
demonstration site operational test series
DSP
digital signal processing
disassemble sequence parameter
distributed system program
distribution system program
double silver plated
drain source protected
dynamic support program
DSPEC
design specification
DSPL
douglas space physics laboratory
DSPLY
display
DSR
data scanning and routing
data set ready
data specification request
digit storage relay
digital stepping recorder
discriminating selector repeater
discrimination selector repeater
distributed state response
dual shift right
dynamic sideband regulator
DSRC
distant space radio center
double sideband reduced carrier
DSRG
data system review group
DSRS
data signalling rate select
DSRV
deep submerge rescue vehicle
deep submergence rescue vehicle
DSS
decision support system
deep space station
digital subset
digital subsystem
digital switching subsystem
direct station selection
director of statistical services
disc storage subsystem
disk storage system
distributed system satellite
division of safety studies
dynamic steady state
dynamic support system
dynamic systems synthesizer

DSSA
  defense security assistance agency
  defense space support agency
DSSB
  data selection and storage buffer
  double single sideband
DSSC
  double sideband suppressed carrier
  double sideband suppressed carrier modulation
  double silk single cotton
DSSCS
  defense special secure communications system
DSSP
  deep submergence systems project
DSSS
  direct sequence spread spectrum
DSSV
  deep submerge search vehicle
DST
  data segment table
  data summary tape
  daylight saving time
  designer specified transformer
  dot sequential transmission
  double summer time
DSTC
  double sideband transmitted carrier
DSTE
  data subscriber terminal equipment
DSU
  data service unit
  data storage unit
  data synchronization unit
  data synchronizer unit
  device switching unit
  digital service unit
  disk storage unit
DSV
  double silk varnish
  douglas space vehicle
  dynamic self verification
DSW
  data status word
  device status word
  door switch
  drum switch
DSX
  distributed systems executive
DT
  dark trace
  data systems engineering
  data terminal
  data transfer
  data translator
  data transmission
  daylight time
  decay time
  deferred telegram
  deuterium tritium reaction
  dialing tone
  difference threshold
  differential time
  digit tube
  digital technique
  digital telemetering
  digroup terminal
  diode transistor
  discharge tube
  disk tape
  disk to tape
  display terminal
  distant transmission
  double throw
  double throw switch
  double track
DTA
  differential thermal analysis
  differential thermoanalysis
  distributing terminal assembly
  double tape armored
DTARS
  digital transmission and routing system
DTAS
  data transmission and switching
DTB
  decimal to binary
  delayed time base
DTC
  data technology corporation
  data transmission center
  data transmission channel
  decision threshold computer
  deposition thickness controller
  desk top computer
  document transformation component
DTCS
  data transmission and control system
  digital test command system
DTDMA
  distributed time division multiple access
DTDS
  digital television display system
DTE
  data ten to eleven
  data terminal equipment
  data transmission equipment
  data transmitting equipment

depot tooling equipment
detroit edison company
digital transmission equipment
digital tune enable
display teter element
DTF
data transmission feature
define the file
drone test facility
DTFA
digital transfer function
analyzer
DTG
data time group
display transmission generator
dynamically tuned gyro
DTI
data team international
department of trade and industry
dial test indicator
digital test indicator
distortion transmission impairment
division of technical information
DTIP
digital tune in progress
DTL
detail
detailed
diode transistor logic
double transistor logic
DTLC
direct coupled transistor logic
DTLZ
diode transistor logic with zener diodes
diode transistor logic, zener
DTM
delay timer multiplier
duration time modulation
DTMF
dual tone multifrequency
DTMFSIGNAL
dual tone multifrequency signal
DTML
diode transistor micrologic
DTMLD
draught moulded
DTMS
defense traffic management service
digital test monitor system
digital test monitoring system
DTN
detection
double tinned
DTNG
detuning

DTO
data take off
digital testing oscilloscope
DTP
data tape punch
data transfer protocol
desk top publishing
directory tape processor
drum timing pulse
DTPA
diethylenetriaminepentaaceticacid
DTPD
divider time pule distributor
DTPL
domain tip propagation logic
DTR
daily transactions reporting
data tape recorder
data telemetry register
data terminal ready
definite time relay
demand totalizing relay
digital telemetering register
document transmittal record
DTRE
defense telecommunications research
establishment
DTREACTION
deuterium tritium reaction
DTRF
data transmittal and routing form
DTRM
dual thrust rocket motor
DTRS
development test requirement specification
DTRSS
data transmission recording subsystem
DTS
data terminal systems
data transfer sequence
data transmission service
data transmission subsystem
data transmission system
diagnostic and test
diagnostic test set
dialog terminal system
digital telemetering system
dix tracking station
double throw switch
DTSC
data transmission subcommittee
DTSG
data transmission study group
DTSM
data base and transaction management

DTSW
  defense telephone service washington
DTT
  design thermal transient
  domain tip technology
DTU
  data transfer unit
  data transformation unit
  digital tape unit
  digital telemetry unit
  digital transmission unit
  digital tuning unit
  display terminal unit
DTUTF
  digital tape unit tape facility
DTV
  digital television display
  digital to television
  drop test vehicle
DTVC
  digital transmission and verification converter
DTVM
  differential thermocouple voltmeter
DTWS
  dial teletypewriter service
DU
  disk unit
DUA
  digitronics users association
DUAL
  dynamic universal assembly language
DUC
  defined user command
DUCO
  duplex controller
DUCS
  display unit control system
DUCY
  duty cycle
DUF
  diffusion under field
  diffusion under film
DUH
  data upper half byte
DUK
  duke power company
DUM
  dummy
DUMC
  deep underwater measuring device
DUMS
  deep unmanned submersibles
DUN
  douglas united nuclear

DUNC
  deep underwater nuclear counter
  deep underwater nuclear counting
DUNS
  data universal numbering system
DUO
  datatron users organization
DUP
  disk utility program
  duplex
  duplicate
  duplicating
  duplication
DUT
  device under test
  diode under test
DUV
  data under voice
DUX
  dat utility complex
  data utility complex
DV
  depended variable
  device
  differential voltage
  direct vision
  direct voltage
  divisor
  double vibration
  dump valve
DVA
  dynamic visual acuity
DVB
  disability veiling brightness
DVCCS
  differential voltage controlled current source
DVD
  detail velocity display
DVESO
  dod value engineering services office
DVET
  data vetting program
DVFO
  digital variable frequency oscillator
DVFR
  defense visual flight rules
DVH
  divide or halt
DVM
  digital voltmeter
  displaced virtual machine
DVMD
  digital voltmeter display
DVOM
  digital volt ohmmeter

DVOR
   doppler vertical omnirange
   doppler very high frequency omnidirectional range
   doppler vhf omnidirectional radio range
DVP
   divide or proceed
DVR
   driver
DVS
   dynamic vertical sensor
DVST
   direct view storage tube
DVT
   design verification test
DVTP
   divide time pulse
DVTVM
   digital vacuum tube voltmeter
DVX
   digital voice exchange
DW
   data word
   dead weight
   dead weight load
   direct writing
   dock warrant
   double word
   drop wire
   drum writer
DWA
   double wire armored
DWAC
   distributed write address counter
DWAT
   dead weight all told
DWB
   daily wireless bulletin
DWBA
   distorted wave born approximation
DWC
   dead weight capacity
DWD
   drum write driver
DWED
   dry well equipment drain
DWF
   daily water flow
   dry weather flow
DWFD
   dry well floor drain
DWG
   drawing
DWI
   descriptor word index

DWICA
   deep water isotropic current analyzer
DWL
   designed waterline
   dominant wavelength
DWS
   development work statement
   double white silk
   double wound silk
DWSGAE
   department of water supply, gas and electricity
DWSMC
   defense weapons systems management center
DWST
   demineralized water storage tank
DWT
   dead weight tonnage
DX
   datex
   duplex
   duplex repeater
DXC
   data exchange control
DXCC
   dx century club
DXCOIL
   direct expansion coil
DXS
   data exchange system
DY
   deflection yoke
   dynamotor
   dynode
   dyprosium
DYANA
   dynamics analyzer
DYB
   dynamic breaking
DYCMOS
   dynamic cmos
   dynamic complementary mos
DYM
   daystrom, incorporated
DYN
   dynamic
   dynamics
   dynamo
   dynamometer
   dynamotor
   dyne
DYNA
   dynamics analyzer programmer

DYNAM
   dynamics
   dynamo
DYNAMO
   dynamo action management operations
DYNAMOWS
   dynamic manned orbital weapon system
DYNASOAR
   dynamic soaring
DYNCM
   dyne centimeter
DYNM
   dynamotor
DYPS
   dynamic programming system
DYSAC
   digitally simulated analog computer
   dynamic storage analog computer
DYSTAC
   dynamic memory and storage analog computer
   dynamic storage analog computer
DYSTAL
   dynamic storage allocation language
DYSYS
   digital dynamic system simulator
DZ
   drop zone
DZA
   drop zone area
DZT
   digit zero trigger
DZTL
   diode zener diode transistor logic

E

E
   early warning
   earth
   east
   echo
   efficiency
   elasticity
   electric field strength
   electric field vector
   electric intensity
   electric potential
   electrical force
   electromotive force
   electron charge
   electronics
   element
   emitter
   enamel
   enamelled
   energy
   equivalent hours
   exponent
EA
   earth
   effective address
   electrical association for women
   electrics association
   electro actuators
   electroacoustic engineers' association
   electronic arrays
   electrostatic analyzer
   elementary assignment
   elliot automation
   enemy area
   equalizing line amplifier
   extended accumulator
   extender amplifier
EAA
   engineer in aeronautics and astronautics
   experimental aircraft association
EAB
   educational activities board
   exclusion area boundry
EABN
   engineer aviation battalion
EABRD
   electrically actuated band release device
EAC
   eastern air command
   electrical apparatus company
   electronic assistance corporation
   electronic autocollimator
   end around carry
   engineering advisory committee
   engineering affairs council
   expected approach clearance
EACC
   error adaptive control computer
EACO
   engineers and architects council of oregon
EACP
   european area communications plan
EACRP
   european american committee on reactor physics
EAD
   evaluation and development
EADA
   electrical appliance dealers' association

EADI
  electronic altitude director indicator
  electronic attitude director indicator
EADIZ
  entering air defense identification zone
EAE
  eastern association of electroencephalographers
  extended arithmetic element
EAEC
  european airlines electronic committee
EAEME
  east african electrical and mechanical engineers
EAES
  european atomic energy society
EAF
  electron arc furnace
  equivalent availability factor
EAFB
  edwards air force base
  eglin air force base
  ellington air force base
  ellsworth air force base
EAG
  equipment advisory group
EAGE
  electrical aerospace ground equipment
EAGER
  electronic audit gager
EAGLE
  elevation angle guidance landing equipment
EAH
  engineering association of hawaii
EAI
  electronic aided instruction
  electronic associates inc
  electronic associates, incorporated
  engineers and architects institute
  equip and install
EAL
  electromagnetic amplifying lens
  electronic associates ltd
  electronic associates, limited
  electronics appointments, limited
  equipment air lock
  equivalent ageing load
  expected average life
EALM
  electronically addressed light modulator
EALP
  east african light and power
EAM
  electric[al] accounting machine
  electrical and mechanical
  electronic accounting machine
  electronic accounting machinery
  electronic automatic machine
  elementary access method
EAN
  effective atomic number
EANDC
  european american nuclear data committee
EANDRO
  electrically alterable non destructive read out
EAOE
  errors and omissions excepted
EAP
  educational activities panel
  electronics and power
  emergency action program
  equipment and parts
  equivalent air pressure
  experimental activity proposal
EAPT
  east african post and telecommunications
EAR
  effective address register
  electronic and aerospace report
  electronic audio recognition
  electronic aural responder
  electronically agile radar
  employee attitude research
  engineering and repair department
  engineering and research
EARB
  electronics and avionics requirements board
  european airlines research bureau
EARC
  extraordinary administrative radio conference
EAROM
  electrically alterable read only memory
  electrically alterable rom
EARS
  elliot automation radar systems
EAS
  electron accelerator system
  electronic automatic switch
  equivalent air speed
  experiment assurance system
  extended area service
  extensive air shower
EASA
  electrical apparatus service association
EASAMS
  elliot automation space and advanced military system
EASB
  electronic area support base

EASCOMINT
    extended air surveillance communications intercept
EASCON
    electronic and aerospace systems convention
    electronics and aerospace systems convention
EASE
    electrical automatic support equipment
    electronic analog and simulation equipment
    engineering automatic system for solving equations
EASI
    electrical accounting for the security industry
    estimate of adversary sequence interruption
EASIAC
    easy instruction automatic computer
EASL
    engineering analysis and simulation language
EAST
    eastern australian standard time
EASTT
    experimental army satellite tactical terminal
EASVC
    e a supervisor call
EASY
    early acquisition system
    efficient assembly system
    engine analyzer system
    exception analysis system
EAT
    earliest arriving time
    east african time
    environmental acceptance test
    expected approach time
EATCS
    european association for theoretical computer science
EATS
    equipment accuracy test station
EAU
    extended arithmetic unit
EAUTC
    engineer aviation unit training center
EAW
    equivalent average word
EAX
    electronic automatic exchange
EB
    electron beam
    emitter base
    encoder buffer
    event block
    executive board
EBAF
    equipment blockage and failure

EBAM
    electron beam accessed memory
    electron beam addressable memory
    electronic beam addressed memory
EBAS
    electronic beam activated switch
EBB
    extra best best
EBC
    educational broadcasting corporation
    enamel bonded single cotton
EBCD
    extended binary coded decimal
EBCDI
    extended binary coded decimals interchange
EBCDIC
    extended binary coded decimal interchange code
EBCSM
    east bay council on surveying and mapping
EBDC
    enamel bonded double cotton
EBDS
    enamel bonded double silk
EBES
    electron beam exposure system
EBFS
    enclosure building filtration system
EBG
    electronics buyer's guide
EBI
    equivalent background input
EBIC
    electron beam induced current
EBICON
    electron bombardment induced conductivity
EBIS
    electron beam ion source
EBJUNCTION
    emitter base junction
EBL
    electron beam lithography
EBM
    electron beam machining
    electron beam melted
    electron beam melting
    electron beam method
    electron beam multiplier
EBMD
    electron beam mode discharge
EBMF
    electron beam microfabricator
EBOR
    experimental beryllium oxide reactor

EBP
  enamel bonded single paper
  etch back process
EBPA
  electron beam parametric amplifier
EBR
  electron beam recording
  electron beam remelting
  electronic beam recorder
  electronic beam recording method
  electronic beam recording system
  epoxy bridge rectifier
  experimental breeder reactor
EBROM
  extended bit rom
EBS
  electric bond and share
  electron beam semiconductor
  electron bombarded semiconductor
  electron bombardment semiconductor
  emergency broadcast system
  enamel bonded single silk
  energy band structure
EBTF
  ecc bypass test facility
EBU
  european broadcasting union
EBUC
  etch back uniformity calculation
EBUJ
  ebu journal
EBW
  electron beam welding
  exploding bridge wire
EBWR
  experimental boiling water reactor
EC
  eddy current
  edge connector
  electric coding
  electric current
  electrical communication
  electrical conductivity
  electrical conductor
  electrically coated
  electricity commission
  electrochromic
  electrocoating
  electron capture
  electron coupled
  electron coupling
  electronic calculator
  electronic calculators
  electronic computer
  electronic conductivity
  emergency call
  emulator control
  enamel covered
  enamel single cotton
  enamelled copper
  enfield cable works
  engineering change
  engineering construction
  engineering council
  engineers club
  engineers corps
  environment condition
  environmental control
  error correcting
  error correction
  european communities
  evaluation center
  execution cycle
  expander cell
  extended control
  extra control
ECA
  economic commission for africa
  electrical contact analyzer
  electrical contactors' association
  electrical contractors association
  electronic control assembly
  electronic corporation of america
  enter control area
  etched card assembly
ECAC
  electromagnetic compatibility analysis center
  european civil aviation conference
ECAD
  error check analysis diagram
ECAFE
  economic commission for asia and the far east
ECAM
  electric controller and manufacturing
ECAP
  electric companies' advertising program
  electronic circuit analysis program
  electronic control assembly pitch
  electronic current analysis program
  error check analysis program
ECAR
  east central area reliability
  east central area reliability coordination agreement
  electronic control assembly roll
ECARS
  electronic coordinatograph read out system
  electronic coordinator graph and read out system

ECAS
electronics cleaning advisory service
ECASS
experimental computer aided shop scheduling
ECAY
electronic control assembly yaw
ECB
electrically controlled birefringence
event control block
ECBC
external call barring circuit
ECC
electric construction company
electric coordinating council
electrical continuous cloth
electrocardiocorder
electron coupled control
electron coupling control
electronic calibration center
electronic components conference
electronic computer center
electronics control corporation
electronized chemicals corporation
emergency control center
emitter coupled circuits
error checking and correction
error correcting circuitry
error correcting code
error correction code
execute control cycle
ECCA
electronic component checkout area
ECCANE
east coast conference on aerospace and navigational electronics
ECCC
european communications coordinating committee
ECCCM
electronic countermeasure
ECCCS
emergency command control communications system
ECCEN
eccentric
ECCHART
electrocardiocharter
ECCL
error checking and correction logic
ECCM
electronic counter countermeasure
ECCNE
electric coordinating council of new england
ECCO
engineers coordinating council of oregon

ECCP
engineering concepts curriculum project
ECCS
emergency core cooling system
ECCSL
emitter coupled current steered logic
emitter coupled current steering logic
ECCT
error correction console technician
ECD
electro chromic display
electronic components division
energy conversion device
energy conversion device electrochromeric display
engineers club of dayton
entry corridor display
estimated completion date
ECDB
electrochemical deburring
ECDC
electrochemical diffused collector
electronic components development committee
ECDM
electrical discharge machining
electrochemical discharge machining
ECDT
electrochemical diffused transistor
ECE
economic commission for europe
electrochemical equivalent
engineering capacity exchange
ECEC
east carolina engineers club
ECEPE
european corporate electronic publishing exhibition
ECET
electronic control engine thrust
ECF
electrochemical forming
emergency cooling functionability
engineers club of fresno
ECFM
eddy current flow meter
ECG
electrocardiogram
electrocardiograph
electrocardiography
electrochemical grinding
electronic components group
ECH
echelon
electrochemical honing

ECHO
  electronic computing hospital oriented
ECI
  electronic communications inc
  electronic communications, incorporated
  electronic components for industry
  electronic counters, incorporated
  emergency cooling injection
  european cooperation in informatics
ECIC
  electric consumers information committee
ECIP
  european cooperation in information processing
ECK
  emergency communications key
ECKC
  engineers club of kansas city
ECL
  eddy current loss
  electrical
  electronic components laboratory
  emitter coupled logic
  endicott computation laboratory
  energy conversion limited
  engineers club of lincoln
  equipment component list
  executive control language
ECLA
  economic commission for latin america
ECLO
  emitter coupled logic operator
ECLS
  environmental control and life support
ECM
  electric cipher machine
  electric coding machine
  electric controller and manufacturing
  electrochemical machining
  electronic counter measure
  engineers club of minneapolis
  extended capacity memory
  extended core memory
  extended core module
ECMA
  electronic computer manufacturers association
  engineering college magazines associated
  european computer makers association
  european computer manufacturers association
ECMC
  electric cable makers' confederation
ECME
  electronic checkout maintenance equipment
  electronic circuit making equipment
  electronic countermeasures
  electronic countermeasures environment
  electronic countermeasures equipment
ECMELINT
  electronic countermeasures electronic intelligence
ECMOB
  electronic countermeasures observer
ECMP
  electronic countermeasures program
ECMR
  effective common mode rejection
ECMRON
  electronic countermeasures squadron
ECMSA
  electronic command meteorological support agency
ECMSN
  electronic countermeasures mission
ECN
  engineering change notice
ECNE
  electric council of new england
ECNM
  engineer's club of northern minnesota
ECO
  electron coupled oscillator
  electronic "contact" operate
  electronic central office
  electronic checkout
  emergency control officer
  engineering changes order
  engineers club of omaha
ECOB
  engineers club of bartlesville
ECOC
  engineering club of oklahoma city
ECOM
  electronic components
  electronic computer oriented mail
  electronic computer originated mail
  electronics command
ECOMA
  european computer measurement association
ECOMS
  early capability orbital manned station
ECON
  electromagnetic emission control
ECOR
  error control register
ECOSOC
  economic and social committee
ECP
  electromagnetic compatibility program
  electronic circuit protector

electronic control products
emitter coupled pair
emulator control program
engineering change proposal
executive control program
the engineer's club of philadelphia
ECPD
engineers council for proffessional development
ECPI
electronic computer programming institute
ECPIP
electric companies' public information program
ECPOG
electrochemical potential gradient
ECPR
electrically calibrated pyroelectric filter
ECPY
electronic control assembly pitch and yaw
ECR
edit, count, recode
electron cyclotron resonance
electronic cash register
electronic conference
electronic control relay
emergency cooling recirculation
engineering change request
error control receiver
estimate change request
excess carrier ratio
executive communication region
extended coverage range
ECRC
electricity council research center
electronic component reliability center
ECRDC
electronic component research and development center
ECREA
european conference of radiotelegraphy experts for aeronautics
ECS
electro convulsive shock
electrocardiogram simulator
electrocardioscanner
electrochemical society
electronic control switch
electronic systems countermeasures
embedded computer system
emergency coolant system
end cell switch
enviromental control shroud
environmental control system
equipment compiler system

error connection servo
etched circuits society
european communication satellite
executive control system
expanded character set
extended core storage
ECSA
european communications security agency
ECSC
electronic countermeasures subcommittee
enamelled single cotton covered
european coal and steel community
ECSF
engineers club of san francisco
ECSG
electronic connector study group
ECSJ
electro chemical society of japan
ECSL
engineers club of st louis
extended control and simulation language
ECSO
european communications satellite organization
ECSP
electronic specialist
ECSS
extendable computer system simulator
extended computer system simulator
ECSW
extended channel status word
ECT
eddy current testing
electroconvulsive therapy
environment control table
ECTA
electronics component test area
ECTL
emitter coupled transistor logic
ECU
electrical conversion unit
electronic computing unit
electronic conversion unit
environmental control unit
ECV
enamel single cotton varnish
ED
edit
editor
education
electrical department
electrical differential
electrical drawing
electro dynamic division
electro static discharge

electrochemical diffused
electrodynamics
electron device
electron diffraction
electronic design
electronic device
electronic differential analyzer
electronic digital analyzer
electronic dummy
electronic systems division
electrostatic storage deflection
end of data
engine drive
engineering division
engineering society of detroit
enhancement depletion
environmental services division
equivalence dosis
error detecting
error detection
evaluation and development
expanded display
expansion deflection
extension shaft disconnect
external delay
external device

EDA
  economic development administration
  effective doubleword address
  electrical development association
  electronic differential analyzer
  electronic digital analyzer

EDAC
  electronic dive angle control
  error detection and correction

EDAM
  electronic design and manufacture

EDAX
  energy dispersion analyzer x-ray

EDB
  educational data bank

EDC
  economic development committee
  electronic desk calculator
  electronic digital computer
  emergency decontamination center
  enamel double cotton
  enamelled double cotton covered
  energy discharge capacitor
  error detection and correction

EDCL
  electric discharge convection laser

EDCO
  editing committee

EDCOM
  editor and compiler

EDCPF
  environmental data collection and processing facility

EDCV
  enamel double cotton varnish
  enamel double cotton varnished

EDCW
  external devise control word

EDD
  electronic data display
  envelope delay distortion

EDDD
  expanded direct distance dialing

EDDF
  error detection and decision feedback

EDDS
  early docking demonstration system

EDE
  electronic defense evaluator
  electronic design engineering
  emergency decelerating
  emitter dip effect
  empire district electric company

EDEW
  enhanced distant early warning

EDF
  electrical discharge forming

EDFP
  engine driven fire pump

EDFR
  effective date of federal recognition

EDG
  emergency diesel generator
  exploratory development goals

EDGE
  electronic data gathering equipment
  experimental display generation

EDHE
  experimental data handling equipment

EDI
  echo doppler indicator
  electron diffraction instrument
  electronic data interchange
  electronic devices, incorporated
  error detection instrument

EDIAC
  electronic display of indexing association and content
  engineering decision integrator and communicator

EDICT
  engineering department interface control technique

EDIS
   engineering document information collection task
   engineering document information collection technique
EDIS
   engineering data information system
EDIT
   editor instruction
   editor program
   error deletion by interactive transmission
EDITAR
   electronic digital tracking and ranging
EDITP
   engineering development integration test program
EDL
   electronics defense laboratory
EDLCC
   electronic data local communications central
   electronic data local communications complex
   electronic data local control center
EDM
   electric dipole moment
   electric[al] discharge machine
   electric[al] discharge machining
   electronic design and manufacture
   electronic drafting machine
EDMOSFET
   enhancement depletion mosfet
EDMS
   engineering data micro reproduction system
   evolutionary data management system
   extended data management system
EDN
   electrical design news
   the electronic engineer's design magazine
EDO
   effective diameter of objective
   engineering duties only
   error demodulator output
   error determination output
   executive director of operations
EDOC
   electrical description of operation chart
EDOS
   extended disk operating system
EDP
   electronic data processing
   electrophoretic display
   experimental dynamic processor
EDPA
   environmental data planning associates, inc

EDPAA
   edp auditors association
EDPACS
   edp audit, control and security
EDPC
   electronic data processing center
EDPD
   electronic data processing device
EDPE
   electronic data processing equipment
EDPEO
   electronic data processing equipment office
EDPI
   electronic data processing institute
EDPM
   electronic data processing machine
EDPOR
   electronic data processing operations research
EDPR
   electronic data performance review
   electronic data processing review
EDPS
   electronic data processing system
EDR
   electrodermal reaction
   electrodermal response
   electronic decoy rocket
   equivalent direct radiation
EDRAW
   erasable direct read after write
EDRCC
   electronic data remote communications complex
EDRI
   electronic distributors' research institute
EDRS
   engineering data retrieval system
EDS
   educational data systems
   electronic data storage
   electronic data switching system
   electronic data system
   electronic data systems corp
   electronic design section
   emergency detection system
   enamel double silk
   environmental data service
   exchangeable disk storage
   exchangeable disk store
EDSAC
   electronic data storage automatic computer
   electronic delay storage automatic calculator
   electronic delay storage automatic computer
   electronic discrete sequential automatic computer

electronic discrete variable automatic computer
EDSD
electronic defense system division
EDSR
electronic digital slide rule
EDST
eastern daylight saving time
elastic diaphragm switch technology
electric diaphragm switch technique
EDSV
enamel double silk varnish
enamel double silk varnished
EDT
eastern daylight time
editor
electric discharge tube
electronic data transmission
energy dissipation tests
engineer design tests
engineering description tape
EDTCC
electronic data traffic control center
electronic data traffic control complex
electronic data transmission communications center
EDTV
extended definition television
EDU
electronic display unit
experimental diving unit
exponential decay unit
EDVAC
electronic digital vernier analog computer
electronic discrete variable automatic calculator
electronic discrete variable automatic computer
EDVAP
electronic digital vernier analog plotter
EDX
event driven executive
EE
echo equalizer
electric electric
electrical engineer
electrical engineering
electrical equipment
electronic engineering
electronics engineer
english electric
equipment engaged
error expected
external environment
EEA
electronic engineering association
error exit address
EEARM
elementary electrical and radio material
elementary electrical and radio material training school
EEB
eastern electricity board
EEC
electrical export corporation
electronic engine control
electronic equipment committee
end of equilibrium cycle
engineered electronics company
english electric company
european economic community
EECL
emitter emitter coupled logic
emitter follower emitter coupled logic
EECMA
european electronic components manufacturers association
EECO
electronic engineering company
EECW
emergency exchanger cooling water
EED
electrical engineering department
electroexplosive device
electromagnetic explosive device
EEDC
electronics economic development committee
EEE
electrical engineering exposition
electronic equipment engineering
EEFAMOS
electrically erasable floating gate avalance injection mos
EEG
electroencephalogram
electroencephalograph
electroencephalography
EEI
edison electric institute
environmental equipment institute
essential elements of information
EEIC
elevated electrode ic
EEL
emitter to emitter coupled logic
engineering electronics laboratory
EELM
english electric leo marconi

EEM
  earth entry module
  electronic engineer's master
  electronic engineers master
  electronic equipment monitoring
  emission electron microscope
EEMJEB
  electrical and electronic manufacturers joint education board
EEMTIC
  electrical and electronic measurement and test instrument conference
EEMUA
  engineering equipment and materials users' association
EEO
  electronic editorial office
  equal employment opportunity
EEOC
  equal employment opportunity commission
EEP
  electroencephalophony
  end to end protocol
EEPD
  energy production and delivery
EEPOL
  electrically erasable programmable logic device
EEPROM
  electrically erasable programmable read only memory
  electrically erasable prom
EER
  envelope elimination and restoration
  explosive echo ranging
EERA
  electrical equipment representatives association
EERI
  earthquake engineering research institute
EERL
  electrical engineering research laboratory
EEROM
  electrically erasable read only memory
EES
  engineering experiment station
EESC
  erie engineering societies council
EESMB
  electrical and electronics standards management board
EET
  east european time
  electrical equipment trailer
  explosive to electric transducer

EETF
  electromagnetic environmental test facility
  electronic environmental test facility
EEUA
  engineering equipment users' association
EEV
  english electric valve
EEVC
  english electronic valve company
EF
  elevation finder
  emitter follower
  engineering foundation
  entire function extra fine
  error factor
  extra fine
EFAL
  electronic flash approach lighting
EFAPP
  enrico fermi atomic power plant
EFAS
  electronic flash approach system
EFB
  electronic feed back
  engineering foundation board
EFBD
  emergency feed baron detector
EFC
  electronic frequency control
EFCL
  error free communication link
EFCOR
  effect corona
EFCS
  emitter follower current switch
EFCV
  excess flow check valve
EFD
  electrofluid dynamic
  engineering flow diagram
EFDA
  european federation of data processing associations
EFDAS
  epsylon flight data acquisition system
EFE
  external field emission
EFET
  enhancement mode fet
EFF
  effective
  efficiency
EFFBR
  enrico fermi fast breeder power reactor

EFFFL
efficiency full load
EFFI
electronic fiber fineness indicator
electronic forum for industry
EFFORPA
elliptic function first order ripple phase approximation
EFG
edge defined film fed growth
EFI
electronic fuel injection
EFICON
electronic financial control
EFIS
electronic flight instruments
EFL
effective focal length
emitter follower logic
equivalent focal length
error frequency limit
EFLA
educational film library association
EFM
eight to forteen modulation
electric field meter
expeditionary force message
EFMG
electric fuse manufacturers' guild
EFNMS
european federation of national maintenance societies
EFO
engineers foundation of ohio
EFOR
equivalent forced outage rate
EFP
electronic field production
EFPD
effective full power days
EFPH
equivalent full power hour
EFPM
effective full power month
EFR
electronic failure report
EFRC
edwards flight research center
EFS
electronic frequency selection
EFSORPA
elliptic function second order ripple phase approximation
EFSR
electronic field seaman recruit

EFT
earliest finish time
electronic funds transfer
EFTA
european free trade association
EFTI
engineering flight test instrumentation
EFTO
encrypt for transmission only
EFTS
electronic funds transfer system
elementary flying training school
EG
electronic guidance
enamel single glass
engine generator
environment generator
executive generator
grid voltage
EGAD
electromagnetic gas detector
electronegative gas detector
EGADS
electronic ground automatic destruct sequencer
EGAL
elevation guidance for approach and landing
EGCR
experimental gas cooled reactor
EGD
electrogasdynamic
EGECON
electronic geographic coordinate navigation
EGGA
european general galvanizing association
EGM
electronic governor module
EGO
eccentric geophysical observatory
eccentric orbiting geophysical observatory
experimental geophysical orbiting
EGPS
electric ground power system
extended general purpose simulator
EGR
electronic governor regulator
exhaust gas recirculation
EGRESS
evaluation of glide reentry structural system
EGRS
electronic and geodetic ranging satellite
EGS
electronics guidance section
electronics guidance station

EGSE
  electronic ground support equipment
EGSMA
  electrical generating systems marketing association
  engine generator set manufacturers' association
EGSP
  electronics glossary and symbol panel
EGTS
  emergency gas treatment system
EH
  electric heater
  electric hoist
  extra high
  extremely high
EHA
  effective halfword address
EHC
  electrical height calculator
  electro hydraulic control
  electronic hardware corporation
EHD
  elastohydrodynamic
  electrohydrodynamic
EHF
  electrohydraulic forming
  engineers hall of fame
  extremely high frequency
EHFA
  electric home and farm authority
EHFB
  electrical historical foundation board
EHMA
  electric hoist manufacturers association
EHP
  effective horsepower
  electric[al] horsepower
  electron hole pair
EHPH
  electric horsepower hour
EHS
  environmental health service
EHSI
  electronic horizontal situation indicator
EHT
  extra high tension
  extremely high tension
EHV
  extra high voltage
  extremely high voltage
EHVA
  electrohydraulic valve actuator
EI
  electrical industries
  electrical insulation
  electro industries
  electro instruments
  electromagnetic interference
  electronic installation
  electronic interface
  electronic interference
  end injection
  engineering idex, inc
  engineering information, inc
  engineering instruction
EIA
  electronic industries association
  energy information administration
  engineering industries association
  environmental impact appraisal
EIAC
  electronic industries association of canada
EIAJ
  electronic industry association of japan
EIAP
  environmental impact assessment project
EIB
  electronics information bulletin
  electronics installation bulletin
EIBA
  electrical industries benevolent association
EIC
  electrical insulation conference
  electromagnetic interference control
  electron induced conduction
  energy industries council
  engineer in charge
  engineering information center
  engineering institute of canada
  equipment identification code
  equipment installation and checkout
EICA
  electronics corporation of america
EICG
  electromagnetic interference control group
EICS
  eletromagnetic intelligence collection system
EID
  electrical inspection directorate
  electronic instrument digest
  electronics intelligence digest
  exposure intensity distribution
EIDLT
  emergency identification light
EIED
  electrically initiated explosive device
EIES
  electronic information exchange system

EIGFET
    enhancement insulated gate fet
    equivalent insulated gate fet
    equivalent insulated gate field effect transistor
EIIA
    electro instrument industry association
EIIS
    energy industry information system
EIK
    extended interaction klystron
EIL
    electron injection laser
EIM
    end of information marker
    excitability inducing material
EIMA
    electrical insulating materials association
EIMAC
    eitel mccullough
EIMAM
    environmental instrumentation measurment and monitoring
EIMB
    electronics installation and maintenance bulletin
EIME
    electronic instrument manufacturers' exhibition
EIMO
    electronic interface management office
EIN
    education information network
    european informatics network
EINI
    electron irradiation and neutron irradiation
EIO
    execute input output
    execute input output system
EIOP
    external input output processor
EIOS
    extended input output system
EIP
    emulator interface program
EIPC
    european institute of printed circuits
EIRP
    effective isotropic radiated power
    effective isotropically radiated power
    equivalent isotropically radiated power
EIS
    economic information systems
    effluent inventory system
    electric induction steel
    electrolyte insulator semiconductor
    electromagnetic intelligence system
    electronic information service
    end interruption sequence
    engineering index service
    environmental impact statement
    environmental information systems
    extended instruction set
EISC
    electronic industry show corporation
EIT
    electrical information test
    electronic installation technician
    engineer in training
EITA
    electric industrial truck association
EITB
    engineering industry training board
EIU
    executive independent utilities
EJ
    electrojet
EJC
    electrical joint compound
    engineers joint council
EJCC
    eastern joint computer conference
    eastern joint computer council
EJCNC
    ejc nuclear congress
EJFET
    enhancement mode junction fet
EK
    eastman kodak
EKC
    eastman kodak company
EKG
    effective kilogram
EKW
    electrical kilowatt
EKY
    electrokymogram
EL
    education level
    electric
    electric life
    electrical
    electrical latching
    electroluminescence
    electroluminescent
    electronics laboratory
    element
    elevation
    energy loss
    etched lead

ELA
electron linear accelerator
ELAC
electroacoustic
ELAN
electrologic language
ELB
electric battery
ELC
electric cable
executive and liaison committee
ELCA
earth landing control area
ELCO
electrolytic capacitor
ELCS
experimental labor control system
ELCT
electronic
electronics
ELCTRM
electronic room
ELD
edge lighted display
electrolytic liquid display
encapsulated light diffusion
extra long distance
ELDC
economic load dispatching computer
ELDO
european launcher development organization
european space vehicle launcher development organization
ELE
electronic launching equipment
ELEC
electric
electrical
electrician
electricity
electro
ELECENGR
electrical engineer
ELECENGRSCOL
electrical engineering school
ELECMECH
electrical mechanical
ELECO
engineering and lighting equipment company
ELECOM
electronic computer
electronic computing
ELECPROC
electrostatic processes

ELECPWRPLNTENGR
electric power plant engineer
ELECT
electric
electrical
electrician
electrolyte
electrolytic
electronics
ELECTCIRDESGNR
electronic circuit designer
ELECTENGR
electronic engineer
ELECTPKGENGR
electronic packaging engineer
ELECTR
electric
electrical
ELECTRL
electrolyte
electrolytic
ELECTRN
electrician
ELECTROCHEMENGR
electrochemical engineer
ELECTROL
electrolysis
ELECTROMECH
electrochemistry
ELECTROMECHENGR
electromechanical engineer
ELECTRON
electronic
electronics
ELECTROPHYS
electrophysical
electrophysics
ELECTROTECHNOL
electro technology electronics
ELECTRPROG
electronic progress
ELECTRWARFARE
electronic warfare
ELECTY
electricity
ELED
edge light emitting diode
ELEMCH
elementary charge
ELEP
electronic converter electric power
ELEV
elevated
elevation
elevator

ELF
electroluminiscent ferroelectric
electromotive force
electronic location finder
extensible language facility
extremely low frequency
ELFA
electric light fittings association
ELFC
electroluminescent ferroelectric cell
ELG
electrolytic grinding
ELGMT
erector launcher, guided missile, transportable
ELIC
electric lamp industry council
ELID
electrostatic latent image development
ELIM
eliminate
elimination
ELINT
electromagnetic intelligence
electronic intelligence
ELIP
electrostatic latent image photography
ELIT
electronics information test
ELLA
european long lines agency
ELLT
electric light
ELM
electrical length measurement
element
element load module
ELMA
electric lamp manufacturers' association
ELMINT
electromagnetic intelligence
ELMT
electronic mechanic technician
ELOD
erasable laser optical disk
ELOISE
european large orbiting instrumentation for solar experiments
ELP
english language programs
ELPC
electroluminescent photoconductive
electroluminescent photoconductor
ELPE
electroluminescent photoelectric

ELPEX
electronic production equipment exhibition
ELPG
electric light and power group
ELPH
elliptical head
ELPHR
experimental low temperature process heat reactor concept
ELPR
electroluminescent photoresponsive
ELR
engineering laboratory report
exchange line relay
ELRAC
electronic reconnaissance accessory
ELRO
electronics logistics research office
ELRSCALE
equal listener response scale
ELS
earth landing system
electronic speciality
error likely situation
exchange line selector
ELSB
edge lighted status board
ELSEC
electronic security
ELSI
extra large scale integration
ELSIE
electronic letter sorting and indicating equipment
electronic location and status indicating equipment
electronic signalling and indicating equipment
ELSS
extravehicular life support system
ELSSE
electronic sky screen equipment
ELT
electrometer
emergency locator transmitter
european letter telegram
ELTEX
electronic time division telex exchange
ELTRC
electric
ELV
electrically operated valve
enclosed frame low voltage
ELVIS
electroluminescent vertical indication system

ELW
  electric weld
ELWLD
  electrical world
EM
  efficiency modulation
  electric charge to mass for particles
  electric machinery
  electrician's mate
  electromagnet
  electromagnetic[al]
  electromechanic[al]
  electron microscope
  electron microscopy
  electronic countermeasure malfunction
  electronic measurements
  emergency maintenance
  emission pump
  end of medium
  end of message
  engineering management
  engineering manual
  engineering memorandum
  environmental management
  epitaxial mesa
  escape motor
  evaluation model
  exact match
  exposure meter
  mechanical efficiency
EMA
  electronic making apparatus
  electronic manufacturers association
  electronic mathematic automation
  electronic measuring apparatus
  electronic missile acquisition
  extended mercury autocode
  extended mercury autocoder
EMAIA
  electrical meter and allied industries association
EMAR
  experimental memory address register
EMATS
  emergency message automatic transmission system
  emergency mission automatic transmission service
EMAV
  electromagnetic relief valve
EMB
  electrical moderization bureau
  electronic maintenance book
  electronic material bulletin
  engineering in medicine and biology

EMC
  eighty meter community
  electromagnetic compatibility
  electronic material change
  electronic mode control
  electronics management center
  electronics marketing corporation
  engineered military circuits
  engineering manpower commission
  engineering military circuits
  engineering mockup critical experiment
  european military communications
  excess minority carriers
EMCA
  electro materials corporation of america
EMCAB
  electromagnetic compatibility advisory board
EMCB
  electrician's mate, construction battalion
EMCBL
  electrician's mate, construction battalion, line and station
EMCCC
  european military communications coordinating committee
EMCGS
  electromagnetic centimeter gram second
EMCO
  electromechanicals company
EMCON
  emery control
  emission control
EMCP
  electromagnetic compatibility program
EMCS
  electromagnetic compatibility stadardization
EMCTP
  electromagnetic compatibility test plan
EMD
  electric motor driven
  engineered magnetics division
  entry monitor display
EMDI
  energy management display indicator
EMDP
  electromotive difference of potential
EMDS
  electronic material data service
EME
  earth moon earth
  electrical and mechanical engineer
  electromagnetic energy
  electromagnetic environment
  engineer electrical and mechanical
  environmental measurements experiment

EMEA
electrical manufacturers export association
electronic maintenance engineering association
EMEB
east midlands electricity board
EMEC
electronic maintenance engineering center
EMERG
emergency
EMESFET
enhancement mode mesfet
EMETF
electromagnetic environmental test facility
EMF
electromotive force
electronic manufacturing facility
EMG
electromyogram
electromyograph
electromyography
EMGE
electronic maintenance ground equipment
EMI
electric music instruments
electrical and musical industries
electrical measuring instrument
electromagnetic impulse
electromagnetic interference
electron magnetic interference
electronic maintenance inspector
electronic memories, incorporated
EMIC
electronic materials information center
EMICE
electromagnetic interference control engineer
EMIDEC
electrical and musical industries data
electronic computer
EMINT
electromagnetic intelligence
EMIP
experimental manned interceptor program
EMIRTEL
emirate telecommunication corporation
EMIS
educational management information system
electromagnetic intelligence system
emission
EMISS
electromolecular instrument space simulator
emission
EML
electromechanical laboratories
emihus microcomponents, limited

emulator
engineering mechanics laboratory
equipment modification list
expected measured loss
EMM
electrical and mechanical maintenance
electromagnetic measurement
electron mirror microscope
electronic manufacturing manual
electronic memories and magnetics
EMMA
electron microscopy and microanalysis
electronic mask making apparatus
eye movement measuring apparatus
EMMCC
erection mechanism motor control center
EMN
electromagnetic moving coil and neutralized winding
equivalent manufacturers' number
EMO
electromechanical optical
EMOD
erasable magneto-optical disk
EMOS
earth mean orbital speed
earth's mean orbital speed
EMP
electromagnetic pulse radiation
electromagnetic power
electromagnetic pulse
electromechanical power
electromechanical pulse
electron microprobe
electronic multiplying punch
EMPIRE
early manned planetary interplanetary round trip experiment
early manned planetary interplanetary round trip expedition
early manned planetary round trip expedition
electronic multipurpose intelligence retalliatory equipment
EMR
electromechanical relay
electromechanical research, incorporated
electromagnetic radiation
electromechanical research
emerson electric manufacturing company
EMRA
electrical manufacturers representatives association
EMRIC
educational media research information center

EMS
   earthquake monitoring system
   electromagnetic surveillance
   electromagnetic susceptibility
   electromotive surface
   electronic mail system
   electronic management system
   electronic medical system
   electronic micro system
   emergency medical services
   emission spectrograph
   emulator monitor system
   engineering management society
   export marketing service
   extended maintenance service
EMSA
   electron microscope society of america
   electronics material support agency
EMSC
   electrical manufacturers standard council
   electronic message service center
EMSL
   electronic material sciences laboratory
EMSR
   electrician's mate, ship repair
EMSRG
   electrician's mate, ship repair, general electrician
EMSRS
   electrician's mate, ship repair, shop electrician
EMSS
   experimental manned space station
EMT
   electrical mechanical tubing
   electrical metal tubing
   electrical metallic tubing
   electrician's mate, telephone
   european mediterranean tropo
EMTA
   electro medical trade association
EMTECH
   electromagnetic technology
EMTED
   electromagnetic test and evaluation data
EMTF
   estimated mean time to failure
EMTI
   edge mounted threaded inserts
EMTTF
   equivalent mean time to failure
EMU
   electromagnetic system of units
   electromagnetic unit
   extravehicular mobility unit

EMUG
   european map user group
EMV
   electromagnetic volume
EMVEC
   eastern massachusetts vermont energy control
EMW
   electromagnetic warfare
   electromagnetic wave
EMWEEK
   electrical merchandising week
EN
   electronics news
   enforcement notification
ENAM
   enamel
   enamelled
ENC
   electron nuclear coupling
   encoder
ENCA
   european naval communication agency
ENCORE
   enlarged compact by response
END
   end of data
   endurance
ENDADR
   end address of main memorix section
ENDOR
   electron nuclear double resonance
ENE
   energize
ENEA
   european nuclear energy association
ENI
   equivalent noise input
ENIAC
   electronic numerical integrator and automatic calculator
   electronic numerical integrator and calculator
ENIC
   voltage negative impedance converter
ENK
   enter key
ENPP
   electronics new product preview
ENQ
   enquiry
   enquiry character
ENR
   equivalent noise ratio
   equivalent noise resistance
   excess noise ratio

ENRZ
  enhanced non return to zero
ENSI
  equivalent noise sideband input
ENSIM
  environmental simulator
ENT
  effective noise temperature
  entry
  equivalent noise temperature
  equivalent noise temperature
ENTAC
  electronic numerical integrator and computer
ENTC
  engine negative torque control
EO
  earth orbit
  elementary operation
  end of operation
  engineering order
  equivalent orifice
  exclusive or circiuit
  executive order
EOA
  end of address
EOAR
  european office of aerospace research
EOARDC
  european office of aerospace research and development command
EOAU
  electrooptical alignment unit
EOB
  electronic order of battle
  end of block
EOC
  electronics operations center
  emergency operating center
  end of card
  end of conversion
  end of cycle
  error of closure
  experimentation operations center
EOCI
  electric overhead crane institute
EOCR
  experimental organic cooled reactor
EOD
  end of data
  explosive ordnance disposal
EODARS
  electrooptical direction and ranging system
EODD
  electrooptic digital deflector

EOE
  edge of earth
  electronic optic electronic
  end of extent
  errors and omissions expected
EOEC
  end of equilibrium cycle
EOEM
  electronic original equipment market
EOF
  earth orbital flight
  end of file
  end of form
EOG
  electrooculogram
  electrooculograph
  electrooculography
EOGO
  eccentric orbiting geophysical observatory
EOI
  end of identity
  end of inquiry
  end or identify
EOIS
  electrooptical imaging system
EOJ
  end of job
EOL
  earth orbit launch
  end of life
  end of line
  end of list
  expression oriented language
EOLM
  electrooptical light modulator
EOLR
  electrical objective loudness rating
EOLT
  end of logic[al] tape
EOM
  earth orbital mission
  end of message
EOMI
  end of message incomplete
EOMS
  earth orbital military satellite
EON
  end of number
  end of number control character
EOO
  electro optics organization
EOOC
  exchange oriented operator control
EOP
  earth orbit plane

end of part
end of program
end output
engineering operating procedure
equipment operating procedure
EOPC
eletro optic phase change
EOPE
end of powered flight
EOQ
economic order quantity
EOR
earth orbit rendezvous
earth orbital flight
end of record
end of reel
end of run
exclusive or
expected operations forecast
explosive ordnance reconnaissance
EORBS
earth orbiting recoverable biological satellite
EOS
earth observation satellite
earth observatory satellites
electrooptical system
end of segment
end of string
equation of state
extended operating system
EOSC
extended operating system card
EOSMD
extended operating system magnetic drum
EOSS
earth orbital space station
EOT
end of tape
end of test
end of text
end of track
end of transmission
end of type
EOTS
electronic[al] optical tracking system
electrooptic tracking system
EOV
electrically operated valve
end of volume
EOVM
end of valid message
EOY
end of year
EP
electric power

electric primer
electrical propulsion
electrically polarized
electronic processing
electronic publishing
electroplate
electropneumatic
elongated punch
emergency preparedness
emulator program
end of program
engineer personnel
english patent
entry point
epitaxial planar
equal
equalizer
equation
equivalent
estimated position
etched plate
explosion proof
extended play
extreme pressure
EPA
electron probe analyzer
environmental protection agency
EPAC
electronic production aids catalog
EPAD
electrical power and distribution
EPAM
elementary perceiver and memorizer
EPBX
electronic private branch exchange
EPC
easy processing channel
edge punched card
educational planning center
electronic program control
engineering change proposals
EPCCS
emergency positive control communications system
EPCI
enhanced programmable peripheral interface
EPCO
emergency power cut off
engine parts coordinating office
EPCU
electrical power control unit
EPD
earth potential difference
electric power distribution
etch pit density

EPDC
  economic power dispatch computer
EPDCC
  elementary potential digital computing component
EPDCE
  elementary potential digital computing element
EPDS
  electrical power distribution system
  electronic parts distributors' show
EPE
  electrical power engineering
  energetic particle explorer
  engineering progress exposition
EPEA
  electrical power engineers' association
EPEC
  emerson programmer evaluator controller
EPG
  edit program generator
  electronic proving ground
  emergency procedure guidelines
EPGCR
  experimental prototype gas cooled reactor
EPH
  electric process heating
  electrochemical plating and honing
EPI
  earth path indicator
  electrical protection
  electronic position indicator
  electronic process inc
  elevation position indicator
  epitaxial
  expanded position indicator
EPIC
  earth pointing instrument carrier
  electric properties information center
  electronic production and inventory control
  epitactic integrated circuit
  epitaxial integrated circuit
EPID
  electrophoretic image display
EPIRB
  emergency position indicating radio beacon
EPL
  electronic products laboratory
  environmental protection limit terminology used in environmental technical specifications
  european program library
  extensible programming language
  extreme pressure lubricant
EPLA
  electronic precedence list agency
EPM
  engineering procedure memos
  environmental project manager
  equivalent per million
  external polarization modulation
EPMA
  electronic parts manufacturers' association
EPMAU
  expected present multi attribute utility
EPN
  effective perceived noise level
EPNDB
  effective perceived noise level decibels
EPNL
  effective perceived noise level
EPNS
  electroplated nickel silver
EPO
  emergency power off
EPP
  electronic packaging and production
EPPI
  electronic plan position indicator
EPPPI
  expanded partial plan position indicator
EPPT
  electrical power production technician
EPPTS
  electrical power production technician specialist
EPR
  electrical pressure regulator
  electron paramagnetic resonance
  electron parametric resonance
  electrophrenic respiration
  engine pressure ratio
  equivalent parallel resistance
  essential performance requirements
  ethylene propylene rubber
EPRA
  electronic production resources agency
EPRD
  electrical power requirements data
EPRDC
  electric power research and development center
EPRI
  electric power research institute
EPROM
  electrically programmable read only memory
  erasabel prom
  erasable programmable read only memory

EPRTCS
   emergency power ride through capability system
EPS
   electric[al] power storage
   electrical power subsystem
   electrical power supply
   electrical power system
   electronic publishing system
   emergency power subsystem
   emergency power supply
   energetic particles satellite
   equilibrium problem solver
   european physical society
   switch mode power supply
EPSA
   electrostatic particle size analyzer
EPSCS
   enhanced private switched communications service
EPSL
   emergency power switching logic
EPSS
   experimental packet switching service
EPSTF
   electrical power system test facility
EPT
   electrostatic printing tube
   environmental proof test
   exciter power logic
EPTA
   expanded program of technical assistance
EPTE
   existed prior to entry
EPU
   electric[al] power unit
   emergency power unit
EPUT
   events per unit time
EPV
   electropneumatic valve
EPWAPDA
   east pakistan water and power development authority
EPWM
   electroplate on white metal
EQ
   equal
   equal to
   equalize
   equalizer
   equation
   equipment
   equivalent

EQCC
   entry querry control console
EQL
   expected quality level
EQP
   equipment
EQPMT
   equipment
EQS
   equatorial scatter
   equivalent to sheathed explosive
EQU
   equate
EQUIP
   equipment
EQUIV
   equivalent
EQV
   equivalence
ER
   echo ranging
   effectiveness report
   electrical resistance
   electrorefined
   electron recording
   electronic reconnaissance
   end of run
   engine relay
   enhanced radiation
   environmental report
   error relay
   established reliability
   expected result
   external register
   external resistance
ERA
   economic regulatory administration
   electric railroaders association
   electrical research association
   electron ring accelerator
   electronic reading automation
   electronic representatives association
   electronic research association
   energy reorganization act
   equal rights amendment
ERAM
   extended range antiarmormunition
ERAS
   electronic reconnaissance access set
ERASER
   elevation radiation seeker rocket
ERB
   edward rocket base
   engineers' registration board

equipment review board
experiment review board
ERBE
    earth radiation budget experiment
ERBM
    extended range ballistic missile
ERC
    electrical research committee
    electronics research center
    equatorial ring current
    error retry count
ERCC
    error checking and correction
    error checking and correction code
ERCOT
    electric reliability council of texas
ERCR
    electronic retina computing reader
    error cause register
ERCS
    emergency rocket command system
    emergency rocket communications
    emergency rocket communications system
ERCW
    emergency raw cooling water
ERD
    electronics research directorate
    exponentially retrograded diode
ERDA
    electrical and radio development association
    electronic resources development agency
    electronics research and development agency
ERDAM
    erda manual
ERDL
    electronic research and development laboratory
ERDR
    earth rate directional reference
ERDT
    electronic research and development technician
ERE
    echor range equipment
EREP
    environmental recording, editing and printing
ERETS
    edwards rocket engine test station
    experimental rocket engine test station
ERF
    error function
ERFA
    european radio frequency agency
ERFC
    error function

error function complement
error function, complementary
ERFPI
    extended range floating point interpretive system
ERFPIS
    extended range floating point interpretive system
ERG
    electron radiography
    electroretinogram
    electroretinograph
    electroretinography
ERGS
    en route guidance system
    experimental route guidance system
ERI
    electronic resources, incorporated
ERIC
    educational research information center
    electronic remote and independent control
    energy rate input controller
ERIS
    emergency resources identification equipment
ERISA
    employee retirement income security act
ERL
    echo return loss
    environmental research laboratory
    equipment revision level
    event record log
ERM
    earth reentry module
    electrical research memorandum
    evaporate rate monitor
ERMA
    electronic reading method of accounting
    electronic recording machine
    electronic recording machine accounting
    electronic recording method of accounting
ERMI
    electronic radio manufacturing industry
ERN
    engineering release notice
ERNIE
    electronic random numbering and indicating equipment
    electronic random numbering and indication equipment
ERO
    emergency repair overseer
    engergy research office
    engineering regional organization
    engineering release order

EROAT
   echo ranging operated acoustic torpedo
EROM
   erasable read only memory
EROS
   earth resources observation satellite
   eleminate range zero system
   experimental reflector orbital shot
ERP
   earth reference pulse
   effective radiated power
   elevated release point
   emergency recorder plot
   emerson radio and phonograph
   emissive radio power
   emitted radio power
   end response
   error recovery procedures
   european recovery program
ERPAL
   electronics repair part allowance list
ERPLD
   extended range phase locked demodulator
   extended range phased locked demodulator
ERPS
   electrolytic reactants production system
ERQ
   end request
ERR
   elk river reactor
   error
ERRC
   error character
ERS
   economic research service
   electronic register sender
   empire radio school
   engineers register study
   entry and recovery simulation
   environmental research satellite
   external regulation system
ERSA
   electronic research supply agency
ERSB
   expendable radio sono buoy
ERSER
   expanded reactance series resonator
ERSI
   electric remote speed indicator
ERSP
   earth resources survey program
ERSR
   equipment reliability status report
ERSS
   earth resources satellite system
   earth resources survey satellite
ERT
   electron ray tube
   environmental research technology
ERTS
   earth resources technology satellite
ERW
   electrical resistance weld
   electronic resistance welding
ERX
   electronic remote switching
ES
   earth switch
   earth to space
   echo sounder
   echo sounding
   edison screw
   electrical radio training school
   electrical section
   electrochemical society
   electromagnetic storage
   electromagnetic switching
   electronic specialty
   electronic standard
   electronic switch
   electrostatic
   electrostatic storage
   electrostatis
   enamel single silk
   engineering societies
   engineering specification
   experimental station
   expert system
   extension station
   external shield
ESA
   ecological society of america
   electrical surge arrester
   electrolysis society of america
   employment standards administration
   european space agency
   executive storage area
ESAIRA
   electronically scanned airborne intercept radar
ESAR
   electronically scanned array radar
   electronically steerable array radar
ESARS
   earth surveillance and rendezvous simulator
   employment service automated reporting system

ESAS
   engineered safeguards actuation system
ESB
   electric storage battery
   electrical simulation of brain
   electric[al] standards board
   electrical systems branch
   engineering societies building
   engineering society of baltimore
   engineering society of buffalo
ESBC
   electronics small business council
ESBO
   electronic selection and bar operating
ESC
   electrical standards committee
   electronic system center
   electrostatic compatibility
   engineering standards committee
   engineers and scientists of cincinnati
   escape
   escape character
   escapement
   escutcheon
ESCA
   electron spectroscopy for chemical analysis
   extended source calibration area
ESCAN
   e scanner
ESCAP
   economic and social commission for asia and the pacific
ESCAPE
   expansion symbolic compiling assembly program for engineering
ESCAT
   emergency security control of air traffic
ESCGS
   electrostatic centimeter gram second
ESCI
   electric space conditioning institute
ESCO
   electrical steel company
ESCOE
   engineering societies commission on energy, inc
ESCOM
   electricity supply commission
ESCS
   electronic spacecraft simulator
ESCWS
   essential service cooling water systems
ESD
   echo sounding device
   electro static discharge
   electronic systems division
   electrostatic storage deflection
   electrostatic storage reflection
   energy storage device
   environmental sensing device
   external symbol dictionary
   external symbol directory
ESDA
   electronic service dealers' association
ESDAC
   european space data center
ESDI
   enhanced small disk interface
   enhanced storage device interface
ESDP
   evolutionary system for data processing
ESDT
   electrostatic storage display tube
ESDU
   event storage and distribution unit
ESE
   east southeast
   electrical support equipment
ESEG
   electronic systems engineering group
ESF
   electrostatic focusing
   engineered safety feature
   extended spooling facility
ESFAS
   engineered safety features actuation system
ESFETA
   empire state federation of electronic technicians' associations
ESFI
   epitaxial silicon film on insulators
   epitaxial silicon film on isolators
ESFIDIODE
   epitaxial silicon film on insulator diode
ESFK
   electrostatically focused klystron
ESG
   electrically suspended gyro
   electrically suspended gyroscope
   electronic sweep generator
   electronically suspended gyro
   electrostatic gyroscope
   electrostatically suspended gyro
   engineers and scientists guild
   engineers' and scientists' guide
   english standard gauge
   expanded sweep generator
ESGA
   electronically suspended gyroscope accelerometer

ESH
  equivalent standard hours
ESHAC
  electric space heating and air conditioning
ESHU
  emergency ship handling unit
ESI
  electro scientific industries
  electromagnetic sciences, incorporated
  emergency stop indicator
  engineering and scientific interpreter
ESIA
  externally specified index adress
ESIS
  electronic store information system
ESJCP
  engineers and scientists joint committee on pensions
ESL
  electro science laboratories
  engineering societies library
  equivalent series inductance
  european systems language
ESLAB
  european space laboratories
ESLO
  european satellite launching organization
  european space launcher organization
ESM
  elastomeric shield material
  electromatic speed meter
  electronic support measure
  electronic warfare support measures
  electronically controlled system with magnetic field coupling
  electronics support measures
  engineers and scientists of milwaukee
  environmental systems monitor
ESMA
  electronic sales and marketing association
ESMC
  engineering societies monograph committee
ESMU
  electronic systems mock up
ESN
  effective segment number
ESNE
  engineering societies of new england, inc
ESO
  economic stabilization office
  electronic supply office
  electronic supply officer
  emergency support organization
ESOC
  european space operations center

ESONE
  european standard of nuclear electronics
  european system of nuclear electronics
ESP
  efficiency speed power
  electronics system plan
  electrosensitive programming
  electrostatic precipitator
  electrostatic probe
  executive and scheduling program
  extrasensory perception
ESPAR
  electronically steerable phased array radar
ESPI
  electronic space products, incorporated
ESPOD
  electronic systems precision orbit determination
ESPOL
  executive system programming oriented language
ESPRIT
  european strategic program for research in information technology
ESPS
  engineering societies personnel service
ESR
  early site review
  effective series resistance
  effective shunt resistance
  effective signal radiated
  electron spin resonance
  electronic scanning radar
  electroslag refining
  equivalent series resistance
  equivalent series resistor
  event storage record
  experimental superheat reactor
ESRANALYSIS
  electron spin resonance analysis
ESRANGE
  european sounding rocket range
  european space range
  european space research range
ESRC
  economic and social research council
ESRIN
  european space research institute
ESRO
  european space research organization
ESRP
  environmental standard review plan
ESRR
  early site review report

ESRS
electronic scanning radar system
ESRSPECTROSCOPY
electron spin resonance spectroscopy
ESRT
electroslag refining technology
ESS
electrical supervisory subassembly
electron spin spectra
electronic science section
electronic sequence switching
electronic speech synthesis
electronic switching system
emplaced scientific station
engineered saftey system
entry survival system
equivalent state subsets
evaluation sage sector
experimental sage sector
expert system shell
ESSA
electronic scanning and stabilizing antenna
emergency safeguards system activation
endangered species scientific authority
environmental science services administration
environmental survey satellite
ESSB
electronic supply support base
ESSBR
electronically scanned stocked beam radar
ESSC
electronic standards subcommittee
ESSCO
electronic space structurers corporation
ESSEX
experimental solid stage exchange
ESSFL
electron steady state fermi level
ESSG
engineer strategic studies group
ESSP
earliest scram set point
engineers society of st paul
ESSPO
electronic support systems project office
electronic supporting systems project office
ESSU
electronic selective switching unit
EST
earliest start time
eastern standard time
eastern summer time
elastic surface transformation
electrolytic sewage treatment
electroshock therapy

electrostatic storage tube
empire social telegram
engineers society of tulsa
enlistment screening test
established
establishment
estimate
estimation
estimator
ESTEC
european space research technical center
european space technology center
ESTF
electronic system test facility
ESTG
estimating
ESTRAC
european space satellite tracking and telemetry network
ESTRACK
european satellite tracking
european space tracking
ESTV
error statistics by tape volume
ESU
electrostatic unit
electrostatics units
ESV
earth satellite vehicle
electrostatic voltmeter
emergency stop valve
enamel single silk varnish
enamel single silk varnished
error statistics by volume
essential service value
ESW
error status word
ESWM
engineering society of western massachusetts
ESWP
engineers society of western pennsylvania
ESWS
emergency service water system
ESY
engineering society of york
ET
eastern standard time
eastern telegraph
edge triggered
effective temperature
electric auto lite company
electric telegraph
electrical technician
electrical time
electrical transcription

electrical typewriter
electronic test
electronic transformers
electronic tube
electronics technician
end of tape
ending tape
energy transfer
engineering test
engineering tests
ephemeris time
escape tower
evaluation test
external tank
ETA
   employment and training administration
   estimated time of arrival
   expected time of arrival
   expected to arrive
ETAC
   environmental technical applications center
ETB
   end of text block
   end of transmission block
   end of transmission block character
ETC
   eastern telegraph company
   electro tech corporation
   electronic temperature control
   electronic tube corporation
   electronic tuning control
   electronic typing calculator
   eleven thirty conversion
   estimated time of completion
   extended text compositor
ETCC
   environmental test control center
ETCG
   elapsed time code generator
ETD
   embedded temperature detector
   estimated time of departure
   expected time of departure
ETDL
   electronics technology and device laboratory
ETE
   effluent thermal effects
   electrical technician electrician
   even transversal electric
   external telecommunications executive
   external test equipment
ETEC
   energy technology eng center
ETES
   european television service

ETF
   electron transfering flavoprotein
   environmental test facility
   estimated time of flight
ETFO
   electronics technical field office
ETFS
   electronic countermeasure transmitter frequency set up
ETG
   electronic training group
ETI
   elapsed time indicator
   electric[al] tool institute
ETIM
   elapsed time
ETL
   eastern trunk line
   electrical testing laboratories
   electronic technology laboratory
   electrotechnical laboratory
   emitter follower transistor logic
   ending tape label
   ericsson telephones, limited
   etching by transmitted light
ETM
   elapsed time meter
   electronic technician's mate
   electronic test and maintencance
   end of tape marker
   enhanced timing module
   even transversal magnetic
   experimental test model
ETMA
   educational television for the metropolitan area
ETMG
   electron tube management group
ETMSR
   electronic technician's mate, ship repair
ETMWG
   electronic trajectory measurements working group
ETN
   electronics technician
   equipment table nomenclature
ETO
   electronic temperature offsetting
   energy technology office
ETOS
   extended tape operating system
ETP
   electrical tough pitch
   electrolytic tough pitch
   electronic tape printer

electronic technical publishing
electronical tough pitch
equivalent top product
experimental test procedure
extended tape processing
extended term plan
ETPY
electronic control assembly thrust vector, pitch and yaw
ETQ
education, training and qualifications committee
ETQAP
education and training in quality assurance practices
ETR
eastern test range
electronics technician, radar
engineering test reactor
engineering test record
estimated time of return
extended temperature range
ETRC
educational television and radio center
engineering test reactor critical assembly
ETS
edwards test station
electronic tandem switching
electronic telegraph service
electronic telegraph system
electronic timing set
environmental technical specifications
environmental test specification
european teleprocessing system
external time sharing
ETSAL
electronics terms for space age language
ETSQ
electrical time superquick
ETSS
electronic tandem switching system
ETT
end of tape test
ETTI
end translation time indicator
ETU
electrical trades union
ETV
educational television
ETVM
electrostatic transistorized voltmeter
ETX
end of text
end of text character

EU
emulator program
energy unit
entropy unit
EUA
electrical utilities applications
EUCLID
easily used computer language for illustrations and drawings
experimental use computer, london integrated display
EUF
equivalent unavailability factor
EUFMC
electric utilities fleet managers conference
EUMAFTN
european mediterranean aeronautical fixed telecommunications network
EURIM
european conference on research of information services and libraries
EUROATOM
european atomic energy community
EUROCAE
european organization for civil aviation electronics
EUROCAEM
european organization for civil aviation electronics manufacturers
EUROCOMP
european computing congress
EUROCON
european convention
EUROCRA
european ocr association
EUROSPACE
european industrial space research group
european space research group
EUROVISION
europa + television
EUT
equipment under test
EUV
extreme ultraviolet
EV
electro voice
electronvolt
enclosed and ventilated
error voltage
expected value
exposure value
EVA
electric vehicle association
electronic velocity analyzer
error volume analysis

    extra vehicular activities
    extra vehicular astronaut
EVAC
    evacuated
    evacuation
EVAL
    evaluation
EVAP
    evaporation
EVAPD
    evaporated
EVAPN
    evaporation
EVATA
    electronic visual auditory training aid
    electronic visual aural training aid
EVBM
    expected value business model
EVC
    electric vehicle council
EVCO
    electronic vibration cut off
EVDL
    electronically variable delay line
EVFM
    ex vessel flux monitor
EVHM
    ex vessel handling machine
EVIST
    ethic and values in science and technology program
EVM
    electronic voltmeter
EVOM
    electronic volt ohmmeter
EVR
    electronic video recorder
    electronic video recording
    electronic video reproduction
EVS
    extra vehicular suit
EVSC
    extra vehicular suit communications
EVSS
    extra vehicular space suit
EVSTC
    extra vehicular suit telecommunications
EVT
    equiviscous temperature
EVTM
    ex vessel transfer machine
EVTS
    engine and vehicle test stand
EW
    early warning

    effective warmth
    electrical world
    electrically welded
    electronic war
    electronic warfare
    electronic wholesalers
    electronics world
EWA
    effective word address
EWASER
    electromagnetic wave amplification by stimulated emission of radiation
EWC
    electric water cooler
EWCAS
    early warning and control aircraft system
EWCL
    electromagnetic warfare and communications laboratory
EWCRP
    early warning control and reporting post
EWCS
    european wideband communication system
EWD
    electronic warfare department
EWEB
    eugene water and electric board
EWF
    electrical wholesalers federation
EWH
    expected working hours
EWL
    equalized ward leonard
EWMU
    enemy wireless monitoring unit
EWO
    electrical and wireless operator
    electronic warfare officer
    engineering work order
EWP
    expected wire phenomenon
EWR
    early warning radar
EWRT
    electrical women's round table
EWS
    early warning signal
    early warning station
    early warning system
EWSC
    electric water systems council
EWST
    elevated water storage tank

EWTAD
   early warning threat analysis display
EWTMI
   european wideband transmission media improvement program
EX
   examined
   example
   except
   exception
   exceptional
   excess
   exchange key
   excluding
   execute
   executed
   execution
   executive
   exit
   experiment
   experimental
   experimental station
   exponent
   extra
   extract
EXACT
   exchange of authenticated electronic component performance test data
EXAPT
   exact automatic programming of tools
   extended subset of apt
   extension of automatically programmed tool language
EXAS
   engineering services and safety
EXC
   excitation
   exciter
EXCH
   exchange
EXCIMER
   excited dimer
EXCIPLEX
   excited state complex
EXCLOR
   exclusive or
EXCO
   executive committee
EXCP
   execute channel program
EXCTR
   exciter
EXD
   external device
EXDAMS
   extended debugging and monitoring system
EXEC
   execute
   executive control system
   executive extension
EXH
   exhibit
   exhibition
   exhibitor
EXMETNET
   experimental meteorological sounding rocket research network
EXOBIOLOGY
   exoterrestrial biology
EXOLLIFE
   exoterrestrial life
EXOR
   exclusive or
EXORGATE
   exclusive or gate
EXP
   expansion
   experiment
   experimental
   exponential
   exponentiation
   exposure
EXPER
   experiment
   experimental
EXPERT
   expanded pert
   extended pert
EXPR
   experiment
   experimental
EXPRESS
   expendable parts record and structures system
EXPTL
   experimental
EXSC
   executive standards council
EXSR
   executive subroutine
EXST
   execute stack
EXSTA
   experimental station
EXT
   extend
   extension
   exterior
   external
   extinguish

    extinguisher
    extra
    extract
    extreme
EXTBAT
    extension battery
EXTCD
    extension cord
EXTEL
    exchange telegraph co ltd
    exchange telegraph company
EXTERRA
    extraterrestrial research agency
EXTFREQ
    extension frequency
EXTM
    extended telecommunications module
EXTRA
    exponentially tapered reactive antenna
EXTRADOP
    extended range doppler
EXTRADOVAP
    extended range doppler velocity and position
EXTRN
    external reference
EXTSN
    extension
EXTTPHONE
    extension telephone
EXTW
    extension wire
EZ
    electrical zero

F

f
    fahrenheit
    farad
    faraday
    feedback
    feet
    fellow
    female
    femto
    fermi
    field
    filament
    file
    filter
    fine
    fixed
    fixed format
    focal length
    following
    foot
    force
    foreground
    formula
    forward
    fraction
    frame
    free energy
    frequency
    function
    fuse
FA
    femtoampere
    field accelerating
    field-accelerating contactor or relay
    field artillery
    final address
    final assembly
    finite automaton
    flat gain amplifier
    floating add
    forced air cooled
    frame antenna
    frequency adjustment
    frequency agility
    friendly aircraft
    fuel air
    fuel assembly
    fully accessible
    fully automatic
    further assemblies
    fused alloy
FAA
    federal aviation administration
    federal aviation agency
FAAAS
    fellow of the american association for the advancement of science
FAAB
    frequency allocation advisory board
FAABMS
    forward area anti ballistic missile system
FAAR
    forward area alerting radar
    forward area alter radar
FABISO
    fabricaton isometric
FABMDS
    field army ballistic missile defense system
FABMIDS
    field anti ballistic missile defense system
    field army ballistic missile defense system

FAC
  facility
  factor
  federal aviation commission
  federation of automatic control
  field accelerator
  file access channel
  first alarm code
  floating point accumulator
  forward air controller
  frequency allocation committee
  fuse arming computer
FACD
  failure and consumption data
  foreign area customer dialing
FACDIR
  failure and consumption data inspection report
FACE
  federation of associations on the canadian environment
  field alterable control element
  field artillery computer equipment
  field artillery computing equipment
FACISCOM
  finance and comptroller information system command
FACP
  fully automated computer program
FACR
  fourier analysis cyclic reduction
FACS
  facsimile
  fine attitude control system
  floating decimal abstract coding system
FACSC
  frequency allocation coordinating subcommittee
FACSI
  fast access coded small images
FACSIM
  facsimile
FACT
  factual compiler
  fairchild assured customer test
  flexible automatic circuit tester
  flight acceptance composite test
  forecast and control technique
  foundation for advanced computer technology
  foundation of advanced computer technology
  fully automatic cataloging technique
  fully automatic compiler translator
  fully automatic compiling technique

FACTOR
  factor operator
  fourteen on one automatically controlled test optimizing routine
FACTS
  facilities administration control and time schedule
FAD
  facilities and design
  findings and determination
  floating add
FADAC
  field artillery digital automatic computer
FADC
  frequency analog to digital converter
FADP
  finnish association for data processing
FAE
  field application engineer
  final approach equipment
FAETU
  fleet airborne electronics training unit
FAETUA
  fleet airborne electronics training unit, atlantic
  fleet airborne electronics unit atlantic
FAETUP
  fleet airborne electronics unit pacific
FAF
  final approach fix
FAGC
  fast automatic gain control
  forward area ground control
FAGM
  field army guided missile
FAGMSS
  field army guided missile system sergeant
FAGS
  federation of astronomical and geophysical services
FAHQMT
  fully automatic high quality machine translation
FAHR
  fahrenheit
FAI
  fail as is
  frequency azimuth intensity
FAIR
  fabrication, assembly and inspection records
  failure analysis information retrieval
  fast access information retrieval
FAIRS
  federal aviation information retrieval system

FAJ
final assembly jig
FAL
financial analysis language
frequency allocation list
FALC
forward acting linear combiner
FALTRAN
fortran to algol translator
FAM
fast access memory
fast auxiliary memory
federal airmail
floating add magnitude
foreign airmail
frequency allocation multiplex
frequency amplitude modulation
frequency and amplitude modulation
frequency assignment model
FAMAS
financial analysis and management system
financial and management accounting system
FAME
ferro acoustic memory
FAMECE
family of military engineer construction equipment
FAMOS
fleet application of meteorological observations for satellites
floating gate avalanche injection metal oxide semiconductor
FAMOST
floating avalanche metal oxide semiconductor transistor
floating gate avalanche injection metal oxide silicon transistor
FAMP
fire alarm monitoring panel
FAMS
forecasting and modelling system
FAO
finish all over
food and agriculture organization
FAP
field application panel
financial analysis program
floating point arithmetic package
fortran assembly program
frequency allocation panel
FAPS
financial analysis and planning system
FAPUS
frequency allocation panel united states
FAPUSMCEB
frequency allocation panel united states military communication electronics board
FAQ
fair average quality
FAR
failure analysis report
farad
faraday
faradic
federal acquisitions regulations
forward acquisition radar
FARADA
failure rate data
FARDRCRM
field artillery radar crewman
FARE
federal acquisition regulations
FARET
fast reactor test assembly
FARGO
fourteen on one automatic report generator operation
FARS
file analysis for random access storage
FAS
federation of american scientists
filtered air supply
flight assistance service
foreign agriculture service
free alongside
frequency assignment subcommittee
FASA
final approach spacing assignment
FASB
fetch and set bit
financial accounting standards board
FASE
fundamentally analyzable simplified english
FASS
forward acquisition sensor
FAST
facility for automatic sorting and testing
fast automatic shuttle transfer
field data applications, systems and techniques
fingerprint access and searching technique
flexible algebraic scientific translator
formal autoindexing of scientific texts
formula and statement translator
four address to soap translator
fuel assembly stability test
fully automatic sorting and testing

FASTAR
  frequency angle scanning, tracking and ranging
FASTFOR
  fast fortran
FASTI
  fast access to system test information
  fast access to systems technical information
  fast access to systems technicals information
FASTROM
  falling sphere trajectory measurement
FASWC
  fleet antisubmarine warfare command
FAT
  factory acceptance test
  fast automatic transfer
  file allocation table
  final assembly test
  formula assembler translator
  fuel and transportation
FATAL
  fadac automatic test analysis language
FATDL
  frequency and time division data link
FATE
  fusing and arming test experiments
FATP
  field assembly test point
FATR
  fixed autotransformer
FATS
  factory acceptance test specifications
  fortran automatic timing system
FAW
  forward area warning
FAWS
  flight advisory weather service
FB
  feedback
  fighter bomber
  film badge
  freight bill
  friction braked landing
  fuse block
  fuse box
FBC
  feedback control
  fiji broadcasting commission
  fluidized bed combustion
  fully buffed channel
  fully buffered channel
FBF
  feedback filter
FBFS
  fuel building filter system

FBFT
  flow bias functional test
FBI
  federal bureau of investigation
FBIS
  foreign broadcast information service
  foreign broadcast intelligence service
FBIT
  fault bit
FBL
  friction braked landing
FBM
  fleet ballistic missile
  four ball machine
  free board measure
FBMP
  fleet ballistic missile program
FBMS
  fleet ballistic missile submarine
  fleet ballistic missile system
FBMWS
  fleet ballistic missile weapon system
FBOE
  frequency band of emission
FBOPERATION
  foreground program background program operation
FBP
  final boiling point
FBR
  fast breeder reactor
  fast breeding reactor
  fast burst reactor
  feedback resistance
  fiber
FBS
  forces broadcasting service
FBT
  facility block table
  flyback transformer
FBU
  field broadcasting unit
FBW
  fly by wire
FBWS
  fly by wire system
FC
  cutoff frequency
  face centered
  fail closed
  faulted circuit
  ferrite core
  field contactor
  file code
  file conversion

finance committee
find called party
find calling party
fine control
fire control
flight control
flow chart
foot-candle
frequency changer
frequency conversion
frequency converter
front connected
fuel cell
fuel cycle
function call
function code
functional checkout
functional code
furnace cooled

FCA
farm credit administration
federal communications act
fire control area
french computing association
frequency control analysis

FCAA
frequency control and analysis

FCAI
fairchild camera and instrument

FCAS
federal civil agencies communications system

FCB
file control block
frequency control board
frequency coordinating board
frequency coordinating body

FCBA
federal communications bar association

FCC
face centered cubic
faced centered cubic crystal
facilities control console
facility communications criteria
federal communication commission
federal communications committee
federal construction council
flight communications center
flight control center
flight control computer
frequency to current converter

FCCN
federal communications commission network
forestry, conservation and communications association

FCCPO
federal contract compliance programs office

FCCTS
federal cobol compiler testing service

FCD
failure and consumption data
failure correction decoding
fine control damper
frequency compression demodulator
frequency control division

FCDF
failure and consumption data form

FCDR
failure and consumption data report
failure cause data report

FCDT
four coil differential transformer

FCE
flexible critical experiment
flight control electronics
frequency converter excitation

FCESR
frequency converter excitation, saturable reactor

FCF
frequency compression feedback
frequency compressive feedback

FCFS
first come first served

FCFT
fixed cost, fixed time

FCI
fluid control institute
flux change per inch
flux changes per inch
fuel coolant interaction
functional configuration identification

FCIC
federal crop insurance corporation

FCIG
field change identification guide

FCIN
fast carry iterative network

FCIR
failure and consumption inspector's report

FCL
feedback control loop
feeder control logic
flight control laboratory

FCLE
forecastle

FCMD
fire command

FCMM
flux change per millimetre

FCMV
  fuel consuming motor vehicle
FCN
  field change notice
FCNI
  flux controlled negative inductance
FCNP
  fire control navigation panel
FCO
  frequency control officer
  functional checkout
FCOC
  facility checkout vehicle
FCOH
  flight controllers operations handbook
FCP
  facility control program
  feed control panel
  file control processor
  file control program
  frequency control products
FCPC
  fleet computer programming center
FCPCP
  fleet computer programming center, pacific
FCPI
  flux change per inch
FCR
  facility change request
  fast ceramic reactor program
  final configuration review
  fire control radar
  forward contactor
  fuse current rating
FCRLSYS
  flight control ready light system
FCS
  facility communication system
  failure consumption sheet
  feedback control system
  fire control system
  flight control system
  frame check sequence
  functional checkout set
FCSC
  foreign claims settlement commission
FCSF
  four conductor, combination, special purpose, flexible
FCST
  federal council for science and technology forcecast
FCT
  field controlled thyristor

  filament center tap
  frequency clock trigger
FCTT
  fuel cladding transient tester
FCU
  file control unit
  flight control unit
FCW
  fast cyclotron wave
  format control word
FCWG
  frequency control working group
  frequency coordination working group
FD
  field
  field decelerating
  field decelerating contactor or relay
  field definition
  field time waveform distortion
  file description
  flange local distance
  floating divide
  floating dollar
  floppy disk
  flow diagram
  flux delta
  focal distance
  focal length to diameter ratio
  focal to diameter ratio
  frame difference
  framed
  frequency distribution
  frequency diversity
  frequency divider
  frequency division
  frequency doubler
  full duplex
FD
  frequency doubler
FDA
  final design approval
  food and drug administration
FDAA
  federal disaster assistance administration
FDACIES
  fort dodge area chapter of iowa engineering society
FDAI
  flight director attitude indicator
FDAS
  frequency distribution analysis sheet
FDAU
  flight data acquisition unit
FDB
  field dynamic braking

FDBK
  feedback
FDC
  floppy disk controller
  fluid digital computer
  form definition component
  functional design criteria
FDCA
  federal defense communications authority
FDCMPT
  fire direction computer
FDCONVERSION
  frequency analog to digital conversion
FDD
  frequency difference detector
FDDI
  fiber distributed data interface
FDDL
  frequency division data link
FDDLPO
  frequency division data link print out
FDE
  field decelerator
  flight data entry
FDEBUG
  fortran symbolic debugging package
FDF
  flight data file
FDG
  fractional doppler gate
FDH
  floating divide or halt
FDI
  field discharge
  flight direction indicator
  flight director indicator
  furnish, deliver and install
FDIC
  federal deposit insurance corporation
FDISTRIBUTION
  fisher's distribution
FDL
  fast development logistic
  ferrit diode limiter
FDLDG
  forced landing
FDM
  frequency diversity multiplex
  frequency division modulation
  frequency division multiplex
  frequency division multiplexer
  frequency division multiplexing
FDMA
  frequency division multiple access
  frequency division multiplex access

FDNET
  fighter direction net
FDNR
  frequency dependent negative resistance
FDOS
  floppy disk operating system
FDP
  factory data processing
  flight data processing
  florida power corporation
  form description program
  forward direction post
  future data processor
FDPM
  frequency domain prony method
FDPSK
  frequency differential phase shift keying
FDR
  fast dump restore
  feeder
  field definition record
  flight data recorder
  frequency diversity radar
  frequency domain reflectometer
  frequency domain reflectometry
  functional design requirements
FDRI
  flight director rate indicator
FDS
  fermi dirac sommerfeld
  fighter data storage
  financial data sciences, inc
  fluid distribution system
  fortran deductive system
  frame difference signal
FDT
  field definition tables
  full duplex teletype
FDTK
  floating drift tube klystron
FDU
  frequencer divider unit
  frequency determining unit
  frequency divider unit
FDX
  full duplex
  full duplex transmission
FE
  ferroelectric
  field engineer
  field engineering
  field engineers
  finite element
  format effector

FEA
   failure effect analysis
   federal energy administration
FEAACSREG
   far east airways and air communications service region
FEAF
   far east air force
FEALD
   field engineering automated logic diagram
FEAP
   fortran executive assembly program
FEAR
   forward firing aerial rocket
FEAT
   frequency of every allowable term
FEB
   function[al] electronic block
FEBA
   forward edge of battle area
FEBC
   far east broadcasting company
FEBS
   functional electronic blocks
FEC
   federal election commission
   federal electronic company
   federal electronic corporation
   forward error control
   forward error correcting
   forward error correction
   frederick electronics corporation
   front end control program
FECB
   file extended control block
FECES
   forward error control electronics system
FED
   field effect diode
   field engineering department
   format element descriptor
FEDAL
   failed element detection and location instruments which detect fuel element failure in power reactors
FEDP
   federal executive development program
FEDPOWCOMM
   federal power commission
FEDSTD
   federal standard
FEE
   failure effects evaluation

FEEDBAC
   foreign exchange eurodollars and branch accounting
FEEL
   ferroelectric electroluminescent
FEFET
   ferroelectric dielectric field effect transistor
FEFO
   first ended, first out
FEFP
   fuel element failure propagation
FEFPL
   fuel element failure propagation loop
FEI
   financial executives institute
   flight error instrumentation
FEIA
   flight engineers international association
FEL
   free electron laser
   frequency engineering laboratories
FEM
   field effect modified
   field effect modified transistor
   field emission microscope
FEMF
   floating electronic maintenance facility
   foreign electromotive force
FEMPX
   front end multiplexer
FEMTRANSISTOR
   field effect modified transistor
FEN
   far east network
   frequency emphasizing network
FEND
   force end
FENG
   flight engineer
FEO
   facility emergency organization
   field engineering order
FEOV
   force of volume
   forced end of volume
FEP
   financial evaluation program
   fluorinated ethylene propylene
   fluorinated ethylene propylene resin
   front end processing
   front end processor
   fuse enclosure package
FERC
   federal energy regulatory commission

FERCON
   ferrule contact
FERD
   fuel element rupture detection
FERROD
   ferrite rod antenna
FES
   far end suppressor
   final environmental statement
   fixed echo suppressor
   florida engineering society
   fundamental electrical standards
FESA
   federation of engineering and scientific associations
FESE
   field enhanced secondary emission
FESR
   finite energy sum rule
FET
   far east time
   field effect transistor
FETE
   federal telecommunication
FETF
   flight engine test facility
FETH
   field effect thyristor
FETS
   field effect transistors
FETT
   field effect tetrode transistor
FETYPELIQUIDCRYSTALDISPLAY
   field effect type liquid crystal display
FEVAC
   ferroelectric variable capacitor
FF
   fast forward
   feedforward
   femtofarad
   first fit algorithm
   fixed focus
   flexi filamented
   flip flop
   flux flow
   form factor
   form feed
   free float
   full field
   full field contactor or relay
   fully fashioned
FFA
   free fatty acid
FFAG
   fixed field alternating gradient
FFAR
   folding fin air rocket
   folding fin aircraft rocket
FFB
   french forces broadcasting
FFC
   fault and facilities control
   feedforward control
   flat field conjugate
   flip flop complementary
FFD
   field forcing, decreasing
FFEC
   field free emission urrent
FFF
   feed forward filter
   flicker fusion frequency
FFI
   field forcing, increasing
   fuel flow indicator
   full field investigation
FFL
   field failure
   first financial language
   fixed and flashing
   front focal length
FFLT
   familiarization flight
FFM
   fuel failure mockup facility
FFNM
   flip flop national module
FFP
   field forcing, protective
   fixed frequency pulsed
FFR
   field forcing, reversing
   folded flow reactor
FFRR
   full frequency range record
   full frequency range recording
   full frequency range records system
FFS
   formated file system
FFSA
   field functional system assembly and checkout
   field functional systems assembly
FFSAC
   field functional systems assembly and checkout
FFSP
   fossil fired steam plant

FFT
fast fourier transform
fast fourier transformation
FFTF
fast flux test facility
FFTFPO
fast flux test facility project office
FFTP
fast fourier transform processor
FFV
field failure voltage
FG
field gain
filament ground
final grid
fission gas
forward gate
function generator
FGAA
federal government accountants association
FGD
fine grain data
FGDAC
function generating digital to analog converter
FGM
fission gas monitor
FGP
foreground program
FGPFL
fixed and group flashing
FGR
floating gate reset
FGSA
fellow of the geological society of america
FHA
federal highway administration
federal housing administration
FHARM
fuel handling and radioactive maintenance
FHB
flat head brass
fuel handling building
FHC
federal housing commissioner
FHD
ferrohydrodynamic
first harmonic distortion
fixed head disk
fixed head disks
FHEO
fair housing and equal opportunity
FHFA
four conductor, heat and flame resistant, armored
FHFTA
four conductor, heat and flame resistant, thin walled, armored
FHG
flat head galvanized
FHLBB
federal home loan bank board
FHMA
frequency hopping multiple access
FHOF
four conductor, heat, oil, and flame resistant
FHP
fractional horsepower
friction horsepower
FHS
flat head steel
format handling system
forward head shield
forward heat shield
FHSR
final hazard summary report
FHT
fast hadamard transform
fully heat treated
FHTE
flight hardware test equipment
FHWA
federal highway administration
FI
fail in place
fan in
farnell instruments
field intensity
fixed interval
flow indicator
free in
friden, incorporated
FIA
factory insurance association
federal insurance administration
flight information area
FIACC
five international associations coordinating committee
FIAD
flame ionization analyzer and detector
FIARE
flight investigation of apollo reentry environment
FIAT
floating interpretative automatic translator
FIB
fortran information bulletin
free into barge
free into bunkers

FIC
  federal information center
  film integrated circuit
  flight information center
  frequency interference control
FICA
  federal insurance contribution act
FICB
  file identification control block
FICC
  frequency interference control center
FICM
  fluidic industrial control module
FICO
  flight information and control operations
FICS
  forecasting and inventory control system
FID
  flame ionization detector
  international federation for documentation
FIDAC
  film input to digital automatic computer
FIDO
  fall out intensity detector oscillator
  functions input, diagnostic output
FIDT
  forced incident destiny testing
FIEI
  farm and industrial equipment institute
FIELDISTOR
  field effect transistor
FIFF
  first in first fit
FIFO
  first in first out
  floating input, floating output
FIG
  figure
  floated integrating gyro
  floated integration gyro
FIGS
  figure shift upper case
  figures
  figures shift
FIIG
  federal item identification guide
FIL
  filament
  filter
  franklin institute laboratories
FILHB
  fillister head brass
FILHS
  fillister head steel

FILO
  first in last out
FILS
  flarescan instrument landing system
FILSUP
  filament supply
FILT
  filter
FIM
  failure indicating module
  field inspection manual
  field intensity meter
  field ion microscope
  flight information manual
FIMATE
  factory installed maintenance automatic test
  factory installed maintenance automatic test equipment
FINAC
  fast interline non activate automatic control
  fast interline non active automatic control
FINAL
  financial analysis language
FINCOM
  finance committee
FIND
  file interrogation of nineteen hundred data
  flight information display electronics
FINQ
  final queue
FINST
  final instruction station
FIOP
  fortran input output package
FIP
  field inspection procedure
  fluorescent indicator panel
FIPS
  federal information processing standard
  flight inspection positioning system
FIR
  far infrared
  finite duration impulse response
  finite impulse response
  flight information region
  food irradiation reactor
  fuel indicator reading
FIRC
  forest industries radio communications
FIRE
  flight investigation reentry environment
FIRETRAC
  firing error trajectory recorder and computer
FIRFILTER
  finite impulse response filter

FIRMS
   forecasting information retrieval of management system
FIRQ
   fast interrupt request
FIRR
   failure and incidents report review
FIRST
   fabrication and inflatable reentry structures for test
   financial information reporting system
FIRTI
   far infrared target indicator
FIS
   field information system
   flight information service
   floating point instruction set
FIST
   fault isolation by semiautomatic techniques
FIT
   failure unit
   fault isolation test
   field installation time
FITAL
   financial terminal application language
FITGO
   floating input to ground output
FITS
   fourteen on one input output tape system
FIU
   forward interpretation unit
   frequency identification unit
FIX
   fixture
FJ
   fixed jack
   fused junction
FJCA
   federal japan communication association
FJCC
   fall joint computer conference
FKB
   function key button
FL
   fetch load
   field loss
   field loss contactor or relay
   filter
   flash
   flight level
   float
   florida
   fluid
   flush
   focal length
   footlambert
FLA
   fluorescent lighting association
   four conductor, lighting, armored
   full load ampere
FLAC
   florida automatic computer
FLAD
   fluorescence activated display
FLAG
   fixed link aerospace to ground
FLAM
   forward launched aerodynamic missile
FLAMR
   forward looking advanced multilobe radar
FLAT
   flight plan aided tracking
FLB
   federal land bank
FLBE
   filter band elimination
   filter band eliminator
   filter for band elimination
FLBH
   filter, band high
FLBIN
   floating point binary
FLBP
   filter band pass
FLCDG
   flow control data generator
FLCR
   fixed length cavity resonance
FLCS
   flight control system
FLD
   field
   fluid
   forming limit diagram
FLDEC
   floating point decimal
FLDL
   field length
FLDO
   final limit, down
FLDS
   fixed length distinguishing sequence
FLEA
   flux logic element array
FLECHT
   full length emergency cooling heat transfer
FLESONARSCOL
   fleet sonar school

FLEX
  flexible
FLF
  final limit, forward
  fixed length field
  flip flop
  follow the leader feedback
FLFNETWORK
  follow the leader feedback network
FLFT
  full load frame time
FLG
  flag
  flange
  flying
  focal length
  following
FLH
  final limit, hoist
FLHP
  filter, highpass
FLI
  flight leader identy
FLICON
  flight control
FLIDAP
  flight data position
FLIDEN
  flight data entry
FLINT
  floating interpretative
  floating interpretive language
FLIP
  film library instantaneous presentation
  flight launch infrared probe
  floating index point
  floating indexed point arithmetic
  floating instrument platform
  floating point interpretative program
  format directed list processor
FLIR
  forward look infrared
  forward looking infrared
  forward looking infrared radar
  forward looking infrared system
FLIRT
  fortran logical information retrieval technique
FLL
  field length for large core memory
  final limit, lower
  frequency locked loop
FLLP
  filter, low pass
FLLS
  fuel low level sensor

FLMTR
  flow meter
FLN
  fluorescence line narrowing
FLOC
  fault locator
FLOCOM
  floating commutator
FLODAC
  fluid operated digital automatic computer
FLOLS
  fresnel lens optical landing system
FLOOD
  fleet observation of oceanographic data
FLOP
  floating octal point
  floating point operation per second
FLORL
  fluorescent runway lighting
FLOSOST
  fluorine one stage orbital space truck
FLOTRAN
  flowcharting fortran
FLOX
  fluorine plus liquid oxygen
FLOZ
  fluid ounce
FLP
  faulty location panel
  floating point
FLPL
  fortran list processing language
FLPRF
  flamenproof
FLPT
  flash point
FLR
  final limit, reverse
  flight line recorder
  flow rate
  forward looking radar
FLRNG
  flash ranging
FLS
  field length for small core memory
  flashing light system
  flow switch
FLSC
  flexible linear shaped charge
FLT
  fault locating test
  fault location technology
  filter
  flight

float
floating
FLTCONT
flight control
FLTG
floating
FLTPG
flight programmer
FLTR
filter
FLU
final limit, up
FLUIDICS
fluid logics
FLUOR
fluorescence
fluorescent
FLVFD
front luminous vacuum fluorescence display
FM
facilities management
fan marker
fast memory
fathom
feedback mechanism
fermium
ferrite metal
field manual
filament mid tap
file maintenance
film microelectronics
finder matrix
fine measurement
floating multiply
frame
frame per minute
frequency modulated
frequency modulation
frequency multiplex
FMA
failure mode analysis
frequency modulation association
frequency modulator altimeter
fundamental mode asynchronous
FMAIN
file maintenance
FMAM
frequency modulation amplitude modulation
FMAP
fan marker approach
FMB
fm broadcasters
FMC
federal maritime commission
final moisture content
FMCS
flight management computer system
FMCW
frequency modulated continuous wave
FMD
fm demodulator
frequency of minimum delay
FMDA
fm development association
FMDR
final missile deviation report
FME
frequency measuring equipment
FMEA
failure mode and effect analysis
FMECANALYSIS
failure mode, effects and critical analysis
FMES
full mission engineering simulator
FMEVA
floating point mean and variance
FMFB
frequency modulated feedback
frequency modulation feedback
frequency modulation with feedback
FMFM
frequency modulation frequency modulation
FMHA
farmers home administration
FMI
fm intercity relay broadcasting
frequency modulation intercity relay broadcasting
FMIC
frequency monitoring and interference control
frequency monitoring interference control
FMKR
fan marker
FML
feedback, multiple loop
front mounting light
FMLS
full matrix least squares
FMN
formation
FMO
frequency multiplier oscillator
FMPE
fast memory parity error
FMPM
frequency modulation phase modulation
FMPROT
fine mesh cover protected

FMPS
  fortran mathematical programming system
  functional mathematical programming system
FMQ
  frequency modulated quartz
  frequency modulated quartz circuit
  frequency modulated quartz oscillator
FMR
  frequency modulated radar
  frequency modulated receiver
FMS
  fathoms
  federation of materials societies
  file management supervisor
  file management system
  flexible manufacturing system
  flux monitoring system
  foreign military sales
FMT
  format
  frequency modulated transmitter
FMTM
  frequency modulation team
FMTR
  florida missile test range
FMTS
  field maintenance test station
FNA
  final approach
FNC
  finnish national committee of the iec
FNH
  flashless nonhygroscopic
FNP
  floating nuclear plant
  front end network processor
  fusion point
FNR
  ford nuclear reactor
FNS
  food and nutrition service
  forever non static
  functional signal
FO
  fail open
  fan out
  fast operate
  fast operating
  fiber optics
  filter output
FOBS
  fractional orbit bombardment system
  fractional orbital bombardment system

FOC
  first operators club
  flight operations center
FOCAL
  formula calculator
  formulating on line calculations in algebraic language
FOCI
  first operational computer installation
FOCOHANA
  fourier coefficient harmonic analyzer
FOCOS
  fordac conversational system
FOCSLE
  forecastle
FOCUS
  financially oriented computer updating service
  forum of contro data users
FOD
  free of damage
FOE
  friend of the earth
FOGWT
  fog wireless telegraph
FOH
  forced outage hours
FOI
  first order interpolator
FOIA
  freedom of information act
FOIL
  file oriented interpretive language
FOIR
  field of interest register
FOL
  facility operating license
  follow
  following
FOLNOAVAL
  following items not available
FOM
  figure of merit
FOMOTCODE
  four mode ternary code
FOODAP
  food distributors application
FOOSP
  fourteen on one statistical program
FOOTL
  foot lambert
FOP
  first order predictor
FOPL
  first order predicate logic

FOPT
  fiber optic photo transfer
FOR
  failure outage rate
  forced outage rate
FORAST
  formular assembler translator
FORATOM
  forum atomique, european association of euratom nuclear industries
FORBLOC
  fortran compiled block oriented simulation language
FORC
  formula coder
FORCAST
  formula assembler translator
FORDAC
  fortran data acquisition and control
FORDIO
  forecast of radio propagation conditions
FORDS
  floating ocean research and development station
FORESDAT
  formerly restricted data
FORGO
  fortran load and go
FORM
  format
  formation
  former
  formula
FORMAC
  formula manipulation compiler
FORMAT
  fortran matrix abstraction techniques
FORMEX
  formal executor
FORTE
  file organization technique
FORTRAN
  formula translation
  formula translator
FORTRUNCIBLE
  fortran style runcible
FOS
  factor of safety
  fiber optical sensor
  finish one side
  floppy operating system
FOSDIC
  film optical scanning device for input to computer
  film optical sensing device for input to computer
FOT
  frequency for optimum traffic
  frequency of optimum operation
  frequency of optimum traffic
FOTA
  fuels open test assembly
FOTS
  fiber optics transmission system
FOV
  field of view
FOW
  oil immersed forced oil cooled with forced water cooler
FOX
  foxboro
FP
  faceplate
  feedback positive
  feedback potentiometer
  file protect
  fireproof
  fixed point
  flagpole
  flame protected
  flameproof
  flash point
  flat pack
  flight position
  flight programmer
  floating point
  foot pound
  forward perpendicular
  frame period
  freezing point
  full period
  fusible plug
FPA
  focal plane array
FPAC
  flight path analysis and command
FPAL
  florida power and light
FPC
  facility power control
  federal pacific electric company
  federal power commission
  fire pump control
  fixed price call
  florida power corporation
  frequency plane correlator
FPCH
  foreign policy clearinghouse

FPCR
    fluid poisson control reactor
FPCS
    fuel pool cooling system
FPCSTL
    fission product control screening test loop
FPD
    flame photometric detector
    full power days
FPDD
    final project design description
FPDI
    food processing development irradiator
FPE
    federal pacific electric
FPF
    feed per foot
FPG
    firing pulse generator
FPGA
    freely programmable gate array
FPH
    floating point hardware
FPI
    faded prior to interception
FPIS
    forward propagation by ionospheric scatter
FPL
    face plate
    field processing language
    file parameter list
    final protective line
    fire plug
    florida power and light
    foxboard programming language
    foxboro programming language
    freelance programmers ltd
    frequency phase lock
FPLA
    field programmable logic array
    field programmable logic arrays
FPLS
    field programmable logic sequencer
FPM
    facility power monitor
    federal personnel manual
    feet per minute
    financial planning model
    frame per minute
    frequency position modulation
    frequency position modulation with phase increments
    functional planning matrices
FPN
    fixed pattern noise

FPO
    fixed path of operation
FPP
    facility power panel
    floating point processor
FPPC
    flight plan processing center
FPPS
    flight plan processing system
FPR
    federal procurement regulations
    fixed program receive
FPRF
    fireproof
FPROM
    field programmable read only memory
FPRS
    forest products radio service
    forest products research society
FPS
    feet per second
    field power supply
    financial planning simulator
    financial planning system
    fixed point system
    flash per second
    floating point system
    flops per second
    fluid power society
    focus projection and scanning
    foot pound second
    frame per second
FPSL
    fission product screening loop
FPSM
    foot pound second magnetic
FPT
    female pipe thread
    file parameter table
    full power trial
FPTF
    fuels performance test facility
FPTS
    forward propagation by tropospheric scatter
    forward propagation tropospheric scatter
FPU
    floating point unit
FPWT
    fire protection water tank
FQA
    field quality audit
FQCY
    frequency
FQE
    free queue element

FQPR
   frequency programmer
FR
   failure rate
   failure report
   fast release
   fast release relay
   federal register
   federal reserve
   field relay
   field reporting
   field resistance
   field reversing
   flash ranging
   flow recorder
   frame
   frequency meter
   frequency response
   front
   full rate
FRA
   federal railroad administration
FRAC
   fractionator reflux analog computer
FRAG
   fragile
   fragmentary
FRAN
   frame structure analysis
FRAP
   fuel rod analysis program
FRAPS
   fuel rod analysis program steady state
FRAPT
   fuel rod analysis program transient
FRAT
   first recorded appearance time
FRB
   federal reserve bank
FRC
   failure recurrence control
   federal radiation council
   federal radio commission
   file research council
   flight research center
   flight rules computer
   flow recorder controller
   functional residue capacity
FRCP
   facility remote control panel
FRCTF
   fast reactor core test facility
FRD
   functional reference device
   functional referenced device

FRED
   fast reference for engineering drawings
   fiendishly rapid electronic device
   figure reader electronic device
   figure reading electronic device
   figute reading device
FREFAL
   floating point regula falsi
FRENA
   frequency and amplitude
FRENAC
   frequency and amplitude coded
FREO
   figure reading electronic device
FREQ
   frequency
FREQCH
   frequency changer
FREQCONV
   frequency converter
FREQIND
   frequency indicator
FREQM
   frequency meter
FREQMULT
   frequency multiplier
FREQN
   frequency
FRES
   fire resistant
FRESCAN
   frequency scanning
FRESCANAR
   frequency scan radar
FRESCANNAR
   frequency scanning radar
FRESCANNER
   frequency scanner
FREWCAP
   flexible reworkable chip attachment process
FRFA
   federal regulatory flexibility act
FRH
   fire resistant hydraulics program
FRIA
   finnish radio industry association
FRINGE
   file and report information processing generator
FRISCO
   fast reaction integrated submarine control
FRL
   fisher radio laboratory
   fractional

FRM
  frame
  frequency meter
FRMC
  frame counter
FRMR
  frame reject
FRNK
  frequency regulation and network keying
FRO
  feed rate override
  flight radio officer
FROB
  flash radar order of battle
FROM
  factory programmable read only memory
  fusible read only memory
FROST
  food reserves on space trip
FRP
  fiber glass reinforced plastic
  fuel reprocessing plant
FRR
  fast recovery rectifier
FRRC
  flow recording ratio controller
FRS
  failure reporting system
  federal reserve system
  fellow of the royal society
  fire research station
  fragility response spectrum
  fragility response system
  frequency response survey
FRSEC
  frame per second
FRT
  federal telephone and radio company
  flow recording transmitter
FRTEF
  fast reactor thermal engineering facility
FRTISO
  floating point root isolation
FRTP
  fraction of rated power
FRTS
  forward propagation by tropospheric scatter
FRTSAP
  federation of radio and television service associations of pennsylvania
FRU
  fleet radio unit
FRUGAL
  fortran rules used as a general applications language

FRUPAC
  fleet radio unit, pacific
FS
  facsimile
  factor of safety
  factor storage
  faraday society
  federal specificatiaon
  federal standard
  feedback
  feedback, stabilized
  female soldered
  field switch
  field separator
  field sequential
  file separator
  final selector
  fire switch
  fiscal service
  flash per second
  float switch
  floating sign
  floating subtract
  flow switch
  foot second
  forest service
  format statement
  forward scatter
  frequency shift
  frequency shifting
  full scale
  functional schematic
  functional selector
  functional symbol
  future computer system
  future series
  future system
FSA
  fine structure analysis
  formatter sense amplifier
  frequency selective amplifier
  full scale accuracy
FSAF
  frequency shift audio frequency
FSAR
  filling storage and remelt system
  final safety analysis report
FSB
  federal specifications board
  floating subtract
  free storage block
FSBLCUT
  fusible cut out
FSBSEM
  free storage block semaphore

FSBW
frame space bandwidth product
FSC
federal supply classification
field support center
final subcircuit
finite state channel
FSCAN
f scanner
FSCEN
flight service center
FSCI
frequency space characteristic impedance
FSCS
flight service communications system
FSCT
floyd satellite communications terminal
FSCW
fast space charge wave
FSD
field support diagram
flying spot device
flying spot digitizer
full scale deflection
FSDC
federal statistical data center
FSE
full screen editor
FSF
flight safety foundation
FSG
flight strip generator
frequency of signal generator
FSGA
four wire shipboard, general use, armored
FSGB
foreign service grievance board
FSI
frame synchronization indication
FSIGBN
field signal battalion
FSIT
flat screen image tube
FSK
frequency shift keying
FSL
formal semantic language
frequency selective limiter
FSLIC
federal saving and loan insurance corporation
FSM
field strength meter
final stage marker
finite state machine
floating subtract magnitude

folded sideband modulation
frequency shift modulation
FSMWI
free space microwave interferometer
FSN
federal stock number
file sequence number
FSO
full scale output
FSP
fleet scheduling program
frequency standard, primary
full screen product
FSPE
the federation of societies of professional engineers
FSPPR
fast supercritical pressure power reactor
FSPT
federation of societies for paint technology
FSQS
food safety and quality service
FSR
feedback shift register
field strength ratio
first soviet reactor
flight safety research
frequency scan radar
frequency shift reflector
full scale range
FSRB
flight safety review board
FSS
federal supply schedule
federal supply services
field sequential system
flight service station
floor service stations
flying spot scanner
FSSRS
fixed step size random search
FSSS
flying spot scanner system
FST
frequency shift transmission
FSTC
foreign science and technology center
FSTV
fast scan television
FSU
field storage unit
full scale unit
FSUC
federal statistics users conference

FSV
   final stage vehicle
   fire service valve
FSVM
   frequency selective voltmeter
FSW
   final status word
   forward swept wing
FSWMA
   fine and speciality wire manufacturers'
   association
FT
   feet
   filing time
   firing tables
   firing temperature
   flame thrower
   flight test
   fluorescent target
   flush threshold
   foot
   frequency and time
   frequency tracker
   full time
   functional test
FTAC
   functional test and calibration
FTB
   for the birds
   frequency time base
   functional test bulletin
FTC
   facility terminal cabinet
   fast time constant
   fast time control
   federal trade commission
   flight test center
   fluid bed thermal cracking
   foot-candle
   frequency time control
FTCD
   foot-candle
FTCELEMENT
   fast time constant element
FTD
   field terminated diode
   florists' telegraph delivery
FTDA
   florists' telegraph delivery
   association
FTE
   factory test equipment
   fftf test engineering
   frame table entry
   functional test equipment

FTF
   flare tube fitting
   flared tube fitting
   forward transfer function
FTFET
   four terminal fet
   four terminal field effect transistor
FTG
   function timing generator
FTI
   frequency time indicator
   frequency time intensity
FTL
   facility tape loading
   fast transient loader
   faster than light
   federal telecommunications laboratories
   foot lambert
   full term license
FTLA
   foot lambert
FTLB
   final turn lead pursuit
FTLP
   fixed term lease plan
FTM
   failed to make
   flight test missile
   frequency time modulation
FTMC
   frequency and time measurement counter
FTML
   folded tape meander line
FTO
   failed to open
   flight test operations
FTP
   fftf test procedure
   field test program
   file transfer packet
   file transfer protocol
   fixed term lease plan
   fixed term plan
   flash temperature parameter
   fuel transfer port
   fuel transfer pump
   functional test procedure
FTPI
   flux transition per inch
FTPMM
   flux transition per millimeter
FTR
   fast test reactor
   federal telephone and radio company
   federal travel regulations

film tracing reproduction
fixed target rejection
fixed target rejection filter
functional test report
functional test requirement
FTRC
federal telecommunications records center
FTRIA
flow and temperature removable instrument assembly
FTS
federal telecommunication system
finish two sides
float switch
foot switch
fourier transform spectroscopy
free time system
frequency and timing subsystem
FTTS
flow through tube sampler
FTU
flight test unit
functional test unit
FTV
flight test vehicle
FTW
federation of telephone workers
FTWG
flight test working group
FTZB
foreign trade zones board
FU
firing unit
functional unit
fuse
FUA
fire unit analyzer
FUD
fire up decoder
FUDR
failure and usage data report
FUFO
fuel fusing option
FUIF
fire unit integration facility
FUN
function
FUNC
function
functional
FUNCTLINE
functional line diagram
FUNDFREQ
fundamental frequency

FUP
follow up
FUR
failure, unsatisfactory or removal
furlong
FUS
fortran utility system
fuselage
fusible
FUSIONTECH
fusion technology
FV
femtovolt
floor valve
frequency to voltage
front view
full voltage
FVA
floor valve adapter
FVC
frequency to voltage converter
FVCONVERTER
frequency to voltage converter
FVD
front vortex back focal distance
FVPB
flight vehicle power branch
FVR
fuse voltage rating
FVS
flight vehicles system
FVSC
flight vehicle system committee
FW
face width
field weakening
field winding
filament wound
firmware
fiscal week
fixed length word
forward wave
foster wheeler corporation
frame synchronization word
full wave
full weight
FWA
first word address
fixed word address
fluorescent whitening agent
forward wave amplifier
FWAC
full wave alternating current
FWB
four wheel brake

FWD
　forward
　free wheeling diode
　front wheel drive
FWDC
　full wave direct current
FWEC
　fort waynie engineers club
　fort worth electronics club
FWG
　french wire gage
FWH
　flexible working hours
FWHM
　full width at half maximum
　full width half maximum
FWL
　fixed word length
FWO
　fleet wireless officer
FWPCA
　federal water pollution administration
　federal water pollution control act
FWQA
　federal water quality administration
FWR
　full wave rectification
　full wave rectifier
FWRU
　full wave rectified unfiltered
FWS
　filter wedge spectrometer
　fish and wildlife service
　fixed wireless station
FWT
　fast walsh transform
FWTM
　full width at tenth maximum
FX
　fixed station
　foreign exchange
FXD
　fixed
FXSTA
　fixed station
FY
　fiscal year
FYDP
　five year defense program
FYI
　for your information
FYIG
　for your information and guidance
FYP
　five year plan

FZ
　float zone
　fuze

G

G
　gage
　gain
　gain of antenna
　gap
　gate
　gauge
　gauss
　general electric
　generator
　giga
　gilbert
　glass
　globular
　government broadcasting
　gram[me]
　gravitational acceleration unit
　gravity
　great
　grid
　gross
　ground
　ground state
　grounded
G TOAGM
　ground to air guided missile
GA
　gain of antenna
　gas amplification
　gauge
　general aircraft
　general assembly
　general atomic company
　general automation
　general average
　georgia
　germanium alloy
　glide angle
　gliding angle
　go ahead
　graphic ammeter
　grapple adapter
　ground to air
　guidance amplifier
GAA
　gyroy and accelerometer panel

GAAES
georgia architectural and engineering society
GAAG
gross actual generation
GAAL
general aircraft and leasing
GAAM
guided anti aircraft missile
GAARD
general automation automatic recovery device
GAAS
gallium arsenide
GAASP
gallium arsenide phosphide
GAC
goodyear aerospace corporation
gross available capacity
guidance and control
GACC
guidance alignment and checkout console
guidance and control coupler
GACEP
guidance and control equipment performance
GACL
guernsey aero club, limited
GACT
greenwich apparent civil time
GAD
galvanized and dipped
germanium alloy diffused
GADL
ground to air data link
GADR
guided air defense rocket
GAEC
greek atomic energy commission
GAFB
griffis air force base
GAG
gross available generation
ground to air to ground
GAHF
grapple adapter handling fixture
GAI
gate alarm indicator
GAINIP
graphic approach to numerical information processing
GAINS
gimballess analytic inertial navigation system
global airborne integrated navigation system
GAL
gallon
general aircraft, limited
generic array logic
guggenheim aeronautical laboratory
GALCIT
guggenheim aeronautical laboratory of california institute of technology
GALLSMIN
gallons per minute
GALV
galvanic
galvanism
galvanized
galvanometer
GALVI
galvanized iron
GALVND
galvanized
GALVO
glavanometer
GAM
graphic access method
guided aircraft missile
GAMA
gas appliance manufacturers association
general aviation manufacturers association
graphics assisted management application
GAMBICA
group of association of manufacturers of british instrumentation, control and automation
GAMBIT
gate modulated bipolar transistor
GAMIC
gamma incomplet
GAMLOGS
gamma ray logs
GAMM
german association for applied mathematics and mechanics
GAMMA
generalized automatic method of matrix assembly
GAN
generalized activity network
generating and analyzing networks
global area network
guidance and navigation
GANIP
graphic approach to numerical information processing
GANS
guidance and navigation system
GAP
general accounting package
general assembly program

GAPA
　ground to air pilotless aircraft
GAPAN
　guild of air pilots and navigators
GAPE
　guyana association of professional engineers
GAPL
　ground assembly parts list
GAPT
　graphical automatically programmed tools
GAR
　growth analysis and review
　guided aircraft rocket
GARD
　gamma atomic radiation detector
　graphic analyzation of resistance defects
GARDE
　gathers, alarms, reports, displays, and evaluates
GARDTRAK
　gamma absorption and radiation detection tracking
GARF
　ground approach radio fuse
GARP
　global atmospheric research program
GAS
　graphics attachment support
GASA
　georgia aeronautics and space administration
GASH
　guanidine aluminium sulfate hexahydrate
GASP
　general activity simulation program
　general analysis of system performance
　generalized academic simulation program
　gevic arithmetic simulation program
　graphic applications subroutine package
　gratis and simulator production system
GASR
　guided air to surface rocket
GASS
　generalized assembly system
GASSAR
　general atomic standard safety analysis report
GAT
　gate associated transistor
　general aviation transponder
　generalized algebraic translator
　georgetown automatic translator
　ground to air transmitter
GATAC
　general accessment tridimensional analog computer

GATB
　general aptitude test battery
GATC
　general american transportation corporation
GATCO
　guild of air traffic control officers
GATE
　garp atlantic tropical experiment
　generalized algebraic translator extended
GATR
　ground to air transmitter receiver
　ground to air transmitting receiving
GATT
　gate assisted turn off thyristor
　ground to air transmitter terminal
GATU
　geophysical automatic tracker unit
GAYIG
　gallium substituted yttrium iron garnet
GB
　gain bandwidth
　gigabyte
　gilbert
　gold bonded
　goodby
　grid bias
　ground beacon
　grounded base
GBA
　give better address
GBH
　group busy hour
GBI
　ground back up instrument
GBO
　goods in bad order
GBP
　gain bandwidth product
GBS
　gigabit per second
　gigabyte per second
GBSAS
　ground based scanning antenna system
GBSR
　graphite moderated boiling and superheating reactor
GBT
　graded base transistor
GBW
　gain bandwidth
GBYTE
　gigabyte
GC
　gain control
　gas chromatograph

gas chromatographic
geiger counter
general cable
general cable corporation
general ceramics
general council
gigacycle
gigacycles per second
global control
great circle
ground control
grounded collector
guidance computer
GCA
  ground communication activities
  ground controlled approach
  group capacity analysis
  guidance coupler assembly
GCAA
  guidance, control and airframe
GCAL
  gram calorie
GCAP
  general circuit analysis program
  generalized circuit analysis program
GCARA
  greater cincinnati amateur radio association
GCB
  great circle bearing
GCC
  ground control center
  guidance checkout computer
GCCA
  graphic communications computer association
GCD
  gain control driver
  gate controlled diode
  great circle distance
  greatest common divisor
GCE
  ground checkout equipment
  ground communications equipment
  ground control equipment
GCF
  greatest common factor
  gross capacity factor
GCFA
  gridded crossed field amplifier
GCFBR
  gas cooled fast breeder reactor
GCFR
  gas cooled fast reactor
GCFRE
  gas cooled fast reactor experiment
GCG
  guidance control group
GCHQ
  government communications headquarters
GCI
  graphics command interpreter
  ground control intercept
  ground controlled interception
GCIIG
  glass and ceramics industry instrumentation group
GCIS
  ground control intercept squadron
  ground controlled interception squadron
GCIT
  ground control interception team
GCL
  gas cooled loop
  ground control landing
  ground controlled landing
  guidance control laboratory
GCLC
  guidance control launch console
GCLPF
  ground capacitor lpf
GCM
  greatest common measure
GCMA
  government contract management association of america
GCMS
  gas chromatography and mass spectroscopy
GCN
  gage code number
  gauge code number
  greenwich civil noon
GCO
  ground cut out
  guidance control officer
GCOE
  ground control operational equipment
GCOS
  general comprehensive operating supervisor
  general comprehensive operation supervisor
GCP
  gain control pulse
  graphic control program
  guidance checkout package
GCR
  gas cooled reactor
  general component reference
  ground control radar
  ground controlled radar
  group code recording

group coded record
group coded recording
GCS
  general computer systems and logic corp
  gate controlled switch
  general computer systems
  gigacycles per second
  guidance and control system
GCSS
  ground control space system
GCT
  general classification test
  greenwich central time
  greenwich civil time
  greenwich conservatory time
  gyro compass trial
GCTS
  ground communications tracking system
GCU
  gyroscope coupling unit
GCW
  general continuous wave
GCWM
  general conference of weights and measures
GD
  gate driver
  general dynamics
  ground
  ground detector
  ground directional
  grown diffused
GDA
  general dynamics astronautics
GDB
  global data base
GDBMS
  generalized data base management system
GDBTG
  generalized data base task group
GDC
  general design criteria
  general dynamics convair
  general dynamics corporation
  gross dependable capacity
  ground digit control
  guidance data converter
GDD
  gas discharge display
GDDM
  graphic data display manager
  graphical data display manager
GDE
  ground data equipment
  group distribution frame
GDES
  governmental department for electrical specifications
GDEU
  guidance digital evaluation unit
GDF
  group distributing frame
  group distribution frame
GDG
  generation data group
  group display generator
GDGS
  guidance digital ground station
GDIG
  gadolinium iron garnet
GDL
  gas dynamic laser
GDM
  grid dip meter
  grid dip modulator
GDMS
  generalized data management system
GDMT
  gemini detailed maneuver table
GDNCE
  guidance
GDO
  grid dip oscillator
GDP
  generalized data processor
GDPS
  generalized disk programming system
  global data processing system
GDR
  group delay response
GDS
  general declassification schedule of executive order 11652
  graphic data system
  graphical display system
GDT
  gas decay tank
  ground delay time
GE
  gas ejection
  gaussian elimination
  general electric
  geoscience electronics
  germanium
  greater than or equal to
  grounded emitter
GEA
  greek electrotechnical association
GEAG
  general electric airborne guidance

GEANPD
   general electric aircraft nuclear propulsion department
GEAP
   general electric atomic products
GEB
   general electric burroughs
GEBCO
   general bathymetric chart of the oceans
GEC
   general electric company
   general electrodynamics corporation
GECAEL
   general electric company applied electronics laboratories
GECEP
   general civil engineering package
GECJ
   general electric company's journal
GECO
   general electric company
GECOM
   general compiler
   generalized compiler
   generalized computer
GECOR
   general communication routine
GECOS
   general comprehensive operating supervisor
   general comprehensive operating system
GECS
   ground environmental control system
GEDA
   goodyear electronic differential analyzer
GEDAN
   general data analyzer
GEDL
   general electric data link
GEE
   general evaluation equipment
GEEIA
   general electronics engineering installation
   ground electronics engineering installation agency
GEEK
   geomagnetic electrokinetograph
GEEP
   general electric electronic processor
GEESE
   general electric electronic system evaluator
GEF
   ground equipment failure
GEFORTRAN
   general electric fortran

GEFS
   general electric flame site
GEGS
   general electric guidance system
GEHME
   general electric heavy military electronics
GEIMS
   general electric inventory management system
GEIS
   generic environmental impact statement
GEISHA
   geodetic inertial survey and horizontal alignment
GEK
   geoelectrokinetograph
   geomagnetic electrokinetograph
GEL
   general electric laboratory
   general electric, limited
GELME
   general electric light military electronics
GEM
   gemini
   general epitaxial monolith
   government electronics market
   ground effect machine
   guidance evaluation of missile
GEMAC
   general electric measurement and control
GEMM
   generalized electronics maintenance model
GEMS
   general education management system
   general electric manufacturing simulator
   general electrical and mechanical systems
GEMSIP
   gemini stability improvement program
GEMSVD
   general electric missile space vehicle department
GEN
   general
   general telephone and electronics corporation
   generator
GENC
   general electric nose cone
GENDA
   general data analysis and simulation
GENDARME
   generalized data reduction, manipulation, evaluation
GENESYS
   general engineering system

GENL
  general
GENR
  generation
  generator
GENTEL
  general telephone and electronics corporation
GENTEX
  general telegraph exchange
GEO
  geoscience electronics
  geostationary orbit
GEODSS
  ground based electro optical deep space surveillance system
GEON
  gyro erected optical navigation
GEOPS
  geodetic estimates from orbital perturbations of satellites
GEORGE
  general organisational environment
  general organization and environment
GEOS
  geodetic earth orbiting satellite
  geodetic satellite
  geosynchronous earth observation system
GEOSAR
  geosynchronous synthetic aperture radar
GEOSCAN
  ground based electronic omnidirectional satellite communications antenna
GEP
  goddard experiment[al] package
  ground environment program
GEPAC
  general electric process automation computer
  general electric programmable automatic comparator
  general electric programmable automatic computer
GEPEXS
  general electric parts explosion system
GEPURS
  general electric general purpose
GEQUIV
  gram equivalent
GERM
  ground effect research machine
GEROS
  general routing optimization system
GERPAT
  german patent
GERSAL
  general electric symbolic assembly language

GERSIS
  general electric range safety instrumentation system
GERT
  graphic[al] evaluation and review technique
GERTIE
  george remote terminal interrogative environment
GERTS
  general electric radio tracking system
  general electric remote terminal supervisor
  general electric remote terminal system
  general remote terminal supervisor
  general remote transmission supervisor
GERV
  general electric reentry vehicle
GES
  gamma european systems
  generalized edit system
  goliath edison screw
  ground electronics system
GESAC
  general electric self adaptive control
GESAL
  general electric symbolic assembly language
  general symbolic assembly language
GESCO
  general electric supply corporation
GESCOM
  general electric scientific color matching
GESMO
  generic environmental statement on mixed oxides
GESOC
  general electric satellite orbit control
GESPL
  generalized edit system programming language
GESSAR
  general electric standard safety analysis report
GET
  germanium transistor
  ground elapsed time
GETAC
  general electric telemetering and control
GETEL
  general electric test engineering language
GETIS
  ground environment team international staff
GETOL
  general electric training operational logic
  ground effect take off and landing

GETR
general electric test reactor
GETS
general track simulation
generalized electronic trouble shooting
GETSS
general electric time sharing system
GETWS
get word from string
GEUEP
general electric utility engineering program
GEV
giga electron volt
GEVIC
general electric variable increment computer
GF
gage factor
galois field
gap filler
gauge factor
generator field
government form
gram force
GFAE
government furnished aircraft equipment
government furnished avionics equipment
government furnsihed aerospace equipment
GFC
gas filled counter
GFCI
ground fault circuit interrupter
GFCRP
gap filler control and reporting post
GFCS
gun fire control system
GFD
gap filler data
GFE
government furnished equipment
GFEAM
government furnished equipment and material
GFI
gap filler input
ground fault interrupter
GFIP
gross fault indicator panel
GFO
gap filler output
GFPM
gated frequency position modulation
GFR
gap filler radar
gas filled rectifier
generator field regulator

GFRC
gas flow radiation counter
general file and recording control
GFRP
gap filler reporting post
glass fiber reinforced plastic
GFS
group final selector
GFT
glass fabric tape
GFV
guided flight vehicle
GFW
glass filament wound
ground fault warning
GFY
government fiscal year
GG
groove gage
ground guidance
ground to ground
GGA
gulf general atomic
GGC
ground guidance computer
GGE
ground guidance equipment
GGG
gadolinium gallium garnet
GGL
gain guided laser
GGM
ground to ground missile
GGS
ground guidance system
GGTS
gravity gradient test satellite
GHA
greenwich hour angle
GHCP
georgia hospital computer group
GHE
ground handling equipment
GHOST
global horizontal sounding technique
GHSE
ground handling and servicing equipment
GHZ
gigahertz
GI
galvanized iron
gastro intestinal
general information
general input
general instrument

general instrument corp
geodesic isotensoid
government and industrial
government initiated
grid interval
ground interception
GIA
general industry applications
GIANT
geneological information and name tabulating system
general instrument advanced nitride technology
geological information and name tabulating system
GIC
general instrument corporation
generalized immittance converter
generalized impedance converter
GICS
global instrumentation control system
ground instrumentation and communications system
GIDEP
government industry data exchange program
GIE
ground instrumentation equipment
GIFS
generalized interrelated flow simulation
guggenheim institute of flight structures
GIFT
gas insulated flow tube
general internal fortran translator
GIGI
gamma inspection of grain integrity
general imaging generator and interpreter
GIGO
garbage in garbage out
GIL
green indicating lamp
GIM
generalized information management
generalized information management system
GIMIC
guard ring isolated molitic ic
GIMRADA
geodesy intelligence and mapping research and development agency
GINO
graphical input output
GIOC
generalized input output controller
GIP
general internal process

GIPS
ground information processing system
GIPSE
gravity independent photosynthetic gas exchanger
GIPSY
generalized information processing system
GIRL
graph information retrieval language
GIRLS
general information retrieval and listing system
generalized information retrieval and listing system
global interrogation, recording, and location system
GIS
general information system
generalized information system
geoscience information society
ground instrumentation system
GISP
general information system for planning
GISVS
generalized information system virtual storage
GIT
graph isomorphism tester
GIU
general intelligence unit
geoballistic input unit
GIUK
greenland iceland united kingdom
GIXU
grain inspection x-ray unit
GJ
grown junction
GJE
gauss jordan elimination
GJP
graphic job processor
GL
gallon
gate leads
glass
glass type loctal tube
glazed
grid leak
ground level
gun laying radar
GLASS
germanium lithium argon scanning system
GLB
glass block

GLC
 gas liquid chromatography
 global loran navigation chart
 ground level concentration
GLCM
 ground launched cruise missle
GLDS
 gemini launch data system
GLE
 gemini laser experiment
GLEAM
 graphic layout and engineering aid method
GLEEP
 graphic low energy experimental pile
 graphite low energy experimental pile
GLEX
 general ledger and expense system
GLF
 general telephone company of florida
GLINT
 global intelligence
GLIPAR
 guide line identification program for anti missile research
GLLD
 ground laser locator designator
GLOAD
 general loader
GLOBECOM
 global communications
GLOCOM
 global communications system
GLOMEX
 global meteorological experiment
GLOP
 gevic logic operation program
 guidance and launch operation
GLOPAC
 gyroscopic low power attitude control
 gyroscopic lower power attitude control
GLOPC
 gyroscopic lower power controller
GLOSS
 global ocean surveillance system
GLOTRAC
 global tracking network
GLOTRACK
 global tracking
GLRC
 geneva lake radio club
GLS
 general ledger system
 general lighting service

GLT
 greetings letter telegram
 guide light
GLV
 gemini launch vehicle
GM
 gaseous mixture
 gated memory
 geiger mueller
 geiger mueller counter
 geiger muller
 general motors
 geometric mean
 glass metall
 good morning
 greenwich meridian
 grid modulation
 group per message
 groupmark
 guided missile
 gun metal
GMAB
 guided missile assembly building
GMAP
 general macro assembler processor
GMAT
 greenwich mean astronomical time
GMB
 guided missile brigade
GMC
 gross maximum capacity
 guided missile committee
 guided missile control
GMCF
 guided missile control facility
GMCM
 guided missile countermeasure
GMCO
 guided missile control officer
GMCP
 guided missile control party
GMD
 geometric mean distance
 guided missile division
GMDEP
 guided missile data exchange program
GME
 general micro electronics
 gimbal module electronics
 group modulation equipment
GMEVALU
 guided missile evaluation unit
GMFCS
 guided missile fire control system

GMG
  gross maximum generation
GMIORG
  guided missile interdepartmental operational requirements group
GMIS
  government management information system
GMLS
  guided missile launching system
GMM
  galvanomagnetic method
GMOCU
  guided missile operations control unit
GMOL
  gram molecule
GMOO
  guided missile operations officer
GMR
  ground mapping radar
GMRD
  guided missile research division
GMS
  gemini mission simulator
  guidance monitor set
  guided missile school
GMSFC
  george marshall space flight center
GMSQUD
  guided missile squadron
GMSRON
  guided missile service squadron
GMSTS
  guided missile service test station
GMSU
  guided missile service unit
GMT
  greenwich mean time
  greenwich meridian time
GMTESTU
  guided missile test unit
GMTU
  guided missile test unit
  guided missile training unit
GMU
  guided missile unit
GMV
  gram molecular volume
  guaranteed minimum value
GMW
  gram molecular weight
GN
  gaussian noise
  generator
  good night
  grid neutralization
  guidance and navigation
GNAC
  guidance, navigation and control
GNAL
  georgia nuclear aircraft laboratory
GNC
  guidance and navigation computer
GND
  ground
  grounded
GNE
  guidance and navigation electronics
GNGC
  guide network group connector
GNMA
  government nation mortage association
GNP
  gross national product
GNR
  guest name record
GNS
  goose norad sector
  guidance and navigation system
GNT
  great northern telegraph
GNTR
  generator
GO
  general output
  geometrical optic
  goniometer
GOAL
  general organization analysis language
  generator for optimized application languages
  ground operations aerospace language
GOAT
  gerber oscillogram amplitude translator
GOBG
  governing
GOCI
  general operator computer interaction
GOCR
  gated off controlled rectifier
GOE
  ground operating equipment
GOES
  geostationary operational environmental satellite
GOL
  general operating language
GOLS
  general on line stack system

GOMAC
   government microcircuit applications conference
GONIO
   goniometer
GOP
   general operational plot
GOR
   gained output ratio
   gas oil ratio
   general operational requirement
GORID
   ground optical recorder for intercept determination
GOS
   global observation[al] system
   global operating system
   graphic operating system
GOSS
   ground operational support system
GOTRAN
   go fortran
GOTS
   graphically oriented time sharing system
GOV
   governor
GOVTEL
   government telegram
GOW
   georgia power company
GP
   gang punch
   gauge pressure
   general purpose
   generalized programming
   german patent
   glide path
   grid pulse
   ground post
   ground protective
   group
GPA
   gate pulse amplifier
   general purpose amplifier
   general purpose analysis
   graphical pert analog
GPATS
   general purpose automatic test set
   general purpose automatic test system
GPC
   gatineau power company
   general peripheral controller
   general precision connector
   general purpose computer

GPCL
   general purpose closed loop
GPCP
   general purpose controller processor
   generalized process control programming
GPCS
   general purpose control system
GPD
   gallon per day
   general purpose data
   generalized pair decomposition
GPDC
   general purpose digital computer
GPDS
   general purpose display system
GPE
   general precision equipment
GPERF
   ground passive electronic reconnaissance facility
GPEXS
   general parts explosion system
GPF
   gas proof
GPFL
   group flashing
GPFLL
   group flashing light
GPG
   gate pulse generator
GPGL
   general purpose graphic language
GPH
   gallons per hour
GPI
   general precision, incorporated
   ground point of intercept
   ground position indicator
GPIA
   general purpose interface adapter
GPIB
   general purpose instrument bus
   general purpose interface bus
GPIS
   gemini problem investigation status
GPL
   general precision laboratory
   general purpose language
   generalized programming language
   grams per litre
GPLP
   general purpose linear programming
GPLS
   glide path landing system

GPM
  gallons per mile
  gallons per minute
  general purpose macrogenerator
  group per message
  group per minute
  groups per minute
GPMS
  general purpose microprogram simulator
GPO
  gemini project office
  general post office
  government printing office
GPOCC
  group occulting
GPOS
  general purpose operating system
GPP
  general plant projects
GPPB
  gemini program planning board
GPR
  general purpose radar
  general purpose register
GPRL
  giant pulse ruby laser
GPS
  gallons per second
  general problem solver
  general purpose simulation
  general purpose simulator
  global positioning system
  graphic programming service
GPSCS
  general purpose satellite communications system
GPSL
  general purpose simulation language
GPSP
  general purpose string processor
GPSS
  general purpose simulation studies
  general purpose simulation system
  general purpose system simulation
  general purpose system simulator
GPSSM
  general purpose surface to surface missile
GPSU
  ground power supply unit
GPT
  gallons per ton
  gas phase titration
  gas power transfer
  general purpose terminal
  guidance position tracking

GPU
  generating power unit
  geopotential unit
  ground power unit
GPWS
  ground proximity warning system
GPX
  generalized programming extent
  generalized programming extended
GQA
  get quick answer
GR
  gas cooled reactor
  gas ratio
  general radio
  general reconnaissance
  general reserve
  generator running
  germanium rectifier
  grade
  grain
  grid resistor
  grid return
  gross
  ground
  group
  grown rate
GRA
  graphic recording ammeter
GRACE
  gamma attenuation reactor physics computer code
  graphic arts composing equipment
  group routing and charging equipment
GRAD
  general recursive algebra and differentiation
  gradient
  graduate resume accumulation and distribution
GRADB
  general remote access data base
  generalized remote access data base
GRADS
  generalized remote access data base system
GRAF
  graphic addition to fortran
GRAMPA
  general analytic[al] model for process analysis
GRAPD
  guard ring avalanche photodiode
GRAPE
  gamma ray attenuation porosity evaluator
GRAPHDEN
  graphic[al] data entry

GRARR
goddard range and range rate
GRAS
generally recognized as safe
GRASER
gamma ray amplification by stimulated emission of radiation
GRASP
general read and simulate program
general resource allocation and scheduling program
generalized retrieval and storage program
graphic service program
GRASS
gamma ray ablation sensing system
GRATIS
generation, reduction and training input system
GRB
geophysical research board
government reservation bureau
GRBM
global range ballistic missile
GRC
green rectifier corporation
GRCSW
graduate research center of the southwest
GRD
general radio discriminator
ground
GRDP
graphic data processing
GRE
graduate reliability engineering
ground radar equipment
GREB
galactic radiation experiment background
galactic radiation experiment background satellite
GREC
grand rapids engineers club
GRED
generalized random extract device
GREMEX
goddard research engineering management exercise
GRF
group repetition frequency
GRFF
general rf fittings
GRFM
general radio frequency meter
GRG
gross reserve generation
GRI
group repetition interval
GRID
graphic interactive display
GRIN
graded index
graphical input
GRINS
general retrieval inquiry negotiation structure
GRIPS
general relation based information processing system
GRIT
graduated reduction in tensions
GRITS
goddard range instrumentation tracking system
GRIVATION
grid variation
GRL
grille
GRLP
ground lamp
GRM
generalized reed muller
generalized reed muller codes
global range missile
guidance rate measuring
GRNC
group not counted
GROM
graphic rom
grommet
GRP
gaussian random process
glass reinforced plastic
glass reinforced polyester
ground relay panel
group
GRR
greek research reactor
guidance reference release
GRRRS
goddard range and range rate system
GRS
gamma ray spectrometry
general radio service
general register set
general retrieval system
generalized retrieval system
gyro reference system
GRSE
guild of radio service engineers
GRSRUBBER
general purpose synthetic rubber

GRT
  general reactor technology
  general recorded tape
  gross register ton
GRTS
  goddard real time system
GRV
  graphic recording voltmeter
  groove
GRVCIC
  graphic reproductions visual communications industries' council
GRW
  graphic recording wattmeter
GRWT
  gross weight
GS
  galvanized steel
  gauss
  general search
  general staff
  general storage
  geological survey
  glide slope
  grain size
  ground speed
  ground station
  ground surface
  group selector
  group separator
  guidance station
  guidance system
  gyroscope
GSA
  general services administration
  genetics society of america
  geological society of america
  guidance system analyst
GSB
  graphic standards board
GSC
  gas solid chromatography
  geodetic space craft
  group switching center
  guidance system console
GSCA
  horizontal sweep circuit analyzer
GSCAN
  g scanner
GSCO
  guidance sustainer cut off
GSCU
  ground support cooling unit
GSD
  general supply depot
  general system development
  general system division
GSDA
  grounded surface distribution apparatus
GSDB
  geophysics and space data bulletin
GSDS
  goldstone duplicate standard
GSE
  ground servicing equipment
  ground support equipment
GSESS
  ground support equipment systems specifications
GSFC
  goddard space flight center
GSI
  gencom systems, inc
  general semiconductor, incorporated
  giant scale integration
  government source inspection
  grand scale integration
  ground speed indicator
GSL
  generalized simulation language
  generation strategy language
GSLD
  group selector long distance
GSMB
  graphic standards management board
GSME
  ground support maintenance equipment
GSO
  ground safety office
GSOP
  guidance systems operation plan
GSP
  general simulation program
  general system of tariff preferences
  geodetic satellite program
  graphic subroutine package
  guidance signal processor
GSPB
  geodetic satellite policy board
GSPO
  gemini spacecraft project office
GSPR
  guidance signal processor repeater
GSR
  galvanic skin resistance
  galvanic skin response
  glide slope receiver
  global shared resources
  ground sensor relay
  group selection register

GSRS
  general support rocket system
GSS
  gamma scintillation system
  general staff support
  geodetic stationary satellite
  geostationary satellite
  global surveillance station
  global surveillance system
  ground support system
GSSC
  ground support simulation computer
GSSL
  general staff support large
GSSLD
  group selector of secondary long distance
GSSM
  general staff support medium
GST
  gate sensitive thyristor
  gemini system trainer
  global space transport
  greenwich sidereal time
GSTA
  ground surveillance and target acquisition
GSTP
  ground system test procedure
GSU
  general service unit
GSUC
  ground stub up connection
GSUG
  gross seasonal unavailable generation
GSV
  governor steam valve
  ground to surface vessel radar
  guided space vehicle
GSVC
  generalized supervisor call
GSW
  galvanized steel wire
GSWR
  galvanized steel wire rope
GT
  antenna gain to noise temperature
  concentration of grams of pu per irradiated short ton of uranium
  game theory
  gas tight
  gas tube
  gate tube
  gemini titan
  glass tube
  glow tube
  greater than
  greetings telegram
  gross ton
  gross tonnage
  ground transmit
  group technology
GTA
  gas tungsten arc
  gemini titan agena
GTAE
  general telephone and electronics
GTAW
  gas tungsten arc weld
GTC
  gain time constant
  gain time control
  gas turbine compressor
  general transistor corporation
GTCC
  group technology characterization code
GTCR
  gate turn off controlled rectifier
GTD
  geometrical theory of diffraction
  graphic tablet display
GTE
  general telephone and electric
  general telephone and electronics
  general television and electronic
  ground test equipment
  ground transport equipment
  group translating equipment
GTEIS
  gte information systems inc
GTF
  general trace facility
  generalized trace facility
  guidance test fixture
GTG
  gas turbine generator
  ground timing generator
GTH
  gas tight high pressure
GTHTGR
  gas turbine high temperature gas cooled reactor
GTI
  general telephone company of illinois
  guidance technology, incorporated
GTL
  gas transport laser
GTM
  ground test missile
GTO
  gate turn off
  goto instruction

GTOL
   ground take off and landing
GTOSCR
   gate turn off silicon controlled rectifier
GTOTHYRISTOR
   gate turn off thyristor
GTOW
   cross take off weight
GTP
   gemini test pad
   general test plan
   golay transform processor
GTR
   ground test reactor
GTRP
   general transpose
GTRR
   georgia institute of technology research reactor
GTS
   general technical services
   general telephone system
   global telecommunications system
   guidance test set
GTTF
   gas turbine test facility experimental, closed cycle, gas turbine system, ft belvoir, virginia
GTU
   guidance test unit
GTV
   gate trigger valve
   gate valve
   general trace facility
   ground test vehicle
   guidance test vehicle
   guided test vehicle
GTW
   gross total weight
GTXT
   generate character text
GU
   guam
   guidance test
GUAR
   guarantee
   guaranteed
GUGA
   grounded unity gain amplifier
GUHA
   general unary hypotheses automation
GUIDE
   guidance for users of integrated data equipment
   guidance for users of integrated data processing equipment
GUISE
   guidance system evaluation
GULP
   general utility library program
GUNFCO
   gulf united nuclear fuels corporation
GUS
   general user system
GUSTO
   guidance using stable tuning oscillations
GUUG
   gross unit unavailable generation
GV
   conductance as a function of voltage
   group velocity
   guard vessel
GVA
   graphic kilovolt ampere meter
GW
   general warning
   gigawatt
   gross weight
   guided weapon
GWC
   global weather central
GWETA
   greater washington educational television association
GWH
   gigawatt hour
GWL
   ground water level
GWMC
   galvanized ware manufacturers' council
GWPM
   gross words per minute
GWS
   guided weapon station
GWSTN
   ground wireless station
GY
   guy
GYRO
   gyro compass
GZ
   ground zero
GZMG
   gradient zone melting
GZTPRD
   ground zero tape read
   nd mechanical system

H

H
  Halt
  hand actuated
  hard
  hardness
  hardware
  head
  heat
  heater
  heavy
  hecto
  height
  henry
  high
  high level
  homing
  honeywell
  horizontal
  hour
  hundredweight

HA
  hand actuated
  half add
  half adder
  hardware
  hectare
  high altitude
  high amplitude
  home address
  hour angle
  hydraulic association

HAA
  heavy anti aircraft
  heavy anti aircraft artillery
  height above airport
  helicopter association of america
  high altitude abort

HAAS
  honeywell automotive accounting system

HAATC
  high altitude air traffic control

HAC
  hot and cold
  hot and cold water
  hughes aircraft company

HACL
  harvard air cleaning laboratory

HACS
  high angle control section

HAD
  half amplitude duration
  high accuracy data
  high accuracy data system
  horizontal array of dipoles

HADC
  holloman air development center

HADEC
  hour angle declination

HADES
  hypersonic air data entry system

HADIOS
  honeywell analog digital input output subsystem

HADS
  hypersonic air data sensor

HADTS
  high accuracy data transmission system

HAF
  high abrasion furnace
  high altitude fluorescence

HAFB
  hill air force base
  holloman air force base

HAFBLCK
  high abrasion furnace black

HAIC
  hearing aid industry conference

HAIRS
  high altitude test and evaluation of infrared sources

HAL
  highly automated logic

HALLED
  hall effect devices

HALO
  high attitude large optics

HALSIM
  hardware logic simulator

HAM
  hardware associative memory
  high activity mode
  high availability manager
  high speed automatic monitor

HAN
  holmes and narver

HANE
  high altitude nuclear effects

HAOSS
  high altitude orbital space station

HAP
  high altitude platform
  high altitude probe

HAPDAR
  hard point demonstration array radar

HAPUG
  harbich, pungs, gerth

HAR
  harmonic
  high altitude recombination energy
HARAC
  high altitude resonance absorption calculation
HARAO
  hartford aircraft reactors area office
HARC
  hudson amateur radio council
HARCO
  hyperbolic area control
  hyperbolic area coverage
  hyperbolic area covering navigation system
HARD
  hardware
HARDTS
  high accuracy radar data transmission system
HARE
  high altitude recombination energy
HARM
  harmonic
  harmony
  high speed antiradiation missile
HARP
  halpern's anti radar point
  high altitude relay point
  high altitude research probe
  high altitude research project
  hitachi arithmetic processor
  holding and reconsignment point
HARS
  heading and attitude reference system
HART
  highway aid by radio truck
HARTRAN
  hardwell fortran
  harwell atlas fortran
HAS
  heading and attitude sensor
  hellenic astronautical society
  high altitude sampler
HASL
  health and safety laboratory
HASP
  high altitude sounding program
  high altitude sounding project
  high altitude sounding projectile
  high altitude space probe
  houston automatic spooling and printing system
  houston automatic spooling priority system
  houston automatic spooling program
HASR
  high altitude sounding rocket
HASSS
  high accuracy spacecraft separation system
HASVR
  high altitude space velocity radar
HAT
  handover transmitter
  hardened and tempered
  hardware acceptance team
  height above touch down
  high altitude test
  high altitude testing
HATRAC
  handover transfer and receiver accept change
HAV
  heating and ventilation
  hot air vulcanizing
HAVE
  height average
HAW
  high level radioactive waste
HAWK
  homing all the way killer
HAYSTAQ
  have you stored answers to questions
HAZ
  heat affected zone
HB
  hard black
  headband
  hexadecimal to binary
  high band
  homing beacon
  honeywell bull
  horizontal baffle
HBC
  high breaking capacity
  honeywell business compiler
  human body counter
  hydrogen bubble chamber
HBR
  ham band receiver
HBW
  hot bridge wire
HBWR
  halden boiling heavy water reactor
HC
  hand carry
  hand control
  hand controlled
  heat coil
  heat controlled
  heater cord
  heating cabinet
  heating coil
  heavy current

heuristic concepts
high capacity
high carbon
high conductivity
high current
holde, crystal
holding coil
home computer
horizontal check
hour circle
hybrid circuit
hybrid computer
HCA
  high conductivity association
HCBS
  host computer basic software
HCD
  high current density
  hollow cathode discharge
  hot carrier diode
  hughes communications division
HCDA
  hypothetical core disruptive accident
HCE
  hollow cathode effect
  human caused error
HCF
  highest common factor
  host command facility
HCG
  horizontal location of center of gravity
HCL
  high, common, low
  horizontal center line
HCLF
  horizontal cask lifting fixture
HCM
  hard core monitor
  hydraulic core mockup
HCMM
  heat capacity mapping mission
HCMOSCIRCUIT
  high speed cmos circuit
HCO
  highly chlorinated oil
HCP
  hemispherical candle power
  hexagonal close packed
  hexagonal closed packed
  horizontal candle power
  host communication processor
HCPS
  hemispherical candle power second
  horizontal candle power second
HCPTR
  helicopter
HCRS
  heritage conservation and recreation service
HCS
  helium circulator seal
HCSA
  house committee on space and astronautics
HCSS
  head compartment support structure
  hospital computer sharing system
  hospital computer user's association
HCT
  heat center tap
HCU
  homing comparator unit
  hydraulic control unit
HCUA
  honeywell computer users' association
HD
  half duplex
  hard drawn
  harmonic distortion
  head
  head diameter
  head driver
  heading
  hearing distance
  heavy duty
  hexadecimal code
  hexadecimal to decimal
  high density
  horizontal drive
HDA
  heavy duty amplifier
  high density acid
HDAM
  hierarchical direct access method
HDATZ
  high density air traffic zone
HDB
  high density bipolar
HDBCODE
  high density bipolar code
HDC
  high duty cycle
HDDR
  high density digital magnetic recording
  high density digital recording
HDDS
  high density data system
HDEP
  high density electronic packaging
HDF
  high frequency direction finder

    high frequency direction finding
    horizontal distributing frame
HDI
    horizon direction indicator
HDL
    harry diamond laboratories
HDLC
    high level data link control
HDLS
    hardware description language system
HDMR
    high density moderated reactor
HDOC
    handy dandy orbital computer
HDP
    high discharge pressure
    horizontal data processing
HDPE
    high density polyethylene
HDR
    header
    hold down and release
HDRSS
    high data rate storage system
HDST
    high density shock tube
HDT
    heat deflection temperature
HDTV
    high definition television
HDV
    halt device
HDW
    hard drawn wire
    hardware
    high pressure demineralized water
    hydrodynamics welding
HDWA
    hardware
HDWE
    hardware
HDX
    half duplex
    half duplex circuit
HDY
    heavy duty
HE
    handling equipment
    heat exchange
    heavy enamel
    heavy enamelled
    heavy equipment
    hemes electronics
    high efficiency
    high energy
    high explosive
    hydroelectric
    hydromagnetic emission
HEALS
    honeywell error analysis and logging system
HEAO
    high energy astronomy observatory
HEAP
    high explosive armor piercing
HEAT
    heating
    high explosive anti tank
HEBDC
    heavy enamel bonded double cotton
HEBDP
    heavy enamel bonded double paper
HEBDS
    heavy enamel bonded double silk
HEBP
    heavy enamel bonded paper
HEBS
    heavy enamel bonded single silk
HEC
    hartford engineers club
    heard european countries
    heavy enamel single cellophane
    heavy enamel single cotton
    hoffman electronics corporation
    hollerith electronic computer
HECATE
    heat exchanger computerized aid for technical engineering
HECLINET
    health care literature information network
HECTOR
    heated experimental carbon thermal oscillator reactor
    hot enriched carbon moderated thermal oscillator reactor
HECV
    heavy enamel single cotton varnish
HED
    horizontal electric[al] dipole
HEDC
    heavy enamel double cotton
HEDCV
    heavy enamel double cotton varnish
HEDL
    hanford engineering development laboratory
HEDS
    heavy enamel double silk
HEDSV
    heavy enamel double silk varnish
HEED
    high energy electron diffraction

HEEP
highway engineering exchange program
HEF
high energy fuel
HEG
heavy enamel single glass
HEHF
hanford environmental health foundation
HEI
heat exchange institute
HEL
harford electric light
HELEN
hydrogenous exponential liquid experiment
HELIOS
heteropowered earth launched interorbital spacecraft
HELP
helicopter electronic landing path
highly extendable language processor
highway emergency locating plan
HELPR
handbook of electronic parts reliability
HEM
heat exchanger method
hybrid electromagnetic
hybrid electromagnetic mode
hybrid electromagnetic wave
HEMT
high electron mobility transistor
HEO
high earth orbit
HEOS
high excentricity orbit satellite
highly excentric orbiting satellite
HEP
header
heterogene element processor
high energy physics
human error probability
HEPA
high efficiency particulate air filters
HEPC
hydro electric power commission
HEPCAPP
hydro electric power commission approved
HEPCO
hydro electric power commission of ontario
HEPDEX
high energy proton detection experiment
HEPL
high energy physics laboratory
HEPO
hydro electric power commission of ontario
HEPP
hoffmann evaluation program and procedure
HER
human error rate
HERA
high explosive rocket assisted
HERALD
harbor echo ranging and listening device
HERF
high energy rate forming
HERFS
high energy rate forming system
HERMES
heavy element and radioactive material electromagnetic separator
HERO
hazard of electromagnetic radiation to ordnance
HERPES
high energy recovery pressure and enthalpy sensor
HES
hanford engineering service
heavy enamel single silk
home entertainment service
hydro electric securities corporation
HESS
houston engineering and scientific society
HEST
high energy shock tunnel
HESV
heavy enamel single silk varnish
HET
high energetic fuel
HETP
height equivalent to a theoretical plate
HETS
height equivalent to a theoretical stage
high environment[al] test system
HEU
highly enriched uranium
hydroelectric unit
HEVAC
heating, ventilating, and air conditioning manufacturers association
HEX
hexadecimal
hexagon
hexagonal
HF
height finder
height finding
high field
high flux
high frequency

HFA
   heat and flame resistant, armored
   high flow alarm
   high frequency amplifier
   high frequency antenna
HFBD
   high frequency band
HFBR
   high flux beam reactor
HFC
   high frequency choke
   high frequency correction
   high frequency current
   high frequency transceiver
HFCF
   half calf
HFDF
   high frequency direction finder
   high frequency direction finding
   high frequency distribution frame
HFE
   human factors in electronics
   human factors in engineering group
HFEF
   hot fuel examination facility
HFG
   heavy free gas
HFI
   high fidelity institute, incorporated
   high frequency input
   high frequency input power
HFIM
   high frequency instruments and measurements
HFIR
   high flux isotope reactor
HFM
   high frequency mode
HFO
   high frequency oscillator
HFORL
   human factors operations research laboratory
HFRDF
   high frequency radio direction finding
   high frequency repeater distribution frame
HFS
   hierarchical file system
   hyperfine structure
HFSA
   human factors society of america
HFSSB
   high frequency single sideband
HFST
   high flux scram trip

HG
   hand generator
   harmonic generator
   head gear
   homing guidance
HGA
   high gain antenna
HGAS
   high gain antenna system
HGE
   handling ground equipment
   hydraulic grade elevations
HGL
   hyperbolic type gas lens
HGR
   head gear receiver
   high group receiving unit
HGRF
   hot gas radiation facility
HGT
   high group transmit
   high group transmitting unit
HH
   heavy hydrogen
HHD
   hogshead
HHF
   hyper high frequency
HHLR
   horace hardy lestor reactor
HHS
   hex head steel
HHSI
   high head safety injection
HI
   high
   high intensity
   high signal
   horizontal interval
   houston instrument corp
   hydraulic institute
HIAC
   high accuracy
HIAD
   handbook of instructions for airplane designers
HIALS
   high intensity approach lighting system
HIBEX
   high impulse boost experiment
   high impulse booster experiment
HIC
   high conversion critical experiment
   high dielectric constant
   hybrid integrated circuit

HICAP
  high capacity
HICAPCOM
  high capacity communication
  high capacity communication system
HID
  high density
  high intensity discharge
HIDAM
  hierarchical indexed direct access method
HIDAN
  high density air navigation
HIDF
  horizontal intermediate distribution frame
HIDM
  high information delta modulation
HIFAM
  high fidelity amplitude modulation
HIFAR
  high flux australian reactor
HIFI
  high fidelity
  high intensity food irradiator
HIFIT
  hardware implemented fault tolerance
  high frequency input transistor
HIG
  hermetical sealed integrating gyroscope
HIIMP
  high impedance
HIIS
  honeywell institute for information science
  honeywell institute of information science
HIL
  high intensity lighting
HILAC
  heavy ion linear accelerator
HILDE
  high level, double ended
HILO
  high low
HIMA
  health industry manufacturers association
HIMAT
  highly maneuverable aircraft technology
HINIL
  high noise immunity logic
HIP
  hanford isotopes plant
  hazard input program
  hot isostatic pressing
HIPAC
  heavy ion plasma accelerator
  hitachi parametron automatic computer

HIPAR
  high power acquisition radar
HIPASS
  high pass
HIPERNAS
  high performance navigation system
HIPFERRITE
  hot isolation pressed ferrite
HIPO
  hierarchy input processing output
  hierarchy plus input, process, output
HIPOT
  high potential
HIPOTT
  high potential test
HIPS
  hyperintense proximal scanning
HIPSA
  hallicrafers incremental power spectrum anaylzer
HIQ
  high quality
HIR
  horizontal impulse reaction
HIRAC
  high random access
HIRDL
  high intensity radiation development laboratory
HIREL
  high reliability
HIRNA
  high precision short range navigation
HIS
  hardware interrupt system
  honeywell information system
HISAM
  hierarchical indexed sequential access method
HISI
  honeywell information system italia
HISS
  high itensity sound simulator
HISTEP
  high speed integrated space transportation evaluation program
HIT
  hierarchical evaluation tool
  high isolation transformer
  high torque
  homing interceptor technology
  hypersonic interference technology
HITAC
  hitachi computer

HITMP
  highest temperature
HIVO
  high voltage
HIVOS
  high vacuum orbital simulation
  high vacuum orbital simulator
HJ
  honest john
HJFET
  heterojunction jfet
HKB
  hard kernel bunch
HKTC
  hongkong telephone company
HL
  half life
  hand lantern
  header label
  heavy loaded
  heavy loading
  hectoliter
  high level
  high low
HLAPCO
  houston lighting and power company
HLAS
  hot line alert system
HLC
  high level center
  high level compiler
HLCV
  hot leg check valve
HLD
  hold
HLDR
  holder
HLFM
  high level flux monitor
HLH
  heavy lift helicopter
HLIV
  hot leg isolation valve
HLL
  hard lunar landing
  high level language
  high level logic
HLLV
  heavy lift launch vehicle
HLLWT
  high liquid level waste tank
HLP
  houston lighting and power
HLRM
  high level radio modulator

HLSE
  high level, single ended
HLSI
  hybrid large scale integration
HLSIGNAL
  high low signal
HLT
  halt
HLTL
  high level transistor logic
HLTTL
  high level transistor transistor logic
HLW
  high level waste
HM
  hallmark
  harmonic mean
  heater mid tap
  heater middle
  heater middle tap
  heavy metal
  hectometer
  hollow metal
  hysteresis motor
HMC
  heading marker correction
HMD
  hot metal detector
  hydraulic mean depth
HMDF
  horizontal main distributing frame
HMEED
  heavy military electronic equipment division
HMES
  heavy military electronics system
HMI
  handbook of maintenance instructions
HMO
  honolulu magnetic observatory
HMOS
  high density metal oxide semiconductor
  high performance metal oxide semiconductor
HMP
  high melting point
HMRB
  hazardous materials regulation board
HMRRO
  human resources research office
HMS
  hanford meteorology surveys
  heavy media separation
  honeywell manufacturing system
HMSO
  honolulu magnetic and seismological observatory

HMSS
  hospital management systems society
HMT
  hand microtelephone
HN
  horn
HNC
  hand numerical control
  higher national certificate
HND
  higher national diploma
HNDR
  heteronuclear double resonance
HNIL
  high noise immunity logic
HNPF
  hallam nuclear power facility
HNR
  handwritten numeral recognition
HNW
  head, nut, and washer
HNYCMB
  honeycomb
HO
  horizontal output
  human operator
  hunting oscillator
HOA
  hands off automatic
HOALM
  holographic optically addressed light modulation
HOB
  homing on offset beacon
HOBO
  homing official bomb
HOBS
  high orbital bombardment system
HOC
  height overlap coverage
  heterodyne optical correlation
HOE
  homing overlay experiment
HOF
  head of form
HOFCO
  horizontal functional checkout
HOJ
  home on jamming
HOL
  high order language
HOLD
  holding
HOLDA
  hold acknowledge

HOM
  high order multiplier
  home
  homing
HOP
  hybrid operating program
HOPE
  hydrogen oxygen primary extraterrestrial
HOR
  horizon
  horizontal
HORAD
  horizontal radar display
HOS
  hardwired operating system
  horizontal obstacle sonar
  hungarian office for standardization
HOST
  harmonically optimized stabilization technique
  houston operations simulation technique
HOT
  handover transmitter
  horizontal output transformer
  horizontal output tube
HOTCE
  hot critical experiments
HOU
  houston lighting and power
HP
  heptode
  hewlett packard
  high pass
  high performance
  high power
  high pressure
  hire purchase
  horizontal polarization
  horsepower
HPA
  hewlett packard associates
  high power amplifier
  holding and positioning aid
  host processor adapter
HPAG
  high performance air to ground
HPBW
  half power bandwidth
HPC
  hard processing channel
HPCC
  high performance control center
HPCI
  high pressure coolant injection

HPCM
   high speed pulse code modulation
HPCS
   high pressure core spray
HPD
   high power density
HPDSN
   hewlett packard distributed systems network
HPF
   high pass filter
   high power field
   highest possible frequency
   highest probable frequency
   highpass filter
   horizontal position finder
   hot pressed ferrite
HPFL
   high performance fuels laboratory
HPG
   high power ground
   high pressure gelatine
HPH
   horsepower hour
HPHR
   horsepower hour
HPI
   height position indicator
   high pressure injection
HPIR
   high probability of intercept receiver
HPIS
   high pressure injection system
HPLC
   high pressure liquid chromatography
HPM
   honeycomb propellant matrix
HPMS
   high performance main storage
HPMV
   high pressure mercury vapor
HPN
   high pass notch
HPOF
   high presure oil filled
HPOFCABLES
   high pressure oil filled cables
HPOT
   helical potentiometer
   heliopotentiometer
   high potential
HPP
   hot processing plant
HPPS
   hewlett packard printer submodule

HPR
   halt and proceed
HPRR
   health physics research reactor
HPS
   hanford plant standards
   health physics society
   high primary sequence
HPSC
   heading per standard compass
HPSIS
   high pressure safety injection system
HPSW
   high pressure service water
HPT
   high point
   high pressure test
   high profile terminal
   horizontal plot table
HQ
   headquarters
   high quality
HQS
   high quality sound
HQTV
   high quality television
HR
   hand radar
   hand reset
   heat resistance
   heat resistant
   heat resisting
   heater
   heavy duty relay
   height range
   hertzsprung russell
   high reflector
   high resistance
   high resistor
   highly resistive
   hot rolled
   hour
   inventory and inspection report
HRAC
   hypersonic research aircraft
HRC
   high rupturing capacity
   horizontal redundancy test
   hypothetical reference circuit
HRCFUSE
   high rupture capacity fuse
HRD
   hertzsprung russel diagram
   high resolution display

HRDANDGR
  harden and grind
HRF
  height range finder
HRI
  height range indicator
  horizon reference indicator
HRIO
  height range indicator operator
HRIR
  high resolution infrared radiometer
  high resolution infrared receiver
  high resolution infrared reception
HRL
  horizontal reference line
HRLEL
  high radiation level examination laboratory
HRM
  high reliability module
HRMS
  height root mean square
HRQ
  hold request
HRRC
  human resources research center
HRS
  heading reference system
  hours
  hydrant refuelling system
HRT
  high rate telemetry
  high resolution tracker
  homogeneous reactor test
HRV
  hypersonic research vehicle
HRZN
  horizon
HS
  half subtractor
  hand set
  head set
  health and safety
  heat shield
  heater shield
  heating surface
  helios semiconductor
  hermetically sealed
  high shock resistant
  high speed
  homing sequence
  horizon scanner
  horizon sensor
  hypersonic
HSA
  heat shield abort
  high speed adapter
  high speed arithmetic
HSAC
  house science and astronautics committee
HSALU
  high speed arithmetic and logic unit
HSAM
  high speed accounting machine
HSAP
  houston automatic spooling and printing system
HSB
  heat shield boost
  horizontal sounding balloon
HSBR
  high speed bombing radar
HSC
  house space committee
  house spacecraft
HSCA
  horizontal sweep circuit analyzer
HSCAN
  h scanner
HSD
  hawker siddeley dynamics
  hierarchically structured data set
  high speed data
  high speed displacement
  horizontal situation display
  hot shut down
HSDA
  high speed data acquisition
HSDC
  high speed data channel
HSDL
  high speed data link
HSE
  health and safety executive
  heat shield entry
  high speed encoder
HSG
  housing
HSGT
  high speed ground transportation
HSI
  hardware software interface
  horizontal situation indicator
HSIIL
  high speed integrated injection logic
HSIMP
  high speed interface message processor
HSJ
  heat shield jettison

HSL
  high speed launch
  high speed logic
HSM
  health services and mental health administration
  high speed measurement
  high speed memory
HSP
  high speed printer
  high speed punch
HSR
  harbor surveillance radar
  heat shield recovery
  high speed reader
HSRO
  high speed repetitive operation
HSS
  hardware specification sheet
  high speed steel
  high speed storage
HSST
  heavy section steel technology
HST
  harmonic and spurious totalizer
  hawaiian standard time
  horizontal seismic trigger
  hypersonic transport
HSTATE
  high state
HSYNC
  horizontal synchronization
HSZD
  hermetically sealed zener diode
HT
  half tone
  handling time
  hawaiian telephone
  hawaiian time
  headset
  heat
  heat treated
  heater tap
  heavy thermoplastic
  height
  high temperature
  high tensile
  high tension
  high torque
  homing transponder
  horizontal tabulation
  horizontal tabulator
  hors tension
HTA
  heavier than air

HTAC
  high tension alternating current
HTB
  high tension battery
HTC
  hughes tool company
HTCAD
  hughes tool company aircraft division
HTD
  hand target designator
HTDC
  high tension direct current
HTGR
  high temperature gas cooled reactor
HTGRCX
  high temperature gas reactor critical experiment
HTIS
  heat transfer instrument system
HTL
  heat transfer loop
  high threshold logic
HTLTR
  high temperature lattice test reactor
HTM
  high temperature
  hypothesis testing model
  hard tube monitor
HTMOS
  high threshold mos
HTN
  heterodyne
HTO
  high temperature oxidation
  horizontal take off
HTOL
  horizontal take off and landing
HTPV
  high temperature power and voltage
HTR
  halt and transfer
  hanford 305 test reactor
  heater
  hitachi training reactor
HTRAC
  half track
HTRB
  high temperature reverse bias
HTRE
  heat transfer reactor experiment, nuclear electronics laboratory, hughes aircraft company
HTRI
  heat transfer research institute

HTRK
  half track
HTS
  hawaii tracking station
  height telling surveillance
  high tensile steel
  high tensile strength
  high tension separation
  high tension supply
  humble timesharig co
HTSF
  high temperature sodium facility
HTSIM
  height simulator
HTSL
  heat transfer simulation loop
HTSS
  honeywell time sharing system
HTST
  high temperature short time
HTSUP
  height supervisor
HTTF
  critical assembly
HTTL
  high power transistor transistor logic
  high speed transistor transistor logic
  high speed ttl
HTU
  heat transfer unit
  height of transfer unit
HTV
  high altitude test vehicle
  homing test vehicle
  hypersonic test vehicle
HUCR
  highest useful compression ratio
HUD
  head up display
HUDFHA
  housing and urban development federal housing administration
HUDWAC
  heads up display weapons aiming comuter
HUKP
  hostile, unknown, faker and pending
HUPCM
  hybrid unidigit pulse code modulation
HUR
  heat up rate
HUT
  held up transient
  hold up tank
HV
  hard valve
  hardware virtualizer
  heavy
  high vacuum
  high velocity
  high voltage
  hypervelocity
HVAC
  heating, ventilation and air conditioning
  high vacuum
  high voltage actuator
HVACC
  high voltage apparatus coordinating committee
HVAR
  high velocity airborne rocket
  high velocity aircraft rocket
HVAT
  high velocity antitank
HVC
  hardened voice channel
  hardened voice circuit
HVCA
  heating and ventilating contractors association
HVD
  half value depth
HVDC
  high voltage direct converter
  high voltage direct current
HVDF
  high frequency and very high frequency direction finder
  high frequency and very high frequency direction finding
HVEC
  high voltage engineering corporation
HVHF
  high very high frequency
HVHMD
  holographic visor helmet mounted display
HVI
  high viscosity index
HVL
  half value layer
HVM
  heterodyne vegetation meter
  high voltage module
HVP
  high video pass
  high voltage potential
HVPC
  high voltage power corporation
HVPS
  high voltage power supply

HVR
  high vaccum rectifier
  high voltage regulator
  high voltage relay
HVRA
  heating and ventilating research association
HVSCR
  high voltage selenium cartridge rectifier
HVT
  half value thickness
  high voltage threshold
HVTR
  home video tape recorder
  home video tape recording
HVY
  heavy
HW
  half wave
  hardware
  heavy water
  hot wire
  howler
HWCTR
  heavy water component test reactor
HWD
  hight, width, depth
HWGCR
  heavy water moderated gas cooled reactor
HWM
  high wet modulus
HWOCR
  heavy water organic cooled reactor
HWP
  harmonic wire projector
  heavy water plant
  hewlett packard company
HWR
  half wave rectifier
  heavy water moderated reactor
  heavy water reactor
HWS
  hanford works standard or specification
HWY
  highway
HX
  half duplex
  heat exchanger
  hexode
HY
  hybrid coul
  hybrid junction
HYB
  hybrid

HYBLOC
  hybrid computer block oriented compiler
HYCOL
  hybrid computer link
HYCOM
  highway communications
HYCOTRAN
  hybrid computer translator
HYD
  hydraulic
  hydrolized
  hydrostatic
HYDAC
  hybrid digital analog computer
  hybrid digital to analog computer
HYDAPT
  hybrid digital analog and pulse time
  hybrid digital analog pulse time
HYDC
  hybrid digital analog computing
HYDROELEC
  hydroelectric
HYDROELECENGR
  hydroelectric engineer
HYDX
  hydroxide
HYFES
  hypersonic flight environmental simulator
HYLO
  hybrid loran
HYPERDOP
  hyperbolic doppler
HYPO
  high power water boiler reactor
HYPREM
  hyper response electric motor
HYPSES
  hydrographic precision scanning echo sounder
HYSCAN
  hybrid scanning
HYSTAD
  hydrofoil stabilization device
HYTRAN
  hybrid translator
HYTRESS
  high test recorder and simulator system
HZ
  hertz
HZMP
  horizontal impulse
HZP
  hot zero power

I

I
iconoscope
illumination
imaginary
inclination
indicating
indicator
inertia
information content
infra
input
instruction
instrumentation
integer
integral
intensity
interceptor
interference
interphone
interpole
inverter
iodine
iso
isotopic
optically inactive

IA
immediate access
immediately available
impendance angle
index array
indirect address
indirect addressing
industry application
information and action
infraaudible
initial apperance
input axis
instrumentation amplifier
interface adapter
international angstrom
iowa

IAA
interim access authorization
international academy of astronautics

IAAB
inter american association of broadcasters

IAAC
international agriculture aviation center
international association for analog[ue] computation

IAAM
international association of aircraft manufacturers

IAAPG
interagency advanced power group

IAARC
international administrative aeronautical radio conference

IAB
instrumentation analysis branch

IABSE
international association for bridge and structural engineering

IAC
immediate access memory
industry advisory committee
information analysis center
installation and checkout
instrumentation and control
integration assembly checkout
interim acceptance criteria
international advisory committee
international algebraic compiler
international analysis code
international association of cybernetics

IACC
industrial analysis and control council

IACP
international association of chiefs of police
international association of computer programmers

IACS
inertial attitude control system
integrated armament control system
intermediate altitude communication satellite
international annealed copperstandard

IAD
immediate action directive
initiation area discriminator
integrated automated documentation
integrated automatic documentation
interface analysis document
international astrophysical decade

IADE
integral absolute delay error

IADIC
integrating analog to digital converter
integration analog to digital converter

IADPC
inter agency data processing committee

IADR
international association for dental research

IAE
information and education

institute for the advancement of engineering
institution of aeronautical engineers
integral absolute error
integral of absolute error
integrated absolute error
international atomic exposition
IAEA
international atomic energy agency
IAEE
international association of earthquake engineers
international automotive engineering exposition
IAEI
international association of electrical inspectors
IAEL
international association of electrical leagues
IAEO
international atomic energy organization
IAES
ieee aerospace and electronic systems
institute of the aeronautical sciences
international association of electrotypers and stereotypers
IAESTE
international association for the exchange of students for technical experience
IAESTECANADA
international association for the exchange of students for technical experience, canada
IAESTEUS
international association for the exchange of students for technical experience, united states
IAF
information and forwarding
international astronautic[al] federation
IAG
ifip administrative data processing group
IAGA
international association of geomagnetism and aeronomy
IAGC
instantaneous automatic gain circuit
instantaneous automatic gain control
IAHR
international association for hydraulic research
IAHS
international association of housing science
IAI
informational acquisition and interpretation
international acquisition and interpretation

IAIE
integral absolute ideal error
IAIG
industrial analytical instrumentation group
IAL
immediate action letter
installation and logistics
international aeradio, limited
international algebraic language
international algorithmic language
investment analysis language
IALA
international association of lighthouse authorities
IALC
instrument approach and landing chart
IALE
integral absolute linear error
IALU
incrementing alu
IAM
impulse amplitude modulation
indefinite admittance matrix
initial address message
institute of aviation medicine
interactive algebraic manipulation
international association of machinists
IAMAP
international association of meteorology and atmospheric physics
IAMC
institute for advancement of medical communication
IAMS
instantaneous audience measurement system
IAMTCT
institute of advanced machine tool and control technology
IANC
international airline navigators council
IANEC
inter american nuclear energy commission
IAO
internal automation operation
international automation operation
IAOR
international abstracts in operations research
IAP
industry application program
inerting and preheating
initial approach
IAPG
interagency advanced power group

IAPMO
   international association of plumbing and mechanical officials
IAPO
   international association of physical oceanography
IAPR
   international association on pattern recognition
IAQ
   international association for quality
IAR
   inspection acceptance record
   instruction address register
   instrument air receiver
   intelligence and reconnaissance
IARU
   international amateur radio union
IAS
   immediate access storage
   immediate access store
   indian astronautical society
   indicated air speed
   industrial applications society
   institute for advanced study
   institute of advanced studies
   institute of aeronautical sciences
   institute of aerospace sciences
   instrument approach system
   integrated analytical system
   interactive application system
   international aircraft standards
   international association of sedimentology
   international association of seismology and physics of the earth's interior
   intrument approach system
IASA
   insurance accounting and statistical association
   international air safety association
IASB
   international aircraft standard bureau
IASE
   installation and service engineering
IASF
   instrumentation in aerospace simulation facilities
IASFC
   instrumentation for aerospace simulation facilities committee
IASH
   international association of scientific hydrology
IASI
   inter american statistical institute

IASPEI
   international association of seismology and physics of the earth's interior
IASTED
   international association of science and technology for development
IAT
   indexible address tag
   individual acceptance test
   information assessment team
   institute for advanced technology
   institute of advanced technology
   institute of automatics and telemechanics
   international accountants society
   international atomic time
IATA
   international air transport association
   is amended to add
IATCR
   international air traffic communications receiving station
IATCS
   international air traffic communications station
   international air traffic communications system
IATCT
   international air traffic communications transmitting station
IATD
   is amended to delete
IATM
   international association for testing materials
IATME
   international association of terrestrial magnetism and electricity
IATR
   is amended to read
IATU
   inter american telecommunications union
IATUL
   international association of technical university libraries
IAU
   infrastructure accounting unit
   international association of universities
   international astronomical union
IAV
   international association of vulcanology
IAVA
   industrial audio visual association
IAVC
   instantaneous automatic video control
   instantaneous automatic volume control

IAWRT
   international association of women in radio and television
IB
   identification beacon
   in bulk
   information bulletin
   information bus
   instruction book
   interface bus
   international broadcasting
IBA
   independent broadcasting authority
   investment bankers association
   ion backscattering analysis
IBAC
   instantaneous broadcast audience counting
IBAM
   institute of business administration and management
IBC
   international broadcasting convention
   international broadcasting corporation
IBCC
   international building classification committee
IBCFA
   injected beam cross field amplifier
IBD
   intermediate block diagram
IBE
   international broadcast engineer
IBEW
   international brotherhood of electrical workers
IBG
   interblock gap
IBI
   insulation board institute
   intergovernmental bureau for informatics
IBIS
   intense bunched ion source
IBL
   intermediate behavioral language
IBM
   international business machines
IBMA
   independent battery manufacturers of america
IBMCE
   ibm customer engineer
IBMOC
   intercontinental ballistic missile operational capability

IBMPC
   ibm personal computer
IBN
   identification beacon
   indexed by name
IBO
   international broadcasting organization
IBOLS
   integrated business oriented language support
IBP
   initial boiling point
IBPA
   international business press associates
IBPCS
   international bureau for physicochemical standards
IBR
   integral boiling reactor
   integrated bridge rectifier
IBRL
   initial bomb release line
IBS
   institute for basic standards
   intercollegiate broadcasting system
   international broadcasters' society
   international business services
IBSAC
   industrialized building systems and components
IBSHR
   integral boiling and superheat reactor
IBSYS
   ibm batch processing system
IBT
   instrument bend test
   instrumented bend test
IBU
   instruction buffer unit
   international broadcasting union
IBUPU
   international bureau of the universal postal union
IBW
   impulse bandwidth
IBWM
   international bureau of weights and measures
IC
   illiac chamber
   image chamber
   image check
   impulse conductor
   in charge
   increase
   index correction
   indication cycle

inductive capacitive
inductive coupling
information center
initial condition
inlet contact
input circuit
input code
inscribed circle
inspected condemned
installation and calibration
installation and checkout
installed capacity
instruction card
instruction code
instruction counter
instruction cycle
instrument and controls
insulated conductors
insulating compound
integrated circuit
integration and checkout
integrator card
interceptor computer
interchange center
intercommunication
interconnection
interface control
interference control
interior communication
intermediate circuit
internal combustion
internal connection
internal conversion
international control
interrupting capacity
interruption code
ion chamber

ICA
ignition control additive
industrial communications agency
industrial communications association
integrated communications adapter
intercompany agreement
international commission on acoustics
international communication association
international congress on acoustics
international council on archives
item change analysis
item control area

ICAA
international committee on aerospace activities

ICAC
instrumentation calibration and checkout

ICAD
integrated control and display
integrated controls and displays

ICAF
international committee on aeronautical fatigue

ICAM
integrated communications access method
integrated computer aided manufacturing

ICAN
international commission for air navigation

ICAO
international civil aviation organization

ICAR
integrated command accounting and reporting
interface control action request

ICARUS
intercontinental aerospacecraft range unlimited system

ICARVS
interplanetary craft for advanced research in the vicinity of the sun

ICAS
intermittent commercial and amateur service
international council of aeronautical sciences

ICB
integrated circuits breadboard
interface control board
interrupt control block
interrupt control center

ICBM
intercontinental ballistic missile

ICBMS
intercontinental ballistic missile system

ICBO
international conference of building officials

ICBR
input channel buffer register

ICBWR
improved cycle boiling water reactor design

ICC
imperial communication committee
information control center
instrumentation control center
intercomputer coupler
international chamber of commerce
international communications corporation
international computation center
international computer center
international computing center
international conference on communication
interstate commerce commission

ICCC
information and computing centers corporation

international chamber of commerce
international council on computer communication
ICCEE
  international classification commission for electrical engineers
ICCF
  interactive computer controlling facility
ICCM
  intercontinental cruise missile
ICCP
  institute for certification of computer professionals
  international conference on cataloging principles
ICCPC
  international computation centers' preparatory committee
ICCS
  intersite control and communications system
ICCT
  initial contact control time
ICCU
  interchannel comparison unit
  intercomputer communication unit
ICD
  interface control dimension
  interface control document
  interface control drawing
  ion controlled diode
ICDC
  international cable development corporation
ICDCP
  interface control drawing change proposal
ICE
  immediate cable equalizers
  in circuit emulation
  in circuit emulator
  in circuit emulgator
  individual compass error
  industrial computer enclosure
  input checking equipment
  institution of civil engineers
  integrated circuit engineering
  integrated cooling for electronics
  intercomputer electronics
  intermediate cable equalizers
  internal combustion engine
ICEA
  insulated cable engineers association
ICECAN
  iceland canada telephone cable
ICED
  interface control envelope drawing
  interface control environment drawing

ICEM
  international council for educational media
ICES
  integrated civil engineering system
  international civil engineering system
  international committee for earth sciences
  international conference of engineering studies
  international council for the exploration of the sea
ICET
  institute for the certification of engineering technicians
ICETK
  international committee of electrochemical thermodynamics and kinetics
ICF
  integrated crystal filter
  intercommunication flip flop
ICFE
  intra collisional field effect
ICG
  interactive computer graphics
ICGM
  intercontinental glide missile
  intercontinental guided missile
ICH
  intermediate chain home
ICHEME
  institution of chemical engineers
ICHENP
  international conference on high energy nuclear physics
ICHMT
  international center for heat and mass transfer
ICI
  intelligent communications interface
  international commission on illumination
ICIASF
  international conference on instrumentation in aerospace simulation facilities
ICIC
  interdepartmental commission on interplanetary communications
ICID
  international commission on irrigation and drainage
ICIP
  international conference on information processing
ICL
  incoming line
  instrument controlled landing
  instrumentation configuration log
  international computer ltd

ICM
   improved capability missile
   instrumentation and communication monitor
   intercontinental missile
   interference control monitor
   inverted coaxial magnetron
ICMA
   independent cable makers' association
ICME
   international conference on medical electronics
ICMP
   interchannel master pulse
ICMST
   international conference on machine searching and translation
ICN
   idle channel noise
   indicator coupling network
   instrumentation and calibration network
ICNF
   irredundant conjunctive normal formula
ICNI
   integrated communications, navigation, and identification
ICO
   installation and checkout
   instrumentation control officer
   inter agency committee on oceanography
   international commission for optics
ICOGRADA
   international conference of graphic design association
ICOLD
   international commission on large dams
ICON
   integrated control
ICONS
   information center on nuclear standards
ICOPAMP
   integrated circuit operational amplifier
ICOR
   intergovernmental conference on oceanic research
ICOS
   integrated cost operation system
ICP
   instrument calibration procedure
   integral circuit package
   intercontinental candle power
   interim compliance panel
   international candle power
   international computer program
   international control plan
   inventory control point

ICPAC
   instantaneous compressor performance analysis computer
ICPC
   interrange communication planning committee
ICPL
   initial control program load
ICPS
   industrial and commercial power systems
   international conference on the properties of steam
   international congress of photographic science
ICR
   input and compare register
   instrumentation control rack
   intelligent character recognition
   interface control register
   international collision regulations
   interrupt control register
   ironcore reactor
ICRA
   interagency committee on radiological assistance
ICRH
   institute for computer research in the humanities
ICRP
   international commission on radiological protection
ICRU
   international commission on radiological units
   international commission on radiological units and measurements
ICS
   information computer systems
   inland computer service
   institute of computer science
   instrumentation checkout station
   integrated communication system
   integrated control system
   interactive counting system
   interagency communications system
   intercarrier sound
   intercommunication system
   internal countermeasures set
   international code of signals
   international computer system
   international connecting set
   interphone control station
   inventory control system
   isolation containment spray

ICSA
  in core shim assembly
ICSAL
  integrated communications system, alaska
ICSC
  interim communications satellite committee
  international communication satellite committee
ICSE
  intermediate current stability experiment
ICSI
  international conference on scientific information
ICSL
  international computer services, limited
  international computing services ltd
ICSR
  interamerican committee for space research
ICST
  institute for computer sciences and technology
ICSU
  international council of scientific unions
ICT
  igniter circuit test
  igniter circuit tester
  incoming truck
  incoming trunk
  inspection control test
  institute of computer technology
  insulating core transformer
  integrated computer telemetry
  interface control tool
  interference compliance test
  internal compool table
  international circuit technology
  international computers and tabulators
ICTASD
  international convention on transistors and associated semiconductor devices
ICTD
  interchannel time displacement
ICTL
  industrial control
ICU
  industrial control unit
  instruction control unit
  instructor's computer utility
  intensive care unit
  interrupt control unit
ICUPLANT
  instructor's computer utility programming language for interactive teaching
ICV
  internal correction voltage

ICW
  interblock communication word
  interrupted continuous wave
  interrupted continuous wave telegraphy
ICWG
  interface control working group
ICWGA
  interface control working group action
ICWM
  international comittee on weights and measures
ICWT
  interrupted continuous wave telegraphy
ICX
  international computer exchange
ICY
  instruction cycle
ID
  idaho
  idem
  identification
  identification data
  identifier
  indicating device
  indicator
  indicator driver
  inductance
  industrial diamond
  information distributor
  inner diameter
  input display
  input division
  inside diameter
  inside dimension
  instruction data
  integral derivative
  intelligence department
  interacitve debugging
  interconnection diagram
  interdigital
  interferometer and doppler
  interior department
  intermediate description
  intermittent duty
  intermodulation distortion
  internal diameter
  item description
  item documentation
IDA
  idaho power company
  industrial development authority
  input data assembler
  institute for defense analysis
  integrated debugging aid
  integrated digital avionics

integrated disc adapter
integro-differential analyzer
interactive debugging aid
intrusion detection alarm
ionospheric dispersion analysis
iterative differential analyzer
IDAC
industrial data acquisition control
IDACON
iterative differential analyzer control
IDAP
iterative differential analyzer pinboard
IDAS
industrial data acquisition system
information displays automatic drafting system
iterative differential analyzer slave
IDAST
interpolated data and speech transmission
IDBPF
interdigital band pass filter
IDC
image dissector camera
information and direction center
instaneous derivation control
instantantaneous deviation control
instrument development company
insulation displacement connector
interceptor distance computer
international data corporation
international diode corporation
IDCC
international data coordinating center
IDCCC
interim data communication collection center
IDCMA
independent data communication manufacturers association
IDCN
interchangeability document change notice
IDCS
image dissector camera system
IDCSP
initial defense communication satellite program
initial defense communications satellite project
IDCSS
initial defense communications satellite system
intermediate defense communications satellite system
IDDD
international direct distance dialing

IDDS
instrumentation data distribution system
international digital data service
IDEA
identify, develop, expose, and action
international directory of engineering analysts
IDEALS
ideal design of effective and logical systems
IDEAS
inquiry data entry access system
integrated design and engineering automated system
IDEEA
information and data exchange experimental activities
IDEF
integrated system definition language
IDENT
identification
IDEP
interagency data exchange program
interservice data exchange program
IDES
information and data exchange system
IDEX
initial defense communication satellite programme experiment
initial defense experiment
IDF
identifier
indicating direction finder
instructional dialogue facility
integrated data file
intermediate distributing frame
intermediate distribution frame
internal distribution frame
international distress frequency
IDFR
identified friendly
IDFT
inverse discrete fourier transform
IDG
inspector of degaussing
IDGIT
integrated data generation implementation technique
IDH
international data highways ltd
IDHS
intelligence data handling system
IDI
improved data interchange
IDIOT
instrumentation digital on line transcriber

IDL
  idle
  instrument development laboratories
  interdiscrepancy laboratory
  international data line
  international date line
IDLT
  identification light
IDM
  instant dimmer memory
  integrated delta modulation
  integrating delta modulation
  interdiction mission
IDMS
  integrated data base management system
  integrated data management system
  isotope dilution mass spectrometry
IDNF
  irredundant disjunctive normal formula
IDO
  idaho operations office
  internal distribution only
IDP
  industrial data processing
  input data processor
  institute of data processing
  integrated data presentation
  integrated data processing
  intermodulation distortion percentage
  internal data processing
  international driving permit
IDPC
  integrated data processing center
IDPG
  impact data pulse generator
IDPI
  international data processing institute
IDQA
  individual documental quality assurance
  individual documented quality assurance
IDR
  industrial data reduction
  infinite duration impulse
  inspection discrepancy report
  intercept during reentry
IDRV
  ionic drive
IDS
  identification section
  input data strobe
  instrument development section
  integrated data storage
  integrated data store
  integrated data system
  interim decay storage
  international data systems
  international dealer systems
IDSC
  international distributed systems center
IDSCP
  interim defense communication program
IDSM
  inertial damped servomotor
IDSP
  initial defence communication satellite program
IDSS
  icam decision support system
IDSTO
  idle used for storage
IDT
  interdigital transducer
  isodensitracer
IDTS
  instrumentation data transmission system
IDTV
  improves definition television
IDU
  indicator drive unit
  industrial development unit
IDUR
  intercept during unpowered rise
IDVID
  immersed deflection vidicon device
IDVM
  integrating digital voltmeter
IE
  id est
  index error
  indicator equipment
  industrial engineer
  information and education
  infrared emission
  initial equipment
  insert extract
  inspection and enforcement
  institute of electronics
  institute of engineers
  instruction execution
  internal environment
  interrogation entry register
IEA
  instruments, electronics and automation
  international electrical association
  international energy agency
  international instruments, electronics and automation exhibition
IEAC
  ieee automatic control

IEAR
  international electronic and aerospace report
IEAUST
  institution of engineers, australia
IEB
  institution of engineers, bangladesh
IEBM
  ieee biomedical engineering
IEC
  industrial electrification council
  infused emitter coupling
  integrated electronic component
  integrated environmental control
  integrated equipment component
  intermittent electrical contact
  international electric company
  international electronic corporation
  international electronical commission
  international electrotechnical commission
  international equipment company
  intrinsic electron conduction
  ion exchange conference
IECEJ
  institute of electrical communication engineers of japan
IECI
  industrial electronics and control instrumentation
IECPS
  international electronic circuit packaging symposium
  international electronics packaging symposium
IECREC
  iec recommendation
IECT
  ieee circuit theory
IED
  individual effective dose
  initial effective data
  institution of engineering designers
  integrated environmental design
IEDU
  ieee education
IEE
  induced electron emission
  industrial electronic engineers
  institute of electrical engineers
  institution of electrical engineers
IEEC
  ieee electronic computers
  industrial electrical equipment council
IEED
  ieee electronic devices
IEEE
  institute of electrical and electronics engineers
  institution of electrical and electronical engineers
  international electrical and electronics exhibition
IEEETEC
  ieee transactions on electronic computer
IEEI
  ieee electrical insulation
  international electronics engineering, incorporated
IEEM
  ieee engineering management
IEETE
  institution of electrical and electronics technician engineers
IEG
  information exchange group
IEGE
  ieee geoscience electronics
IEI
  international electronic industries
  institution of engineers of ireland
IEIM
  ieee instrumentation and measurement
IEIT
  ieee information theory
IEL
  information exchange list
  iowa electric light and power company
IEM
  interim examination and maintenance
  ion exchange membrane
IEMC
  ieee electromagnetic compatibility
  industrial equipment manufacturers council
  international electronics manufacturing company
IEMCELL
  interim examination and maintennace cell
IEME
  inspectorate of electrical and mechanical engineering
IEMTF
  interim examination and maintennace training facility
IEOS
  integrated electronical office system
IEP
  instrument for evaluation of photographs
IEPG
  independent european program group

IEPR
  ieee proceedings
IEQE
  ieee quantum electronics
IER
  industrial equipment reserve
  institutes for environmental research
IERC
  industrial equipment reserve committee
  international electronic research corporation
IERE
  institution of electronic and radio engineers
IERI
  illuminating engineering research institute
IES
  illuminating engineering society
  illuminating engineers society
  institute of environmental sciences
  integral error squared
  intrinsic electric strength
  iowas engineering society
IESA
  indiana electronic service association
IESD
  instrumentation and electronic systems division
IESL
  institution of engineers, sri lanka
IESP
  ieee spectrum
IESU
  ieee sonics and ultrasonics
IET
  initial engine test
IETF
  initial engine test facility
  initial engine test firing
IEU
  instruction execution unit
IEV
  international electrotechnical vocabulary
IEX
  instruction execution
IF
  image frequency
  image to frame ratio
  information collector
  infrared
  instruction field
  instrument flight
  interface
  intermediate frequency
  inverter and cathode follower

IFA
  integrated file adapter
  intermediate frequency amplifier
IFAC
  inter firm accounting project
  international federation for automatic control
  international federation of automatic control
IFALPA
  international federation of airline pilots' associations
IFAN
  international federation for the application of standards
IFATCA
  international federation of air traffic control associations
  international federation of air traffic controllers association
IFB
  interrupted feedback line
  invitation for bid
IFC
  in flight calibrator
  infantaneous frequency correlation
  instantaneous frequency correlation
  integrated fire control
  interface clear
  international finance corporation
  international formulation committee
IFCF
  integrated fuel cycle facilities
IFCN
  interfacility communication network
IFCR
  interface control register
IFCS
  in flight checkout system
  international federation of computer sciences
IFCU
  interface control unit
IFD
  incipient fire detector
  instantaneous frequency discriminator
  international federation for documentation
IFDC
  integrated facilities design criteria
IFDR
  interface data register
IFE
  internal field emission
IFEMS
  international congress of the federation of electron microscope societies
  international federation of electron microscopes societies

IFES
  international field engineering service
IFETCH
  instruction fetch
IFF
  identification friend or foe
  if and only if
  ionized flow field
IFFT
  inverse fast fourier transform
IFI
  industrial fasteners institute
IFIP
  international federation for information processing
  international federation of information processing
IFIPC
  international federation of information processing congress
IFIPS
  international federation of information processing societies
IFIS
  infrared flights inspection system
  integrated flight instrument system
  integrated flight instrumentation system
IFL
  induction field locator
  institute of fluorescent lighting
  integrated fuse logic
IFLA
  international federation of library associations
IFLC
  international frequency list committee
IFLOT
  intermediate focal length optical tracker
IFLTT
  intermediate focal length tracking telescope
IFMBE
  international federation for medical and biological engineering
IFME
  international federation for medical electronics
IFN
  information
IFO
  identifiable flying object
IFOR
  interactive fortran
IFORS
  international federation of operational research science
  international federation of operational research societies
  international federation of operations research societies
IFP
  imperial and foreign post
  international fixed public
IFPI
  international federation of the phonographic industry
IFPM
  in flight performance monitoring
IFPP
  imperial and foreign parcel post
IFPS
  in flight performance signal
  interactive financial planning system
IFR
  increasing failure rate
  inflight refueling
  infrared
  instantaneous frequency measuring receiver
  instantaneous frequency receivers
  instrument flight rules
  intermediate frequency range
  internal function register
  international federation of radio telegraphists
  interrupt flag register
IFRB
  international frequency registration board
IFRU
  interference frequency rejection unit
IFS
  identification friend of foe switching circuit
  integrated flight system
  intermediate frequencies
  intermediate frequency strip
  international federation of surveyors
  interrelated flow simulation
  ionospheric forward scatter
IFSEM
  international federation of societies for electron microscopy
IFSS
  international flight service station
IFT
  institute of food technologists
  intermediate frequency transformation
  intermediate frequency transformer
  international foundation for telemetering
  international frequency table
IFTC
  international federation of thermalism and climatism
  international film and television council

IFTM
   in flight test and maintenance
IFTS
   in flight test system
IFTSSS
   in flight test system scan select
IFVME
   inspectorate of fighting vehicles and mechanical equipment
IG
   imperial gallon
   impulse generator
   inert gas welding
   inertial guidance
   instantaneous grid
IGA
   intergranular attack
IGAAS
   integrated ground airborne avionics system
IGAUP
   interceptor generation and umpiring program
IGC
   institute for graphic communication
   internal gain control
   international geophysical committee
IGCS
   integrated guidance and control system
IGE
   institution of gas engineers
   instrumentation ground equipment
   international general electric
IGEC
   international general electric company
IGEOSA
   international general electric operations sa
IGESUCO
   infrastruction ground environment subcommittee
IGFET
   insulated gate fet
   insulated gate field effect transistor
   isolated gate fet
   isolated gate field effect transistor
IGFM
   internal gamma flux monitor
IGGS
   interactive geo facilities graphic support
IGI
   inner grid injection
IGIA
   interagency group on international aviation
IGJ
   international geophysical year
IGL
   index guided laser

IGMF
   inertial guidance maintenance facility
IGMOSFET
   insulated gate metal oxide semiconductor fet
IGN
   ignited
   ignition
IGNDET
   ignition detector
IGO
   impulse governed
   impulse governed oscillator
   intergovernmental organization
IGOR
   instrument ground based optical recording
   intercept ground optical recorder
IGORITT
   intercept ground optical recorder tracking telescope
IGP
   inertial guidance platform
IGPP
   institute of geophysics and planetary physics
IGRAP
   inert gas receiving and processing
IGS
   improved gray scale
   inertial guidance system
   integrated graphics system
   interactive graphics system
IGSCC
   inter granular stress corrosion crack
IGSS
   inertial guidance system simulator
IGT
   insulated gate tetrode
IGTC
   inertial guidance test center
IGTM
   interial guided tactical missile
IGU
   international gas union
IGV
   inlet guide vane
IGY
   international geophysical year
IH
   indirect heating
   indirectly heated
   interrupt handler
IHACE
   international heating and air conditioning exposition

IHAS
  integrated helicopter avionics system
  international high altitude station
IHB
  international hydrographic bureau
IHD
  international hydrological decade
IHE
  institution of highway engineers
IHF
  independent high frequency
  inhibit halt flip flop
  institute of high fidelity
IHFBC
  international high frequency broadcasting conference
IHFM
  institute of high fidelity manufacturers
IHP
  indicated horsepower
  inner helmholtz plane
IHPH
  indicated horsepower hour
IHPHR
  indicated horsepower hour
IHS
  infrared homing system
IHSBR
  improved high speed bombing radar
IHTS
  intermediate heat transport system
IHX
  intermediate heat exchanger
IHXGV
  intermediate heat exchanger guard vessel
II
  image intensifier
  infrared industries
  instruction and inspection
  inventory and inspection
IIA
  information industry association
  invert indicator from accumulator
IIAS
  international institute of administrative services
IIBTT
  ion implanted base transistor technology
IIC
  integrated interface circuit
  international institute of communications
IID
  indenpendent identically distributed
IIF
  infinite duration impulse response

IIGA
  ieee industry and general applications
IIIL
  isoplanar integrated injection logic
IIL
  integrated injection logic
  invert indicators of the left half
IILS
  international institute for labor studies
IINSE
  international institute of nuclear science and engineering
IIOP
  integrated input output processor
IIPBM
  index of individually planned bills of material
IIR
  infinite duration impulse response
  infinite impulse response
  integrated instrumentation radar
  intermediate infrared
  international institute of refrigeration
  inventory and inspection report
  invert indicators of right half
IIRFILTER
  infinite impulse response filter
IIRS
  institute for industrial research and standards
IIS
  institute of industrial supervisors
  institute of information scientists
  integrated instrument system
  interactive instructional system
  invert indicators form storage
IISL
  international institute of space law
IISO
  institution of industrial safety officers
IIT
  illinois institute of technology
IITF
  in core instrument test facility
IITR
  illinois institute of technology research
IITRAN
  illinois institute of technology translator
IITRI
  illinois institute of technology research institute
  international telegraph and telephone research institute
IIW
  international institute of welding
IJ
  internal junctor

IJAJ
  intentional jitter anti jaming
  international jitter anti jam
IJCAI
  international joint conference on artificial intelligence
  international joint council on artificial intelligence
IJCS
  integrated joint communication system
IJJU
  intentional jitter anti jam unit
  international jitter jammer unit
  international jitter jamming unit
IJP
  internal job processing
IKBS
  intelligent knowledge based system
IKEC
  indiana kentucky electric corporation
IKOR
  immediate knowledge of results
IL
  idle
  illinium
  illium
  illustration
  indicating lamp
  indicating light
  individual line
  inertial laboratory
  installation and logistics
  instrumental landing
  instrumentation laboratory
  insulation level
  insulator
  intensity level
  intermediate language
ILA
  instrument low approach
  instrument landing approach
  international language for aviation
ILAAS
  integrated light attack avionic system
  integrated light avionics attack system
ILAAT
  interlaboratory air to air missle technology
ILABC
  inter laboratory committee
ILAR
  institute of laboratory animal resources
ILAS
  instrument landing approach system
  instrument low approach system
  interrelated logic accumulating scanner

ILB
  independent lateral band
  inner lead bonding
ILC
  input language converter
  instruction length code
  instruction length counter
ILCC
  integrated launch control and checkout
ILCCS
  integrated launch control and checkout system
ILCF
  inter laboratory committee on facilities
ILCO
  instantaneous launch control officer
ILD
  injection laser diode
  instructional logic diagram
  intermediate level diagram
ILE
  integral linear error
  interface latch element
ILEED
  inelastic low energy electron diffraction
ILF
  inductive loss factor
  infralow frequency
ILG
  instrument landing guidance
ILIP
  in line instrument package
ILIR
  in house laboratories independent research
ILIXCO
  international liquid xtal corp
ILJM
  illinois journal of mathematics
ILL
  illumination
  illuminator
  intensity of illumination
  international lunar laboratory
ILLCS
  intralaunch facility and launch control facility cabling subsystem
ILLIAC
  illinois integrator and automatic computer
ILLIAD
  illinois algorithmic decoder
ILLUM
  illuminant
  illumination
ILLUMENGR
  illumination engineer

ILM
  independent landing monitor
  information logic machine
ILMA
  incandescent lamp manufacturers association
ILN
  internal line number
ILO
  injection locked oscillator
  international labor office
  international labor organization
ILP
  intermediate language processor
ILPF
  ideal low pass filter
ILRT
  integrated leak rate test
  intermediate level reactor test
ILS
  ideal liquidus structures
  instrument landing system
  integrated logistic support
  international lunar society
ILSAP
  instrument landing system approach
ILSE
  interagency life sciences supporting space research and technology exchange
ILSF
  intermediate level sample flow
ILSRO
  interstate land sales registration office
ILSTAC
  instrument landing system and tactical air navigation
ILSW
  interrupt level status word
ILTS
  institute of low temperature science
ILU
  illinois university
ILW
  intermediate level wastes
IM
  ideal modulation
  imaginary
  impulse modulation
  industry motion picture
  inner marker
  inspection manual
  installation and maintenance
  institute of metrology
  instrument measurement
  instrumentation
  instrumentation and measurement
  intensity measuring
  intensity modulation
  interactive mode
  interceptor missile
  intermediate missile
  intermediate modulation
  intermodulation
IMA
  industrial medical association
  international mineralogical association
  invalid memory address
  ion microspectroscope analysis
IMACS
  international association for mathematics and computers in simulation
IMAG
  ieee magnetics
  imaginary
IMAGE
  information management by application generation
IMAPCT
  inventory management program and control techniques
IMARE
  institute of marine engineers
IMAS
  industrial management assistance survey
IMC
  image motion compensation
  initial moisture content
  institute of measurement and control
  instrument meteorological conditions
  integrated maintenance concept
  integrated microelectronic circuitry
  intensity millicurie
  international micrographic congress
  international microwave corporation
IMCC
  integrated mission control center
IMCD
  input marginal checking and distribution
IMCI
  intensity millicurie
IMCO
  intergovernmental maritime consultative organization
  international metered communications
IMCPM
  improved capability missile
IMCU
  intensity millicurie
IMCV
  input media conversion

IMD
   interactive map definition
   intercept monitoring display
   intermodulation distortion
IMDC
   interceptor missile direction center
IME
   indiana and michigan electric
   institute of makers of explosives
   institute of mining engineers
   institution of mechanical engineers
IMEA
   incorporated municipal electrical association
IMEAPWA
   institute for municipal engineering american public works association
IMECHE
   institute of mechanical engineers
IMEP
   indicated mean effective pressure
IMF
   image matched filter
   interactive mainframe facility
   internal magnetic focus
   internal magnetic focus tube
   international monetary fund
IMFI
   industrial mineral fiber institute
IMFSS
   integrated missile flight safety system
IMG
   improved measurement group
IMGCN
   interceptor missile ground control network
IMH
   institute of materials handling
IMHEP
   ideal man helicopter engineering project
IMI
   improved manned interceptor
   information management inc
   innoviative management inc
   intermediate manned interceptor
   irish management institute
IMIR
   interceptor missile interrogation radar
IMIS
   integrated management and information system
   integrated management information system
   integrated municipal information system
   interim maneuver identification system
IMITAC
   image input to automatic computers

IMK
   identification mark
IMKR
   inner marker
IML
   initial microprogram load
IMM
   institute of mathematics machines
   institute of mining and metallurgy
   integrated magnetic memory
   integrated maintenance management
IMMAC
   inventory management and material control
IMMITTANCE
   impedance and admittance
IMMS
   international material management society
IMMSNJC
   international material management society, new jersey chapter
IMOS
   interactive multiprogramming operating system
   ion implant metal oxide semiconductor
   ion implanted metal oxide semiconductor
IMP
   impedance
   impression
   impulse
   inflatable micrometeoroid paraglider
   injection into microwave plasma
   integrated manufacturing planning
   integrated microprocessors
   integrated microwave package
   integrated microwave products
   interface message processor
   inter industry management program
   intermessage processor
   interplanetary monitoring platform
   interplanetary monitoring probe
   intrinsic multiprocessing
IMPA
   initialized moore probabilistic automation
IMPAC
   industrial multilevel process analysis and control
IMPACT
   implementation, planning and control technique
   integrated management planning and control technique
   integrated managerial programming analysis control technique
   inventory management program and control technique

IMPATT
  impact avalanche and transit time
  impact avalanche transit time
IMPATTDIODE
  impact avalanche and transit time diode
IMPE
  impregnate
IMPGEN
  impulse generator
IMPI
  international microwave power institute
IMPL
  initial microprogram load
  initial microprogram loading
IMPRESS
  interdisciplinary machine processing for research and education in the social sciences
IMPS
  integrated master programming and scheduling
  interface message processors
  international meter pulse sender
  interplanetary measurement probes
  interplanetary monitoring probes
IMPTS
  improved programmer test station
IMR
  institute for materials research
  interrupt mask register
  isolation mode rejection
IMRA
  infrared monochromatic radiation
  international marine radio association
IMRADS
  information management retrieval and dissemination system
IMRAMN
  international meeting on radio aids to marine navigation
IMRAN
  international marine radio aids to navigation
IMRC
  international marine radio company
IMRR
  isolation mode rejection ratio
IMS
  image motion simulator
  industrial management society
  industrial mathematics society
  information management system
  institute of management sciences
  institute of marine science
  institute of mathematical statistics
  instructional management system
  integrated meteorological system
  interceptor missile
  international management system corp
  international metallographic society
  interplanetary measurement satellite
  interplanetary monitor satellite
  inventory management and simulator
IMSA
  international municipal signal association
IMSL
  international mathematical and statistical libraries
IMSSS
  interceptor missile squadron supervisors' station
IMT
  impulse modulated telemetering
  impulse modulated telemetry
IMTS
  improved mobile telephone service
IMU
  inertial measurement unit
  inertial measuring unit
  interference mock up
  international mailers union
  international mathematical union
IN
  inch
  indiana
  indium
  inertial navigation
  input
  institute of navigation
  insulate
  insulating
  insulator
  interference to noise ratio
  item number
INA
  international normal atmosphere
INACS
  interstate airways communiations station
INACTV
  inactivation
  inactive
INAS
  inertial navigation and attack systems
INC
  inclosure
  included
  increased
INCA
  integrated communications agency
  integrated navigation and communication, automatic

INCAND
incandescent
INCE
institute of noise control engineering
INCH
independent channel handler
integrated chopper
INCL
inclosure
included
inclusive
incoming line
INCO
information and control
international chamber of commerce
INCOH
incoherent
INCOM
indicator compiler
input compiler
INCOMEX
international computer exhibition
INCOPAC
international consumer policy advisory committee
INCOR
intergovernmental conference on oceanographic research
israeli national committee for oceanographic research
INCOSPAR
indian national committee for space research
indian national committee space research
INCR
increase
increment
interrupt control register
INCUM
indiana computer users meeting
IND
independent
independent run
index
indicative
indicator
inductance
induction
industrial
industry
INDEP
independent
INDEX
inter nasa data exchange
INDN
indication

INDR
indicator
INDREG
inductance regulator
induction regulator
INDS
incore nuclear detection system
INDTR
indicator transmitter
INDUC
inductance
induction
INEA
international electronics association
INEL
idaho nuclear engineering laboratory
international exhibition of industrial electronics
INEWF
indian national electricity workers' federation
INF
infinity
infra
irredundant normal form
irredundant normal formula
INFANT
iroquois night fighter and night tracker
INFCO
information committee
INFO
information
information network and file organization
information system
INFOCEN
information center
INFOL
information oriented language
INFOR
information
interactive fortran
INFORMS
information organization reporting and management system
information, organization, and management system
INFRAL
information retrieval automatic language
ING
inertial navigation and guidance
inertial navigation gyro
intense neutron generator
INGO
international nongovernmental organization
INIC
current negative immittance converter

current negative impedance converter
inverse negative impedance converter
INIS
international nuclear information system
INJ
injection
INL
internal noise level
INLC
initial launch capability
INLR
item no longer required
INMARSAT
international martitime satellite system
INMM
institute of nuclear materials management
INN
intermediate network node
INOP
inoperative
INP
indium phosphide
inert nitrogen protection
INPC
impulse noise performance curve
INPH
interphone
INPO
institute of nuclear power operations
INPT
input
INQ
inquire
INR
interference to noise ratio
INREQ
information on request
INS
inches
inertial navigation sensor
inertial navigation system
institute for naval studies
institute for nuclear study
insulated
insulation
insulator
internation simulation
interstation noise suppression
ion neutralization spectroscopy
INSAC
interstate airways communications
INSACS
interstate airways communications station
interstate airways communications system

INSATRAC
interception by satellite tracking
INSCAIRS
instrumentation calibration incident repair service
INSMETLS
international shielding metals
INSOL
insoluble
INSPEC
information service in physics, electrotechnology and control
INSPECT
integrated nationwide system for processing entries from customs terminals
INSRADMAT
inspector of radio material
INSSCC
interim national space surveillance control center
INST
instant
instantaneous
institute
instruction
instrument
instrumentation
INSTA
instruments authorized
INSTAR
inertialess scanning, tracking and ranging
INSTARS
information storage and retrieval system
INSTCTL
instrumentation control
INSTEE
institution of electrical engineers
INSTLN
installation
INSTR
instructor
INSTRE
institute of radio engineers
INSTREF
instrument reference
INT
initial
integral
integration
integrator
interior
intermediate
internal
interogative
interphone

interpreter
interrupt
interrupter
interruption
intersection
interval
INTAMP
   intermediate amplifier
INTC
   industrial nuclear technology conference
INTCO
   international code of signals
INTCOM
   international liaison committee
INTCON
   internal connection
INTCP
   intercept
INTEC
   intercontinental electronics corporation
INTEG
   integration
INTEL
   intelligence
INTELSA
   international telecommunications satellite consortium
INTELSAT
   international telecommunications satellite
   international telecommunications satellite consortium
INTER
   intermediate
INTERCOM
   intercommunication
   intercommunications monitor
INTERCON
   international convention
INTERDICT
   interference detection and interdiction countermeasures team
INTERDOC
   integrated terminology document management system
INTERF
   interference
INTERGALVA
   international galvanizing conference
INTERMAG
   international conference on magnetics
   international magnetics conference
INTERMEC
   interface mechanism
INTERVISION
   international television

INTF
   internal frosted
INTFC
   interference
INTG
   interrogator
INTGR
   integrate
   integrating
   integrator
INTIPS
   integrated information processing system
INTLK
   interlock
INTMED
   intermediate
INTMT
   intermittent
INTOP
   international operations simulation
INTPHTR
   interphase transformer
INTRAN
   input translator
INTREPT
   intelligence report
INTREX
   information transfer experiment
INTRPT
   interruption
INTSUM
   intelligence summary
INTUG
   international communications user group
INU
   inertial navigation unit
INV
   invalid
   inverse
   inverted
   inverter
INVAL
   invalid
INVCURR
   inverse current
INVTR
   inverter
INVV
   inverse voltage
IO
   image orthicon
   input output
   input output devices
   international octal
   international octal base

interpretive operation
iterative operation
IOA
    input output adapter
IOAU
    input output access unit
    input output arithmetic unit
IOB
    input output block
    input output buffer
IOC
    in out converter
    initial operational capability
    input output channel
    input output control
    input output controller
    input output converter
    integrated optic[al] circuit
    intergovernmental oceanographic commission
    iterative orbit calculator
IOCC
    input output control center
    input output control command
IOCD
    input output under count control and disconnect
IOCP
    input output control program
    input output under count control and proceed
IOCR
    input output control routine
IOCS
    input output computer services
    input output control service
    input output control system
IOCU
    international organization of consumers unions
IOD
    industrial operations division
    input output dump program
IODC
    input output delay counter
IODD
    ideal one dimensional device
IOE
    institution of electronics
    intake opposite exhaust
IOF
    international oceanographic foundation
IOGE
    integrated operational ground equipment
IOL
    instantaneous overload

IOM
    input output multiplexer
IOMP
    input output message processor
ION
    ionosphere and aural phenomena advisory committee
IONO
    ionosphere
IOP
    input output processor
    input output pulse
IOPKG
    input output package
IOPL
    intermittent operating life
IOPN
    in operation
IOPS
    input output programming system
IOR
    input output register
IOREG
    input output register
IORP
    input output of a record and proceed
IORT
    input output of a record and transfer
IORU
    international commission on radiological unit and measurements
IOS
    input output selector
    input output skip
    input output supervisor
    input output system
    instrumentation operations station
    interceptor operator simulator
    international organization for standardization
IOSP
    input output under signal and proceed
IOST
    input output under signal and transfer
IOT
    initial orbit time
    input output test
    input output transfer
    interoffice trunk
IOTA
    information overload testing aid
    information overload testing apparatus
    instant oxide thickness analyzer
IOTAE
    initial operation test and evaluation

IOTESTPROGRAM
   input output test program
IOU
   immediate operational use
   input output unit
   input output utilizer routine
IOUBC
   institute of oceanography, university of british columbia
IOVST
   international organization for vacuum science and technology
IP
   identification of position
   ignition point
   index of performance
   induced polarization
   induced polarization method
   industrial production
   information pool
   information processing
   inhouse publishing
   initial phase
   initial point
   inland postage
   input
   institute of petroleum
   instruction pulse
   instrument panel
   instrumentation plan
   interactive processing
   interelement protection
   interface process
   interface programmable
   intermediate pressure
   international programming
   isoelectric point
   italian patent
   item processing
IPA
   image power amplifier
   information processing architecture
   information processing association
   integrated photodetection assemblies
   intermediate power amplifier
   international patent agreement
   international phonetic alphabet
   international psychoanalytical association
   isopropyl alcohol
IPADAE
   integrated passive action detection acquisition equipment
IPAM
   inter partition access method

IPAS
   ieee power apparatus and systems
IPB
   illustrated parts book
   illustrated parts breakdown
   interconnection and programming bay
IPBM
   interplanetary ballistic missile
IPC
   illinois power company
   illustrated parts catalogue
   industrial process control
   industrial progammable controller
   information processing center
   information processing code
   institute of printed circuits
   integrated peripheral channel
   integrated process control
   intermittent positive control
   interplanetary communications
   interprocess communication
   interprocess communication module
   interprocessor channel
IPCCS
   information processing in command and control systems
IPCEA
   insulated power cable engineers association
IPCR
   institute of physical and chemical research
IPCS
   interproject control station
IPD
   insertion phase delay
IPDD
   initial project design description
IPDH
   in service planned derated hours
IPDP
   intervals of pulsations of diminishing period
IPDS
   intelligent printer data stream
IPDTAS
   interim point defense target acquisition system
IPE
   industrial production equipment
   information processing equipment
   initial portable equipment
   institute of power engineers
   institution of production engineers
IPET
   independent professional electronic technicians

IPF
   inches per foot
IPFM
   integral pulse frequency modulation
IPG
   internal problem generator
IPH
   inches per hour
IPI
   identified friendly prior to interception
   impregnated paper insulated
IPL
   identified parts list
   indianapolis power and light
   information processing language
   information processing letters
   initial program load
   initial program loader
   initial program loading
   instrument pool laboratory
IPLV
   information processing language v
IPM
   impulse per minute
   inch per minute
   inches penetration per month
   incidental phase modulation
   information processing machine
   institute for practical mathematics
   interference prediction model
   internal polarization modulation
   interruption per minute
IPMP
   ieee parts, materials and packaging
IPO
   installation planning order
IPOEE
   institute of post office electrical engineers
IPOT
   inductive pontential divider
   inductive potentiometer
IPP
   imaging photopolarimeter
   integrated plotting package
IPPJ
   institute of plasma physics japan
IPPS
   institute of physics and the physical society
IPR
   inches per revolution
   inspection planning and reliability
IPRE
   incorporated practical radio engineers
   institute of practical radio engineers

IPRL
   interceptor pilot research laboratory
IPRO
   international patent research office
IPROP
   ionic propulsion
IPS
   inch per second
   information processing system
   installation performance specification
   instruction per second
   instrument power supply
   instrument power system
   instrumentation power supply
   instrumentation power system
   integrated project support
   interceptor pilot simulator
   interim policy statement
   intermodulation products
   internal plate screen
   international pipe standard
   interpretative programming system
   interruption per second
   ionospheric prediction service
   iron pipe size
   item processing system
IPSJ
   information processing society of japan
IPSOC
   information processing society of canada
IPSSB
   information processing system standards board
IPST
   international practical scale of temperature
IPT
   improved programming technologies
   inch per tooth
   inplant transporter
   internal pipe thread
   international planning team
   international production technology
   interphase transformer
   interplanetary travel
IPTC
   international press telecommunications committee
   international press telecommunicaton center
IPTERMINAL
   input terminal
IPTP
   inplant test program
IPTS
   international practical temperature scale

IPU
  instruction processing unit
  instruction processor unit
IPW
  interstate power company
IPY
  iches penetration per year
IQ
  instrument quality
IQC
  international quality center
IQEC
  international quantum electronics conference
IQF
  interactive query facility
IQI
  image quality indicator
IQL
  interactive query language
IQRP
  interactive query and report processor
IQS
  interactive query system
IQSU
  international year of the quiet sun
IQSY
  international quiet sun year
IR
  incident report
  index register
  indicator reading
  indicator register
  industrial relations
  industrial robot
  information request
  information retrieval
  infrared
  infrared radiation
  infrared reflectance
  inside radius
  insoluble residue
  inspection record
  inspection request
  inspector's report
  instantaneous relay
  instantaneous release
  institute of radio engineers
  instruction register
  instrument reading
  instrument register
  instrumentation requirement
  insulation resistance
  intelligence and reconnaissance
  interchangeability and replaceability
  intermediate infrared
  intermediate register
  internal register
  internal resistance
  interrogator responder
  interrogator responser
  interrupt register
  interrupt request
  irradiance
IRA
  input reference axis
IRAC
  interdepartment radio advisory council
  interdepartmental radio advisory committee
IRACQ
  infrared acquisition
  instrumentation radar and acquisition
IRAD
  independent research and development
IRAH
  infrared alternate head
IRAM
  integrated random access memory
IRAN
  inspect and repair as necessary
  inspection and repair as necessary
IRAP
  interagency radiological assistance program
IRAR
  impulse response area ratio
IRASA
  international radio air safety association
IRASER
  infrared amplification by stimulated emission of radiation
IRB
  interrupt request block
IRBM
  intermediate range ballistic missile
IRBO
  infrared homing bomb
IRBT
  intelligent remote batch terminal
IRC
  industrial reorganization corporation
  infrared countermeasure
  international record carrier
  international record carriers
  international rectifier corporation
  international reply coupon
  international resistance company
  item responsibility code
IRCA
  international radio club of america
IRCC
  international radio consultative committee

IRCCD
infrared sensitive charge coupled device
IRCCM
infrared counter countermeasure
IRCM
infrared countermeasure
IRCS
inertial reference and control system
intercomplex radio communications system
IRD
international resource development
IRDOME
infrared dome
IRDP
information retrieval data bank
IRDS
integrated reliability data system
IRE
institute of radio engineers
IREC
international rectifier
IRED
infrared emitting diode
IREE
institute of radio and electrics engineers
IREP
integrated reliability evaluation program
interim reliability evaluation program
IREX
international research and exchanges board
IRF
impedance reduction factor
international rectifier corporation
interrogation recurrence frequency
IRFB
international radio frequency board
IRFC
intermediate range function test
IRFNA
inhibited red fuming nitric acid
IRG
inertial rate gyro
international register
interrange instrumentation group
interrecord gap
IRGAR
infrared gas radiation
IRGPG
interrange global planning group
IRHD
international rubber hardness degrees
IRI
industrial research institute
infrared industries, incorporated
institution of the rubber industry
integrated range instrumentation
IRIA
infrared information and analysis
IRIG
inertial rate integrating gyroscope
inertial reference integrating gyroscope
interrange instrumentation group
IRIS
infrared information symposium
infrared interferometer spectrometer
instand response information reconnaissance intelligence system
instant response information system
interactive real time information system
IRL
information retrieval language
IRLS
interrogation, recording of location subsystem
interrogation, recording, and location system
IRM
information resource management
information resources management
infrared measurement
integrated range mission
interactive request modification
intermediate range monitor
iodine radiation monitor
isothermal remanent magnetization
IRMA
information revision and manuscript assembly
infrared miss distance approximator
IRMC
inter services radio measurements committee
IRMP
infrared measurement program
inter services radiation measurement program
IRMS
information retrieval and management system
IRN
interface revision notice
IROAN
inspect and repair only as needed
IROD
instantaneous readout detector
IRON
infrared optical noise
IROS
increased reliability of system
IRP
inertial reference package
initial receiving point
interrupt processor

IRPA
  international radiation protection association
IRPI
  individual rod position indicator
IRPL
  inter services radio propagation laboratory
IRPP
  industrial readiness planning program
IRQ
  interrupt request
IRR
  institute for reactor research
  instrumentation revision record
  integral rocket ramjet
  intelligence radar reporting
  internal rage of return
  irregular
  israeli research reactor
IRRAD
  infrared ranging and detection
  infrared ranging and detection equipment
IRRD
  international road research documentation
IRREG
  irregular
IRRI
  international rece research institute
IRRMP
  infrared radar measurement program
IRRS
  infrared reflection spectroscopy
IRS
  information retrieval system
  input read submodule
  inquiry and reporting system
  inspector of radio services
  integrated records system
  internal revenue service
  irish standard
  isotope removal service
IRSC
  international radium standard commission
IRSCANNER
  infrared scanner
IRSM
  institute of radio service men
IRSP
  infrared spectrometer
IRSS
  instrumentation and range safety system
  integrated range safety system
IRT
  in reactor thimble
  infrared tracker
  initialize reset tape
  institute of reprographic technology
  interrogator responder transducer
  interrogator responder transponder
IRTE
  institute of road transport engineers
IRTF
  intermediate range task force
IRTO
  international radio and television organization
IRTP
  integrated reliability test program
IRTS
  interim recovery technical spec
  international radio and television society
IRTWG
  interrange telemetry working group
IRU
  inertial reference unit
IRVB
  india rubber vulcanized, braided
IRX
  interactive resource executive
IS
  ignition and separation
  impact switch
  incomplete sequence
  index sequential
  indian standard
  information science
  information separator
  information service
  information system
  infrared spectroscopy
  infrasonic
  installation of systems
  instruction sheet
  instrument
  instrumentation summary
  instrumentation system
  insulating sleeve
  integrated satellite
  intelligence service
  interference suppressor
  internal shield
  international standard
  interstage shielding
  interval signal
  iowa state university of science and technology
  irish standard
ISA
  ignition and separation assembly
  instrument society of america
  instrument sub assembly

interactive survey analysis
international federation of the national standardizing associations
international society of acupuncture
international standard atmosphere
international standardization association
ion scattering analysis
ISAD
integrate sample and dump
ISADAC
interim standard airborne digital computer
ISADS
innovative strategic aircraft design studies
ISAL
information system access line
ISALIS
indian society for automation and information sciences
ISAM
index sequential access method
indexed sequential access method
integrated switching and multiplexing
ISAP
information sort and predict
interactive survey analysis package
ISAR
information storage and retrieval
ISAS
institute of space and aeronautical science
integrated spacecraft avionics system
ISB
independent sideband
intelligence systems branch
international society of biometeorology
ISBB
international society of bioclimatology and biometeorology
ISBD
information services business division
ISBMODULATION
independent sideband modulation
ISBN
international standard book number
ISC
industrial security committee
information society of canada
instruction staticizing control
instrument system corporation
insulated signal coupler
integrated storage control
interactive sciences corp
international space congress
international standards council
interstellar communications
intrasite cabling

ISCA
international standards steering committee for consumer affairs
international symposium on computer architecture
intersociety committee on methods for ambient air sampling and analysis
ISCAN
i scanner
inertialess steerable communication antenna
ISCAS
international symposium on circuits and systems
ISCB
interallied staff communications board
ISCC
inter society committee on corrosion
internal service coordination center
interstate solar coordination council
ISCS
international scientific cooperative service
ISCT
inner seal collar tool
institute of science technology
ISCTC
inter services components technical committee
ISD
ibm standard data
induction system deposit
information services division
informations systems division
interactive screen definition
internal symbol dictionary
internal symbolic dictionary
international subscriber dialing
ISDE
integral square delay error
ISDF
intermediate sodium disposal facility
ISDN
integrated services digital network
ISDOS
information systems design and optimization system
ISDS
integrated ship design system
ISE
illogical sequence error
institution of structural engineers
integral square error
international software enterprise
international standard electric corp
interpret sign error
ion sensitive electrode

ISEA
   industrial safety equipment association
ISEC
   international standard electric corporation
ISEE
   international sun earth explorer
ISEEP
   infrared sensitive element evaluation program
ISEIG
   inter services electronic identification group
ISEP
   international standard equipment practice
   international telegraph and telephone standard equipment practice
ISEPS
   international sun earth physics satellites program
ISER
   integral systems experimental requirements
ISES
   international solar energy society
ISEUNA
   international stereotypers and electrotypers union of north america
ISF
   international science foundation
ISFET
   ion sensitive field effect transistor
ISFMS
   index sequential file management system
   indexed sequential file management system
ISFSI
   independent spent fuel storage installation
ISG
   interconnected systems group
   interservice group
ISHM
   international society for hybrid microelectronics
ISI
   in service inspection
   indian standards institute
   indian standards institution
   information storage
   institute for scientific information
   integral systems incusa
   internal specified index
   internally specified index
   international safety institute
   international statistical institute
   intersymbol interference
   iron and steel institute
   israel standards institute
ISIB
   interservice ionospheric bureau
ISIC
   intersymbol interference corrector
ISIE
   integral square ideal error
ISIM
   inhibit simultaneity
   inventory simulation
ISIR
   international satellite for ionospheric research
ISIRS
   international scorption information retrieval systems
ISIS
   infratest software informations service
   instant sales inidicator system
   integral service information system
   integrate strike and interceptor system
   integrated set of information systems
   integrated ship instrumentation system
   integrated strike and interception system
   internally switched interface system
   international satellite for ionospheric studies
   international science information service
   international scientific information service
   intratest software information service
ISIWM
   incorporated society of inspectors of weights and measures
ISJM
   israeli journal of mathematics
ISK
   insert storage key
ISL
   information search language
   information system language
   initial spare parts list
   initial system load
   injection coupled synchronous logic
   institute of space law
   instructional systems language
   integrated schottky logic
   interactive simulation language
   isolated signal lines
   item study listings
ISLE
   integral square linear error
ISLIC
   israel society for special libraries and information centers
ISLS
   interrogation path sidelobe suppression
ISM
   industrial scientific and medical
   industrial, scientific, medicalus
   information system for management

initial segment membrane
international standards method
interpretive structural modeling
ISMFREQUENCY
industrial scientific medial frequency
ISMI
improved space manned interceptor
ISMIS
improved sage manned intercept system
ISML
intermediate system mock up loop
ISMMP
international standard methods for measuring performances
ISMOD
index sequential module
indexed sequential module
ISN
initial sequence number
internal statement number
international society for neurochemistry
ISO
individual system operation
intermediate station operation
international organization for standardization
international science organization
international standards organization
isometric
isotropic
ISODIS
iso draft international standard
ISOL
isolated
ISOLDE
isotopic low weight device
ISOMITE
isotope miniature thermionic electric
ISONE
international standard of nuclear electronics
ISONET
iso network committee
ISOPAR
improved symbolic optimizing assembly routine
ISOPLANAR
isolation oxide planar
ISOR
iso recommendation
ISOSC
iso subcommittee
ISOTC
iso technical committee
ISP
imperial smelting process
industrial security plan

international society of photogrammetry
italian society of physics
ISPEC
insulation specification
ISPEMA
industrial safety personal equipment manufacturers association
ISPL
initial spare parts list
ISPO
instrumentation ships project office
international statistical programs office
ISQC
indian society for quality control
ISR
information storage and retrieval
input select and reset
institute of semiconductor research
instrumentation status report
interrupt service routine
interrupt status register
intersecting storage ring
intersection stockage ring
ISRAC
international telegraph and telephone secure ranging and communication
ISRB
inter services research bureau
ISRCSC
inter services radio components standardization committee
ISRFCTC
inter services radio frequency cables technical committee
ISRSM
international symposium on rocket and satellite meteorology
ISRSYSTEM
information storage and retrieval system
ISRU
international scientific radio union
ISS
ideal solids structures
inertial subsystem
information storage system
input subsystem
institute of strategic studies
integrated switch stick
interface signal simulator
ion silicon system
ISSAIME
iron and steel society of aime
ISSCC
international solid state circuits conference

ISSE
international sight and sound exposition
ISSEM
information system security evaluation methodology
ISSM
initialized stochastic sequential machine
ISSMB
information systems standards management board
ISSN
international standard serial number
ISSS
installation service supply support
international society of soil science
IST
incompatible simultaneous transfer
incredibly small transistor
indian standard time
information science and technology
input stack tape
integrated switching and transmission
integrated system test
integrated system transformer
integrated systems technology
integrated systems test
international standard thread
interstellar travel
ISTAR
image, storage translation and reproduction
ISTC
international switching and testing center
ISTIM
interchange of science and technical information in machine language
interchange of scientific and technical information in machine language
ISTM
international society for testing materials
ISTMC
instrumentation section test and monitor console
ISTP
information system theory project
ISTS
international symposium on space technology and science
ISU
instruction stream units
interface surveillance unit
interface switching unit
interference suppression unit
international scientific union
ISW
initial status word
internal status word
ion switch
ISWC
international short wave club
ISWG
imperial standard wire gauge
ISX
information switching exchange
IT
identification and traceability
industry telephone maintenance
industry transistor
information theory
input translator
inspection and test
instant transactions
institute of technology
instrument test
instrument transformer
instrument tree
insulating transformer
integral time
intelligent terminal
intensity of telephone interference
interfacial tension
interfering transmitter
internal translator
interrogator transponder
interval timer
irradiation time
isometric transition
isothermal transformation
italic
italic type
item transfer
ITA
independent television authority
industry and trade administration
institute for telecom and aeronomy
international tape association
international typographic association
ITAAP
inspection test and analysis plan
ITAE
integral of time multiplied absolute error
integral of time multiplied absolute value of error
integrated time and absolute error
ITAECRITERION
integral of time multiplied absolute value of error criterion
ITAL
italic
italic type

ITAR
  international traffic in arms regulations
ITAS
  integrated transport accounting system
ITAT
  international telegraph and telephone
ITB
  initial temperature difference
  instantaneous trip block
  international time bureau
ITBE
  interchannel time base error
ITC
  integral tube components
  intelligent telecommunications controller
  intermediate toll center
  international telecommunication convention
  international telemetering conference
  international trade commission
  intertropical convergence
  investment tac credit
  ionic thermoconductivity
ITCC
  international technical communications conference
ITCIS
  integrated telephone customer information system
ITCS
  integrated target command system
  integrated target control system
ITD
  initial temperature difference
  interchannel time displacement
ITDD
  integrated tunnel diode device
ITDE
  interchannel time displacement error
  intertrack time displacement error
ITE
  institute of telecommunications engineers
  institute of traffic engineers
  institute of transportation engineers
  inverse time element
ITELEMENT
  integral time delay element
ITEM
  interference technology engineer's master
  internal thermal environment management
ITEWS
  integrated tactical electronic warfare system
ITF
  integrated test facility
  interactive terminal facility

ITFO
  international trade fairs office
ITFS
  instructional television fixed services
ITFTRIA
  instrument tree flow and temperature removal instrument assembly
ITI
  indian telephone industries, limited
  inspection and test instruction
  inspection test instruction
  interactive terminal interface
  international technology institute
ITIRC
  ibm technical information retrieval center
ITL
  ignition transmission line
  integrate transfer launch
  integrated transfer launch
  integration, test, and launch
  inverse time limit
ITLC
  integrated transfer launch complex
ITM
  inch trim moment
  interceptor tactical missile
  intercommunication teleprocessing monitor
ITMA
  irradiation test management activity
ITN
  independent television news
  interamerican telecommunications network
  international telecomputer network
ITO
  indium tin oxide
ITOR
  intercept target optical reader
ITP
  integrated test program
  international television program
  italian patent
ITPA
  independent telephone pioneer association
ITPS
  integrated teleprocessing system
  internal teleprocessing system
ITR
  incore thermionic reactor
  initial trouble reports
  instrument test rig
  integrated thyristor rectifier
  inverse time relay
  isolation test routine

ITRIA
   instrument tree removable instrument assembly
ITRL
   instrument test repair laboratory
   international telephone and radio laboratories
ITS
   idaho test station
   independent television service
   inertial timing switch
   informatics teaching system
   institute for telecommunication sciences
   instrumentation and telemetry system
   insulation test specification
   integrated termination system
   integrated tracking sysem
   integrated trajectory system
   interim table simulation
   intermarket trading system
   internal time sharing
   international telecommunications service
   international temperature scale
   international time sharing corp
ITSA
   institute for telecommunication sciences and aeronomy
ITSL
   international translator
ITSP
   instrument tree spool piece
ITT
   incoming teletype
   incoming trunk terminal
   international telegraph and telephone
   international telephone and telegraph corporation
   intertoll trunk
ITTA
   indianapolis television technicians' association
ITTC
   international telegraph and telephone corporation
ITTCS
   international teleraph and telephone communications systems
ITTE
   international telegraph and telephone, europe
ITTF
   international telegraph and telephone, federal division
ITTFL
   international telegraph and telephone, federal laboratories
ITTK
   international telegraph and telephone laboratories kellogg switchboard and supply company
ITTL
   international telegraph and telephone laboratories
ITTPMD
   interpretative trace and trap programm plus modifications
ITTWC
   itt world communications, inc
ITU
   international telecommunication union
ITV
   independent television
   industrial television
   instructional television
ITVAC
   industrial transistor value automatic computer
ITVB
   international television broadcasting
ITVSDA
   independent television service dealers' association
IU
   information unit
   input unit
   instrument unit
   instrumentation unit
   interference unit
   international unit
IUA
   inertial unit assembly
   iomec users association
IUAPPA
   international union of air pollution prevention associations
IUB
   international union of biochemistry
IUBS
   international union of biological sciences
IUCAF
   inter union commission on allocation of frequencies
IUCN
   international union for the conservation of nature and natural resources
IUCR
   international union of crystallography
IUDH
   in service unplanned derated hours
IUE
   international union for electroheat

international union of electrical, radio, and machine workers
IUERMW
international union of electrical, radio and machine workers
IUGG
international union of geodesy and geophysics
IUGS
international union of geological sciences
IUNDH
in service unit derated hours
IUP
installed user program
IUPAC
international union of pure and applied chemistry
IUPAP
international union of pure and applied physics
IUS
inertial upper stage
initial upper state
installed user system
IUSR
international union of scientific radio
IUTAM
international union of theoretical and applied mechanics
IUVSTA
international union for vacuum science techniques and applications
IV
current voltage
increased value
independent variable
initial velocity
insurance value
intermediate voltage
interval
inverter
iodine value
IVA
intervehicular activity
IVALA
integrated visual approach and landing aid
IVAM
interorbital vehicle assembly mode
IVC
international video corporation
IVD
image velocity detector
inductive voltage divider
interpolated voice data

invalid decimal
ion vapor deposition
IVDP
initial vector display point
IVDS
independent variable depth sonar
IVHM
in vessel handling machine
IVHMEM
in vessel handling machine engineering model
IVMU
inertial velocity measurement unit
IVP
interface verification procedure
IVR
instrumented visual range
integrated voltage regulator
IVS
in vessel storage
integrated versaplot software
IVSI
inertial lead vertical speed indicator
instantaneous vertical speed indicator
IVSM
in vessel storage module
IVT
intervalve transformer
IVV
instantaneous vertical velocity
IVVI
instantaneous vertical velocity indicator
IW
impulse weight
index word
indirect waste
industry power
inside width
instruction word
interior width
isotopic weight
IWAHMA
industrial warm air heater manufacturers association
IWCAC
international wireless communications advisory committee
IWCB
internal web channel bus
IWCS
integrated wideband communication system
international wire and cable symposium
IWCSSEA
integrated wideband communications system southeast asia

IWDS
  international world day service
IWE
  instantaneous word encoder
IWFNA
  inhibited white fuming nitric acid
IWFS
  industrial waste filter system
IWG
  implementation working group
  iowa illinois gas and electric
  iron wire gage
  iron wire gauge
IWLS
  iterative weighted least squares
IWP
  interim working party
IWRC
  independent wire rope core
IWRMA
  independent wire rope manufacturers' association
IWS
  institute of work study
IWSA
  international water supply association
IWSC
  institute of wood science
IWSIT
  integrated work sequence inspection traveler
IWSP
  institute of work study practitioners
IWT
  institute of wireless technology
IWTS
  integrated wire termination system
IX
  index
  ion exchange
IZ
  isolation zone

J

J
  jack
  job
  joint
  joule
  jumper
  junction

JA
  joint account
  jump address
JACC
  joint automatic control conference
JACCP
  joint airborne communication and command post
JACE
  joint alternate command element
JACM
  journal of the association for computing machinery
JACOUSTSOCAMER
  journal of the acoustical society of america
JACSPAC
  joint air communications of the pacific
JACTRU
  joint air traffic control radar unit
JADE
  japan area defence environment
JAEC
  japan atomic energy commission
JAEIP
  japan atomic energy insurance pool
JAERI
  japanese atomic energy research institute
JAIEG
  joint atomic information exchange group
JAIF
  japan atomic industrial forum
JAM
  jamming
  jet age malfunction
JAMTRAC
  jammers tracked by azimuth crossings
JAN
  joint army navy
JANAF
  joint army navy air force
JANAIR
  joint army navy aircraft instrument research
JANAP
  joint army navy air force publications
JANARS
  joint army navy air force radiotelephone system
JANCOM
  joint army navy communications
JANS
  joint army navy specification
JANSPEC
  joint army navy specification
JANTAB
  joint army navy technical aeronautical board

JAPAN GCR
 japan gas cooled reactor
JAPC
 joint air photo center
JAPP
 japanese patent
JAPPLPHYS
 journal of applied physics
JAPT
 journal of approximation theory
JARRP
 japan association for radiation research on polymers
JARS
 joliet amateur radio society
JASA
 journal of the american society of acoustics
JASDA
 juilie automatic sonic data analyzer
JASG
 joint advanced study group
JASIS
 journal of the american society for information science
JASTOP
 jet assisted stop
JATO
 jet assisted take off
JB
 junction box
JBBL
 jamming of beacons and blind landing
JBCOUNT
 job account
JBMMA
 japanese business machine makers association
JBRITIRE
 jornal of the british institute of radio engineers
 journal of the british institution of radio engineers
JBS
 japan broadcasting system
JBTO
 joint bromarc test organization
JC
 jack connection
 junction
 junction center
JCA
 joint commission on accreditation of universities
 joint communications agency

JCAM
 joint commission on atomic masses
JCAR
 joint commission on applied radioactivity
JCB
 job control block
 joint communications board
JCC
 joint communications center
 joint computer conference
 joint conference committee
 joint coordination center
JCCOMNET
 joint coordination center communications network
JCCRG
 joint command and control requirements group
JCEB
 joint council on educational broadcasting
JCEC
 joint communications electronics committee
JCENS
 joint communications electronic nomenclature system
JCET
 joint council on educational television
JCFI
 job control file internal
JCFS
 job control file source
JCG
 joint coordinating groups
JCGS
 joint center for graduate study
JCI
 joint communications instructions
JCL
 job control language
JCLOT
 joint closed loop operations test
JCNPS
 joint committee on nuclear power standards
JCP
 jersey central power and light
 job control program
 joint committee on printing
JCPOA
 joint council of post office associations
JCS
 job control system
JCSS
 journal of computer and system sciences
JCT
 job control table

journal control table
junction
JCTN
   junction
JCTPT
   junction point
JDC
   job description card
   just discriminable change
JDEG
   joule per degree
JDM
   journal of data management
JDREMC
   joint departmental radio and electronics measurements committee
JDS
   job data sheet
JEA
   jordan engineers association
JEC
   japanese electrotechnical committee
JECA
   joint engineers council of alabama
JECC
   japan electronic computer center
   japan electronic computer company
   joint electronic components conference
JECNS
   joint electronic communications nomenclature system
JECS
   job entry central service
   job entry peripheral services
JEDEC
   joint electronic device engineering council
JEEP
   joint establishment experimental pile
JEIA
   japanese electronic industries association
   joint electronics information agency
JEIDA
   japan electronic industry development association
JEIPAC
   japan electronic information processing automatic computer
   jicst electronic information processing automatic computer
JEMC
   joint engineering manangement conference
JEOCN
   joint european operations communications network

JEOL
   japan electron optics laboratory
JEOLCO
   japan electron optics laboratory company
JEPIA
   japan electronic parts industry association
JEPS
   job entry peripheral service
JERC
   joint electronic research committee
JES
   japanese export standard
   job entry subsystem
   job entry system
JESA
   japan engineering standard association
   japanese engineering standards association
JESC
   joint electronics standardization committee
JESM
   japan society of electrical discharge machining
JETEC
   joint electron tube engineering council
JETR
   japan engineering test reactor
JETS
   job executive and transput satellite
   junior engineering technical society
JF
   junction frequency
JFAC
   joint flight acceptance component testing
JFAP
   joint frequency allocation panel
JFET
   junction fet
   junction field effect transistor
JFKSC
   john f kennedy space center
JFL
   joint frequency list
JFP
   joint frequency panel
JG
   joules gram
JGEOPHYSRES
   journal of geophysical research
JGFET
   junction gate field effect transistor
JGN
   junction gate number
JHG
   joule heat gradient

JHU
  john hopkins university
JI
  josephson interferometer
  junction isolation
JIAC
  joint automatic control conference
JIB
  job information block
JIC
  joint industrial council
  joint industry conference
JICST
  japan information center of science and technology
JIE
  junior institution of engineers
JIEE
  japanese institute of electrical engineers
  journal of institution of electrical engineers
JIFDATS
  joint in flight data acquisition and transmission system
  joint in flight data transmission system
JIFTS
  joint in flight transmission system
JILA
  joint institute for laboratory astrophysics
JIMS
  journal of the indian mathematical society
JINR
  joint institute for nuclear research soviet bloc countries research center
JINSTNAV
  journal of the institute of navigation
JIP
  joint input processing
JIR
  job improvement request
JIS
  japanese industrial standard
  job information station
JISC
  japanese industrial standards committee
JISTECHNIQUE
  junction insulated schottky technique
JIT
  job instruction training
JJ
  josephson junction
JK
  jack
  joule per degree kelvin
JKG
  joule per kilogram

JKT
  junction
JLC
  joint logistics committee
JLMS
  journal of the london mathematical society
JLPC
  joint logistics planning committee
JLRSE
  joint long range strategic estimate
JM
  jettison motor
JMA
  japan meteorological agency
  japan microphotography association
JMC
  joint meteorological committee
JMD
  joint monitor display
JMED
  jungle message encoder decoder
JMET
  journal of meteorology
JMKU
  journal of mathematics of kyoto university
JMM
  joint man machine
  joint man machine system
JMOS
  joint mos
JMRP
  joint meteorological radio propagation
JMSAC
  joint meteorological satellite advisory committee
JMSJ
  journal of the mathematical society of japan
JMSPO
  joint meteorological satellite program office
JMT
  job method tranining
JMTG
  joint missile task group
JN
  junction
JNA
  joint navy
JND
  just noticeable difference
JNSC
  joint navigation satellite committee
JNWP
  joint numerical weather prediction
JNWPU
  joint numerical weather prediction unit

JOBNO
  job no
JOBTICS
  job and time control system
JOD
  journal of development
JOHNNIAC
  john von neumann integrator and automatic computer
JOHNNIC
  john von neumanns integrator and automatic computer
JOL
  job orientation language
JOLA
  journal of library automation
JOR
  jet operations requirements
JORSJ
  journal of the operations research society of japan
JOVIAL
  jules' own version of internal algebraic language
  jules' own version of international algebraic language
JP
  jet pilot
  jet propellant
  jet propelled
  jet propulsion
  jones plug
  jumper
  jute protected
  jute protected cable
  jute protection
JPA
  job pack area
JPB
  joint planning board
JPC
  jet propulsion center
JPDR
  japan power demonstration reactor
JPL
  jet propulsion laboratory
JPOS
  journal of the patent office society
JPRS
  joint publications research service
JPTO
  jet propelled take off
JPW
  job processing word

JQE
  journal of quantum electronics
JRATA
  joint research and test activity
JRB
  joint radio board
JRDOD
  joint research and development objectives document
JRG
  junction register
JRIA
  japan radioisotope association
JRR
  japanese research reactor
JRS
  junction relay set
JS
  jam strobe
  jamming to signal
JSA
  japanese standards association
JSC
  japan science council
JSD
  jackson system development
JSEC
  joint services electrical and electronics committee
JSEM
  japan society for electron micrsocopy
JSI
  job step index
JSIA
  joint service induction area
JSM
  job stream manager
JSMPTE
  journal of the society to motion picture and television engineers
JSRCSC
  joint services radio component standardization committee
JSRU
  joint speech research unit
JSSC
  joint services standardization committee
JSSPG
  job shop simulation program generator
JST
  jamming station
  japanese standard time
JT
  job table

joint
junction transistor
JTAC
joint technical advisory committee
joint telecommunications advisory committee
JTC
joint telecommunications committee
joint transform correlator
JTCP
jovial test control program
JTDS
joint track data storage
JTE
junction tandem exchange
JUB
job unit block
JUDGE
judged utility decision generator
JUG
joint users group
junction gate
JUGFET
junction gate fet
junction gate field effect transistor
JUNC
junction
JWG
joint working group
JWGM
joint working group meeting

K

K
karat
kelvin
key
kilo
kiloohm
klystron
knot
KA
keyed address
kiloampere
KAAE
kaiser aircraft and electronics
KAC
kennedy administrative complex
KAEDS
keystone association for educational data systems
KAFB
keesler air force base
kirkland air force base
KAH
kiloampere hour
KAHL
kahl power reactor
KAI
kuybyshev aviation institute
KALDAS
kidsgrove algol digital analog simulation
KAN
kansas power and light
KAPL
knolls atomic power laboratory
KAPSE
kernel ada programming support environment
KB
keyboard
kilobar
kilobit
kilobyte
kilobaud
knowledge base
KBES
knowledge-based expert system
KBPS
kilobits per second
KBS
kilobit per second
kilobytes per second
KBYTE
kilobyte
KC
kilocurie
kilocycle
kilocycle per second
KCAL
kilocalorie
KCC
keyboard common contact
KCG
key calling
KCHS
kilo characters per second
KCI
kilocurie
KCL
kirchhoff's current law
KCPS
kilocycles per second
KCS
kilocharacters per second
kilocycles per second

KCSC
kidde computer services co
KCU
kilocurie
KDD
kokusai denshin denwa
KDE
keyboard data entry
KDH
key depression per hour
KDK
knit de knit texturing
KDP
known datum point
KDR
keyboard data recorder
keyboard date recorder
KE
keuffel and esser company
kinetic energy
KEAJ
kanto electric association of japan
KEAS
knot equivalent air speed
KEC
kingston engineers club
KERG
kiloerg
KES
kansas engineering society
KEV
kiloelectronvolt
KEWB
kinetic experiment on water boilers, small test reactors
KFLOP
kiloflop
KG
kilogauss
kilogram
kilogram force
KGCAL
kilogram calorie
KGE
kansas gas and electric company
KGF
kilogram force
KGM
kilogram meter
KGPS
kilogram per second
KGR
klydonograph type gradient recorder
KGS
known good system

KGWT
kilogram weight
KH
kelsey hayes company
krohn hire
KHAI
kharkov aviation institute
KHZ
kilohertz
KIA
kobus interface adapter
KIAS
knots indicated airspeed
KIC
kollsman instrument corporation
KIFIS
kollsman integrated flight instrument system
kollsman integrated flight instrumentation system
KIM
keyboard input matrix
KIMCODE
kimble method for controlled devacuation
KINE
kinescope
KIP
kilopound
KIPO
keyboard input printout
KIPS
kilowatt isotope power system
knowledge information processing system
KIS
keyboard input simulation
KISNOPI
keyboard input simulation noise problem input
KISS
keep it simple stupid
keep it simple, sir
KJ
kilojoule
KL
key length
key lever
kiloliter
KLA
klystron amplifier
KLO
klystron oscillator
KM
kilomega
kilometer

KMC
 kilomegacycle
KMCP
 kodak metal clad plate
KMCS
 kilomegacycle per second
KMER
 kodak metal etch resist
KMHZ
 kilomegahertz
KMPH
 kilometers per hour
KMPS
 kilometers per second
KMRF
 keyswitch magic relay finder
KMS
 keyset multiple selector
KN
 knot
KNSW
 knife switch
KO
 kick off
KOHM
 kiloohm
KOR
 kodak orthoresist
KORSTIC
 korea scientific and technological information center
KOS
 kent on line system
KP
 key pulsing
 key punch
 kilopond
KPC
 kiloparsec
KPE
 key point error
KPH
 kilometers per hour
KPIC
 key phrase in context
KPO
 keypunch operator
KPR
 kodak photoresist
KPSM
 klystron power supply modulator
KR
 knowledge representation
KRF
 knowledge of results feedback

KRFT
 knowledge of results feedback task
KRR
 kansai research reactor
KS
 kansas
 keystone
 knowledge source
KSC
 kagoshime space center
 kennedy space center
KSCAN
 k scanner
KSE
 kuwait society of engineers
KSL
 keyboard simulated lateral telling
KSPM
 klystron power supply modulator
KSR
 keyboard send receive
 keyboard send receive set
KSS
 kellog switchboard and supply
 kilosample per second
KST
 known segment table
KSTR
 kema suspension test reactor
KSU
 key service unit
KTFR
 kodak thin film resist
KTI
 keyboard training, incorporated
KTP
 kronos transaction processing
KTR
 keyboard typing reperforator
KTS
 key telephone system
 kodak tracking station
KTSA
 kahn test of symbol arrangement
KU
 kentucky utilities
KUTD
 keep up to date
KV
 karnaugh veitch
 kilovolt
 kinematic viscosity

KVA
  kilovolt ampere
KVAC
  kilovolt alternating current
KVAH
  kilovolt ampere hour
KVAHM
  kilovolt ampere hour meter
KVAM
  kilovolt ampere meter
KVAR
  kilovar
  kilovolt ampere reactive
KVARH
  kilovar hour
KVCP
  kilovolt constant potential
KVDC
  kilovolt direct current
KVP
  kilovolt peak
KW
  kilowatt
KWE
  kilowatt of electrical energy
KWESTEVDET
  key west test and evaluation detachment
KWFT
  kilowatt foot
KWHM
  kilowatt hour meter
KWHR
  kilowatt hour
KWHRMETER
  kilowatt hour meter
KWIC
  keyword in context
  keyword in context indexing
KWIPS
  kilo whetstone instructions per second
KWIT
  keyword in title
KWOC
  keyword out of context
KWOCREGISTER
  keyword out of context register
KWORD
  kiloword
KWOT
  keyword out of title
KWR
  kilowatt reactive
KY
  kentucky
  keying device

L

L
  label
  lambert
  lamp
  large
  lateral
  latitude
  launch
  launcher
  left
  length
  length of operand
  lengthwise
  level
  level zero
  light
  line
  line motion control
  line telegraphy
  link
  listening
  liter
  load
  locus
  long
  longitude
  loran
  lost
  low
  low level
  low power
  lumen
  luminance
LA
  lag angle
  lambert
  laser gyro axis
  launch analysis
  launch area
  lead angle
  level amplifier
  light artillery
  light wire armored
  lightning arrester
  line adapter
  link address
  link allotter
  load address
  load adjuster
  look ahead
  louisiana

low altitude
low angle
LAAR
liquid air accumulator rocket
LAB
label
labelling
laboratory
low altitude bombing
low amplitude bombing
LABIL
light aircraft binary information link
LABPIE
low altitude bombing position indicator equipment
LABROC
laboratory rocket
LABS
low altitude bombing system
LAC
launch analysis console
load accumulator
lockheed aircraft corporation
lunar atlas chart
LACBPE
los angeles council of black professionals engineers
LACBWR
la crosse boiling water reactor
LACCIRCUIT
look ahead carry circuit
LACE
launch angle condition evaluator
launch automatic checkout equipment
liquid air cycle engine
local automatic circuit exchange
london airport cargo edp
london airport cargo edp scheme
LACES
los angeles council of engineering societies
LACIE
landsat oriented large area crop inventory experiment
large area crop inventory experiment
LACONIQ
laboratory computer on line inquiry
LACR
low altitude coverage radar
LACS
listener active state
LACT
lease automatic custody transfer
LAD
load address

location aid device
logic and adder
logic and adder module
logical aptitude device
low accuracy data
low accuracy designation
LADAPT
lookup dictionary adapter
lookup dictionary adaptor program
LADAR
laser detection and ranging
LADC
laser advanced development center
LADIZ
leaving air defense identification zone
LADOG
low altitude drive on ground
LADS
linear analysis and design of structures
listener addressed state
low altitude detection system
LADSIRLAC
liverpool and district scientific, industrial and research library advisory council
LADWP
los angeles department of water and power
LAE
left arithmetic element
LAEC
los angeles electric club
LAEDP
large area electronic display panel
LAG
lagging
load and go assembler
LAGS
laser activated geodetic satellite
LAH
logical analyzer of hypothesis
LAL
langley aeronautical laboratory
lower acceptance level
LALO
low altitude observation
LAM
lambert
laminated
lamination
load accumulator with magnitude
look at me
loop adder and modifier
LAMA
local automatic message accounting

LAMBDA
   language for manufacturing business and distribution activities
LAMCIS
   los angeles multiple corridor identification system
LAMCS
   latin america communication system
LAMIS
   local authority management information system
LAMP
   low altitude manned penetrator
   lunar analysis and mapping program
LAMPF
   los alamos meson physics facility
LAMPRE
   los alamos molten plutonium reactor experiment
LAMPS
   light airborne multipurpose system
LAMSAC
   local authority management services and computer committee
LAN
   landing aid
   langley
   local apparent noon
   local area network
LANAC
   laminar navigation and anti collision
   laminar navigation and anti collision system
LANG
   language
LANNET
   large artificial nerve net
   large artificial neuron network
LANS
   loran airborne navigation system
LAP
   launch analysis panel
   lesson assembly program
   linear arithmetic processor
   link access procedure
   list assembly program
   list assembly programming
LAPB
   link access procedure balanced
LAPD
   limited axial power distribution
LAPDOG
   low altitude pursuit drive on ground
LAPES
   low altitude parachute extraction system

LAPPES
   large power plant effluent study
LAPUT
   light activated programmable unijunction transistor
LAQ
   lacquer
LAR
   local acquisition radar
   low angle reentry
LARA
   line addressed random access memory
LARAM
   line addressable random access memory
LARC
   large automatic research calculator
   library automation research and consulting services
   life accurance rates calculation
   life insurance rates calculation
   livermore atomic research computer
   livermore automatic research calculator
   livermore automatic research computer
LARCT
   last radio contact
LARD
   load adjuster reference datum
LARDS
   low accuracy radar data transmission system
LARF
   low altitude radar fusing
LARGOS
   laser activated reflecting geodetical satellite
LARIAT
   laser radar intelligence acquisition technology
LARK
   ladies' amateur radio club
LARR
   large area record reader
LARS
   laser angular rate scanner
   laser angular rate sensor
   low altitude radar tracking and target recongnition system
LARV
   low altitude research vehicle
LAS
   large astronomical satellite
   launch auxiliary system
   logical compare accumulator with storage
   loop actuating signal
   low altitude satellite
LASA
   large aperture seismic array

LASCR
  light activated silicon controlled receiver
  light activated silicon controlled rectifier
  light activated silicon controlled switch
LASER
  light amplification by stimulated emission of radiation
LASH
  laser anti tank semi active homing
LASHUP
  land air synergic homogeneous ultra processor
LASL
  los alamos scientific laboratory
LASO
  low altitude search option
LASP
  local attached support processor
LASR
  letter writing with automatic send receive
LASRM
  low altitude supersonic research missile
LASS
  laser activated semiconductor switch
  library automated service systems
  light activated silicon switch
  line amplifier and super synchronization
LASSO
  landing and approach system spiral oriented
  laser search and secure observer
  laser synchronization from stationary orbit
  light aviation special support operations
LASV
  low altitude supersonic vehicle
LAT
  lateral
  latitude
  local apparent time
LATAR
  laser augmented target acquisition and recognition
LATCC
  london air traffic control center
LATCH
  local attached support processor
LATINCON
  latin american convention
LATREC
  laser acoustic time reserval, expansion and compression
  laser acoustic time reversal, expansion and compression
LATS
  long acting thyroid stimulator

LAU
  lower arithmetic unit
LAVA
  linear amplifier for various applications
LAW
  local air warning
LAWB
  los alamos water boiler
LAWRC
  limited air weather reporting certificate
LAYDET
  layer detection
LB
  launch bunker
  letter box
  light bombardment
  line buffer
  local batch
  local battery
  low band
LBA
  laser beam analyzer
  linear bounded automaton
  local battery apparatus
LBD
  logic block diagram
LBE
  live better electrically
LBF
  load bit field
LBL
  label
  lawrence berkeley laboratory
LBM
  laser beam machining
  load buffer memory
  lunar breaking module
LBMCTX
  local battery magneto call telephone exchange
LBO
  line build out
  line building out
LBP
  length between perpendiculars
  line binder post
LBR
  laser beam recorder
  laser beam recording
LBS
  launch blast simulator
  least significant bit
  local battery signaling
  local battery supply
  local battery switchboard

local battery system
low speed breaker
LBT
  listen before talk
  local battery telephone
  low bit test
LBTS
  local battery telephone set
  local battery telephone switchboard
LBTX
  local battery telephone exchange
LBU
  launcher booster unit
LBWBUZCALTX
  local battery with buzzer calling telephone exchange
LBWMABCTX
  local battery with magneto and buzzer calling telephone exchange
LBWOC
  level bombing wind offset computer
LBX
  lexar business exchange
LC
  landing craft
  last card
  launch complex
  launch conference
  launch control
  launch corridor
  launch countdown
  lead coated
  lead covered
  letter card
  level control
  library of congress
  light case
  light current
  line carrying
  line circuit
  line connection
  line connector
  line construction tools
  line of communication
  line of contact
  link circuit
  liquid crystals
  load carrier
  load cell
  load center
  load compensating
  localization code
  location counter
  logic cell
  logic corporation
  logistics command
  loop check
  loss of contact
  loss of contract
  loud and clear
  low carbon
  lower case
  lower characters
  lower control
  luminosity class
LCA
  launch control analyst
  launch control area
  load carrying ability
  local communication adapter
  local communication area
  logic cell array
  logistics control area
  low complement of address
LCAO
  linear combination of atomic orbitals
LCARC
  lake county amateur radio club
LCB
  launch control building
  line control block
  longitudinal position of center of buoyancy
LCC
  landing craft control
  launch control center
  launch control console
  leadless chip carrier
  life cycle costing
  liquid crystal cell
  listeners century certificate
  local communications complex
  local communications console
  london communication committee
LCCA
  load current contact aiding
  low cost computer attachment
LCCFC
  launch control complex facilities console
LCCS
  launch checkout and count down system
  launch control and checkout system
LCD
  launch control design
  least common denominator
  liquid crystal display
  lowest common denominator
LCDS
  low cost development system

LCDTL
  load compensated diode transistor logic
  low current diode transistor logic
LCE
  launch complex engineer
  launch complex equipment
  load circuit efficiency
LCES
  least cost estimating and scheduling
LCF
  language central facility
  launch control facility
  least common factor
  local cycle fatigue
  lowest common factor
LCFC
  launch complex facilities console
LCFS
  last come first served
  launch control facility simulator
LCG
  logistics control group
  look ahead carry generator
LCH
  load channel
LCHR
  launcher
LCHS
  large component handling system
LCI
  launch complex instrumentation
LCL
  less than carload
  lifting condensation level
  linkage control language
  local
  localizer
  lower control limit
LCLU
  landing control and logic unit
LCLV
  liquid crystal light valve
LCLZR
  localizer
LCM
  large capacity memory
  large core memory
  launch control monitor
  launch crew member
  least common multiple
  low cost module
  lowest common multiple
LCMM
  life cycle management model

LCMS
  launch control and monitoring system
LCN
  load classification number
  local civil noon
LCNT
  link celestial navigation trainer
LCO
  launch control officer
  launch control operation
  limited condition of operability
  limiting conditions for operation
LCOC
  launch control officer's console
LCOP
  launch control officer's panel
LCOSS
  launch control officer's safety system
LCP
  language conversion program
  large computer project
  last card program start
  launch control panel
  left-hand circular polarization
  left-handed circular polarization
  loading control program
LCR
  inductance capacitance resistance
  level crossing rate
  load complement register
  log count rate
  low compression ratio
LCRE
  lithium cooled reactor experiment
LCRM
  launch control room
LCS
  large capacity core store
  large capacity storage
  large core storage
  large core store
  lateral channel stop
  launch control sequence
  launch control simulator
  launch control system
  line coding storage
  loop control system
  loudness contour selector
LCSB
  launch control support building
LCSCU
  launch coolant system control unit
LCSLT
  low cost solid logic technology

LCSO
    launch complex safety officer
    launch control safety officer
LCSRM
    loop current stp response method
LCSS
    launch control and sequencer system
    launch control system simulator
LCSW
    latch checking switch
LCT
    last card total
    launch control trailer
    linkage control table
    local civil time
    location, command, and telemetry
LCTB
    launch control training building
LCTI
    large components test installation
LCTL
    large components test loop
LCTSU
    launch control transfer switching unit
LCU
    launch control unit
    line control unit
    local control unit
LCV
    low calorific value
LCVD
    least coincidence voltage detection
LCVIP
    licensee contractor vendor inspection report
LCVP
    landing craft vehicle personnel
LCWF
    launch complex work flow
LCX
    launch complex
LCZR
    localizer
LD
    label definition
    lamp driver
    landing distance
    launching division
    lead
    leading edge delay
    leak detector
    length to diameter
    lethal dose
    level discriminator
    lift to drag
    lift to drag ratio
    light driver
    line drawing
    line driver
    line of departure
    line time waveform distortion
    linear decision
    load
    load draught
    loader
    logarithmus dualis
    logic driver
    long-distance
    loss and damage
    low drag
    lunar docking
    luminescence diode
LD:
    lethal dose
LDA
    line driving amplifier
    linear dynamic analyzer
    locate drum address
LDB
    light distribution box
    load b register
    local data buffer
LDBLC
    low drag boundary layer control
LDC
    lasa data center
    laser discharge capacitor
    latitude data computer
    light direction center
    line drop compensator
    linear detonating cord
    long-distance call
    lower dead center
LDCC
    large diameter component cask
LDCS
    long-distance control system
LDDC
    long-distance dialing center
LDDL
    logical data definition language
LDDS
    low density data system
LDE
    laminar defect examination
    linear differential equation
    long delay echo
LDEF
    long duration exposure facility
LDEP
    lancashire dynamo electronic products

LDF
  linear discriminant function
  load factor
LDFSTN
  landing direction finding station
LDG
  loading
LDGE
  lunar excursion module dummy guidance equipment
LDI
  load indicator
  lossless digital integrator
LDL
  landing direction light
  logical data language
LDM
  linear delta modulation
  load distribution matrix
LDMOST
  lateral double diffused most
LDNA
  long distance navigation aid
LDNS
  doppler navigation system
LDO
  launch division officer
  light diesel oil
LDP
  laboratory data processor
  language data processing
  local data package
LDPE
  low density polyethylene
LDR
  level distribution recorder
  light dependent resistor
  low data rate
LDRI
  low data rate input
LDRS
  lem data reduction system
  lunar excursion module data reduction system
LDRT
  low data rate
LDS
  large disk storage
  last data sample
LDSU
  local distribution service unit
LDT
  level detector
  linear differential transformer
  logic design translator
  logical device table
  long-distance transmission
LDTEL
  long distance telephone
LDX
  long distance xerography
LE
  launch equipment
  launch escape
  launching equipment
  leading edge
  left end
  length
  less than or equal to
  light equipment
  limit of error
  linkage editor
  low explosive
LEAA
  law enforcement assistance administration
LEAD
  lens electronic automatic design
LEADS
  law enforcement automated data system
LEANS
  lehigh analog simulator
LEAP
  language for the expression of associative procedures
  leading edge airborne panoramic
  lift off elevation and azimuth programmer
  logistics evaluation and analysis program
LEAR
  logistics evaluation and review
  logistics evaluation and review technique
LEARSYN
  logistics evaluation and review synchronization
LEAS
  lease electronic accounting system
  lower echelon automatic switchboard
LEB
  london electricity board
  lower equipment bay
LEC
  langage exploitation condense
  launch escape control
  liquid encapsulated czochralski
  list execution condition
  lockheed electronics company
LECA
  launch escape control area
LECC
  linear error correcting code

LECO
  london engineering congress
LECOS
  lunar environment construction and operations simulator
LECS
  launching equipment checkout set
LED
  light emitting diode
LEDT
  limited entry decision table
LEED
  laser energized explosive device
  low energy electron diffraction
LEER
  low energy electron reflections
LEF
  light emitting film
LEFM
  linear elastic fracture mechanics
LEG
  logistics evaluation group
LEGM
  low energy gamma monitor
LEID
  low energy ion detector
LEIT
  light emission via inelastic tunnelling
LEL
  large engineering loop
  lower explosion level
LELTS
  lightweight electronic locating and tracking system
LELU
  launch enable logic unit
LEM
  antenna effective length for magnetic field antennas
  launch enclosure maintenance
  launch escape motor
  lower explosion limit
  lunar excursion module
  lunar exploration module
LEMES
  low energy magnetic electron spectrum
LEMT
  lunar excursion module track
LEMUF
  limit of error on material unaccounted for
LEO
  low earth orbit
  lyon's electronic office

LEP
  lower end plug
  lowest effective power
LEPS
  launch escape propulsion system
LER
  licensee event report
  london electric railway
LERA
  limited employee retirement account
LERT
  lockheed emergency reset timer
LES
  launch enable system
  launch escape system
  lilliput edison screw
  limited early site
  lincoln experimental satellite
  local engineering standards
  loop error signal
  louisiana engineering society
LESA
  lunar exploration system for apollo
LESC
  launch escape system control
LESCS
  launch escape stabilization and control system
LESR
  limited early site review
LESS
  launch escape system simulator
  least cost estimating and scheduling
  least cost estimating and scheduling system
LESSC
  louisville engineering and scientific societies' council
LEST
  launch enable system turret
LET
  laboratory electronics technician
  launch escape tower
  life environmental testing
  lincoln experimental terminal
  linear energy transfer
  logical equipment table
LETCS
  launch escape tower canard system
LETS
  law enforcement teletype system
  law enforcement teletype writer service
LEU
  launch enable unit
  low enriched uranium

LEV
  level
  lunar escape vehicle
  lunar excursion vehicle
LF
  launch facility
  leapfrog
  leapfrog configuration
  line feed
  line finder
  listener function
  load factor
  low frequency
LFA
  last field address
  low flow alarm
  low frequency amplifier
LFBR
  liquid fluidized bed reactor concept
LFBRCX
  liquid fluidized bed reactor critical experiment
LFC
  laminar flow control
  level of free convection
  load frequency
  load frequency control
  low frequency choke
  low frequency correction
  low frequency current
LFCR
  low frequency out off ratio
LFD
  least fatal dose
LFDF
  low frequency direction finder
LFE
  laboratory for electronics
LFF
  low frequency filter
LFIM
  low frequency instruments and measurement
LFINT
  low frequency intersection
LFM
  launch first motion
  linear frequency modulator
  low powered fan marker
LFMF
  low frequency, medium frequency
LFO
  low frequency oscillator
LFQ
  light foot quantizer

LFR
  line frequency rejection
LFRD
  lot fraction reliability definition
  lot fraction reliability deviation
LFS
  launch facility simulator
  logical file system
  loop feedback signal
LFSR
  linear feedback shift register
LFT
  late finish time
LFTA
  low frequency timing assembly
LFU
  least frequently used
LFUNCTION
  listener function
LFV
  lunar flying vehicle
LFVLF
  low frequency very low frequency
LG
  landing gear
  landing ground
  leg
  length
  level gage
  line generator
  line to ground
  long
  loop gain
LGA
  light gun amplifier
LGB
  laser guided bomb
LGC
  lunar excursion moudle guidance computer
  lunar gas chromatograph
LGE
  lunar excursion moudle guidance equipment
LGG
  light gun pulse generator
LGL
  logical left shift
LGN
  line gate number
LGO
  lamont geological laboratory
LGP
  librascope general precision
LGR
  logical right shift
  low group receiving unit

LGT
 low group transmit
 low group transmitting unit
LGTH
 length
LGWS
 laser guided weapons systems
LH
 latent heat
 left hand
 left handed
 lighthouse
 liquid helium
 liquid hydrogen
 low high
 low noise high output
 low to high
 luteinizing hormone
LHA
 local hour angle
LHC
 left-hand circular
LHCJEA
 london and home counties joint electricity authority
LHCP
 left-hand circular polarization
 left-hand circular polarized
LHCTL
 left-hand control
LHD
 left-hand drive
LHDC
 lateral homing depth charge
LHE
 liquid helium
LHEDGE
 low high edge
LHFEB
 left-hand forward equipment bay
LHGR
 linear heat generation rate
LHM
 left-hand circularly polarized mode
 loop handling machine
LHP
 left half plane
LHPS
 lead hydrogen purge system
LHR
 lowe hybrid resonance
 lower hybrid resonance
 lumen hour
LHRH
 left hand right hand

LHS
 left-hand side
 loop handling system
LHSC
 left-hand side console
LHSI
 low head safety injection
LHT
 lighthouse tube
LHTR
 lighthouse transmitter receiver
LHW
 lef half word
 left-hand word
LI
 intensity level
 lee insulator
 level indicator
 linear interpolator
 link
 location identifier
 low intensity
 lubricity index
 lyons instruments
LIA
 laser industries association
 laser industry association
LIBOL
 litton business oriented language
LIC
 last instruction cycle
 launcher interchange circuit
 line integrated circuit
 linear integrated circuit
 loop insertion cell
LICOF
 land line communication facilities
LICOR
 lightening correlation
LID
 leadless inverted device
 locked in device
LIDAR
 laser infrared radar
 laser intensity direction and ranging
 light detecting and ranging
 light detection and ranging
 light radar
LIDF
 line intermediate distributing frame
LIDS
 listener idle state
LIE
 limited information estimation

LIEMC
  long island electronics manufacturers' council
LIFE
  logistics evaluation and review integrated flight equipment
LIFMOP
  linearly frequence modulated pulse
  linearly frequency modulated pulse
LIFO
  last in first out
LIFT
  logically integrated fortran translator
LIGHT
  lighting
  lightning
LIL
  lunar international laboratory
LILCO
  long island lighting company
LILO
  link loader
LILOC
  light line optical correlation
LIM
  limit
  limit value
  limiter
  limiting
  linear induction motor
LIMAC
  large integrated monolithic array computer
LIMB
  liquid metal breeder, heterogeneous, thorium breeder, graphite moderated and lead cooled reactor
LIMDAT
  limiting date
LIMFAC
  limiting factor
LIMP
  lunar interplanetary monitoring probe
LIMSW
  limit switch
LIN
  linear
  liquid nitrogen
LINAC
  linear accelerator
LINAS
  laser inertial navigation attack system
LINC
  laboratory instrument computer
LINCOMPEX
  linked compressor and expander

LINCOS
  lingua cosmica
LINFT
  linear foot
LINLOG
  linear logarithmic
LINMH
  linear meters per hour
LINS
  laser inertial navigation system
  lightweight inertial navigation system
  loran inertial system
LINUS
  lexington integrated utility system
LIOCS
  logical input output control system
LIP
  launch in process
LIPL
  linear information processing language
LIPS
  logic inference per second
LIQ
  liquid
LIQFRKT
  liquid fuel rocket
LIR
  line integral refractometer
  load indicating relay
  load indicating resistor
LIROC
  last instruction read out cycle
LIS
  language implementation system
  launch instant selector
  line isolation switch
  loop input signal
  low inductance stripline
LISA
  library systems analysis
  linear system analysis
LISARDS
  library information search and retrieval data system
LISN
  line impedance stabilization network
LISP
  laser isotope separation program
  list processing
  list processing language
  list processor
LIST
  library and information services, tees side
LIT
  liquid injection technique

LITASTOR
light tapping storage
LITE
legal information through electronics
LITR
low intensity test reactor
LITS
laboratory for information transmission systems
LITVC
liquid injection thrust vector control
LIVCR
low input voltage conversion regulation
LIVE
lunar impact vehicle
LIW
loss in weight
LIWB
livermore water boiler
LJE
local job entry
LJP
local job processing
LK
link
lock
looking
LKG
leakage
linking
LKGABKG
leakage and breakage
LKGE
linkage
LL
land line
light line
light load
light loading
lightly loaded
limited liability
lincoln laboratory
line link
line to line
linking loader
liquid limit
live load
loudness level
low level
lower limit
LLBAM
lincoln laboratory boolean algebra minimizer
LLC
liquid liquid chromatography
logical link control

LLCO
loudspeaker and leaflet company
LLD
low level detector
LLDL
low level differential logic
LLE
left lower extremity
LLF
link line frame
LLFM
land-line frequency modulation
low level flux monitor
LLFPB
lumped, linear, finite, passive, bilateral
LLG
line to line to ground
LLL
lawrence livermore laboratory
low level logic
low light level
LLLLLL
laboratories low level linked list language
LLLTV
low light level television
LLLWT
low level liquid waste tank
LLM
lunar landing mission
lunar landing module
LLO
local lock out
LLP
live load punch
lunar landing program
LLPL
low low pond level
LLPN
lumped, linear, parametric network
LLR
line of least resistance
load limiting resistor
long length record
LLRES
load limiting resistor
LLRM
low level radio modulator
LLRR
lowest level remove replace
LLRT
low level reactor test
LLRV
lunar landing research vehicle
LLS
liquid level switch

long left shift
lunar landing simulator
lunar logistic system
LLSAGW
low level surface to air guided weapon
LLSS
long life space system
LLSV
lunar logistics supply vehicle
LLT
land-line teletype
long island lighting
LLTT
land-line teletype
LLTV
lunar landing training vehicle
LLV
lunar landing vehicle
lunar logistics vehicle
lunar logistics vessel
LLW
low level waste
LM
landing module
landsdale microelectronics
large memory
laser machine
latch magnet
library material
light maintenance
limit
line mark
lines per minute
load module
loss margin
lumen
lunar excursion module
lunar module
LMAFS
lookout mountain air force station
LMAL
langley memorial aeronautical laboratory
LMC
land mobile communications
launch monitor console
launch monitor control
LMCC
land mobile communications council
LMCD
liquid metal cooled demonstration
LMCN
launch maintenance conference network
launch missile control network
LMDE
lunar module descent engine

LME
launch monitor equipment
LMEC
liquid metal engineering center
LMEE
light military electronic equipment
LMF
linear matched filter
low and medium frequency
LMFBR
liquid metal fast breeder reactor
LMFR
liquid metal fuel reactor concept
liquid metal fuelled reactor
LMFRE
liquid metal fuel reactor experiment
LMG
light machine gun
LMGEN
load module generator
LMH
lumen hour
LMHR
lumen hour
LMHX
liquid metal heat exchanger
LMI
logistics management institute
LML
lookout mountain laboratories
LMLR
load memory lock out register
LMM
locator middle marker
LMN
lineman
LMO
lens modulated oscillator
linear master oscillator
LMP
light metal products
liquid metal plasma valve
LMPRT
locally most powerful rank test
LMR
library maintenance routine
linear multiple regression
liquid metal reactor
LMRS
lunar module rendezvous simulator
LMS
least mean square
level measuring set
local missile selector
logic metadata system

london mathematical society
lumen second
LMSC
   lockheed missile and space company
LMSD
   lockheed missile and space division
LMSEC
   lumen second
LMSEO
   labor management standards enforcement office
LMSQFT
   lumen per square foot
LMSS
   lunar mapping and survey system
LMT
   length, mass, time
   local mean time
   logic master tape
LMTD
   logarithmic mean temperature difference
LMU
   line monitor unit
LMW
   lumen per watt
LN
   line
   liquid nitrogen
   low noise
LNA
   low noise amplifier
LNAP
   ladder network analysis program
   low non essential air pressure
LNCHR
   launcher
LNDIS
   land intermediate station
LNDL
   least negative down level
LNG
   length
   liquefied natural gas
   liquid natural gas
   liquified natural gas
LNKEDT
   linkage editor
LNP
   least newtonian path
   loss of normal power
LNR
   low noise receiver
LNT
   launch network test
   liquid nitrogen temperature

LNTWA
   low noise traveling wave amplifier
LNTWTA
   low noise traveling wave tube amplifier
LNVT
   launch network verfication test
LO
   launch operations
   launch operator
   level off
   lift off
   liquid oxygen
   local
   local oscillator
   lock on
   lock out
   locked open
   locked oscillator
   logical operation
   longitude
   longitudinal optical
   lorenz
   low
   low ordinary
   lubrication oil
   lunar orbiter
LOA
   length overall
   line of assurance
LOAC
   low accuracy
LOADS
   low altitude defense system
LOAMP
   logarithmic amplifier
LOAP
   list of applicable publications
LOAS
   lift off acquisition system
LOB
   line of balance
   line of bearing
LOBAR
   long baseline radar
LOC
   large optical cavity
   launch operations center
   launch operations complex
   launch operator's console
   line of communication
   local
   localizer
   location

LOCA
   loss of coolant accident
   low cost computer attachment
LOCAL
   laboratory program for computer assisted learning
LOCATE
   library of congress automation techniques exchange
   loran omega course and track equipment
LOCATS
   lockheed optical communications and tracking system
LOCC
   launch operations control center
LOCCIT
   loco citato
LOCF
   loss of coolant flow
LOCI
   logarithmic computing instrument
LOCMOS
   local oxidation complementary mos
   local oxidation of metal oxide semiconductor
LOCMOSTECHNIQUE
   local oxidation of metal oxide silicon technology
LOCO
   long core
LOCOS
   local oxidation of silicon
LOCOSTECHNIQUE
   local oxidation of silicon technique
LOCP
   launch operation control panel
   loss of coolant protection
LOCS
   librascope operation control system
   logic and control simulator
LOCT
   lockheed command and tracking
LOCTRACS
   lockheed tracking and control system
LOD
   launch operations directorate
   launch operations division
   leading ones detector
   length of day
   low density
LODESMP
   logistics data element standardization and management process
LODESTAR
   logically organized data entry, storage and recording

LOERO
   large orbiting earth resources observatory
LOF
   local oscillator frequency
   loss of flow
   lowest operating frequency
LOFA
   loss of flow accident
LOFAR
   low frequency acquisition and ranging
   low frequency analysis and recording
LOFI
   low fidelity
LOFT
   loss of flow test
   loss of fluid test
   loss of fluid test facility
   low frequency radio telescope
LOFTI
   low frequency transionospheric satellite
   low frequency transionospheric
LOFW
   loss of feedwater
LOG
   logarithm
   logging
   logic
   logical
LOGAIRNET
   logistics air network
LOGALGOL
   logical algorithmic language
LOGAMP
   logarithmic amplifier
LOGANDS
   logical commands
LOGBALNET
   logistics ballistics network
LOGCOMD
   logistical command
LOGDEC
   logarithmic decrement
LOGEL
   logic[al] generating language
LOGFTC
   logarithmic fast time constant
LOGIFAMP
   logarithmic intermediate frequency amplifier
LOGIPAC
   logical processor and computer
LOGIT
   logical inference tester
   logical interference tester
LOGLAN
   logical language

LOGMTD
   logarithmic mean temperature difference
LOGR
   logistical ratio
LOGRAM
   logical program
LOGTAB
   logic[al] tables
LOGTANBG
   logarithm tangent bearing
LOH
   light observation helicopter
LOHAP
   light observation helicopter avionics package
LOI
   loss of ignition
   lunar orbit injection
LOKTAL
   locked octal
LOL
   lenght of lead
   limited operating life
LOLA
   layman oriented language
   low level oil alarm
   lunar orbit and landing approach
   lunar orbit landing approach
LOLITA
   language for the on-line investigation and transformation of abstractions
LOLP
   loss of load probability
LOM
   locator at the outer marker
   locator outer marker
   lunar orbital mission
LOMAC
   logical machine corp
LOMAR
   logistics, maintenance and repair
LOMB
   lockheed missile beacon
   lockheed modified beacon
LOMGEN
   lombard general
LOMIS
   locator map in source
LOMOR
   long-distance medium frequency omni range
LOMSHIP
   lombard shipping
LOMUSS
   lockheed multiprocessor simulation system

LONG
   longitude
   longitudinal
LONO
   low noise
LOP
   launch operations panel
   launch operator's panel
   line of position
   logic processor
   lookout post
   loss of offsite power
   lunar orbit plane
LOPAC
   load optimization and passenger acceptance control
LOPAD
   logarithmic on-line data processing system for analog data
   logarithmic outline processing for analog data
   logarithmic outline processing system for analog data
LOPAIR
   long path infrared
LOPAR
   low power acquisition radar
LOPASS
   low pass
LOPC
   lunar orbital photocraft
LOPI
   loss of pipe integrity
LOPO
   low power water reactor
LOPP
   lunar orbital photographic project
LOPPLAR
   laser doppler radar
LOPROBLEM
   linear optimization problem
LOPS
   lunar orbital photographs omnirange
   lunar orbiter photographs
LOPT
   line output transformer
LOQG
   locked oscillator quadrature grid
LOR
   low frequency omnirange
   lunar orbit[al] rendezvous
LORAC
   long-range accuracy
LORAD
   long-range active detection

LORADAC
    long-range active detection and communications
LORAN
    long-range air navigation
    long-range navigation
LORAPH
    long range passive homing
LORC
    lockheed radio command
LORDS
    licensing on-line retrieval data system
LOREC
    long-range earth current
LORL
    large orbital research laboratory
    large orbiting research laboratory
LORPGAC
    long-range proving ground automatic computer
LORSAC
    long-range submarine communications
LORV
    low observable reentry vehicle
LOS
    launch operation station
    launch optional selector
    line of sight
    local operating system
    loop output signal
    loss of signal
    lunar orbiter spacecraft
LOSARP
    line of sight and repeater placement program
LOSAT
    language oriented system analysis table
LOSOS
    local oxidation of silicon on sapphire
LOSOSTECHNIQUE
    local oxidation of silicon on sapphire technique
LOSP
    loss of offsite power
    loss of system pressure
LOSS
    landing observer signal system
    large object salvage system
    lunar orbital survey system
LOSSYS
    landing oberserver's signal system
LOST
    let down storage tank
LOT
    laminated overlay transistor
    large orbital telescope
    list on tape
LOTIS
    logic, timing and sequencing
LOTS
    launch optical trajectory system
    logistic over the shore vehicle
    logistics over the shore
LOU
    louisville gas and electric
LOV
    limit of visibility
LOX
    liquid oxygen
    liquid oxygen explosive
LOZ
    liquid ozone
LP
    launch pad
    launch platform
    lighting panel
    limit of proportionality
    line printer
    linear printer
    linear programming
    list processing
    load point
    log periodic
    long persistance
    long play
    long playing
    longitudinal parity
    loop
    low pass
    low pass filter
    low point
    low power
    low pressure
    low primary
    lower panel
    lunar and planetary
LPA
    launcher plant assembly
    link pack area
    log periodic antenna
    low pressure alarm
LPAC
    launching programs advisory committee
LPARM
    liquid propellant applied research motor
LPC
    laboratory precision connector
    linear power control
    linear power controller
    linear predictive coder

linear predictive coding
lockheed propulsion company
longitudinal parity check
loop control
loop preparation cask
low power channel
lower pump cubicle
LPCH
local process control host
LPCI
low pressure coolant injection
LPCP
launching preparations control panel
LPCRS
low pressure coolant recirculation system
LPCS
low pressure core spray
LPCVD
low pressure chemical vapor deposition
LPD
language processing and debugging
linear power density
log periodic dipole
log periodic dipole antennas
low power difference
low pressure difference
LPDA
log periodic dipole array
LPDR
local public document room
LPDTL
low power diode transistor logic
LPDTML
low power diode transistor micrologic
LPE
liquid phase epitaxial
liquid phase epitaxy
lopp preparation equipment
LPERE
linear phase with equal ripple error
LPF
logically passive function
low pass filter
LPFGEN
linear programming file generator
LPFPRINT
linear programming file print
LPG
liquefied petroleum gas
liquid petroleum gas
list program generator
LPGAS
liquefied petroleum gas

LPGE
lunar excursion module partial guidance equipment
LPGS
liquid pathway generic study
LPGTC
liquified petroleum gas industry technical committee
LPHB
low pressure heating boiler
LPI
lightning protection institute
lines per inch
list per inch
low power injection
LPIA
liquid propellant information agency
LPICBM
liquid propellant intercontinental ballistic missile
LPIS
low pressure injection system
LPL
linear programming language
list processing language
louisiana power and light
low power logic
LPLM
lowest planned level of maintenance
LPM
laser precision microfabrication
licensing project manager
line per minute
local processor memory
LPMA
loose parts monitor assembly
LPMATGEN
linear programming matrix generation
LPMOSS
linear programming mathematical optimization subroutine system
LPO
low power output
lunar program office
LPP
launching preparation panel
low power physics
LPPC
load point photocell
LPPT
low pressurization pressure test transmitter
LPR
late position report
line printer
liquid propellant rocket

long-playing record
loop position regulator
lynchburg pool reactor
LPRD
  launch program requirement document
LPREPORT
  linear programming report
LPRINT
  lookup dictionary print program
LPRM
  low power range monitor
LPRS
  local power recirculation system
LPS
  laboratory peripheral system
  launch phase simulator
  line program selector
  linear programming system
  lines per second
  liters per second
  low power schottky
  low pressure scram
LPSD
  logically passive self dual
LPSOL
  linear programming solution
LPSTTL
  low power schottky transistor transistor logic
LPSW
  load program status word
  low pressure service water
LPT
  low power test
LPTD
  long-play talkdown
LPTF
  low power test facility
LPTR
  livermore pool type reactor
LPTTL
  low power transistor transistor logic
LPTV
  large payload test vehicle
LPUL
  least positive uplevel
LPV
  light pen value
  log periodic v antenna
LPW
  lumen per watt
LPWG
  lunar and planetary working group
LPZ
  low population zone

LQ
  letter quality
  limiting quality
  liquid
LQG
  linear quadratic gaussian
LQP
  letter quality printer
LQT
  linear quantizer
LR
  last record
  left + right
  left right
  left to right parse
  level recorder
  limited recoverable
  line relay
  liquid rocket
  lloayd's register
  load ratio
  load register
  load resistor
  lock range
  locus of radius
  logical record
  long-range
  low register
  low resistance
  low resistive
  low resistor
  lower
  lunar rendezvous
LRA
  laser gyro reference axis
  locked rotor amperes
LRAS
  lunar module replaceable assemblies
LRASV
  long-range air to surface vessel
  long-range air to surface vessel radar
LRB
  load request block
  local reference beam
LRBM
  long-range ballistic missile
LRBR
  long-range ballistic rocket
LRC
  langley research center
  level recorder controller
  level recording controller
  lewis research center
  load ratio control

longitudinal redundancy check
longitudinal redundancy check character
LRCO
  limited radiocommunication outlet
  limited remote communication outlet
LRCR
  longitudinal redundancy check register
LRD
  long-range data
LRDP
  long-range development program
LRDSB
  left minus right double sideband
LRE
  liquid rocket engine
  lossless reciprocal embedding
LRG
  long-range
LRHS
  large radioisotope heat source
LRI
  left right indicator
  long range radar input
LRIA
  level removable instrument assembly
LRIM
  long-range input monitor
LRIR
  low resolution infrared radiometer
LRL
  lawrence radiation laboratory
LRLM
  lower reject limit median
LRLTRAN
  lawrence radiation laboratory translator
LRM
  limited register machine
  long-range missile
  lunar reconnaissance module
  lunar rendezvous mission
LRML
  long-range missile launcher
LRN
  long-range navigation
LRP
  launch reference point
  logical record processor
  long-range path
  long-range plan
  long-range planning
LRPL
  liquid rocket propulsion laboratory
LRR
  long-range radar

LRRP
  lowest required radiated power
  lowest required radiating power
LRS
  linguistic research system
  liquid radwaste system
  long-range search
  long right shift
LRSAGW
  long range suface to air guided weapon
LRSM
  long range seismograph measurements
LRSS
  long-range survey system
LRT
  load ratio transformer
  local leak rate test
LRTF
  long-range technical forecast
LRTM
  long-range training mission
LRU
  last recently used
  least recently used
  least replaceable unit
  line replaceable unit
LRV
  lunar roving vehicle
LRY
  latching relay
LS
  laboratory system
  landing ship
  language specification
  lapped seam
  laser system
  late scramble
  launch and service
  launch sequencer
  launch simulator
  launch site
  least significant
  left side
  left sign
  letter service
  level switch
  light source
  lightning arrester
  limit stop switch
  limit switch
  line sequential
  line stretcher
  lobe switching
  local storage
  local sunset

long shot
longitudinal staggering
loudspeaker
low secondary
low speed
LS SEARCH
 limited semantics search
LSA
 limited space accumulation
 limited space charge accumulation
 line sensing amplifier
 logic state analyzer
 low cost solar array
 low specific activity
 low speed adapter
LSAOSCILLATOR
 limited space charge accumulation oscillator
LSAR
 local storage address register
LSB
 launch service building
 launcher support building
 least significant bit
 least significant byte
 lower sideband
LSBR
 large seed blanket reactor
LSC
 least significant character
 least square center
 legal services corporation
 limit signaling comparator
 linear sequential circuit
 linear shaped charge
 lines of communications
 liquid scintillation counter
 liquid solid chromatography
LSCAN
 l scanner
LSCC
 line sequential color composite
LSCL
 limit switch closed
LSCR
 launch sequence and control equipment
LSD
 launch systems data
 least significant decade
 least significant digit
 light sensing device
 limit switch down
 limited space charge drift
 linkage system diagnostic
 low speed data
 lysergic acid diethylamide

LSDF
 large sodium disposal facility
LSE
 launch sequencer equipment
 launch support equipment
 london stock exchange
 longitudinal section electric
 low speed encoder
 lunar support equipments
LSECS
 life support and environmental control system
LSES
 life support and environmental system
LSF
 limit switch forward
 loss factor
LSFFAR
 low spin folding fin aircraft rocket
LSFT
 low steamline flow test
LSH
 light ship
LSHI
 large-scale hybrid integration
LSHV
 laminated synthetic high voltage
LSI
 large-scale integrated
 large-scale integration
 launch success indicator
LSIA
 lamp and shade institute of america
LSIC
 large-scale integrated circuit
LSIG
 least significant
LSIT
 large-scale integration technology
LSL
 ladder static logic
 logistics system laboratory
 low speed logic
LSLI
 large-scale linear integration
LSM
 launch site maintenance
 launcher status multiplexer
 linear synchronous motor
 longitudinal section magnetic
 low speed modem
LSMU
 lasercom space measurement unit
LSN
 line stabilization network
 linear sequential network

LSO
 landing signal officer
 launch safety officer
LSOC
 logistical support operations center
LSOP
 limit switch open
LSP
 launch sequence plan
 line spectrum pair
 logical signal processor
 low speed printer
 lunar survey probe
LSPBP
 large solid propellant booster program
LSPC
 lewis space flight center
LSPF
 least squares polynomial fit
LSR
 large ship reactor
 launch signal responder
 laurence, scott and electromotors
 limit switch reverse
 line source range
 load shifting resistor
 local shared resources
 local sunrise
 lynchburg source reactor
LSS
 language support system
 large scale standard
 large space structure
 large space system
 launcher status summarizer
 life support system
 limited storage site
 linking segment subprogram
LSSD
 level sensitive scan design
LSSM
 local scientific survey module
 logical scientific survey module
LSSP
 latest scram set point
LSSS
 limiting safety system setting
LSST
 lead sheathed steel taped
LST
 landing ship for tanks
 late start time
 left store
 local sidereal time
 local standard time
 local summer time
LSTATE
 low state
LSTN
 light station
LSTTL
 low power schottky transistor transistor logic
LSU
 limit switch up
 line selector unit
 local storage unit
 louisiana state university
LSV
 lunar survey viewfinder
LSW
 limit switch
 line switch
LT
 laboratory test
 language translation
 language translator
 launcher test
 left
 less than
 letter telegram
 level transmitter
 level trigger
 light
 light test
 ligic theory
 limit
 line telecommunications
 line telegraphy
 line telephony
 line terminator
 link trainer instructor
 local time
 logic[al] theory
 long ton
 low temperature
 low tension
 lug terminal
LTA
 lem test article
 lighter than air
 logic time analyzer
LTAL
 lower transition altitude
LTALT
 light alternating
LTB
 last trunk busy
 low tension battery

LTBO
  linear time base oscillator
LTC
  launch vehicle test conductor
  line terminal control
  line traffic coordinator
  linear transformation converter
  load tap changing
  long time constant
  low tension current
LTDS
  laser target designation system
  launch trajectory data system
LTDSTD
  limited standard
LTE
  local thermodynamic equilibrium
LTF
  life time function
LTFRD
  lot tolerance fraction reliability deviation
LTG
  lighting
LTH
  lighthouse
LTHA
  long term heat ageing
LTHO
  lighthouse
LTI
  light transmission index
  linear time invariant
LTL
  lot truck load
LTM
  load ton miles
LTN
  line terminating network
  listen
LTNG
  lightning
LTNGARR
  lightning arrester
LTP
  library technology project
  line type processor
  low temperature polymer
  lower trip point
LTPD
  lot tolerance percent defective
LTPL
  long-term procedural language
LTPR
  light proof
LTPS
  lincoln tube process specification
LTR
  lattice test reactor
  left test register
  letter
  licensing technical review
  lockheed training reactor
LTROM
  linear transformer memory
  linear transformer read only memory
LTRS
  letter shift
  letter shift lower case
LTS
  laser triggered switch
  lateral test simulator
  launch telemetry station
  launch tracking system
  lift off transmission subsystem
  long-term stability
  lunar touchdown system
LTSR
  line trunk scanner register
LTSW
  light switch
LTTL
  low power transistor transistor logic
LTU
  line termination unit
LTV
  large test vessel
  launch test vehicle
  long tube vertical
  lunar excursion module test vehicle
LTW
  low tension winding
LU
  line unit
  load unit
  logical unit
  lumen
LUB
  logical unit block
  lubricant
LUCID
  language for utility checkout and instrumentation development
  language used to communicate information system design
  loughborough university computerized information and drawings
LUCOM
  lunar communication

LUCS
london university computer services
LUCUS
london university computer services
LUE
left upper extremity
LUF
lowest usable frequency
LUH
lumen hour
LUHF
lowest usable high frequency
LUI
logical unit of information
LUL
left upper lobe
LULS
lunar logistic system
LUM
luminous
LUMAS
lunar mapping system
LUME
light utilization more efficient
LUMF
lockheed underwater missile facility
LUS
load, update, subset
LUSI
lunar surface inspection
LUSTER
lunar dust and earth return
LUT
launch umbilical tower
launcher umbilical tower
LV
laser vision
launch verification
launcher vehicle
legal volt
low voltage
luncheon voucher
LVA
logarithmic video amplifier
low voltage avalanche
LVAP
launch vehicles and propulsion
LVAS
launch vehicle alarm system
LVCD
least voltage coincidence detection
least voltage coincidence detector
LVD
low velocity detonation
low voltage drop

LVDA
launch vehicle data adapter
LVDC
launch vehicle digital computer
LVDT
linear variable differential transformer
linear velocity displacement transformer
LVE
launch vehicle engine
LVEA
leligh valley electronic association
LVHF
low very high frequency
LVHV
low volume high velocity
LVI
low viscosity index
LVL
level
LVM
launch vehicle monitor
LVMOST
laterl v groove depletion most
LVN
low voltage neon
LVO
launch vehicle operation
LVOD
launch vehicle operations division
LVOR
low power very high frequency
omnidirectional range
LVP
light valve projector
low voltage protection
LVPG
launch vehicle planning group
LVPS
low voltage power supply
LVR
long vertical right
longitudinal video recorder
low voltage relay
low voltage release
LVRS
launch vehicle recovery system
LVS
launch vehicle simulator
low velocity scanning
LVSE
launch vehicle system engineer
LVSG
launch vehicle study group
LVSSTS
launch vehicle safety system test set

LVST
longitudinal velocity sorter tube
LVT
landing vehicle tracked
linear velocity transducer
LVTC
launch vehicle test conductor
LVTR
low power very high frequency transmitter receiver
LW
last word
leave word
light wall
light warning
light weight
lime wash
long wave
low frequency
lumen per watt
LWA
limited work authorization
LWB
long wheel base
LWBR
light water breeder reactor
LWC
liquid water content
LWD
larger word
laser welder driller
launch window display
LWGCR
light water moderated, gas cooled reactor
LWGR
light water cooled, graphite moderated reactor
LWIR
long wave infrared
LWL
load water line
low water line
LWM
liquid waste monitor
low water mark
LWOP
leave without pay
LWOST
low water ordinary spring tide
LWR
light water reactor
liquid waste release
lower
LWRU
light weight radar unit

LWS
light warning set
LWSR
light weight search radar
LWST
light waste storage tank
LWT
listen while talk
local winter time
LWTT
liquid waste test tank
LWW
launch window width
LX
logabox
lux
LXA
load index from address
LXD
load index from decrement
LXS
lux second
LY
langley
light year
LYR
layer
LYRIC
language for your remote instruction by computer
LZ
left-hand zero leading zero
left zero
live zero
LZCC
landing zone control center
LZP
left zero print
LZT
local zone time

M

M
mach
mach number
magnaflux
magnet
magnetic
magnetic moment
magnetic vector
magnetization
magnetron

main
main channel
maintenance
manual
mark
marker
marker beacon
mass
maxwell
mean
measure
mechanical
medium
medium mega
meg
mega
megaohm
melts at
member
memory
merge
meridian
meta
metallic
meter
micro
microphone
middle
mile
milli
millimicrometer
millimicron
million
minute
mired
mist
mobile
mode
model
modem
modulation
modulator
module
modulus
molecular weight
moment
moment of force
monitor
monochrome
motor
motor starter
movement
mutual
MA
  mach

magnetic amplifier
main alarm
maintenance agreement
maintenance analysis
maritime administration
massachusetts
mast aerial
mathematical association
mechanical advantage
mechanoacoustic
megaampere
memory address
memory available
mercury atlas
message assembler
metric association
microalloy
microwave associates
milliammeter
milliampere
modify address
multiple access
myria
MAA
  material access area
  mathematical association of america
MAAC
  mid atlantic area council
  milliampere alternating current
MAARC
  magnetic annular arc
MAB
  missile assembly building
MABLE
  miniature autonetics base line equipment
MAC
  machine aided cognition
  maintenance allocation chart
  maintenance and construction
  malfunction circuitry trainer
  man and computer
  maximum admissible concentration
  maximum allowable concentration
  mcdonnel aircraft corporation
  mean aerodynamic chord
  measurement and control
  memory access command
  memory access controller
  memory address counter
  merchant aircraft carrier
  mineralogical association of canada
  monitor and control
  monthly availability charge
  multi access computing
  multi analyzer configuration

multi application computer
multiple access computer
multiple access computing
multiple address code
multiple address computer
MACC
modular alter and compose console
MACCS
manufacturing and cost control system
manufacturing cost collection system
MACCT
multiple assembly cooling cask test
MACE
management applications in a computer environment
master control executive
MACL
minimum acceptable compliance level
MACMIS
maintenance and construction management information system
MACON
matrix connector punched card programmer
MACP
macro control processor
MACS
management and computer services
medium altitude communications satellite
missile air conditioning system
multipurpose acquisition and control system
multipurpose acquisition control system
MACSMB
measurement and automatic control standards management board
MACSRRPCC
maxwellian averaged cross section reactor physics computer code
MACSS
medium altitude communication satellite system
MACSYMA
mac symbolic mathematics
MACU
monitor and control unit
MAD
magnetic airborne detector
magnetic anomaly detection
magnetic anomaly detector
maintenance analysis diagram
maintenance, assembly, disassembly
man magnetic airborne detector man
material analysis data
material analysis department
mean absolute deviation
memory access director

michigan algorithm decoder
michigan algorithmic decoder
morse automatic decoder
multi aperture device
multiple access device
multiply and add
music associated's detector
MADA
multiple access discrete address
MADAEC
military application division of the atomic energy commission
MADAM
moderately advanced data management
multipurpose automatic data analysis machine
MADD
multichannel analog to digital data decoder
MADAP
maastricht automatic data processing
MADAR
malfunction analysis detection and recording
malfunction and data recorder
MADBLOC
mad language block oriented computer
MADC
milliampere direct current
MADCAR
management data charting and review
MADDAM
macromodule and digital differential analyzer machine
multiplexed analog to digital, digital to analog multiplexed
MADDDIA
magnetic drum digital differential analyzer
MADE
microalloy diffused electrode
minimum airborne digital equipment
multichannel analog to digital data encoder
MADGE
microwave aircraft digital guidance equipment
MADIS
millivolt analog to digital instrumentation system
MADM
manchester automatic digital machine
MADP
main air display plot
MADRE
magnetic drum receiving equipment
martin automatic data reduction equipment
MADREC
malfunction and detection recording
malfunction detection and recording

MADS
   machine aided drafting systm
   missile attuned determination system
   multiple access digital system
MADT
   microalloy diffused base transistor
   microalloy diffused transistor
MADW
   military air defense warning
MADWN
   military air defense warning network
MAE
   maine association of engineers
   maryland association of engineers, inc
   master electric
   material and equipment
   mean absolute error
   mechanical and electrical
   men absolute error
   missile airborne equipment
   missile assembly equipment
MAECON
   mid-america electronics conference
   mid-american electronics conference
MAEE
   marine aircraft experimental establishment
MAELU
   mutual atomic energy liability underwriters
MAER
   mechanical and electrical room
MAERU
   mobile ammunition evaluation and reconditioning unit
MAES
   manufacturing and engineering support
   mexican american engineering society
MAESTRO
   machine assisted educational system for teaching by remote operation
MAET
   missile accident emergency team
MAF
   major academic field
   manpower authorization file
   minimum audible field
   missile assembly facility
   moisture and ash free
   multiple access facility
MAFB
   malstrom air force base
MAFD
   minium acquisition flux density
MAG
   magazine
   magnet
   magnetic
   magnetics
   magnetism
   magneto
   magnetron
   magnitude
   marine aircraft group
   maximum available gain
   military advisory group
MAGAMP
   magnetic amplifier
MAGCI
   magnetic cast iron
MAGCS
   magnetic cast steel
MAGG
   modular alphanumeric graphics generator
MAGGS
   modular advanced graphics generation system
MAGI
   mathematical applications group, inc
   multi array gamma irradiator
MAGIC
   machine for automated graphics interface to a computer
   machine for automatic graphics interface to a computer
   matrix algebra general interpretive coding
   michigan automatic general integrated computation
   midac automatic general integrated computation
   modified action generated input control
MAGIS
   marine air ground intelligence system
MAGLATCH
   magnetic latch
MAGLOC
   magnetic logic computer
MAGMOD
   magnetic modulator
MAGN
   magnetic
   magnetic airborne detection
   magnetism
MAGNETTOR
   magnetic modulator
MAGPIE
   machine automatically generating production inventory evaluation
MAGREC
   magnetic recording

MAGTRAC
  magnetic amplifier, transistorized, automatic target tracker
MAI
  management assistance inc
  micanite and insulators
  multiple access interface
MAID
  maintenance automatic integration director
  monrobot automatic internal diagnosis
  multiple aircraft identification display
MAIDS
  multipurpose automatic inspection and diagnostic system
MAIN
  mid-american interpool network
MAINT
  maintenance
MAIP
  matrix algebra interpretive program
MAIR
  modular airborne intercept radar
  molecular airborne intercept radar
MAJAC
  maintenance antijam console
  monitor, antijam, and control
MAJCON
  major aid command controlled
MAL
  macroassembly language
  material allowance list
MALE
  multiaperture logic element
MALLAR
  manned lunar landing and return
MALT
  mnemonic assembly language translator
MAM
  management and administration manual
  medium altitude missile
  microwave attenuator monitor
  milliampere minute
  missile assembly and maintenance
  multiple access to memory
MAMB
  missile assembly and maintenance building
MAMBO
  minuteman assembly and maintenance building
MAMI
  machine aided manufacturing information
MAMIE
  magnetic amplification of microwave integrated emissions
  minimum automatic machine for interpolation and extrapolation
MAMMAX
  machine made and machine assisted index
MAMOS
  marine automatic meteorological observing station
MAMP
  meter ampere
  milliampere
MAMS
  missile assembly and maintenance shop
  missile assitance maintenance structure
MAN
  maintenance alert network
  mantisse
  manual
  manufactured
  metropolitan area network
  microwave aerospace navigation
MANAV
  manoeuvring and navigation system
MANDA
  management and administration
MANDRO
  mechanically alterable nondestructive read out
MANIAC
  mathematical analyzer numerical integrator and computer
  mechanical and numerical integrator and calculator
  mechanical and numerical integrator and computer
MANIP
  manual input
MANL
  manual
MANOP
  manual of operations
  manual operation
  manually operated
MANOVA
  multivariate analysis of variance
MANTRAC
  manual angle tracking capability
MAO
  maintenance and operation
MAOS
  metal alumina oxide silicon
  metal aluminia oxide semiconductor
  metal aluminia oxide silicon
MAP
  macroarithmetic processor
  macroassembly program

main arithmetic processor
maintenance analysis procedure
maintenance analysis program
management and programming
manifold absolute pressure
manufacturing automation protocol
master operational recording tape assembly program
mathematical analysis without programming
maximum a posteriori probability
mercury all position
message acceptable pulse
message acceptance pulse
missile assignment program
model and program
module analysis processor
multiple address processing system
multiple aimpoint
multiple allocation procedure

MAPCHE
mobile automatic programmed checkout equipment

MAPI
mitsubishi atomic power industries

MAPID
machine aided program for preparation for instruction data
machine aided program for preparation of instruction data

MAPLHGN
maximum average planar linear heat generator

MAPLHGR
maximum average planar linear heat generation rate
maximum average planar linear heat generator

MAPORD
methodology approach to planning and programming air force operational requirements, research and development

MAPP
masking parameter print out
materials and plate protection
midcontinent area power planner

MAPRAT
maximum power ratio

MAPS
management aids program suite
master activation phasing schedule
multicolor automatic projection system
multiple address processing system
multivariate analysis and prediction of schedules

MAPSE
minimal ada programming support environment

MAR
malfunction array radar
manufacturing assembly report
master angular reference
memory address register
mercury arc rectifier
microanalytical reagent
miscellaneous apparatus rack
multifunction array radar
multiple array radar

MARC
machine readable catalog
machine readable cataloging
machine readable code
magnetic abrasion resistant coating
material accountability recoverability code
monitor and results computer
mortage account report compiler
multi axial radial circuit

MARCA
midcontinent area reliability coordination agreement

MARCAS
maneuvering reentry control and ablation studies
maneuvering reentry control and ablation system

MARCEP
maintainability and reliability cost effectiveness program

MARCIA
mathematical analysis of requirements for career information apraisal

MARCOM
microwave airborne communications relay

MARCONIREV
marconi review

MARDAN
marine differential analyzer

MARGEN
management report generator

MARI
motivator and response indicator

MARIDAS
maritime data system

MARINETRANSP
marine transportation

MARLIS
multi aspect relevance information system
multi aspect relevance linkage information system

multiple aspect relevance linkage
information system
MAROTS
maritime orbital test satellites
MARS
machine retrieval system
magnetic airborne recording system
management analysis reporting service
management analysis reporting system
manned astronautical research station
manned astronomical research station
marconi automatic relay system
marketing activities reporting system
martin automatic reporting system
master attitude reference system
memory address register storage
military affiliate radio system
military affiliated radio system
military airborne radar system
military amateur radio system
millimeter wave amplification by resonance saturation
mobile atlantic range system
multi access retrieval system
multi aperture reluctance switch
multiple apertured reluctance switch
MART
maintenance analysis review technique
mean active repair time
missile automation radiation tester
MARTAC
martin automatic rapid test and control
MARTEC
martin thin film electronic circuit
MARTI
maneuverable reentry technology investigation
MARTINI
massive analog recording technical instrument for nebulous indications
MARTOS
multi access real time operating system
MARV
maneuverable anti radar vehicle
MAS
maintenance and supply
management advisory services
marine acoustical services
material application service
metal alumina semiconductor
metal alumina silicon
microprogram automation system
military agency for standardization
milliampere second
modular accounting system
modular applications system
multiaspect signaling
MASC
multilayer aluminium oxide silicon dioxide combination
MASCOT
meteorological auxiliary sea current observation transmitter
mobile air transportable satellite communications terminal
motorola automatic sequential computer operated tester
MASD
mach aids to surface to air missile development
MASER
microwave amplification by stimulated emission of radiation
MASFET
metal alumina silicon fet
metal alumina silicon field effect transistor
MASIS
management and scientific information system
managment and scientific information system
mercury abort sensing instrumentation system
MASK
maneuvering and sea keeping
masking
MASR
memory address select register
MASRT
marine air support radar teams
MASRU
marine air support radar unit
MASS
michigan automatic scanning system
monitor and assembly system
multiple access sequential selection
MASSDAR
modular analysis, speed up, sampling, and data reduction
MASSPO
manned spaceflight support project office
MASSTER
mobile army sensor system test evaluation and review
MAST
magnetic annular shock tube
missile automatic supply technique
MASTAP
master system tape
MASTER
matching available student time to educational resources

multiple access share time executive routine
multiple access shared time executive routine
MASTIF
multiaxis spin test inertia facility
MASW
master switch
MAT
machine aided translation
machine analysis table
master operational recording tape address table
mechanical aptitude test
mechanical assembly technique
microalloy transistor
military aircraft types
missile acceptance team
missile acceptance test
mobile aerial target
modular assembly technique
multiple address telegram
MATA
michigan aviation trades association
MATABE
multiple weapon automatic target and battery evaluator
MATCALS
marine air traffic control and landing system
MATCH
multi element assured tracking chopper
MATCON
microwave aerospace terminal control
MATCU
marine air traffic control unit
MATD
mine and torpedo detector
MATE
measuring and test equipment
modular automatic test equipment
multiple access time division equipment
multiple access time division experiment
multisystem automatic test equipment
MATH
mathematical
MATHPAC
mathematical package
MATIC
multiple area technical information center
MATICO
machine applications to technical information center operations
MATLAN
matrix language
MATPS
machineaided technical processing system

MATRS
miniature airborne telemetry receiving station
MATS
mechanical accounting for telephone service
mechnical accounting for telephone service
multi purpose automatic test system
multiple access time sharing
MATTS
multiple airborne target trajectory system
MATU
marine air traffic unit
MATV
master antenna television
MAU
maintenance analysis unit
maintencance analysis unit
multiattribute utility
MAUDE
morse automatic decoder
MAULT
manual automatic ultrasonic laboratory test
MAV
milliampere per volt
MAVAR
microwave amplification by variable reactance
mixer amplification by variable reactance
modulating amplifier using variable reactance
modulating amplifier using variable resistance
MAVIN
machineassisted vendor information network
MAW
marine aircraft wing
mission adaptive wiring
MAWCS
mobile air weapons control system
MAX
maximum
maxwell
MAXSECOM
maximum security communications
MAXUPO
maximum undistorted power output
MAZH
missile azimuth heading
MAZO
missile azimuth orientation
MB
magnetic bearing
magnetic brake
main battery
make break
megabyte

memory buffer
metric board
millibar
mobile base
MBA
master of business administration
material balance area
MBAR
millibar
MBAUD
megabaud
MBB
make before break
MBC
main beam clutter
manual battery control
marine broadcasting company
miniature bayonet cap
multiple board computer
MBCU
mobile bombardment communications unit
MBD
magnetic bubble device
MBE
missile borne equipment
molecular beam epitaxy
MBEO
minority business enterprise office
MBF
modulator band filter
MBK
make break keying
multiple beam klystron
MBL
main battle line
marine biological laboratory
miniature button light
mobile
MBM
magnetic bubble memory
metal-barrier-metal
MBMS
model base management software
MBO
management by objectives
monostable blocking oscillator
MBOS
missile base operations supervisor
MBPS
megabit per second
megabits per second
MBR
marker beacon receiver
material balance report
mechanical buffer register

memory buffer register
mini badge reader
MBRE
memory buffer register even
MBRO
memory buffer register odd
MBRUU
may be retained until unserviceable
MBRV
maneuverable ballistic reentry vehicle
MBS
magnetron beam switching
main bang suppressor
main buffer storage
management by system
megabit per second
megabit pro second
megabyte pro second
multibit shifter
multiple business system
municipal broadcasting system
mutual broadcasting system
MBSG
missouri basin system group
MBSP
main bang synchronization pulse
MBSS
main beach signal station
MBST
multiple beam switching tube
MBT
memory block table
metal base transistor
MBTS
missile battery test set
MBTWK
multiple beam traveling wave klystron
MBV
minimum breakdown voltage
MBWO
m type backward wave oscillator
microwave backward wave oscillator
MBX
management by exception
MBY
make busy
MBYTE
megabyte
MBZ
must be zero
MC
machine
machine certificate
machine check

machine code
machine cycle
magnetic card
magnetic clutch
magnetic core
magnetic course
major component
making capacity
managment committee
manhole cover
manual control
manufacturing change
marginal check
marginal checking
maritime commission
master change
master clock
master clods
master control
mathematical center
maximum count output
medical corps
medium curing
megacycle
megacycles per second
mercury contact
message composer
metal clad
metercandle
metric carat
microcomputer
microminiature circuit
microphase corporation
midcourse correction
military computer
millicurie
millicycle
missile code
missile command
missile control
mode change
mode control
modem controller
molded components
momentary contact
morse code
motor car
motor contact
motor converter
moulded component
moving coil
multichip
multicomputing
multicomputing system
multiple contact
multiplex channel
mutual coupling

MCA
main console assembly
manufacturing chemists association
massachusetts computer associates
material control and accountability
material coordinating agency
maximum credible accident
military coordinating activity
ministry of civil aviation
model cities administration
multichannel analyzer
multiplexing channel adapter
multiprocessor communications adapter

MCAA
mechanical contractors association of america

MCAD
mechanical computer aided design

MCAP
ministry of civil aviation publication

MCAR
machine check analysis recording

MCAS
minuteman configuration accountability system

MCAT
midwest council on airborne television

MCBETH
military computer basic environment for test handling

MCBF
mean cycles between failures

MCC
main communications center
main control circuit
maintenance control center
maintenance control circuit
maintenance of close contact
management control center
management controls corporation
manual control center
master control console
memory control circuit
mercury control center
miniature center cap
minuteman change committee
missile checkout console
missile command coder
missile control center
mission control center
modified close control
modulation with constant coefficient
monitor control console
motor control center

movement control center
multi chip carrier
multi component circuit
multiple communication control
multiple computer complex
MCCB
minuteman change commitment board
MCCD
meander channel ccd
multiplexed ccd
MCCH
mission control center houston
MCCU
multiple communication control unit
MCD
marginal checking and distribution
measurement control devices
millicurie destroyed
months for cyclical dominance
MCDF
missile defense alarm system control and display facilities
MCDP
microprogrammed communication data processor
MCDS
management control data system
MCE
master of civil engineering
MCEB
military communications electronics board
MCESS
milwaukee council of engineering and scientific societies
MCEU
mobile civil emergency unit
MCF
mean carrier frequency
military computer family
monolithic crystal filter
MCG
magnetocardiography
man computer graphics
master control gage
microwave command guidance
mobile command guidance
MCGCIS
marine corps ground controlled interceptor squadron
MCGS
microwave command guidance system
MCH
machine check handler
machine check interruption

MCHFR
minimum critical heat flux rates
MCI
machine check interruption
malleable cast iron
microwave communications, incorporated
millicurie
MCID
multipurpose concealed intrusion detector
MCIS
maintenance control information system
materials compatibility in sodium
multiple corridor identification system
MCIT
management controls and information technology
MCL
master clear line
microwave cavity laboratories
mini circuits laboratory
miniature cartridge light
minority carrier lifetime
MCM
magnetic card memory
magnetic core memory
merged charge memory
microcomputer machines
microwave circuit module
milli circular mil
missile carrying missile
monte carlo method
moving coil motor
MCMA
mains cables manufacturers' association
MCMB
multiple conductor, marker buoy
MCMJ
michigan mathematical journal
MCMM
management control material management
MCMOS
motorola cmos
MCMS
multichannel memory system
MCN
midcourse navigation
MCNC
microcomputer numerical control
MCO
missile control officer
MCOM
mathematics of computation
MCONTACT
make contact

MCP
  master control program
  memory centered processor
  message control program
  microchannel plate
  multichannel communication program
MCPDP
  meander channels plasma display panel
MCPF
  multichannel peak factor
MCPR
  maximum critical power ratio
MCPS
  megacycles per second
MCQ
  memory call queue
MCR
  magnetic character reader
  magnetic character recognition
  main control room
  master change record
  master control routine
  maximum continuous rating
  military compact reactor
  minuteman change request
  multi contact relay
MCROA
  marine corps reserve officers association
MCRR
  machine check recording and recovery
MCRU
  mobile control and reporting unit
MCRWV
  microwave
MCS
  main control station
  maintenance control section
  maintenance control system
  management control system
  master control set
  master control system
  medical computer services
  megacycle per second
  megacycles per second
  message control system
  method of constant stimuli
  microcomputer system
  microwave carrier supply
  missile calibration station
  missile checkout station
  mobile calibration station
  mobile checkout station
  mobile communications systems
  modular computer system
  modulation controlled synchronization
  monte carlo simulation
  motor circuit switch
  multichannel communication system
  multichannel switch
  multichannel system
  multiple column selector
  multiple console support
  multiprogrammed computer system
  multipurpose communications and signaling
MCSP
  multiple conductor, shielded, pressure resistant
MCSS
  military communications satellite system
MCSWGR
  metal clad switchgear
MCT
  magnetic card and tape unit
  magnetically coupled transformer
  mechanical comprehension test
  mercury cadmium telluride
  movable core transformer
MCTG
  model change training guide
MCTI
  metal cutting tool institute
MCTR
  message center
MCTS
  master central timing system
MCU
  machine control unit
  machine tool control unit
  magnetic card unit
  main control unit
  maintenance control unit
  measurement control unit
  microprogram control unit
  millicurie
  miniature command unit
  minicomputer unit
MCUG
  military computer users group
MCV
  magnetic cushion vehicle
  movable closure valve
MCW
  modified continuous wave
  modulated carrier wave
  modulated continuous wave
MCX
  minimum cost expediting
MCY
  machinery

MD
  macro directory
  macrodata
  magnet driver
  magnetic disk
  magnetic drum
  main drum
  maintenance documentation
  manual data
  maryland
  maximum demand
  mean deviation
  measured discard
  medical department
  medium duty
  message data
  message per day
  mine disposal
  missile division
  modulation demodulation
  modulator
  modulator demodulator
  monitor displays
  months' date
  moulded depth
  movement derective
  multiple divide
  multiply divide

MDA
  mechanically despun antenna
  minimum descent altitude
  monoalythic design automation
  multidimensional analysis
  multidocking adapter

MDAA
  mutual defense assistance act

MDAC
  mcdonnell douglas astronautics corporation
  multiplying digital to analog converter

MDAP
  morphological dictionary adapter program
  mutual defense assistance program

MDB
  microsecond delay blasting

MDC
  main display console
  maintenance data collection
  maintenance dependency chart
  master direction center
  maximum dependable capacity
  missile development center
  missile direction center
  mobile distress calling
  multiple device controller
  multistage depressed collectors

MDCS
  maintenance data collection system
  master data control system
  master digital command system

MDCT
  mechanical draft cooling tower

MDCU
  mobile dynamic checkout unit

MDDPM
  magnetic drum data processing machine

MDDR
  minimum distance decoding rule

MDE
  magnetic decision element
  main distribution equipment
  mission display equipment
  modular design of electronics

MDEFWP
  motor driven emergency feedwater pump

MDF
  main distributing frame
  main distribution frame
  manual direction finder
  master data file
  medium frequency direction finder
  microcomputer development facilities
  mild detonating fuse

MDI
  magnetic detection indicator
  magnetic direction indicator
  manual data input
  miss distance indicator
  monopulse display improvement
  multiple display indicator

MDIC
  microwave dielectric integrated circuit

MDIE
  mother daughter ionosphere experiment

MDIF
  manual data input function

MDIRT
  miss distance indicator radioactive test

MDIS
  manual data input system

MDIU
  manual data insertion unit

MDL
  macrodescription language
  main defence line
  microwave development laboratories
  mine defense laboratory
  miniature display light
  minimum detectable level
  mohawk data language

MDM
    magnetic drum memorex
    maximum design meter
    metal dielectric metal
    microdensitometer
    multiprocessing diagnostic monitor
MDMS
    miss distance measuring system
MDNC
    mechanical drafting numerical control
MDO
    main dial office
MDP
    main display panel
    maintenance diagnostic program
    microprocessor debugging program
    motorola data processor
MDR
    magnetic document reader
    magnetic field dependent resistor
    manual data room
    mark document reader
    memory data register
    mission data reduction
    multichannel data recorder
MDRS
    manpower data relay station
MDS
    magnetic disk storage
    magnetic disk store
    magnetic drum storage
    magnetic drum store
    main device scheduler
    maintenance data system
    malfunction detection system
    manual data supervisor
    marine distress signal
    market data system
    master drum sender
    memory disk system
    message dropping station
    meteoroid detection satellite
    microcomuter development system
    microprocessor development system
    minimum detectable signal
    minimum discernible signal
    modern data system
    mohawk data sciences
    multiple deployment system
    multiple distribution service
    multipoint distribution systems
MDSMB
    medical devices standards management board
MDSR
    malay district signal regiment
MDSS
    meteorological data sounding system
    mission data support system
MDST
    mountain daylight saving time
MDT
    manual data technician
    mci data transfer corporation
    mean down time
    medium data technique
    modified data tag
    mountain daylight time
MDTA
    man power development and training act
MDTL
    modified diode transistor logic
    motorola diode transistor logic
MDTS
    megabit digital to troposcatter subsystem
    modular data transaction system
    modular data transmission system
MDU
    message decoder unit
    mine disposal unit
    missile design unit
    montana dakota utilities
MDV
    map and data viewer
    minimum domain velocity
MDW
    mars departure window
    multipair distribution wire
MDZ
    middle zero
ME
    maine
    manufacturing engineering
    master element
    master of engineering
    measurement engine
    measuring element
    mechanical efficiency
    mechanical electrical
    mechanical engineer
    mechanical engineering
    megacycle
    microelectronic
    military engineer
    mining engineer
    modular electronics
    molecular electronics
MEA
    minimum en route altitude
    minimum en route instrument altitude

MEACON
   masking beacon
MEACONING
   measuring and confusing
MEAL
   master equipment allowance list
   master equipment authorization list
MEAM
   municipal electric association of massachusetts
MEAR
   maintenance engineering analysis record
   maintenance engineering analysis report
MEAS
   measurable
   measure
   measuring
MEASCAL
   measure calibrate
MEATR
   materials, engineering, and advanced test reactor
MEB
   microelectronics bulletin
   midlands electricity board
MEBES
   manufacturing electron beam exposure system
MEC
   manual emergency control
   marshalltown engineers club
   metrology engineering center
   microwave electronics company
   microwave electronics corporation
   milgo electronics corporation
   minimum energy curve
MECA
   maintainable electronic component assembly
   matsushita electric corporation of america
   multielement component array
   multivalued electronic circuit analysis
MECCA
   mechanized catalog
MECL
   motorola emitter coupled logic
   multiemitter coupled logic
   multiemitter coupled transistor logic
MECNY
   municipal engineers of the city of new york
MECOMSAG
   mobility equipment command scientific advisory group
MECR
   maintenance engineering change report
MED
   medium
   microelectronic device
MEDAC
   medical equipment display and conference
MEDACS
   medical administrative control system
MEDAL
   micromechanized engineering data for automated logistics
MEDALS
   modular engineering drafting and library system
MEDDA
   mechanical defense decision anticipation
   mechanized defense decision anticipation
MEDI
   missile error data integration
MEDIA
   magnavox electronic data image apparatus
   missile error data integration analysis
MEDIC
   medical electronic data interpretation and correlation
MEDIS
   message diversion relay system
MEDIUM
   missile error data integration, ultimate method
MEDLARS
   medical literature analysis and retrieval system
MEDLOC
   mediterranean lines of communication
MEDS
   medical evaluation data system
MEDSERV
   medical service corps
MEDSMB
   medical standards management board
MEDSPECC
   medical specialist corps
MEDUSA
   multiple element directional universally steerable antenna
MEE
   minimum essential equipment
   mission essential equipment
MEECN
   minimum essential emergency communications network
MEED
   mechanical and electrical engineering division

MEETAT
  maximum improvement in electronics effectiveness through advanced techniques
MEFV
  maintenance equipment floor valve
MEG
  megaohm
  message expediting group
  miniature electronic gyro
MEGA
  megaampere
MEGAFLOP
  mega floating point
MEGC
  megacycle
  megacycle per second
MEGV
  megavolt
MEGW
  megawatt
MEGWH
  megawatt hour
MEHP
  mean effective horsepower
MEI
  manual of engineering instructions
MEIEC
  metropolitan electronic industry education council
MEIM
  minuteman engineering instruction manual
MEIU
  mobile explosives investigation unit
MEKTS
  modular electronic kay telephone system
MEL
  many element laser
  marine engineering laboratory
  master equipment list
  materials evaluation laboratory
  microenergy logic
MELABS
  microwave engineering laboratories
MELBA
  multipurpose extended lift blanket assembly
MELEC
  microelectronics
MELEM
  microelement
MELF
  metal electrode face bonding
MELH
  missile elevation heading
MELI
  master equipment list identification
MELS
  microwave and electronic systems
MELVA
  military electronic light valve
MEM
  mars excursion mission
  mars excursion module
  memory
  midland electric manufacturing
  minimum essential medium
MEMA
  microelectronics modular assembly
  motor and equipment manufacturers association
MEMB
  membran
MEMCARD
  memory card
MEMMA
  mining electro mechnanical maintenance association
MEMMDLE
  memory module
MEMO
  memorandum
MEMTEST
  membrain test language
MEN
  multiple earthed neutral
  multiple event network
MENG
  master of engineering
MEO
  major engine overhaul
MEOL
  manned earth orbiting laboratory
MEP
  mean effective pressure
  mean probable error
  moon earth plane
MEPDP
  meander electrodes plasma display panel
MEQ
  milliequivalent
MER
  minimum energy requirement
  most economical rating
MERA
  molecular electronics for radar application
MERDC
  mobility equipment research and development center
MERDL
  medical equipment research and development laboratory

MERL
  massachusetts institute of technology electronic research laboratories
MERMUT
  mobile electronic robot manipulator and underwater television
MERRECT
  mercury rectifier
MERS
  mobility environmental research study
MES
  manual entry subsystem
  michigan engineering society
  miniature edison screw
  miscellaneous equipment specification
  mission event sequency
MESA
  manned environmental system assessment
  medium power switching application
  miniature electrostatic accelerometer
  mining enforcement and safety administration
MESCO
  message electronic switching computer
MESFET
  metal schottky fet
  metal semiconductor field effect transistor
MESG
  maximum experimental safe gap
  microelectrostatically suspended gyro
MESROM
  materials evaluation subcaliber rocket motor
MESS
  maximum efficiency structural system
  monitor event simulation system
MESTIND
  measurement standards instrumentation division
MESUCORA
  measurement, control regulation, and automation
MET
  management engineering team
  meteorological broadcast
  missile electrical technician
  missile environment tape
  modified expansion tube
  motorola environmental telemetry
META
  maintenance engineering training agency
  methods of extracting text automatically
  metropolitan educational television association
METALINDY
  metal indusry
METAPLAN
  methods of extracting text automatically programming language
  methods of extracting text autoprogramming language
METCO
  meteorological coordinating committee
METRIC
  multiechelon technique for recoverable item control
METROC
  meteorological rocket
METROVICK
  metropolitan vickers
METS
  missile electrical technician specialist
METSAT
  meteorological satellite
METU
  marine electronic technical unit
MEV
  megaelectron volts
  million electronvolts
MF
  machine finish
  machine finished
  medium frequency
  middling fair
MFLOP
  million floating point operations per second
MFSK
  multiple frequencies frequency shift keying
MG
  machine glazed
  main generator
  metal glass
  motor generator
MGSET
  motor generator set
MHHW
  mean higher high water
MHWI
  mean high water lunitidal interval
MHWS
  mean high water spring tide
MI
  marconi instruments
  microinch
  moment of inertia
MICC
  mineral insulated copper covered
MICR
  magnetic ink character reader
  magnetic ink character recognition

MKWHR
   mils per kilowatt hour
ML
   manual local
   metallic longitudinal
MLHCP
   mean lower hemispherical candlepower
MLLW
   mean lower low water
MLS
   microwave landing system
MLWI
   mean low water lunitidal interval
MLWS
   mean low water spring tide
MM
   magnetic mechanic
   maximum and minimum
MMA
   meter manufacturers' association
MMAU
   millimass unit
MMF
   magnetomotive force
MMIIL
   multiinput-multioutput integrated injection logic
MMUPS
   massachusetts general hospital multi programming system
MNR
   mean neap rise
MO
   manned and operational
   master oscillator
MOJMRP
   meteorological office joint meteorological radio propagation
MPA
   modulated pulse amplifier
MPD
   magnetoplasma dynamics
MPE
   multiple protective earthing
MPL
   maximum permissible level
MR
   mate's receipt
   maximum range
   molar refraction
   moment of resistance
MRAT
   mobile radiation tester
MRE
   mean radial error
MRIA
   magnetic recording industry association
MRU
   material recovery unit
   message retransmission unit
   microwave relay unit
   mobile radio unit
MRUA
   mobile radio users' association
MRV
   maneuvering reentry vehicle
   multiple reentry vehicle
MRWC
   multiple read write computer
MRX
   memorex
   memorex corporation
MS
   machine screw
   machine selection
   machine steel
   machinery steel
   machining system
   macromodular system
   magnetic south
   magnetic storage
   magnetic strip
   magnetic synchro
   magnetic synchron
   magnetostatic
   magnetostriction
   main stage
   main storage
   main store
   main switch
   manned station
   margin of safety
   mark sensing
   marker switch
   mass spectrometry
   mast section
   master of sciences
   master scheduler
   master sequencer
   master slave
   master switch
   material specification
   maximum stress
   mean square
   medical survey
   medicine and surgery
   medium scale
   medium soft
   medium speed
   megasecond

memory system
message switching
metallurgical society
metric size
mild steel
milestone
military specification
military standard
millisecond
millisiemens
missile station
mission sequencer
mississippi
mobile service
morse tape
most severe
motor starting
motorship
multiplexer storage

MSA
management science america
material surveillance assembly
mechanical signature analysis
meteorological satellite activities
mine safety appliances
mineralogical society of america
mines safety appliance
missile systems analyst

MSAC
missile system analyst console
moore school automatic computer

MSAR
mines safety appliance research

MSAT
missile systems analyst technician

MSB
main switchboard
mine sweeping boat
mining standards board
missile storage bunker
missile support base
most significant bit
most significant byte

MSBE
molten salt breeder experiment

MSBLK
mild steel, black finish

MSBR
molten salt breeder reactor

MSBRT
mild steel, bright finish

MSC
macroselection compiler
main storage controller
manned spacecraft center
manned spacecraft flight center
manpower services commission
marine corps school
mass storage control
mass storage controller
mean spherical candles
mediumscale computer
meteorological satellite center
microsoftware catalog
microwave semiconductor corporation
mile of standard cable
miles of standard cable
missile and space council
missile system checkout
most significant character
motor speed changer

MSCA
mixed spectrum critical assembly

MSCC
manned spaceflight control center
military space surveillance control center

MSCE
main storage control element
mobile systems checkout equipment

MSCI
molten steel coolant interaction

MSCLE
maximum space charge limited emission

MSCOP
missile systems checkout program

MSCP
mean spherical candlepower
missile systems checkout programmer
motor short circuit protector

MSCS
mass storage control system

MSCU
multistations control unit

MSD
marine signal detachment
master standard data
mean solar day
mean square difference
missile system division
missile systems development
mission systems data
most significant decade
most significant digit

MSDC
manual slave direction center

MSDR
main storage data register
multiplexer storage data register

MSDS
    message switching data service
    missile static development site
MSDT
    maintenance strategy diagramming technique
MSE
    mask superposition error
    mean square error
    minimum size effect
    missile support equipment
MSEC
    megasecond
    millisecond
MSEF
    missile systems evaluation flight
MSEM
    mission status and evaluation module
MSEMPR
    missile support equipment manufacturers planning reports
MSEQ
    master sequencer
MSF
    manned space flight
    mass storage facility
    matched spatial filter
    medium standard frequency
MSFC
    marshall space flight center
MSFET
    metal on silicon field effect transistor
    metal schottky gate field effect transistor
    metal semiconductor fet
MSFF
    master slave flipflop
MSFL
    manned space flight laboratory
MSFN
    manned space flight network
MSFP
    manned space flight program
MSFS
    main steam feddwater system
    manned space flight system
    marshall space flight system
MSFX
    master fixture
MSG
    mapper sweep generator
    mapping supervisor gap filler
    maximum stable gain
    message
    miscellaneous simulation generation
    miscellaneous simulation generator
    modular steam generator
    multiplicand select gate
MSGDPU
    message dropping and picking up
MSGWTG
    message waiting
MSHA
    mine safety and health administration
MSHI
    mediumscale hybrid integration
MSI
    man system integration
    manned satellite inspector
    marketing systems inc
    mediumscale integrated
    mediumscale integration
    middlescale integration
    missile subsystem integration
MSICIRCUIT
    mediumscale integrated circuit
MSIG
    most significant
MSIO
    mass storage input output
MSIS
    manned satellite inspection system
MSIV
    main steam isolation valves
MSIVLCS
    main steam line isolation valve leakage control system
MSK
    mask
    minimum shift keying
MSL
    manned space laboratory
    maximum service life
    maximum service limit
    mean sea level
    measurement standards laboratory
    meteorological satellite laboratory
    missile
MSLCOM
    missile command
MSLD
    mass spectrometer leak detector
MSLS
    maneuverable satellite landing system
MSLT
    military solid logic technology
MSM
    master slave manipulator
    meal semiconductor metal
    missile standards manual
    modified source multiplication

MSMB
   mechanical standards management board
MSMP
   multiple source moire patterns
MSMV
   monostable multivibrator
MSN
   microwave system news
MSOL
   manned scientific orbital laboratory
MSOR
   missile systems operational report
MSOS
   mass storage operating system
MSP
   maintenance surveillance procedure
   manual switching position
   military space program
   mode select panel
   modular system program
   most significant position
   motorized set point
MSPG
   magnetic shock pulse generator
MSPR
   master spares positioning resolver
MSPS
   megasample per second
MSR
   magnetic shift register
   magnetic stripe reader
   mark sheet reader
   mass storage resident
   material status report
   mean spring rise
   mean square root
   merchant ship reactor
   missile sight radar
   missile site radar
   mode status register
   molten salt reactor
MSRE
   molten salt reactor experiment
MSRM
   main steam radiation monitor
MSRS
   main steam radiation system
MSRT
   missile system readiness test
MSS
   main steam system
   main support structure
   management science systems
   manned space system
   manual safety switch
   manufacturers standardization society of the valve and fittings industry
   mass storage system
   master surveillance station
   mean solar second
   megasample per second
   metastable states
   meteorological satellite section
   missile safety set
   missile subsystem
   mission status summary
   mode selector switch
   monolithic systems technology
   multispectral scanner
MSSB
   missile servicing and storage building
MSSC
   main storage stock control
MSSCE
   mixed spectrum superheater critical experiment
MSSCS
   manned space station communications system
MSSN
   mean square signal to noise
MSSR
   mars soil sample return
   mixed spectrum superheat reactor concept
MSSS
   maintenance and service subsystem
   manned static space simulator
MST
   marconi self tuning
   mean solar time
   measurement
   microsecond trip
   missile systems test
   monolithic systems technology
   mountain standard time
MSTM
   missile service test model
MSTS
   missile static test site
   multisubscriber time sharing
   multisubscriber time sharing system
MSU
   maintenance service unit
   middle south utilities
   mobile signal unit
   modem sharing unit
MSV
   manned space vehicle
   mean square voltage

    meteor simulation vehicle
    mobile surface vehicle
MSVC
    mass storage volume control
MSVD
    missile and space vehicle department
MSVI
    mass storage volume inventory
MSVO
    missile and space vehicle office
MSW
    master switch
    microswitch
MT
    machine translation
    magic tee
    magnetic particle test
    magnetic tape
    magnetic tube
    mail transfer
    maintenance technician
    mark trunk
    mast
    material test
    maximum torque
    mean time
    measurement
    measuring transformer
    mechanical translation
    mechanical transport
    megaton
    mode transducer
    montana
    morse tape
    morse taper
    motor terminal
    motor transport
    multiasking
    multiple transfer
    multiple twin
    multitasking
MTA
    missile transfer area
    modified tape armored
    modified tape armored cable
    motion time analysis
    multiple terminal access
    multiterminal adapter
    my talk address
MTAC
    mathematical tables and other aids to computation
MTB
    maintenance of true bearing
    materials transportation bureau

MTBE
    mean time between failure
MTBF
    mean time before failure
MTBM
    mean time between maintenance
MTBMA
    mean time between maintenance action
MTBO
    mean time between overhauls
MTBR
    mean time between repair
    mean time between replacement
MTBS
    mean time between stops
MTBUF
    mean time between undetected failures
MTC
    machine tool conference
    machine tool control
    machine tool controller
    magnetic tape cassette
    magnetic tape command
    magnetic tape control
    magnetic tape controller
    maintenance time constraint
    master table of contents
    master tape control
    master taper control
    memory test computer
    micrographic technology corporation
    missile test center
    mission and traffic control
    moderator temperature coefficient
    mutating transformation converter
MTCA
    multiple terminal communication adapter
MTCABLE
    multiple twin cable
MTCF
    mean time to catastrophic failure
MTCON
    microwave aerospace terminal control
MTCU
    magnetic tape control unit
MTCV
    main turbine control valve
MTD
    mean temperature difference
    minimal toxic dose
    mounted
MTDA
    marine tactical data

MTDS
　manufacturing test data system
　marine tactical data system
　missile trajectory data system
MTE
　magnetic tape encoder
　maximum tracking error
　multisystem test equipment
　multiterminal emulator
　multithreshold element
MTECP
　maintenance test equipment certification procedure
MTECR
　maintenance test equipment certification requirement
MTF
　mean time to failure
　mechanical time fuse
　mississippi test facility
　modulation transfer
　modulation transfer function
MTFF
　mean time to first failure
MTFR
　mean time for repair
MTG
　microsyn torque generator
　mounting
　multiple trigger generator
MTGBKT
　mounting bracket
MTH
　magnetic tape handler
MTI
　machine tools industry
　materials technology institute
　mechanical technology, incorporated
　moving target indication
　moving target indicator
MTIC
　moving target indicator coherent
MTICFAR
　moving target indicator constant false alarm rate
MTL
　master tape loading
　material
　mean tidal level
　merged transistor logic
　microelectronic testing laboratory
　mobiltherm light
MTLP
　master tape loading program

MTM
　method of time measurement
　method time measurement
　methods time measurement association
　mission test module
　mobile transfer method
MTMA
　methods time measurement association
MTMAUK
　methods time measurement association united kingdom
MTMOD
　magnetic tape module
MTMS
　memorex tape management system
MTN
　multinational trade negotiations
MTNS
　metal thick nitride semiconductor
　metal thick nitride silicon
　metal thick oxide nitride silicon
MTO
　master terminal operator
　missile test operator
MTOP
　molecular total overlap population
MTOS
　magnetic tape operating system
　magnetic tape operation system
　metal thick oxide semiconductor
　metal thick oxide silicon
　metal thin oxide silicon
MTOSFET
　metal thick oxide semiconductor field effect transistor
MTP
　mechanical thermal pulse
MTPF
　maximum total peaking factor
　minimal total processing time
MTPS
　magnetic tape programming system
MTR
　magnetic core transistor relay
　magnetic tape recorder
　master tool record
　material testing reactor
　material transfer recorder
　materials testing report
　mean time to restore
　milliammeter
　missile track radar
　missile tracking radar
　motor
　moving target reactor

multiple track radar range
multiple track range radar
MTRE
magnetic tape recorder end
missile test and readiness equipment
missile test and readiness evaluation
MTRG
metering
MTRI
missile test range instrumentation
MTRS
magnetic tape record start
magnetic tape recorder start
MTS
magnetic tape station
magnetic tape storage
magnetic tape store
magnetic tape subsystem
magnetic tape system
magnetic type system
marine technology society
mass terminations system
master timing system
message toll service
message traffic study
michigan terminal system
michigan time sharing
missile test station
missile tracking station
missile tracking system
mobile telephone service
module testing system
module tracking system
money transfer system
motor operated transfer switch
MTSC
magnetic tape selective composer
MTSE
magnetic trap stability experiment
MTSMB
material and testing standards management board
MTSS
manned test space station
military test space station
MTST
magnetic tape selective typewriter
MTT
magnetic tape terminal
magnetic tape transport
microwave theory and techniques
multi terminal pro task
MTTD
mean time to diagnosis

MTTE
multithreshold threshold element
MTTF
mean time to failure
MTTFF
mean time to first failure
MTTL
motorola transistor transistor logic
MTTPO
mean time to planned outage
MTTR
maximum time to repair
maximum time to replace
mean time to repair
mean time to replace
mean time to restore
MTTUO
mean time to unplanned outage
MTU
magnetic tape unit
master terminal unit
master trigger unit
multiplexer and terminal unit
MTUBE
metal tube
MTV
marginal terrain vehicle
missile test vehicle
motor test vehicle
MTVAL
master tape validation
MTWP
multiplier traveling wave phototube
MTWTV
multithreshold weight threshold vector
MTX
matrix
MU
machine unit
maintenance unit
make up
memory unit
message unit
mobile unit
mock up
modular unit
multiple unit
multiple use
MUART
microprocessor universal asynchronous receiver transmitter
MUC
multiple use counter
MUCHFET
multichannel fet

MUDAID
   multivariate, univariate, and discriminant analysis of irregular data
MUDDC
   multiunit direct digital control
MUDPAC
   melbourne university dual package analog computer
MUDPIE
   museum and university data processing information exchange
MUDWNT
   make-up demineralizer waste neutralizer tank
MUF
   material unaccounted for
   maximum usable frequency
MUG
   maximum unilateralized gain
   maximum usable gain
MUL
   multiply
MULDOS
   mulby disk operating system
MULS
   mobile unit launch site
MULT
   multiple
   multiplier
MULTEWS
   multiple electronic warfare surveillance
MULTI
   multivibrator
MULTICS
   multiplexed information and computing service
   multiplexed information and computing system
MULTIV
   multivibrator
MULTR
   multimeter
MUM
   multiuse mnemonics
   multiple unit message
MUMS
   mobile utility module system
MUOD
   mean unplanned outage duration
MUPF
   modified ultrapherical polynominal filter
MUPO
   maximum undistorted power output
MUR
   management update and retrieval system
   mock up reactor

MURCIRCUIT
   module ultrarapid circuit
MURCODE
   modular ultrarapid code
MURG
   machine utilization report generator
MURLOGIC
   modular ultrarapid logic
MURS
   machine utilization reporting system
MUSA
   multiple unit steerable antenna
MUSASYSTEM
   multiple unit steerable antenna system
MUSAT
   multipurpose uhf satellite
MUSC
   multi unit supervisory control
MUSE
   machine user symbolic environment
   modular utilities for systems education
MUSIC
   machine utilisation statistical information collection
   maryland university sectored isochronous cyclotron
   military, university union, science, industrial complex
   multi sensor intelligence correlation
MUSS
   missile unit support system
   mobile unit support system
MUST
   message user service transcriber
MUT
   modular universal terminal
   module under test
MUW
   music wire
MUX
   multiplex
   multiplexer
MUXARC
   multiplexing automatic error correction
MUXER
   multiplexer
MV
   maintenance version
   mean value
   mean variation
   mean voltage
   measured value
   medium voltage
   megavolt
   megavoltage

mercury vapor
millivolt
minimal variant
motor vehicle
multiconverter vector
multivibrator
muzzle velocity
MVA
megavolt ampere
millivolt ampere
MVAC
megavolt alternating current
millivolt alternating current
MVAR
megavar
megavolt ampere reactive
MVARH
megavar hour
MVAU
maximum volt ampere utilization
MVB
multivibrator
MVC
manual volume control
multi variant counter
multiple variate counter
MVCM
millivolt per centimeter
MVD
map and visual display
motor voltage drop
MVDC
megavolt direct current
millivolt direct current
MVDF
medium and very high frequency direction finder
MVI
medium viscosity index
MVM
millivolt per meter
MVP
manpower validation program
mechanical vacuum pump
multivariable programming
MVPS
medium voltage power supply
MVS
magnetic voltage stabilizer
minimum visible signal
multiple virtual storage
MVT
multiprocessing with a variable number of tasks
multiprogramming with a variable number of tasks
multiprogramming with variable tasks
multitasking with a variable number of tasks
MVTR
moisture vapour transmission rate
MW
machine word
manual word
medium wave
megawatt
megawatts
microwave
milliwatt
modulated wave
molecular weight
multiwire
multiwiring
music wire
MWAEFILTER
minimum weighted absolute error filter
MWARA
major world air route area
MWARC
maritime world administrative radio conference
MWCM
milliwatt per square centimeter
MWD
megawatt day
MWDDEA
mutual weapons development data exchange agreement
MWDT
megawatt days per ton
MWE
megawatt, electric[al]
MWG
meteorological working group
missile working group
music wire gage
MWH
megawatt hour
megawatt, heat
MWI
message waiting indicator
MWL
master warning light
milliwatt logic
MWMRTL
milliwatt motorola resistor transistor logic
MWMSE
minimum weighted mean square error
MWMSEFILTER
minimum weighted mean square error filter

MWO
  maintenance work order
MWP
  maximum working pressure
MWPC
  multiwire proportional chamber
MWR
  mean width ratio
MWRAILS
  micorowave remote area instrument landing system
MWS
  microwave station
MWSC
  minimum wage study commission
MWSD
  missile and weapons systems division
MWT
  make up water treatment
  marconi's wireless telegraph
  molecular weight
MWTH
  megawatt thermal
MWV
  maximum working voltage
MWYE
  megawatt year of electricity
MX
  matrix
  maxwell
  multiplex
  multiplexer
MXD
  mixed
MXM
  matrix memory
MXR
  mask index register
  mixer
MY
  mean year
  million years
MYLTR
  my letter
MYMSG
  my message
MZ
  minus zero
MZCP
  mean zonal candlepower
MZPI
  microwave zone position indicator

N

N
  nano
  navigation
  negative
  negatively conducting
  neper
  neutral
  neutron
  neutron number
  newton
  nichrome
  nitrogen
  no
  noise
  normal
  normal concentration
  normality
  north
  nuclear
  number
  number of bits
  number of revolutions
  number of turns
  nylon
NA
  nanoampere
  naval aviator
  navigation aid
  navy aircraft
  neutral axis
  next assembly
  not applicable
  not assigned
  not available
  numerical aperture
NAA
  national aeronautic association
  national association of accountants
  neutron activation analysis
  north american aviation
NAACD
  north american aviation columbus division
NAAFI
  navy, army and air force institute
NAAL
  north american aerodynamics laboratory
NAALA
  north american aviation los angeles
NAAM
  national association of architectural metal manufacturers

NAAPS
nozzle actuator auxiliary power supply
NAARD
north american aviation rocketdyne division
NAAS
naval auxiliary air station
NAASC
north american aviation science center
NAASD
north american aviation space division
NAATS
national association of air traffic specialists
NAB
national acoustics board
national alliance of businessmen
national association of broadcasters
national association of broadcasting
naval air base
navigational aid to bombing
nuclear assembly building
nut and bolt
NABCE
national association of black consulting engineers
NABE
national association of business education
NABER
national association of business and educational radio
NABET
national association of broadcast employees and technicians
NABS
nuclear armed bombardment satellite
NABUG
national association of broadcast unions and guilds
NAC
national agency checks
NACA
national advisory commission for aeronautics
NACAA
national association of computer assisted analysis
NACACP
national cash register's applied cobol packages
NACATS
north american clear air turbulence tracking system
NACC
national automatic control conference
north america control committee
NACCAM
national coordinating committee for aviation meteorology
NACE
national advisory committee for electronics
national association of corrosion engineers
national association of country engineers
NACEIC
national advisory council on education for industry and commerce
NACEO
national advisory council on economic opportunity
NACME
national action council for minorities in engineering
NACOA
national advisory committe on oceans and atmosphere
NACOM
national communications
NACOS
national communications schedule
NACS
northern area communications system
NACTCU
northeast area communications traffic control unit
NAD
no-acid descaling
noise amplitude distribution
nuclear accident dosimetry
NADAC
navigation data assimilation computer
NADC
naval air development center
NADGE
nato air defense ground environment
NADS
north american data systems
NADU
naval air development unit
NADWARN
natural disaster warning system
NAE
national academy of engineering
NAEB
national association of educational broadcasters
NAEC
national aerospace education council
national association of electric companies
national aviation education council
NAECON
national aerospace electronic conference

national aerospace electronics conference
national aerospace electronics convention
NAED
national association of electrical distributors
NAEDS
national association of educational data systems
NAEF
naval aeronautical engineering facility
NAER
north american electronic research
NAES
naval air experimental station
NAESU
navy airborne electronics service unit
NAET
national association of educational technicans
NAF
naval air facility
naval avionics facility
NAFEC
national administrative facilities experimental center
national aviation facilities experimental center
NAFI
naval avionics facility, indianapolis
NAFLI
natural flight instrument system
NAFMB
national association of fm broadcasters
NAG
naval advisory group
naval astronautics group
navigation and guidance
NAI
no-address instruction
northrop aircraft, incorporated
NAIC
national astronomy and ionosphere center
nippon atomic industry group
NAIOP
navigation aid inoperative for parts
navigational aid inoperative for parts
NAIPRC
netherland's automatic information processing research center
NAIR
narrow absorption infrared
NAK
negative acknowledge
negative acknowledgement
negative acknowledgment character

NAL
national accelerator laboratory
native assembly language
NALLA
national long lines agency
NALOG
natural logarithm
NAM
national association of manufacturers
network access machine
NAMA
national automatic merchandising association
NAMDAR
north american data airborne recorder
NAMFI
nato missile firing installation
NAMI
naval aerospace medical institute
NAMIS
nitride barrier avalanche injection metal insulator semiconductor
NAMISCEN
naval missile center
NAMISTESTCEN
naval air missile test center
NAMRISME
north american manufacturing research institution of sme
NAMTC
naval air missile test center
NANAC
national aircraft noise abatement council
NAND
not and
not and circuit
NANEP
naval air navigation electronics project
NANWEP
navy numerical weather prediction
NAOGE
national association of government engineers
NAOR
naval aviation observer radar
NAOS
north atlantic ocean station
NAP
noise abatement procedure
north american philips
NAPA
national automotive parts association
NAPALM
national automatic data processing program for army material command logistics management

NAPCA
national air pollution control administration
NAPE
national association of power engineers
NAPHCC
national association of plumbing heating cooling contractors
NAPL
national association of photolitographers
NAPM
national association of photographic manufacturers
national association of purchasing management
NAPPE
network analysis program using parameter extractions
NAPS
nimbus automatic programming system
numerical analysis problem solving system
NAPSIC
north american power system interconnection committee
NAPSS
numerical analysis problem solving system
NAPUS
nuclear auxiliary power unit system
NAQF
north atlantic quality figures
NAR
national association of rocketry
net assimilation rate
north american rockwell
numerical analysis research
NARAD
naval air research and development
NARATE
navy automatic radar test equipment
northrop automatic radar test system
NARBA
north american regional broadcasting agreement
NARBEC
north american regional broadcasting engineering committee
NARC
ninth area radio club
NARCOM
north atlantic relay communication system
NARDA
national appliance and radio tv dealers' association
NARDIS
navy automated research and development information system

NAREC
naval research electronic computer
NARETU
naval air reserve electronics training unit
NARF
nuclear aerospace research facility
nuclear aircraft research facility
NARM
national association of relay manufacturers
NARND
national association of radio news directors
NARS
national accommodations reservation services
national archives and records service
NARTB
national association of radio and television broadcasters
NARTC
naval air rocket test center
NARTEL
north atlantic radio telephone
NARTS
national association of radio telephone systems
naval aeronautic test station
naval air rocket test station
NARUC
national association of regulatory and utilities commissions
national association of regulatory utility commissioners
NARUCE
national association of regulatory utility commission engineers
NAS
national academy of sciences
national advanced systems
national aerospace standard
national aircraft standard
national airspace system
naval air station
network administrative station
NASA
national aeronautics and space administration
NASAERC
nasa electronic research center
NASAP
network analysis and systems application program
NASARR
north american search and range radar
north american search and ranging radar
NASB
navigational aid support bases

NASC
  national aeronautics and space council
  national aerospace standards committee
NASCO
  national academy of sciences' committee on oceanography
NASCOM
  nasa communications
NASD
  national association of securities dealers
NASDAQ
  national association of security dealers automated quotation
  national association of securities dealers automated quotation system
NASIS
  national association for state information systems
NASKWEST
  naval air station, key west
NASL
  naval applied sciences laboratory
NASOPT
  network analysis system with optimization facility
NASPA
  national society of public accountants
NASRR
  north american search and ranging radar
NASS
  naval air signal school
NASTART
  normal after start
NASTRAN
  nasa structural analysis
NASU
  national association of system 3 users
NASW
  national association of science writers
NAT
  natural unit
  network analysis technique
  normal allowed time
NATA
  national association of testing authorities
  north american telephone association
NATC
  noval air test center
NATCOM
  national communications symposium
NATCS
  national air traffic control service
NATE
  neutral atmosphere temperature experiment

NATEC
  naval air technical evaluation center
NATESA
  national alliance of television and electronics service associations
NATFREQU
  natural frequency
NATIS
  national association for trade and technical schools
NATIV
  north american test instrument vehicle
NATIVE
  north american test instrument vehicle
NATLSEMICON
  national semiconductor
NATMC
  national advanced technology management conference
NATTS
  national association of trade and technical schools
NAU
  network addressable unit
NAUS
  national air space utilization system
NAV
  navigation
  navigational
NAVA
  national audio visual association
NAVAIDS
  navigational aids
NAVAPI
  north american voltage and phase indicator
NAVAR
  navigation air radar
  navigation and ranging
NAVARHO
  navigation and radio homing
NAVBM
  navy ballistic missile
NAVBMC
  naval ballistic missile committee
NAVCM
  navigation countermeasure
  navigation countermeasures and deception
NAVCOM
  naval communications
NAVCOMMFAC
  naval communications facility
NAVCOMMSTA
  naval communications station
NAVCOMMSYS
  naval communications system

NAVCOMMU
   naval communications unit
NAVCOMMUNR
   naval communications unit, naval reserve
NAVCOR
   navigation computer corporation
NAVDAC
   navigation[al] data assimilation center
NAVELEC
   naval electronics system command
NAVFAC
   naval facilities
   naval facilities engineering command
NAVGMU
   naval guided missile unit
NAVIC
   naval information center
NAVL
   navigational light
NAVMISFAC
   naval missile facility
NAVRADSTA
   naval radio station
NAVSAT
   navigational satellite
NAVSEC
   naval ship engineering center
NAVSPASUR
   naval space surveillance
NAVSTA
   naval station
NAVSTAR
   navigation system using timing and ranging
NAW
   night adverse weather
NAWAPA
   north american water and power alliance
NAWAS
   national attack warning system
   national warning system
NAXSTA
   naval air experimental station
NB
   narrow band
   nebraska
   nitrobenzene
   no bias
   nominal bore
NBA
   narrow band allocation
NBAA
   national business aircraft association
NBC
   narrow band conducted
   national broadcasting company
   noise balancing circuit
   noise balancing control
NBCD
   natural bcd
NBCDCODING
   normal binary coded decimal coding
NBCV
   narrow band coherent video
NBDL
   narrow band data line
   narrow band data link
NBDN
   nuclear blast detector network
NBER
   national bureau of engineers registration
NBFM
   narrow band frequency modulation
NBFU
   national board of fire underwriters
NBH
   network busy hour
NBL
   new brunswick laboratory
NBMG
   navigation bombing and missile guidance
NBMGS
   navigation bombing and missile guidance system
NBO
   network buildout
NBR
   narrow band radiated
   nitrile butadiene rubber
NBS
   national broadcasting service
   national broadcasting system
   national bureau of standards
   new british standard
NBSD
   night bombardement short distance
NBSFS
   nbs frequency standard
NBSR
   nbs reactor
NBT
   negative balance test
   null ballance transmissometer
NBTDR
   narrow band time domain reflectometry
NBTL
   naval boiler and turbine laboratory
NBTS
   new boston tracking station

NBVM
  narrow band voice modulation
NC
  national center
  national coarse
  network control
  network controller
  neutralization capacitor
  nickel cadmium
  nickel clad copper
  nixdorf computer
  no coil
  no connection
  noise criteria
  noise criterion
  noncoin
  nonconversational
  nonlinear capacitance
  normally closed
  north carolina
  northrop corporation
  nuclear capability
  numeric[al] control
  numerically controlled
NCA
  national coal association
  naval communications annex
  new communities administration
  northwest computing association
  nothern counties civil engineers and land surveyors
NCAE
  national conference on airborne electronics
NCAEI
  national conference on the application of electrical insulation
NCAG
  nixdorf computer ag
NCAR
  national center for atomsphere research
NCASI
  national council for air and stream improvement
NCB
  national coal board
  national college of business
  naval communications board
NCBEE
  national council of state boards of engineering examiners
NCBR
  near commercial breeder reactor
NCC
  national computer conference and exposition
  national computing center
  network control center
  norad control center
  normally closed contact
NCCAT
  national committee for clear air turbulence
NCCCC
  naval command, control and communications center
NCCCHE
  national certification commission in chemistry and chemical engineering
NCCK
  noncoherent carrier keying
NCCLS
  national committee for clinical laboratory standards
NCCOMPUTER
  numeric[al] control computer
NCCONTACT
  normally closed contact
NCCS
  national command and control system
NCD
  new component design
NCDC
  new community development corporation
NCDEAS
  national committee of deans of engineering and applied science
NCEA
  north central electric association
NCEE
  national council of engineering examiners
NCER
  national conference on electromagnetic relays
NCET
  national council for educational technology
NCF
  naval communications facility
  nominal characteristics file
NCFEA
  north carolina federation of electronic associations
NCFMF
  national committee for fluid mechanics films
NCFSK
  noncoherent frequency shift keying
NCGA
  national computer graphics association
NCGG
  national committee for geodesy and geophysics
NCH
  n channel
  no charge

NCHEML
national chemical laboratory
NCHMOS
n channel metal oxide semiconductor
NCHS
national center for health statistics
NCHVRFE
national college for heating, ventilating, refrigeration, and fan engineering
NCI
national computer institute
national computing industries
netherlands center for informatics
northeast computer institute
NCIC
national crime information center
NCL
national central library
national chemical laboratory
network control language
NCLO
naval communication liaison officer
NCM
numerically controlled machine
numerically controlled machine tool
NCMACHINE
numerically controlled machine
NCMC
numerically controlled machining center
NCMCE
national council of minority consulting engineers
NCMES
numerically controlled measuring and evaluating system
NCMI
national country maintenance index
NCMS
numerically controlled machining system
numerically controlled manufacturing system
NCMT
numerically controlled machine tool
NCN
nixdorf communication network
NCO
number controlled oscillator
numerically controlled oscillator
NCOR
national committee for oceanographic research
NCP
network control program
noncarbon paper
normal circular pitch
numerically controlled peripheral

NCPAMA
noise control products and materials association
NCPAS
national computer program abstract service
NCPI
national computer program index
NCPM
noncritical phase matching
NCPS
national commission on product safety
NCPTWA
national clearinghouse for periodical title word abbreviations
NCPVS
network control program virtual storage
NCQR
national council for quality and reliability
NCR
national cash register
national cash register company
no carbon required
NCRL
national citizens radio league
NCROBOT
numerically controlled robot
NCRP
national committee on radiation protection
NCRUCE
national conference of regulatory utility commission engineers
NCS
national communication system
national computer systems
naval communication system
naval communications station
net control station
netherlands computer society
network communication system
network control station
nuclear components spares
numeric[al] control society
numeric[al] control system
NCSAG
nuclear cross section advisory group
NCSBCS
national conference of states on building codes and standards
NCSC
national council of schoolhouse construction
NCSE
north carolina society of engineers
NCSL
national conference on standards laboratories

NCTA
  national cable television association
  national community television association
NCTC
  naval communications training center
NCTM
  national council of teachers of mathematics
NCTS
  national council of technical schools
NCTSI
  national council of technical service industries
NCU
  naval communications unit
  network control unit
NCUA
  national credit union administration
NCUR
  national committee for utilities radio
NCV
  no commercial value
ND
  natural dosis
  negative declaration
  no detect
  nondelay
  nondeterministic
  nondirectional
  nondirector
  norsk data
  north dakota
NDAC
  no data accepted
NDB
  nautical directional beacon
  nondirectional beacon
  nondirectional radio beacon
NDC
  national documentation center
  noise dose count
NDCCC
  national defense communications control center
NDCT
  natural draft cooling tower
NDE
  nondestructive evaluation
  nondestructive examination
  nonlinear differential equation
NDF
  nonrecursive digital filter
NDI
  numerical design index
  numerical designation index

NDL
  network definition language
NDM
  negative differential mobility
NDN
  new data network
NDP
  normal diametral pitch
  numeric data processor
NDPS
  national data processing service
NDR
  net difference report
  network data reduction
NDRO
  nondestructive read only
  nondestructive read out
NDRS
  nuclear definition and reporting system
NDRW
  nondestructive read and write
NDS
  network development system
  neutron doped silicon
  nominal detectable signal
  nondestructive test
  nuclear detection satellite
NDT
  nil ductility transition
  nondestructive test
  nondestructive testing
  nuclear detection test
NDTC
  nondestructive testing center
NDTP
  nuclear data type program
NDTT
  nil ductility transition temperature
NE
  national electronics
  noise equivalent
  north east
  northern electric
  not equal to
NEA
  national education association
  national electronics association
  negative electron affinity
  nuclear energy agency
  nuclear engineering associates
NEAC
  nippon electric automatic computer
NEACP
  national emergency airborne command post

NEADAI
   national education association department of audiovisual instruction
NEAT
   national cash register electronics autocoding technique
   ncr electronic autocoding technique
NEB
   noise equivalent bandwidth
NEBSS
   national examinations board in supervisory studies
NEC
   national electric[al] code
   national electronics conference
   national electronics council
   national engineering consortium
   netherlands electrotechnical committee
   nippon electric co
   nippon electric company
   no-error check
   nuclear energy center
NECA
   national electrical contractors association
NECAP
   nutmeg electric companies atomic project
NECCO
   national electric coil company
NECG
   national executive committee on guidance
NECIES
   northeast coast institution of engineers and shipbuilders
NECO
   nippon electric company
NECODE
   national electrical code
NECPUC
   new england conference of public utility commissioners
NECS
   national electrical code standards
   nationwide educational computer service
NECSS
   nuclear energy center site survey
NECTA
   national electrical contractors trade association
NED
   navigation error data
   normal equivalent deviate
NEDA
   national electronic distributors association
NEDETS
   nuclear detection system

NEDSA
   nonerasing deterministic stack automation
NEEB
   northeastern electricity board
NEEDS
   new england educational data systems
NEEP
   nuclear electronic effect program
NEES
   naval engineering experimental station
   new england electric system
NEF
   national extra fine
   noise equivalent flux
NEFD
   noise equivalent flux density
NEFO
   national electronics facilitis organization
NEG
   negation
   negative
NEGIT
   negative impedance transistor
NEHA
   national environmental health association
NEI
   new england instrument
   noise equivalent input
   noise equivalent intensity
NEIL
   neon indicating light
NEIPG
   national electronic industries procurement group
NEIS
   national electrical industries show
   national engineering information system
NEL
   national engineering laboratory
   naval electronics laboratory
   navy electronics laboratoy
   neon light
NELA
   national electric light association
NELABS
   northern electric laboratories
NELAT
   naval electronics laboratory assembly tester
NELC
   naval electronics laboratory center
NELCO
   new england laminates company
NELCON
   new zealand electronics convention

NELCONNZ
national electronics conference, new zealand
NELIA
nuclear energy liability insurance association
NELIAC
naval electronics laboratoy international algebraic compiler
navy electronic laboratory international algol compiler
navy electronics laboratory international algebraic compiler
navy electronics laboratory international algorithmic language compiler
NELPA
northwest electric light and power association
NELPIA
nuclear energiy liability property insurance association
NEMA
national electrical manufacturers' association
NEMAG
negative effective mass amplifiers and generators
NEMI
national elevator manufacturing industry
NEMP
nuclear electromagnetic pulse
NEP
never ending program
newton extrapolation polynominal
noise equivalent power
NEPA
national environmental policy act
nuclear energy for propulsion of aircraft
NEPAME
national exposition of power and mechanical engineering
NEPCON
national electronic packaging conference
national electronic production conference
NEPD
noise equivalent power density
NEPEX
new england power exchange
NEPIA
nuclear energy property insurance association
NEPTUNE
north eastern electronic peak tracing unit and numerical evaluator
NER
national engineers register
NERA
national economic research associates
NERC
national electric reliability council
national electronic research council
natural environment research council
nuclear energy research center
NEREM
northeast electronics research and engineering meeting
NERHL
northeastern radiological health laboratory
NERV
nuclear emulsion recovery vehicle
NERVA
nuclear engine for rocket vehicle application
NES
national engineering service
national estimating society
near end suppressor
new england electric system
noise equivalent signal
NESA
national electric sign association
NESC
national electric[al] safety code
national environmental satellite center
naval electronic systems command
newcastle electric supply company
NESCOM
new standards projects committee
NESP
national environmental studies project
NEST
naval experimental satellite terminal
NESTEV
naval electronics system test and evaluation
NESTOR
neutron source thermal reactor
NESTS
nonelectric stimulus transfer system
NESW
non-essential service water relay pump
NET
national educational television
network
noise equivalent temperature
noise evaluation test
NETA
northwest electronic technical association
NETFS
national educational television film service
NETR
nuclear engineering test reactor
NETRC
national educational television and radio center

NETS
  network electrical technique system
  network techniques
NETSET
  network synthesis and evaluation technique
NEUT
  neutral
NEW
  national electrical week
  national engineers week
  navy early warning
NEWA
  national electrical wholesalers association
NEWRADS
  nuclear explosion warning and radiological data system
NEWS
  naval electronic warfare simulator
NEWWA
  new england water works association
NEXUS
  numerical examination of urban smog
NF
  nanofarad
  national fine
  national standard fine
  night fighter
  night frequency
  noise factor
  noise figure
  noise frequency
  noise fuse
  normal form
NFAH
  national foundation for the arts and the humanities
NFB
  negative feedback
NFC
  no further consequences
  not favorably considered
NFCA
  nonfuel core array
NFDC
  national flight data center
NFE
  nearly free electron
NI
  noninductive
NL
  navigation localizer
NLD
  naval electrical department
NLOGM
  naval liaison officer for guided missiles

NLP
  natural language processing
  nonlinear programming
NLQ
  near letter quality
NLR
  noise load ratio
  noise loading ratio
  nonlinear resistance
  nonlinear resistor
NLRB
  national labor relations board
NLREG
  nonlinear regression
NLS
  no-load speed
  nonlinear smoothing
  nonlinear systems
NLSMA
  national lamp and shade manufacturers' association
NLT
  negative line transmission
  noise limiter
NM
  nanomemory
  nanometer
  nautical mile
  new mexico
  night message
  no message
  noise margin
  noise meter
  nonmagnetic
  not measured
  nuclear magnetron
  nuclear measurements
NMA
  national management association
  national microfilm association
NMAA
  national machine accountants association
NMAP
  national metric advisory panel
NMC
  national meteorological center
  naval missile center
NMCC
  national mass communications council
NMCL
  naval missile center laboratory
NMCS
  national military command system
  nuclear materials control system

NME
  noise measuring equipment
NMEL
  navy marine engineering laboratory
NMF
  naval missile facility
NMFHAWAREA
  naval missile facility, hawaiian area
NMFPA
  naval missile facility, point arguello
NMG
  numerical master geometry
NMI
  nasa management instruction
  national maintenance index
  nautical mile
  nonmaskable interrupt
NMIC
  national missile industry conference
NMIS
  nuclear materials information system
  nuclear materials inventory system
NMK
  niagara mohawk power
NML
  nuclear magnetic logging
NMM
  network measurement machine
NMMSS
  nuclear materials management and safeguards system
NMOP
  national mission operating procedure
NMOS
  n channel metal oxide semiconductor
NMP
  national meter programming
NMPC
  national minority purchasing council
NMR
  national missile range
  naval missile range
  normal mode rejection
  nuclear magnetic resonance
NMRM
  nuclear magnetic resonance measurement
NMRS
  national mobil radio system
NMS
  naval meteorological service
  network management services
  neutron monitoring system
  nuclear medical science
NMSS
  national meteorological satellite system
  national multipurpose space station
NMST
  new materials system test
NMTBA
  national machine tool builders' association
NMTC
  naval missile test center
NN
  nearest neighbour
  neutralization number
NNC
  nonnumerical control
NNCSC
  national neutron cross section committee
NNE
  north northeast
NNEC
  national nuclear energy commission
NNI
  nonnuclear instrumentation
NNP
  nonnegation property
NNR
  new nonofficial remedies
NNRDC
  national nuclear rocket development center
NNSDD
  newport news shipbuilding and dry dock company
NNSS
  navy navigational satellite system
NNW
  north northwest
NO
  normally open
NOA
  national oceanography association
NOAA
  national oceanographic and atmospheric administration
NOALA
  noise operated automatic level adjustment
NOC
  normally open contact
  normally open contract
NOD
  network out dialing
  night observation device
NODAC
  naval ordnance data automation center
NODC
  national oceanographic data center

NODE
noise diode
NOEV
norad operational evaluation
NOF
national optical fond
ncr optical font
NOGAD
noise operated gain adjusting device
NOHP
not otherwise herein provided
NOI
notice of inquiry
NOIBN
not otherwise indexed by name
NOL
naval ordnance laboratory
normal operating losses
NOLMDI
naval ordnance laboratory miss distance indicator
NOMA
national office management association
NOMAD
naval oceanographic meteorological automatic device
navy oceanographic and meteorological automatic device
NOMAG
nonmagnetic
NOMSS
national operational meteorological satellite system
NOMTF
naval ordnance missile test facility
NONCOHO
noncoherent oscillator
NONLIN
nonlinear
NONMAGCI
nonmagnetic cast iron
NONP
negative on positive
NONRSNT
nonresonant
NOOOA
norad office of operational analysis
NOOP
no operation
NOP
near object probe
no operation
nonoperational
normed programming

NOR
not or
not-or circuit
NORAC
no radio contact
NORAD
north american air defense command
north american air defense system
NORC
naval ordnance research computer
NORCUS
northwest college and university association for science
NORDO
no radio
NORM
not operationally ready maintenance
NORMCLSD
normally closed
NORMOPN
normally open
NORS
not operationally ready supply
NORVIPS
northrop voice interruption priority system
NOS
national operational satellite
network operating system
night observation system
not otherwise specified
number of stops
NOSMO
norden optics setting, mechanized operation
NOSS
national oceanic satellite system
national orbiting space station
nimbus operational satellite system
NOT
not logic element
number of turns
NOTAM
notes to airmen
notice to airmen
NOTS
naval ordnance test station
NOTU
naval operational training unit
NP
nameplate
national pipe
naval publication
neap
neaper
neuropsychiatric
nominal horsepower

nondeterministic polynomial time
nonprint
nonprinting
nonprocurable
normal pressure
number of primary turns
NPA
normal pressure angle
number plan area
numerical production analysis
NPAR
negative positive acknowledgement and retransmission
NPC
nasa publication control
national power conference
nonprinting character
nuclear propulsion committee
nucleonic products company
NPCC
northeast power coordinating council
NPCODE
nonprint code
NPD
national power demonstration
network protective device
nuclear power demonstration
NPDES
nuclear pollution discharge elimination system
NPDS
nuclear particle detection system
NPE
nuclear power engineering
NPEC
nuclear power engineering committee
NPEX
normal priority exit
NPF
nuclear power facility
nuclear problems forum
NPFO
nuclear power field office
NPG
nonprocessor grant
normalized programming generator
nuclear power group
NPGS
nuclear power generating station
NPL
national physical laboratory
new process line
new processor line
new product line
new programming language
nonprogramming language
NPLG
navy program language group
NPM
narda microline precision miniature
NPN
negation permutation negation
negative positive negative
NPO
negative positive zero
nuclear power operator
NPOLE
north seeking pole
NPP
new product planning
nuclear power plant
NPPS
nebraska public power system
NPPSO
naval publications and printing service office
NPR
new production reactor for plutonium and electricity production
noise power ratio
nonprocessor request
numerical position readout
NPRC
nuclear power range channel
NPRCG
nuclear public relations contact group
NPRDS
nuclear plant reliability data system
NPRF
northrop pulse radiation facility
NPRFCA
national petroleum radio frequency coordinating association
NPRS
negative poll response state
NPS
network processing supervisor
nuclear and plasma sciences
numerical plotting system
NPSA
new program status area
NPSH
net positive suction head
NPSK
n phases pulse shift keying
NPSRA
nuclear powered ship research association of tokyo
NPSS
nuclear and plasma sciences

NPT
- american national briggs pipe thread
- american standard national taper pipe thread
- national taper pipe thread
- network planning technique
- nonproliferation treaty
- normal pressure and temperature

NPV
- nitrogen pressure valve

NQAA
- nuclear quality assurance agency

NQR
- nuclear quadrupole resonance

NR
- n operand register
- natural rubber
- navigational radar
- negative resistance
- new range
- nitrile rubber
- noise ratio
- noise reduction
- nonlinear resistance
- nonreactive
- nonrecoverable
- nonreversing
- norad region
- nuclear reactor
- number
- reynolds' number

NRA
- naval radio activity
- network resultion area
- nuclear reaction analysis

NRAO
- national radio astronomy observatory

NRC
- national research council
- national rocket club
- naval radio compass
- noise reduction coefficient
- nuclear regulatory commission

NRCI
- national radio company, incorporated

NRCST
- nationl referral center for science and technology

NRD
- national range division
- naval radio direction finder
- negative resistance diode

NRDB
- nonreversing, dynamic braking

NRDC
- national research and development corporation
- national research development council
- national resources defense council

NRDL
- naval radiological defense laboratory

NRDS
- nuclear rocket development station

NRDSCG
- naval research and development satellite communications group

NRE
- negative resistance element
- nuclear rocket engine

NRECA
- national rural electric cooperative association

NRF
- nuclear resonance fluorescence

NRFD
- not ready for data

NRI
- national radio institute

NRIP
- number of rejected initial pick ups

NRIS
- national resource information system
- natural resource information system

NRL
- national reference library for science and invention
- naval research laboratory

NRM
- natural remanent magnetization
- normal response mode
- normalization
- normalize

NRMA
- national retail merchant association

NRMC
- north american rockwell microelectronics company

NRMEC
- north american rockwell microelectronics company

NRMS
- nominal root mean square

NRNY
- navy radio, new york

NRP
- normal rated power

NRR
- negative radial rake
- negative resistance repeater
- office of nuclear reactor regulation

NRRE
    netherlands radar research establishment
NRRO
    naval radio research observatory
NRRS
    naval research station
NRS
    national radio station
    naval radio station
    naval rocket society
NRSEP
    national roster of scientific and engineering personnel
NRSL
    navy radio and sound laboratory
NRSTP
    national register of scientific and technical personnel
NRSW
    nuclear river service water
NRT
    net register ton
    net registered tonnage
    nonrequestor terminal
NRTDAS
    non real time data automation system
NRTS
    national reactor test station
NRTSC
    naval reconnaissance and technical support center
NRU
    nonreplaceable unit
NRW
    nuclear radwaste
NRWO
    nuclear radwaste operator
NRX
    nuclear engine reactor experiment
NRZ
    no return to zero
    nonreturn to zero
NRZC
    no return to zero change
    nonreturn to zero change
    nonreturn to zero with change
NRZI
    no return to zero invert
    nonreturn to zero invert
    nonreturn to zero inverted
NRZL
    nonreturn to zero level
NRZM
    nonreturn to zero mark

NS
    nanosecond
    national semiconductor
    national standard
    naval station
    new style
    new system
    no scramble
    noise sensitivity
    nonsequenced
    nonshorting
    nonskew
    north south
    not specified
    nuclear ship
    number of secondary turns
NSA
    national security agency
    national shipping authority
    national standards association
    netherlands society for automation
    nonsequenced acknowledgement
    nuclear science abstracts
NSAC
    nuclear safety analysis center
NSAF
    national sanitation foundation
NSAG
    n channel self aligned gate
NSB
    national science board
    nonsustained breakdown
    nuclear standards board
NSC
    national safety council
    national security council
    national space council
    network switching center
    noise suppression circuit
    numerical sequency code
NSCC
    nuclear services closed cooling
NSCR
    nuclear science center reactor
NSD
    network status display
    nonsequential disk
NSDM
    national security decision memorandum
NSEC
    nanosecond
    naval ships engineering center
NSEF
    navy security engineers facility

NSEIP
  norwegian society for electronic information processing
NSF
  national science foundation
  nuclear science foundation
NSFH
  north south fine, hundreds
NSFT
  north south fine, tens
NSFU
  north south fine, units
NSGN
  noise generator
NSHEB
  north of scotland hydro electric board
NSI
  nonsatellite identification
  nonsequenced information
  nonstandard item
NSIA
  national security industrial association
NSIC
  nuclear safety information center
NSIF
  near space instrumentation facility
NSIL
  nonsaturating inverter logic
NSL
  national science laboratories
  national science library
  national semiconductor laboratories
  naval supersonic laboratory
  northrop space laboratory
  nuclear safety line
NSM
  network space monitor
  network status monitor
  nondeterministic sequential machine
NSMB
  nuclear standards managment board
NSMSES
  naval ship missile systems engineering station
NSP
  nasa support plan
  national space program
  navigational satellite program
  network services protocol
  network support plan
  nonseries parallel
  nonstandard part
  northern state power
NSPA
  national society of public accountants

NSPAC
  national standards policy advisory committee
NSPE
  national society of professional engineers
NSPI
  national society for programmed instruction
NSPP
  nuclear safety pilot plant
NSQCRE
  national symposium on quality control and reliability in electronics
NSR
  neutron source reactor
  nitrile silicone rubber
  noise to signal ratio
NSRB
  national security resources board
NSRC
  national stereophonic radio committee
NSRDS
  national standard reference data system
NSRFI
  national symposium on radio frequency interference
NSRQC
  national symposium in reliability and quality control
NSRS
  naval supply radio station
NSRW
  nuclear service raw water
NSRWP
  nuclear service raw water pump
NSS
  national space station
  navy secondary standards
  navy shore station
  new system simulator
  nuclear steam system
NSSCC
  national space surveillance control center
NSSL
  national severe storms laboratory
NSSM
  national security study memorandum
NSSS
  national space surveillance system
  naval space surveillance system
  nuclear steam supply system
NST
  n digit selected ternary
  national standard taper
  network support team
  newfoundland standard time

NSTA
   national science teachers association
NSTF
   neutron sensor testing facility
NSTIC
   naval scienific and technical information center
NSTL
   national space technology laboratories
NSTP
   nuffield science teaching project
NSV
   nonautomatic self verification
NSW
   national software works
NT
   night telegram
   nit
   no tool
   no transmission
   normal temperature
   not tested
   numbering transmitter
NTA
   national technical association
   norwegian telecommunication administration
   nuclear test aircraft
NTAG
   nuclear technical advisory group
NTC
   national telecommunications conference
   national telemetering conference
   national transformers committee
   negative temperature coefficient
   negative thermal coefficient
NTCRESISTOR
   negative temperature coefficient resistor
NTD
   neutron transmutation doping
NTDS
   naval tactical data system
   naval technical data system
   navy tactical data system
NTE
   navy teletypewriter exchange
   network terminating equipment
NTER
   normalized transmission energy requirement
NTEXIS
   north texas interconnected system
NTF
   nuclear test facility
NTFC
   national television film council

NTI
   noise transmission impairment
NTIA
   national telecommunication and information administration
NTIS
   national technical information service, department of commerce
NTL
   natural thermo luminescence
   nonthreshold logic
   nonuniform transmission line
NTO
   nonorthogonal timing error
NTOL
   normal take off and landing
NTP
   near time processing
   normal temperature and pressure
   nuclear test plant
NTPC
   national technical processing center
NTR
   negative true rake
   noise temperature ratio
   nothing to report
   nuclear test reactor
NTS
   navigation technology system
   negative torque signal
   nevada test site
   not to scale
NTSA
   national technical services association
NTSB
   national transportation safety board
NTSC
   national television system committee
NTT
   new england telegraph and telephone
   nippon telegraph and telephone
NTTPC
   nippon telegraph and telephone public corporation
NTU
   network terminating unit
   number of transfer units
NTV
   nippon television network corporation
NTVA
   nondeterministic time variant automation
NTWS
   nontrack while scan
NTWT
   net weight

NTX
   national teletypewriter exchange
   naval teletype exchange
   navy teletype exchange
NU
   number unobtainable
NUA
   network user address
NUCOM
   numerical contouring mechanism
NUDAC
   nuclear data center
NUDE
   nuclear experimental data evaluation reactor physics computer code
NUDETS
   nuclear detection system
NUE
   national union electric
NUI
   network user identification
NULACE
   nuclear liquid air cycle engine
NUM
   numeral
   numerator
   numeric
NUMAR
   nuclear magnetic resonance
NUMEC
   nuclear materials and equipment corporation
NUMS
   nuclear materials security
NUPAD
   nuclear powered active detection
NUPPS
   nonuniform progressive phase shift
NURE
   national uranium resource evaluation program
NUS
   nominal ultimate strength
   non uniformly spaced
   nuclear utility service
NUSAR
   nuclear sweep and radar
NUSL
   naval underwater sound laboratory
   navy underwater sound laboratory
NUSRL
   naval underwater sound reference laboratory
NUSUM
   numerical summary message
NUTL
   nonuniform transmission line
NUTSEQUAMIR
   national universal television sequential encoded, quadrature and amplitude modulated and intermittently reserved
NV
   nanovolt
   neutralization value
   nevada
   nonvolatile
NVA
   no voltage amplification
NVACP
   neighborhoods voluntary associations and consumer protection
NVDM
   network virtual data manager
NVLAP
   national voluntary lab accreditation pro
NVM
   nonvolatile matter
NVR
   no voltage release
NVRAM
   nonvolatile random access memory
NVSD
   night vision system development
NVSM
   nonvolatile semiconductor memory
NVT
   neutron velocity time
NW
   naked wire
   nanowatt
   network
   northwest
NWAC
   national weather analysis center
NWAHACA
   national warm air heating and air conditioning association
NWB
   national wiring bureau
NWCM
   milliwatt per square centimeter
NWDS
   number of words
NWDSEN
   number of words per entry
NWE
   narrow width effect
NWEB
   northwestern electricity board
NWFAL
   nation wide fall out

NWG
national wire gage
NWL
normal water level
NWMA
national woodwork manufacturers association
NWP
numerical weather prediction
NWPOG
numerical weather prediction operational grid
NWRC
national weather records center
NWRF
naval weather research facility
NWS
national weather service
nosewhell steering
NWSC
national weather satellite center
NWSSG
nuclear weapons system satellite group
NXDO
nike x development office
NY
new york
NYADC
new york air defense center
NYAP
new york assembly program
NYARTCC
new york air route traffic control center
NYCBAN
new york center beacon alpha numerics
NYPP
new york power pool
NYR
not yet required
not yet returned
NYSE
new york stock exchange
NYSEVAB
new york stock exchange voice answer back
NYSPA
new york state power authority
NZ
nike zeus
NZBC
new zealand broadcasting corporation
NZBS
new zealand broadcasting service
NZE
north zenith east system
NZIE
new zealand institution of engineers
NZJCB
new zealand joint communications board
NZSI
new zealand standards institute
NZT
non zero test
NZTJWG
nike zeus target joint working group
NZTV
nike zeus target vehicle

O

O
ohm
ortho
oscillator
output
overall
overall readability
oxygen
OA
office of applications
omnirange antenna
open architecture
operating assembly
operating authorization
operational analysis
operations analysis
operations analyst
operator access
output axis
overall
OAC
office automation conference
operations and checkout
OACO
operation and checkout
OAD
operational availability date
ordering and distribution
overall dimensions
OAI
office of aeronautical intelligence
or accumulators to indicators
OAIDE
operational assistance and instructive data equipment
OAL
operational applications laboratory
operations and logistics
ordnance aerophysics laboratory

OAM
   office of aerospace medicine
   operation and maintenance
   organization and methods
OAME
   orbital altitude and maneuver electronics
   orbital altitude and maneuvering electronics
OAMP
   optical analog matrix processing
OAMS
   orbital altitude and maneuvering system
   orbital attitude and maneuvering system
OANS
   occupied areas news service
OAO
   orbiting astronomical observatory
OAP
   operations and procedures
OAPM
   optimal amplitude and phase modulation
OAR
   office of aerospace research
   office of analysis and review
   operand address register
   operation[al] address register
   operations ananlysis report
   operations and regulations
   optical automatic ranging
   overhaul and repair
OARAC
   office of air research automatic computer
OARC
   ordinary administrative radio conference
OART
   office of advanced research and technology
OAS
   old age and survivors' insurance
   open hearth acid steel
   operational announcing system
   organizational accounting structure
OASF
   orbital astronomy support facility
   orbiting astronomical support facility
OASIS
   ocean all source information system
   on-line administrative information system
   operational automatic scheduling information system
OASM
   office of aerospace medicine
OASR
   office of aeronautical and space research
OASV
   orbital assembly support vehicle

OAT
   operating ambient temperature
   outer atmospheric temperature
   outside air temperature
   overall test
OATC
   overseas air traffic control
OATP
   operational acceptance test procedure
OAV
   operational aerospace vehicle
   output available
OB
   octal to binary
   old boy
   on board
   output buffer
   outside broadcast
   outside broadcasting
   overseas broadcast
   overseas broadcasting
OBA
   octave band analyzer
   oxygen breathing apparatus
OBAWS
   on-board aircraft weighting system
OBC
   on-board computer
   oversea broadcasting
OBD
   omnibearing distance
OBE
   operating basis earthquake
   operating basis event
OBF
   output buffer full
OBGS
   orbital bombardement guidance system
OBI
   omnibearing indicator
   open back inclinable
OBK
   open breaker keying
OBM
   ordnance bench mark
OBN
   out of band noise
OBO
   orbital bomber
OBOE
   observed bombing of enemy
OBP
   on-board processor
OBR
   outboard recorder

OBRA
  overseas broadcasting representatives association
OBS
  observation
  observed
  observer
  obstacle
  omnibearing selector
  open hearth basic steel
  operand buffering system
OBSNFL
  observation flight
OBSV
  observation
  observatory
  observer
OBW
  observation window
OC
  office of controls
  on call
  on center
  on course
  open circuit
  open circuited
  open closed
  open collector
  operating characteristic
  operating coil
  operating curve
  operation code
  operation control
  operational computer
  operations conductor
  operations control
  operator circuit
  outlet contact
  outside circumference
  overcurrent
OCA
  operational control authority
OCAL
  on- line cryptanalytic aid language
OCAS
  on line cryptanalytic aid system
OCB
  oil circuit breaker
  operated circuit breaker
  operations coordinating board
  outgoing calls barred
  override control bits
OCBR
  output channel buffer register

OCC
  occupied
  open circuit characteristic
  operational computer complex
  operational control center
  ordnance command converter
OCCA
  overseas communications cooperation association
OCCM
  office of commercial communications management
OCDM
  office of civil defense mobilization
OCDRE
  organic cooled dueterium reactor experiment
OCDU
  optics coupling display unit
OCEAN
  oceanographic coordination evaluation and analysis network
OCEANS
  omnibus conference on experimantal aspects of naval missile range spectroscopy
OCF
  operational control facility
OCI
  open circuit inductance
  optical coupled isolator
  optically coupled insulator
  oxide control and indication
OCK
  operation control key
OCL
  operation control language
  operational check list
  operational control level
  operator control language
  outgoing correspondence log
OCM
  ohm centimeter
  optical countermeasures
OCNS
  oklahoma city nord sector
OCO
  open close open
  operation capability objectives
  operational checkout
OCOUTPUT
  open collector output
OCP
  operating control procedure
  operational checkout procedure
  operator's control panel
  orbital control program

order code processor
output control pulse
OCPROTECTIONEQUIPMENT
overcurrent protection equipment
OCR
oil circuit recloser
optical character reader
optical character reading
optical character recognition
organic cooled reactor
overcurrent relay
overhaul component requirement
overhead component requirement
OCRA
optical character recognition font a
OCRB
optical character recognition bar
optical character recognition font b
OCRIT
optical character recognizing intelligent terminal
OCRUA
optical character recognition users association
OCS
office computer system
office of cataloguing and standardization
office of communications systems
open circuit stable
operational call sign
operations control system
operator's connection set
optical character scanner
optical computer system
outer continental shelf
OCST
overcast
OCSTAGE
open collector stage
OCT
octagon
octahedral
octal
octave
output clock trigger
OCTL
open circuited terminating line
open circuited transmission line
OCU
office channel unit
operational control unit
OCVD
open circuit voltage decay
OCW
orange cyan wideband

OD
omnidirection
on demand
open drain
operations directive
optical density
original design
outer diameter
output data
output display
outside diameter
outside dimension
oven dry
ODA
operational data analysis
operational design and analysis
ODAP
operation data analysis program
ODB
output display branch
output to display buffer
ODD
operator distance dialing
optical data disk
ODG
operational design group
ODINPUT
output disable input
ODLRO
off diagonal longrange order
ODM
one day mission
orbital determination module
ODN
own doppler nullifier
ODOP
offset doppler
orbital doppler
ODP
operational development phase
operational development program
original document processing
output to display buffer
output to display parity error
ODR
omnidirectional range
ODS
operational data summary
output data strobe
ODSTAGE
open drain stage
ODT
octal debugging technique
on-line debugging technique

operational development team
outside diameter tube
ODTF
operational development test facility
ODU
output display unit
ODVAR
orbit determination and vehicle attitude reference
OE
office of education
omission excepted
open end
operating engineer
operational efficiency
oregon electric railway company
output enable
OEAP
operational error analysis program
OEC
office of electronics and control
ohio edison company
OECD
organization for economic cooperation and development
OECO
outboard engine cut off
OEEC
office of european economic cooperation
OEG
operations evaluation group
OEI
overall efficiency index
OEM
original equipment maker
original equipment manufacturer
original equipment market
other equipment manufacturer
OEMI
other equipment manufacturers information
OEN
oak electronetics
OEP
oil extended polymer
OER
operational equipment requirement
OERC
optimum earth reentry corridor
OES
operations and equipment section
order entry systems
OESLA
office of engineering standards liaison and analysis

OETP
operations experimental test plan
OF
oil filled
oil fuel
optional feature
orbital facilities
oscillator frequency
output factor
outside face
oxygen free
OFA
oil immersed forced air cooled
OFACS
overseas foreign aeronautical communications station
OFARS
overseas foreign aeronautical receiver station
OFATS
overseas foreign aeronautical transmitter station
OFB
oil forced blast
operational facilities branch
OFC
one flow cascade cycle
operational flight control
orthonormal function coding
OFDM
optical frequency division multiplexing
OFG
optical frequency generator
OFHC
oxygen free hard copper
oxygen free high conductivity
oxygen free high conductivity copper
OFIR
oceanic flight information region
OFM
observation file maintenance
OFMC
operational fixed microwave council
OFO
office of flight operations
OFP
operating force plan
oscilloscope face plane
OFPP
office of federal procurement policy
OFR
on-frequency repeater
operational failure report
ordering function register
overfrequency relay

OFSD
  operating flight strength diagram
OFSM
  operational flight safety monitor
OFT
  operational flight trainer
OFTMS
  output format table modification submodule
OG
  ogee
  or gate
  outer gimbal
  outgoing
  output gate
OGA
  outer gimbal axis
OGAE
  oklahoma gas and electric
OGE
  oklahoma gas and electric
  operating ground equipment
  operational ground equipment
OGI
  outer grid injection
OGL
  outgoing line
OGM
  office of guided missile
OGMC
  ordnance guided missile center
OGMS
  ordnance guided missile school
OGO
  orbiting geophysical observatory
OGR
  outgoing repeater
OGRL
  outgoing rural line
OGRS
  outgoing relay set
  outgoing rural selector
OGS
  outgoing secondary switch
OGSE
  operational ground support equipment
OGSEL
  operational ground support equipment list
OGST
  overthread guide sleeve tool
OGT
  outgoing trunk
OGTC
  outgoing toll center
  outgoing toll circuit
OGU
  outgoing unit
OH
  octal to hexadecimal
  ohio
  ohmic heating
  oil hardened
  on hand
  open hearth
  operational hardware
  oval head screw
  overhead
OHA
  office of hearings and appeals
  outside helix angle
OHC
  occupational health center
  overhead camshaft
OHD
  one hour duty
  over the horizon detection
OHDETS
  over horizon detection system
OHF
  occupational health facility
OHM
  ohmmeter
OHMCM
  ohm centimeter
OHMR
  office of hazardous materials regulations
OHP
  ohio power
  outer helmholtz plane
  oxygen at high pressure
OHS
  open hearth steel
OHSGT
  office of high speed ground transportation
OHV
  overhead valve
OI
  oil immersed
  oil immersion
  oil insulated
  operating instructions
OIA
  optics inertial analyzer
OIB
  operating impendance bridge
  operations integration branch
OIC
  on-line instrument and control program
  operations instrumentation coordinator
  optical integrated circuit

OICO
  office of integration and checkout
OID
  order initiated distribution
OIDPS
  overseas intelligence data processing system
OIFC
  oil insulated fan cooled
OIG
  optically isolated gate
OIL
  orange indicating lamp
OIP
  operating internal pressure
OIS
  office information system
  office of information service
  operating information system
  operational instrumentation system
  operational intercommunication system
  ounce inches per second
OISA
  office of international science activities
OISC
  oil insulated self cooled
OIST
  operator integration shakedown test
OIWC
  oil immersed water cooled
  oil insulated water cooled
OJT
  on the job training
OK
  oklahoma
  okonite
  oktal
OKITAC
  oki transistorized computer
OL
  oil lighter
  on line
  only loadable
  open loop
  operating level
  operating license
  operating location
  operations and logistics
  oscillating limiter
  overhead line
  overlap
  overload
  overload relay
OLB
  outer lead bonding
OLBM
  orbital launched ballistic missile
OLC
  on-line computer
  outgoing line circuit
OLCA
  on-line circuit analysis
OLD
  on-line debug
  open loop damping
OLERT
  on-line executive for real time
OLF
  orbital launch facility
OLIP
  on-line instrument package
OLIVER
  on-line interactive variable editing reporter
OLM
  on-line monitor
OLO
  orbital launch operation
OLP
  oxygen lance powder
OLPARS
  on-line pattern analysis and recognition system
OLPS
  on-line programming system
OLR
  objective loudness rating
  overload relay
OLRT
  on-line real time
OLS
  operational launch station
  optical landing system
OLSA
  off-line selectric analyser
OLSC
  on-line scientific computer
OLSS
  on-line software system
OLTEP
  on-line test executive program
OLTP
  on-line transaction processing
OLTS
  on-line mainframe testing system
  on-line test system
  on-line transaction system
OLU
  outdoing line unit

OLV
  open frame low voltage
  orbital launch vehicle
OLVP
  office of launch vehicle and propulsion
  office of launch vehicle programs
OM
  old man
  operational maintenance
  operational management
  operational monitor
  operations manager
  optical master
  orbit modification
  organic matter
  outer marker
  oxygen to metal ratio
OMA
  orderly marketing agreement
OMAT
  office of manpower and automation training
  office of manpower automation and training
OMB
  office of management and budget
  outer marker beacon
  outer marking beacon
OMBE
  office of minority business enterprise
OME
  office of management engineer
OMEC
  optimized microminiature electronic circuit
OMETA
  ordnance management engineering training agency
OMF
  object module file
  operation and maintenance facilities
OMI
  optical measurement instrument
OMIBAC
  ordinal memory inspecting binary automatic computer
OMKR
  outer marker
OML
  orbiting military laboratory
  ordnance missile laboratories
OMLIT
  one man live interception test
OMM
  operation and maintenance manual
OMNITENNA
  omnirange antenna

OMP
  operating maintenance panel
  operating maintenance procedures
  output make up
OMPR
  optical mark page reader
OMPRA
  one man propulsion research apparatus
OMPT
  observed man point trajectory
OMR
  operational modification report
  optical mark read
  optical mark reader
  optical mark reading
  optical mark recognition
  organic moderated reactor
OMRCA
  organic moderated reactor critical assembly
OMRE
  organic moderated reactor experiment
OMRR
  ordnance material research reactor
OMRV
  operational maneuvering reentry vehicle
OMRW
  optical maser radiation weapon
OMS
  office management system
  operational monitoring system
  orbital maneuvering subsystem
  output multiplex synchronizer
  output per manshift
  ovonic memory switch
OMSF
  office of manned space flight
OMT
  orthomode transducer
  orthogonal mode transducer
OMTS
  organizational maintenance test station
OMU
  optical measuring unit
ON
  octane number
  oil immersed natural cooled
  oxidation number
ONAL
  off net access line
ONFD
  on-line file display
ONI
  office of naval intelligence
  operator number identification

ONOF
  office of network operations and facilities
ONR
  office of naval research
OOA
  optimum orbital altitude
OOF
  office of the future
OOK
  on-offkeying
OOL
  operator oriented language
OOLR
  overall objective loudness rating
OOO
  out of order
OOP
  out of print
OOPS
  off-line operating simulator
OOS
  out of service
  out of sight
OOSC
  out of sight controller
OOSCC
  out of sight control center
OOU
  out of use
OP
  object program
  observation post
  office products
  operand
  operating point
  operating procedure
  operation
  operation code
  operation part of instruction
  operation register
  operational priority
  operations plan
  operator
  oppenheimer phillips process
  orbital period
  out of print
  outer panel
  output
  over pressure
  overproof
  oxygen pressure process
OPA
  optoelectronic pulse amplifier
OPADEC
  optical partial decoy
  optical particle decoy
OPADR
  operand address
OPAI
  office of patents and inventions
OPAL
  optical platform alignment linkage
OPAMP
  operational amplifier
OPC
  office of program coordination
  open printed circuit
  operational control
  operations code
  operations control
OPCIT
  opus citatum
OPCODE
  operation code
OPCOM
  operations communications
OPCON
  operations and control system
  optimizing control
OPCONCEN
  operations control center
OPCONCENTER
  operations control center
OPCTR
  operation counter
OPD
  office products division
OPDAC
  optical data converter
OPDAR
  optical detection and ranging
OPDARS
  optical detection and ranging system
OPE
  operations project engineer
OPEC
  organization of petroleum exporting countries
OPEN
  opening
OPEP
  orbital plane experimental package
OPER
  operating
  operation
  operator
OPERA
  operational analysis

ordnance pulses experimental research assembly
out of pile expulsion and reentry apparatus
OPERG
  operating
OPFM
  outlet plenum feature model
OPI
  office of public information
  oll pressure indicator
OPIC
  overseas private investment corporation
OPIM
  order processing and inventory monitoring
OPL
  old product line
OPLAN
  operation plan
OPLE
  omega position location experiment
  omega positioning and location experiment
OPM
  operations per minute
  operator programming method
OPN
  open
  operation
OPND
  operand
OPNL
  operational
OPNS
  operations
OPO
  optical parametric oscillator
  orbiting planetary observatory
  other programmed operations
OPP
  octal print punch
  operator preparation program
  opposite
  oriented polypropylene
  out of print at present
OPPCE
  opposite commutator end
OPPD
  omaha public power district
OPPE
  office of program planning and evaluation
OPPG
  office of propulsion and power generation
OPPN
  - opposition

OPPOSIT
  optimization of a production process by an ordered simulation and iteration technique
OPPS
  over pressurization protection switch
  over pressurization protection system
OPR
  offsite procurement request
  operating
  operator
  optical page reading
OPREG
  operation register
OPRIC
  operator in charge
OPRNL
  operational
OPS
  offshore power systems
  oil pressure switch
  on-line process syntheziser
  operational paging system
  operational protection system
  operations
  operations per second
  optical power spectrum
OPSCAN
  optical scannings users group
OPSCON
  operations control
OPSEC
  operations per second
OPSF
  orbital propellent storage facility
OPSP
  office of product standards policy
OPSR
  office of pipeline safety regulations
OPT
  optic
  optical
  optics
  optimum
  optional
  output transformer
OPTA
  optimal performance theoretically attainable
OPTAG
  optical pick off two axis gyroscope
OPTAR
  optical automatic ranging
OPTEVFOR
  operational test and evaluation force

OPTI
  office of productivity, technology, and innovation
OPTIM
  order point technique for inventory management
OPTS
  on-line program testing system
OPTUL
  optical pulse transmitter using laser
OPUR
  object program utility routines
OPUS
  optimization program for unstable secondaries
OPV
  ohms per volt
OPW
  operating weight
  orthogonalized plane wave
OPX
  off-premise extension
OQ
  oil quench
  oil quenched
OQC
  outside quality control
OQL
  observed quality level
  on-line query language
  outgoing quality level
OR
  omnidirectional range
  operating reactor
  operation register
  operational equipment requirement
  operational readiness
  operational research
  operations requirements
  operations research
  operator
  or circuit
  or operation
  ordering register
  oregon
  out of range
  output register
  outside radius
  overall resistance
  overhaul and repair
  overload relay
  overrange
  oxygen enhancement ratio
ORA
  or to accumulator
  output reference axis
  output register address
ORACLE
  oak ridge automatic computer and logical engine
  optimum record automation for court and law enforcement
  optional reception of announcements by coded line electronics
ORAN
  orbital analysis
ORATE
  ordered random access talking equipment
ORB
  ocean research buoy
  omnidirectional radio beacon
ORBIS
  orbiting radio beacon ionospheric satellite
ORBIT
  oak ridge binary internal translator
  on-line retrieval of bibliographic data
  on-line, real time branch information transmission
  oracle binary internal translator
  orbit ballistic impact and trajectory
ORBS
  orbiting rendezvous base system
ORC
  on-line reactivity computers
  operations research center
  ordnance rocket center
ORD
  operation readiness
  operational readiness date
  operational ready date
  optical rotary dispersion
ORDENG
  ordnance engineering
ORDIR
  omnirange digital radar
ORDIS
  optical reading direct input system
ORDRAT
  ordnance dial reader and translator
ORDS
  observation requirement data sheet
ORDVAC
  ordnance variable automatic computer
ORE
  occupational radiation exposure
  opertional research establishment
OREG
  operation register

OREM
  objective reference equivalent measurement
ORF
  orifice
ORFM
  outlet region feature model
ORG
  operations research group
  organ
  organic
  organization
  organize
ORGDP
  oak ridge gaseous diffusion plant
ORGN
  organization
ORI
  octane requirement increase
  operational readiness inspection
ORIA
  office of regulatory and information affairs
ORIC
  oak ridge isochronous cyclotron
ORIENT
  orientation
ORIG
  origin
  original
ORINS
  oak ridge institute of nuclear studies
ORIT
  operational readiness inspection team
  operational readiness inspection test
ORL
  orbital research laboratory
  ordnance research laboratory
ORLY
  overload relay
ORM
  operator's reference manual
  operators reference manual
ORNL
  oak ridge national laboratory
ORO
  oak ridge operations office
  operations research office
OROM
  optical read only memory
ORP
  orbital rendezvous procedure
ORPICS
  orbital rendezvous positioning, indexing, and coupling system
ORR
  oak ridge research reactor

  omnidirectional radar range
  omnidirectional radio range
  orbital rendezvous radar
ORRAS
  optical research radiometrical analysis system
  optical research radiometrical system
ORRT
  operational readiness and reliability test
ORS
  octahedral research satellite
  operational research society
  optimal real storage
  or to storage
  output shift register
  overrange station
ORSORT
  oak ridge school of reactor technology
ORT
  operating room technician
  operational readiness test
  orbital rendezvous technique
  overland radar technology
ORTAG
  operations research technical assistance group
ORTAI
  orbit to air intercept
ORTF
  organization radio television france
ORTT
  over reaching transfer trip
ORTUEL
  organized reserve training unit, electronics
ORU
  optimal replaceable unit
ORUNIT
  owens rutherford unit
ORV
  orbital reentry vehicle
  orbital rescue vehicle
  orbital return vehicle
ORZ
  omnirange zero
OS
  odd symmetric
  off scale
  oil switch
  old style
  on sale
  on station
  one shot multivibrator
  one side
  operating system
  operational sequence
  operator's set

ordnance surveying
organizations system
oscilloscope
out of stock
outsize
overseas
OSA
  optical society of america
  overall system attenuation
OSAM
  overflow sequential access method
OSAR
  overhead systems appearance research
OSAY
  outside screw and yoke
OSB
  orbital solar observation
OSBA
  outlet and switch box association
OSC
  on-site safety committee
  operation switching cabinet
  oscillation
  oscillator
  oscillograph
  oscilloscope
  own ship's course
OSCAR
  optimum systems covariance analysis results
  orbiting satellite carrying amateur radio
  overnight statewide customer accounting reporting
OSCILLOSC
  oscilloscope
OSCP
  oscilloscope
OSD
  operational sequence diagram
  operational systems development
OSE
  operational support equipment
  orbital sequence of events
  overall system effectiveness
OSFD
  office of space flight development
OSFM
  office of spacecraft and flight missions
OSFP
  office of space flight programs
OSG
  operand select gate
OSHA
  occupational safety and health act
  occupational safety and health administration
OSI
  office of scientific intelligence
  on-line software international
  open system interconnection
  or storage to indicators
OSIC
  optimization of subcarrier information capacity
OSIGO
  office of the signal officer
OSIS
  office of science information service
  office of scientific information service
OSL
  orbital space laboratory
  orbiting space laboratory
OSM
  operating system monitor
OSMV
  one shot multivibrator
OSMVT
  operating system for multiprogramming with a variable number of tasks
OSO
  office of space operations
  orbital solar observatory
  orbiting satellite observer
  orbiting scientific observatory
  orbiting solar observatory
OSP
  office of statistical policy
  on station position
  oribtal support plan
OSPCP
  operating system, primary control program
OSPE
  organizational spare parts and equipment
OSQ
  operating system q
OSR
  office of scientific research
  operational support requirement
  optical scanning recognition
  output shift register
OSRD
  office of standard reference data
OSS
  ocean surveillance satellite
  office of space science
  office of space systems
  office of statistical standards
  operational storage site
  optical surveillance system
  orbital space station

orbital space station study
ordnance safety switch
OSSA
　office of space science and applications
OSSP
　outer solar system probe
OSSRS
　optimum step size random search
OSSS
　optical space surveillance system
　orbital space station system
　orbital space surveillance subsystem
OST
　objectives strategy tactics
　office of science and technology
　on-shift tests
　on-site test
　operational suitability test
　operational system test
OSTA
　office of space and terrestrial applications
OSTAC
　ocean science and technology advisory committee
OSTF
　operational suitability test facility
　operational systems test facility
OSTI
　office for scientific and technical information
OSTP
　office of science and technology policy
OSTRUMPE
　operating system trump extended
OSTS
　office of state technical service
OSUR
　ohio state university reactor
OSURF
　ohio state university research foundation
OSURO
　ohio state university radio observatory
OSV
　ocean station vessel
　office of space vehicles
　orbital support vehicle
OSWG
　optical systems working group
OT
　oil tempered
　oil tight
　on truck
　operating temperature
　operating time
　operational tiros
　oscillation transformer

　output
　overall test
　overland telegraph
　overtime
OTA
　office of technology assessment
　open test assembly
　operational transconductance amplifier
　operational transductance amplifier
　other talk address
OTB
　off-track betting
　off-track betting corp
OTBD
　outboard
OTC
　office of telecommunications
　office of transport and communications
　old timers club
　once through cooling
　operational test center
　oregon technical council
　originating toll center
　originating toll circuit
　originating trunk center
　over the counter
　overseas telecommunications commission
OTCCC
　open type control circuit contacts
OTD
　orbit test direction
　orbit test directive
　orbital test direction
　orbital test directive
OTDA
　office of tracking and data acquisition
OTDT
　over temperature delta t
OTE
　odd transversal electrical
　operational test equipment
OTEC
　ocean thermal energy conversion
　oriental telephone and electric company
OTF
　optimum traffic condition
　optimum traffic frequency
OTH
　over the horizon
OTHR
　over the horizon radar
OTI
　optimum time invariant
OTIS
　oregon total information system

OTL
  output transformerless
OTM
  odd transversal magnetic
  office of telecommunications management
OTN
  operational teletype network
OTO
  optical tracker operator
OTP
  office of telecommunication policy
  operational test procedure
OTPNL
  outer panel
OTR
  optical tracking
  organic test reactor
  overload time relay
OTRAC
  oscillogram trace reader
OTS
  office of technical services
  office of technological service
  optical technology satellite
  optical tracking satellite
  optical transport systems
  orbital test satellite
OTSA
  oregon television service association
OTSG
  once through steam generator
OTSGS
  once through steam generating system
OTSR
  once through superheat reactor
OTT
  one terminal per task
  one time tape
  outgoing trunk terminal
OTTO
  optical to optical
OTU
  office of technology utilization
  operational test unit
  operational training unit
OTV
  operational television
  operational test vehicle
  optimum time varying
  orbit transfer vehicle
OTVSA
  oregon television service association
OU
  output unit
OUF
  optimal usual frequency
  optimum usual frequency
OUO
  official use only
OUPT
  output
OURAD
  our radiogram
OUT
  output
OUTLIM
  output limiting facility
OUTRAN
  output translator
OUTREG
  output register
OV
  orbiting vehicle
  over
  overvoltage
OVAM
  orbital vehicle assembly mode
OVBD
  overboard
OVE
  orator verbis electric
OVEC
  ohio valley electric corporation
OVERS
  orbital vehicle reentry simulator
OVF
  overvoltage factor
OVFLO
  overflow
OVHD
  overhead
OVHL
  overhaul
OVL
  overlap
  overlay
OVLBI
  orbital very long baseline interferometer
OVLD
  overload
OVLY
  overlay
OVM
  orbiting velocity meter
OVRD
  override
OVRO
  owens vallay radio observatory

OVSP
  overspeed
OVV
  overvoltage
OW
  oil immersed water cooled
  oil in water
  open wire
  order wire
  order writing
OWF
  optimum working frequency
  orbital, weightless flight
OWGL
  obscure wire glass
OWL
  over water line
OWM
  office of weights and measures
OWP
  one way polar
OWPR
  ocean wave profile recorder
OWR
  omega west reactor
OWRR
  office of water resources research
OWS
  ocean weather ship
  ocean weather station
  operational weather support
  orbital weapon system
  orbital workshop
OX
  oxygen
OXIM
  oxide isolated monolith
OXIS
  oxide isolation
OXY
  oxygen
OZ
  ounce
  ozone
OZFT
  ounce foot
OZIN
  ounce inch

P

P
  microprocessor
  pair
  para
  parity
  peg
  pentode
  perforateur honeywell bull
  period
  permeance
  peta
  philips electrologica
  phone
  phosphorous
  pick
  pico
  plate
  plug
  poise
  polar
  polarization
  pole
  polynomial time
  portable
  positive
  positively conducting
  power
  pressure
  primary
  prismatic joint
  probe
  proceedings
  processor
  program
  proportional
  proton
  prototype
  pulse
  pump
  punch
PA
  packet adapter
  pad abort
  particular average
  pascal
  pending availability
  pendulous axis
  per annum
  performance analysis
  phase angle
  picoampere
  pilot amplifier
  pilotless aircraft
  planetary atmosphere
  point of aim
  polar to analog
  position approximate

post adjutant
power amplifier
precision angle
preliminary amplifier
pressure angle
privacy act
probability of acceptance
process automation
product analysis
product analyst
product assurance
program access
program address
program analysis
program attention
program authorization
programmed arithmetic
protected area
public address
pulsating arc
pulse amplifier
PAA
pan american airways
professional and administrative
PAAC
program analysis adaptable control
PAAR
precision approach airfield radar
PAB
potter and brumfield
PABLA
problem analysis by logical approach
PABX
private automatic branch exchange
PAC
pacific
packaged assembly circuit
packed memory
pedagogic automatic computer
personal analog computer
pilotless aircraft
portland cement association
procurement and contracts
professional acitivities committee
program address counter
program allocation checker
program for automatic coding
project analysis and control
purchasing and contracting
PACAACS
pacific area airways and air communications service
PACC
pert associated cost control
product administration and contract control
PACCS
post attack command and control system
PACCT
pert and cost correlation technique
PACE
packaged card random access memory executive
performance and cost evaluation
phase array control electronics
plant acquisition and construction equipment
precision analog computing equipment
preflight acceptance checkout equipment
prelaunch automatic checkout equipment
processing and control element
producers of associated components for electronics
professional activities committees for engineers
program analysis control and evaluation
programming analysis consulting education
projects to advance creativity in education
PACED
program for advanced concepts in electronic design
PACELV
preflight acceptance checkout equipment launch vehicle
PACER
process assembly case evaluator routine
program assisted console evaluation and review
program of active cooling effects and requirements
PACESC
preflight acceptance checkout equipment for spacecraft
PACIR
practical approach to chemical information retrieval
PACK
packing and allocation for a compool
kaleidoscope
PACM
parts and components manual
pulse amplitude code modulation
PACMETNET
pacific meteorological network
PACMISRAN
pacific missile range
PACOB
propulsion auxiliary control box
PACOR
passive correlation and ranging
passive correlation and ranging station

PACORE
  parabolic corner reflector
PACP
  propulsion auxiliary control panel
PACS
  pacific area communications service
PACT
  pay actual computer time
  print active computer tables
  production action control technique
  production analysis control technique
  program for automatic coding techniques
  programmed analysis computer transfer
  programmed automatic circuit tester
  progress in advanced component technology
  project for the advancement of coding techniques
PACTA
  packed tape assembly
PACTCU
  pacific area communications message traffic control unit
PAD
  packed assembler disassembler
  packed assembly disassembly
  polyaperture device
  poor acquisition data
  port of aerial debarkation
  post activation diffusion
  post alloy diffused
  post alloy diffusion
  power amplifier driver
  propellant actuated device
PADAR
  passive detection and ranging
PADL
  pilotless aircraft development laboratory
PADLOC
  passive active detection and location
  passive detection and location
PADLOCC
  passive detection and location of countermeasures
PADRE
  patient automatic data recording equipment
  portable automatic data recording equipment
PADS
  passive active data simulation
  precision antenna display system
PADT
  post alloy diffused transistor
  post alloy diffusion technique
  post alloy diffusion transistor
PAE
  planning and estimating
  port of aerial embarkation
  post accident environment
  progress aerospace enterprises
PAEM
  program analysis and evaluation model
PAF
  printed and fired
PAFB
  patrick air force base
PAFEC
  program for automatic finite element calculation
PAG
  program advisory group
PAGE
  pert automated graphical extension
PAGEOS
  passive geodetic earth orbiting satellite
PAHEP
  plasma and high energy physics
PAHO
  pan american health organization
PAHR
  post accident heat removal
PAI
  piping and instrumentation
  place accumulator in indicators
  precise angle indicator
  process automation interface
  production acceptance inspection
  programmer appraisal instrument
PAIC
  public address intercom system
PAID
  piping and instrumentation diagram
  piping and instrumentation drawing
PAIGH
  pan american institute of geography and history
PAIL
  post attack intercontinental link
PAIP
  public affairs and information program
PAIR
  performance and integration retrofit
PAKSI
  pakistan standards institute
PAL
  pacific aerospace library
  parcel air lift
  patent associated literature
  pedagogic algorithmic language
  permanent artificial lighting
  permissive action link
  phase alternation line

philippine air lines
philips assembler language
power and light
process assembly language
production and application of light
profit and loss
programmable array logic
programmed application library
programmer assistance and liaison
prototype application loop
psychoacoustic laboratory
PALASM
programmable array logic assembler
PALC
passenger acceptance and load control
point arguello launch complex
PALD
phase alternation line delay
PALIS
property and liability information system
PALS
permissive action link systems
point arguello launch site
precision approach and landing system
PALX
private automatic loudspeaking exchange
PAM
panvalet access method
plasma arc machine
pole amplitude modulation
portable alpha monitor
post accident monitoring
primary access method
process automation monitor
pulse amplitude modulation
PAMA
polyalkylmethacrylate
pulse address multiple access
PAMAC
parts and material accountability control
PAMFM
pulse amplitude modulation frequency modulation
PAMI
personnel accounting machine installation
PAMPER
practical application of mid-points for exponential regression
PAMS
pad abort measuring system
paging area memory space
portable acoustic monitoring system
proceedings of the american mathematical society

PAMUX
parallel addressable multiplexer
PAN
panoramic
PANAGRA
pan american grace airways
PANAR
panoramic radar
PANE
performance analysis of electrical networks
PANOR
panoramic
PANS
procedures for air navigation services
PANSDOC
pakistan national scientific and technical documentation
PANSMET
procedures for air navigation services meteorology
PAO
parts assembly order
pulsed avalanche diode oscillator
PAP
pacific automation products
paper
periphery access processor
plant air package
port a punch
PAPA
programmer and probability analyzer
PAPE
photoactive pigment electrophotography
PAPI
pacific automation products, incorporated
pulp and paper industry
PAPP
pull and push plate
PAQR
polyacene quinone radical
PAR
paragraph
parallax and refraction
parallel
parameter
peak accelerometer recorder
peak to average ratio
pennsylvania advanced reactor
performance analysis and review
performance appraisal report
perimeter acquisition radar
perimeter array radar
positive acknowledgement and retransmission
precision approach radar
problem analysis and response program

production automated riveting
program activity recording
program appraisal and review
progressive aircraft rework
project authorization request
publication analysis report
pulse acquisition radar
purchasing approval request
PARA
parabolic
parachute
parachutist
paragraph
paramagnetic
problem analysis and recommended action
PARAB
parabola
PARABOL
parabolic
PARADE
passive active range determination
passive active ranging and determination
PARAM
parametric
PARAMI
parsons active ring around miss distance indicator
PARAMP
parametric amplifier
PARAS
parasitic
PARASYN
parametric synthesis
PARC
progressive aircraft reconditioning cycle
PARD
parts application reliability data
periodic and random deviation
pilotless aircraft research division
precision annotated retrieval display
PARDOP
passive ranging doppler
PARE
programmed analysis and resources evaluation
PARESEV
paraglider research vehicle
PARIS
postal address reader indexter system
pulse analysis recording information system
PARL
prince albert radar laboratory
PARM
program analysis for resource anagement

PAROS
passive ranging on submarines
programmed automated replenishment ordering system
PARR
pakistan atomic research reactor
post accident radioactivity removal
procurement authorization and receiving report
PARS
passenger airline reservation system
perimeter acquisition radars
pilotless aircraft research station
programmed airline reservation system
PARSAC
particle size analog computer
PARSAVAL
pattern recognition system application evaluation
PARSEC
parallax second
PARSECS
program for astronautical research and scientific experiments concerning space
PARSET
precision askania range system of electronic timing
PARSEV
paraglider research vehicle
PARSIM
plant appropriation request simulation
PARSYN
parametric synthesis
parametric synthesis part partial
PART
partial
particle
PARTAC
precision askania range target acquisition and control
PARTEI
purchasing agents of the radio, television, and electronics industries
PARTNER
proof of analog results through numerically equivalent routine
PAS
passing aid system
passive
philadelphia astronautical society
polish astronautical society
primary alert system
privacy act statement
professional activity study
professor of air science

program address storage
program alternative simulation
public address system
PASAT
particle accelerator science and technology
PASC
pacific area standards congress
pan american standards committee
PASCAL
philips automatic sequence calculator
PASE
power assisted storage equipment
PASS
passage
passivate
position and surveyance system
production automated scheduling system
program aid software system
program alternative simulation system
PAT
parametric artificial talking device
patent
patented
pattern
pattern analysis test
performance acceptance test
performance alignment tester
peripheral allocation table
personalized array translator
polaris acceleration test
position adjusting type
power ascension testing
procedure for automatic testing
production acceptance test
production assessment test
program altitude test
program attitude test
programmer aptitude test
programmer aptitude tester
proportional to absolute temperature
pseudo adder tree
PATA
pneumatic all terrain amphibian
PATC
position, attitude, trajectory control
professional, administrative, technical, and clerical
PATCO
professional air traffic controllers association
professional air traffic controllers organization
PATD
parts and tool disposition
patented
PATE
programmed automatic test equipment
PATH
performance analysis and test histories
PATI
passive airborne time difference intercept
PATN
pattern
PATRIC
pattern recognition and information correlation system
pattern recognition, interpretation, and correlation
PATS
precision altimeter techniques
precision altimeter techniques study
PATSU
patrol aircraft service unit patt pattern
PATT
project for the analysis of technology transfer
PATTERN
planning assistance through technical evaluation of relevance numbers
PAU
pilotless aircraft unit
PAUDGET
photometer, automated universal distribution gonioelectric type
PAV
phase angle voltmeter
position and velocity
position and velocity tracking
PAVE
programmed analysis for value engineering
PAVR
pure and vulcanized rubber
pure and vulcanized rubber insulation
PAVT
position and velocity tracking
PAW
powered all the way
pratt and whitney
PAWBP
pension and welfare benefit programs
PAWOS
portable automatic weather observable station
PAWS
phased array warning system
programmed automatic welding system
PAX
place address in index register
private automatic exchange
PAYE
pay as you earn

PB
   packard bell
   parity bit
   peripheral buffer
   phonetically balanced
   piggy back
   pipe break
   playback
   plot board
   plug board
   plugging back
   power boiler
   power box
   process bulletin
   program block
   proportional band
   publications bulletin
   pulse beacon
   pulse bonded
   push button
PBA
   printed board assembly
PBAC
   program budget advisory committee
PBAPS
   pipe break air piping system
   pipe break automatic protective system
PBB
   polybrominated biphenyl
PBC
   periodic bond chain
PBCC
   packard bell computer corporation
   palen beach computer consultants
PBCH
   primitive bose chaudhuri hocquenghem
PBDG
   push button data generator
PBE
   prompt burst experiments
PBF
   pilot bypass filter
   power burst facility
PBHP
   pounds per brake horsepower
PBHTGR
   peach bottom high temperature gascooled reactor
PBI
   process branch indicator
PBIP
   pulse beacon impact predictor
PBIT
   parity bit

PBJ
   paper braided jute
PBL
   parachute braked landing
   planetary boundary layer
   public broadcast laboratory
PBM
   potential barrier method
   pulse burst modulation
PBO
   plotting board operator
PBP
   plotting board plot
   push button panel
PBPS
   post boost propulsion system
PBR
   plum brook reactor
   procedure base register
PBRE
   pebble bed reactor experiment
PBRF
   plum brook reactor facility
PBS
   pressure boundary subsystem
   program buffer storage
   project breakdown structure
   public broadcasting service
   push button switch
PBSE
   philadelphia baltimore stock exchange
PBSTA
   push button station
PBT
   parity bit test
   piggy back twistor
PBTF
   pump bearing test facility
PBV
   post boost vehicle
PBW
   parts by weight
   proportional bandwidth
PBX
   private branch exchange
PBXFS
   private branch exchange final selector
PC
   paper core
   parameter checkout
   parity check
   parsec
   part card
   path control
   per cent

per compass
percent
percentage
permeance coefficient
personal computer
phase coherent
phase control
phase controlled
philco corporation
photo cell
photo circuits corporation
photo conductive
photo conductor
picocoulomb
picocurie
picture
pitch circle
pitch control
plastic core
plate circuit
plug card
plug compatible
plugable card
plugboard controlled
point of curve
polar crane
polar to cartesian
port call
portland cement
positve column
post chlorinated
power cartesian
power circuit
power components
power contactor
power conversion
predictable computation
prime level code
print command
print cycle
print of curve
printed card
printed circuit
process control
process controller
processor controller
production control
professional communication
program card
program check
program coordination
program counter
programmable control
programmable controller
programmable logic controller

programmable machine control
programmed check
pseudocode
pulsating current
pulse code
pulse comparator
pulse compression
pulse controller
pulse counter
punched card
purified concentrate
pyrocarbon

PCA
polar cap absorption
pole cap absorption
pool critical assembly
portland cement association

PCAC
partially conserved axial current
partially conserved axial vector current

PCAE
parts control and expediting

PCAM
punch[ed] card accounting machine

PCAS
primary central alarm station

PCAT
procedures for the control of air traffic

PCB
page control block
polychlorinated biphenyl
polychlorinated biphenylene
power circuit breaker
printed circuit board
process control block
program communication block
propulsion control box

PCBA
pioneer citizens band association

PCBC
partially conserved baryon current

PCBOARD
printed circuit board

PCBS
positive control bombardment system

PCC
partial crystal control
phase correction circuit
photoelectric counter chronometer
planning coordination committee
point of compound curve
polarity coincidence correlator
positive control communications
process control computer
processor control cards

program control card
program controlled computer
protocol converter concentrator
PCCA
power and communication contractors association
PCCARD
plug compatible card
printed circuit card
PCCD
peristaltic charge coupled device
PCCN
part card change notice
PCCP
preliminary contact change proposal
PCCR
procurement code change request
PCCS
photographic camera control system
positive control communication system
PCD
photoconductive decay
pitch circle diameter
plasma coupled devices
power control and distribution
power control device
PCDP
punched card data processing
punched card data processing equipment
PCDS
power conversion and distribution system
PCE
parameter checkout engineer
peripheral control element
photocell emitter
pool control error
process control element
processor control element
punched card equipment
pyrometric cone equivalent
PCEA
pacific coast electric[al] association
PCEM
process chain evaluation model
PCETF
power conversion equipment test facility
PCF
pound per cubic foot
power cathode follower
pulse to cycle fraction
PCFM
production control file manager
production file manager
PCFP
predicted comparative failure probability

PCG
phonocardiogram
planning and control guide
printed circuit generator
PCH
p channel
punch
PCHANNELMOS
p channel metal oxide semiconductor
PCHG
punching
PCHMOS
p channel metal oxide semiconductor
PCI
packet communication, inc
panel call indicator
pattern correspondence index
peripheral command indicator
photon coupled isolator
pilot controller integration
planning card index
polycrystal isolation
prestressed concrete institute
printed circuits, incorporated
product configuration identification
program check interruption
program controlled interrupt
program controlled interruption
programmable communication interface
PCIS
primary containment isolation system
PCK
peck
PCL
permissible contamination limits
printed circuit lamp
process control language
PCLA
power control linkage assembly
PCLDI
prototype closed loop development installation
PCLS
prototype closed loop system
PCLT
prototype closed loop test
PCM
passive countermeasure
performance capability measure
peripheral computer manufacturer
pitch control motor
plug compatible mainframer
plug compatible manufacturer
plug compatible module
power cooling mismatch

primary code modulation
process communication monitor
pulse code modulation
pulse code modulator
pulse count modulation
punch[ed] card machine
PCMD
pulse code modulation digital
PCMDHS
pulse code modulation data handling system
PCME
pulse code modulation event
PCMFM
pulse code modulation frequency modulation
PCMFSKAM
pulse code modulation frequency shift keying amplitude modulation
PCMI
photochromic microimage
photochromic microimage system
PCMP
packed computational
PCMPM
pulse code modulation phase modulation
PCMPN
pulse code modulation pseudonoise
PCMPS
pulse code modulation phase shift
PCMS
punched card machine system
PCMSIGNAL
pulse code modulated signal
PCMTS
pulse code modulation telemetry system
PCN
pacific communications network
prelaunch channel number
PCNE
protocol converter for native equipment
PCNS
polar coordinate navigation system
PCO
picture control oscilloscope
port communications office
post central office
primary communications oriented
program coordination office
PCOS
primary communications oriented system
PCP
parallel cascade processor
parallel circular plate
peripheral control program
photon coupled pair
post construction permit

preassembled cable in pipe
primary control program
primary coolant pump
procesor control program
process control processor
program change proposal
program control plan
project control plan
pulse comparator
punched card punch
PCPCN
part card procurement change notice
PCPS
program change packages
PCQ
production control quantometer
productivity criteria quotient
PCR
partial carriage return
peer code review
per call rate
perpheral control routine
photoconductive relay
plant control room
power control room
procedure change request
production control record
program change request
program control register
program counter
programmer in charge of records
pulse compression radar
punched card reader
PCRCA
pickled, cold rolled and close annealed
PCRS
primary control rod system
PCRV
prestressed concrete reactor vessel
PCS
peripheral computer system
permanent change of station
planning control sheet
pointing control system
polymer clad silica
position control system
position, course, speed
power conversion system
precedence charting system
primary coolant system
print contrast scale
print contrast signal
process communication supervisor
process control system
production control system

program counter store
project control system
punched card system
PCSC
  power conditioning, switching and control
PCSE
  pacific coast stock exchange
PCSIR
  pakistan council of scientific and industrial research
PCSP
  program communications support
  program
PCT
  paper crepe tape
  per cent
  perfect crystal technology
  photon coupled transistor
  pitch centering torque
  planning and control techniques
  point contact transistor
  portable camera transmitter
  potential current transformer
  pressure concentration temperature
  program control table
  program counter timer
PCTE
  portable common tools environment
PCTF
  plant component test facility
  power conversion test facility
PCTFE
  polychloride trifluorethylene
  polychlorotrifluoroethylene
  polymonochlorotrifluoroethyle
PCTM
  pulse count modulation
PCTR
  physical constant test reactor
PCU
  peripheral control unit
  pound centigrade unit
  power control unit
  power conversion unit
  pressure control unit
  primary control unit
  print control unit
  progress control unit
PCUA
  power controller unit assembly
PCUSEQ
  pressure control unit sequencer
PCV
  pollution control valve
  pressure control valve

PCVC
  partially conserved vector current
PCW
  pulsed continuous wave
PD
  paid
  partial discharges
  passive detection
  pedestal
  period
  periodic duty
  peripheral device
  periscope depth
  phase discriminator
  phelps dodge
  physical development
  pitch circle diameter
  pitch diameter
  planned derating
  plasma deposited
  plate dissipation
  plotting display
  polar distance
  port dues
  port of debarkation
  position doubtful
  positive displacement
  potential difference
  power distribution
  power divider
  power doubler
  power driven
  preliminary design
  pressure difference
  pressure distillate
  primer driver
  priority directive
  proability of detection
  probability density
  process descriptor
  project documentation
  projected decision date
  projected display
  propellant dispersion
  proportional derivative
  proportional plus derivative
  proton donar
  proximity detector
  pulley drive
  pulse doppler
  pulse driver
  pulse duration
PDA
  patient data automation
  peak distribution analyzer

percent defective allowable
post acceleration
post deflection accelerated
post deflection acceleration
post deflection accelerator
precision drive axis
predocketed application
preliminary design approval
preliminary design authorization
probability discrete automata
probability distribution analyzer
proposed development approach
pulse distribution amplifier
push down automation
PDAID
    problem determination aid
PDAIDS
    problem determination aids
PDB
    public debt bureau
PDC
    parallel data communicator
    parts distribution center
    performance data computer
    performance development corporation
    power distribution control
    predetection combining
    premission documentation change
    pressure die casting
    prevention of deterioration center
    probability of detection and conversion
    professional development committee
    pure direct current
    single paper double cotton
PDCL
    provisioning data check list
PDCS
    performance data computer system
PDD
    physical damage division
    premodulation processor deep space data
    program design data
    prospective decision date
PDE
    partial differential equation
PDES
    preliminary draft environmental statement
PDF
    plant design factor
    point detonating fuse
    probability density function
    probability distribution function

PDG
    production development group
    programs development group
PDGDL
    plasma dynamics and gaseous discharge laboratory
PDGS
    precision delivery glides system
PDH
    planned derated hours
PDI
    panel data interface
    pictorial deviation indicator
    pilot direction indicator
    power dissipation index
PDIO
    photodiode
PDIS
    pressure differential switch
    proceedings of the national symposia
PDL
    poundal
    procedure definition language
    programmable digital logic
    protocol description language
PDM
    phase displacement
    precedence diagramming method
    pulse data modulation
    pulse delta modulation
    pulse duration modulation
PDME
    precision distance measuring equipment
PDMMS
    product design minuteman mechanical system
PDN
    production
PDO
    program directive operations
    publications distribution office
PDP
    plasma display panel
    polysilicon dielectric polysilicon
    positive displacement pump
    power distribution plan
    process data processing
    process development pile
    program definition phase
    program development plan
    programmable data processor
    programmed data processing
    programmed data processor
    project development plan
PDPC
    position display parallax corrected

PDPS
   parts data processing system
   program data processing section
   program data processing system
PDQ
   programmed data quantizer
PDR
   periscope depth range
   power directional relay
   precision depth recorder
   predetection recording
   preliminary data report
   preliminary design review
   pressurized deuterium reactor
   priority data reduction
   process dynamics recorder
   processed data recorder
   processing data rate
   program discrepancy report
   program drum recording
   public document room
PDRC
   pressure difference recording controller
PDRP
   program data requirement plan
PDS
   pacific data systems
   partitioned data set
   personal data system
   personnel data system
   phot document sensor
   power density spectra
   power distribution specification
   power distribution system
   procedures development simulator
   production data sheet
   program data source
   propellant dispersion system
PDSMS
   point defense surface missile system
PDST
   pacific daylight saving time
PDSTATION
   pick-up discharge station
PDT
   pacific daylight time
   panoramic design technique
   peripheral data transfer
   post alloy diffused transistor
   programmable data terminal
PDU
   pilot display unit
   power distribution unit
   pressure distribution unit

PDV
   premodulation processor deep space voice
PE
   pacific electric
   parity error
   peclet number
   pentaeythrol
   peripheral equipment
   perkin elmer
   permanent echo
   permanent error
   phase encoded
   phase encoding
   philadelphia electric
   philips electric
   photoelectric
   photoelectron
   pilot equalizer
   pioneer electric and research
   polyethylene
   polyethylene potentia energy
   port of embarkation
   potential energy
   power equipment
   precipitation evaporation
   premissible error
   probable error
   processing element
   professional engineer
   program element
   project engineer
PEA
   pennsylvania electric association
   push effective address
PEAC
   photoelectric alignment collimator
   photoelectric autocollimator
PEACU
   plastic energy absorption in compression unit
PEARL
   process and experiment automation real time language
PEAT
   programmer exercised autopilot test
PEB
   program element breakdown
   pulsed electron beam
PEC
   packaged electronic circuit
   peripheral equipment corp
   photoelectric cell
   pilot error correction
   production equipment code
   program environmental control

PECAN
  pulse envelope correlation air navigation
PECBI
  professional engineers conference board for industry
PECONTROLSYSTEM
  phase encoding control system
PECOS
  program environmental checkout system
  project evaluation and cost optimisation system
PECR
  program error correction report
PECS
  portable environmental control system
PED
  pedestal
  personal equipment data
  personnel equipment data
  phosphorus enhanced diffusion
  power equipment division
  program execution directive
  proton enhanced diffusion
PEDMS
  portable and extensible data management system
PEDN
  planned event discrepancy notification
PEDRO
  pneumatic energy detector with optics
  pneumatic energy detector with remote optics
PEDS
  protective equipment decontamination section
PEEP
  pilot's electronic eye level presentation
PEF
  physical electronics facility
  powerhouse exhaust facility
  pulse eliminating filter
PEG
  principle of the equivalent generator
  public service electric and gas
PEGE
  program for evaluation of ground environment
PEI
  preliminary engineering inspection
PEIC
  periodic error integrating controller
PEIR
  performance evaluation and information reduction
  process evaluation and information reduction
PEK
  phase exchange keying

PEL
  picture element
  proportional elastic limit
PELSS
  precision emitter location strike system
PEM
  photoelectromagnetic
  processing element memory
  processing element module
  processor element memory
  production engineering measure
  program element monitor
PEN
  pentode
  program error note
PENA
  primary emission neuron activation
PENCIL
  pictorial encoding language
PENNSTAC
  pennsylvania state automatic computer
PENT
  penetrate
  penetration
  pentode
PEO
  patrol emergency officer
PEOS
  propulsion and electrical operating system
PEP
  patent examinig procedure
  peak envelope power
  planar epitaxial passivated
  planetary ephemeris program
  platform electronic package
  princeton electronic products
  princeton experimental package
  program evaluation procedure
  program evaluation program
  programmable extension package
  proton electron positron colliding beams
  proton electron proton
  pulse effective power
PEPAG
  physical electronics and physical acoustics group
PEPCO
  potomac electric power company
PEPP
  planetary entry parachute program
PEPPRE
  photo electric portable probe reader
PEPR
  precision encoding and pattern recognition
PEPSS

programmable equipment for personnel
subsystem simulation
PER
  preliminary engineering report
  probable error radial
  program event recording
PERA
  production engineering research association
PERCOS
  performance coding system
PERCY
  photoelectric recognition cybernetics
PERECORDING
  phase encoded recording
PEREF
  propellant engine research environmental facility
PERF
  perfection
  perforate
  perforated
  perforator
  performance
PERFR
  perforator
PERGO
  project evaluation and review with graphic output
PERIOD
  periodical
PERM
  permanent
  permeability
  program evaluation for repetitive manufacture
PERP
  perpendicular
PERS
  personal computer
  program for evaluation of rejects and substitutions
PERT
  performance evaluation review technique
  pert program evaluation and review technique
  program estimation revaluation technique
  program evaluation and review technique
  program evaluation research task
PERTCO
  program evaluation review technique with cost
PERTTIME
  program evaluation and review technique time
PERU
  production equipment records unit

PES
  passive electromagnetic system
  photoelectric scanner
  pueblo engineers society
PESD
  program execution subdirective
PEST
  parameter estimation by sequential testing
  project engineer scheduling technique
PESY
  pheripheral exchange synchronisation
PET
  patterned epitaxial technology
  performance evaluation test
  peripheral equipment tester
  personal electronic transactor
  philco epoxy transistor
  physical equipment table
  polyethylene
  polyethylene terephthalate
  position event time
  process evaluation tester
  production environmental test
  production environmental testing
  production experimental test
PETA
  portable electronic traffic analyzer
PETC
  polyethylene tetrachloride
PETE
  pneumatic end to end
PETN
  pentaerythritol tetranitrate
PETS
  pacific electronics trade show
PEV
  peak envelope voltage
PEWS
  parts early warning system
PEX
  private electronic exchange
PEXRA
  programmed electronic x-ray automatic diffractometer
PEXRAD
  programmed electronic x-ray automatic diffractometer
PF
  page fault
  partial function
  penetration fracture
  permanent force
  picofarad
  pilot stop filter
  pole fittings

pore free
power factor
preflight
probability of failure
program function
protection factor
proximity fuse
pulse feedback
pulse frequency
punch off
PFA
pulverized fuel ash
pure fluid amplification
pure fluid amplifier
PFAM
programmed frequency amplitude modulator
PFB
preformed beam
provisional frequency board
PFC
power factor capacitor
PFCC
power factor corrector capacitor
PFCS
primary flow control system
PFD
phase frequency distortion
preferred
primary flash distillate
PFE
photoferroelectric effect
plenum fill experiment
primary feedback element
PFES
proposed final environmental statement
PFG
primary frequency generator
PFI
photon flow integrating
photon flux integration
power factor indicator
PFIMODE
photon flow integrating mode
PFL
propulsion field laboratory
PFLL
phase and frequency locked loop
PFM
power factor meter
pulse frequency modulation
PFN
pulse forming network
PFNS
position fixing and navigation system

PFP
probability of failure, performance
PFR
parts failure rate
plug flow reactor
polarized field frequency relay
power fail recovery
power fail recovery system
preliminary flight rating
programmable film reader
prototype fast reactor
pulse frequency
punch feed read
PFRS
portable field recording system
PFRT
preliminary flight rating test
PFS
peripheral fixed shim
physical file system
propellent field system
PFT
page frame table
paper, flat tape
prime factor transform
PFV
peak forward voltage
PG
page
pilot generator
power gain
power generation
precision ground
pressure gage
pressure gauge
program generator
program guidance
programmer group
protective ground
proving ground
pulse generator
PGA
pin grid array
power generation assembly
professional group audio
programmable gain amplifier
PGAC
professional group automatic control
PGAE
pacific gas and electric
PGANE
professional group aeronautical and navigational electronics
PGAP
professional group antennas and propagation

PGBTR
    professional group broadcast and television receivers
PGBTS
    professional group broadcast transmission systems
PGC
    polynomial generator checker
    power generation committee
    programmed gain control
PGD
    program for geographical display
PGE
    pennsylvania gas and electric
    portland general electric
PGEC
    professional group on electronic computers
PGECP
    professional group electronic component parts
PGED
    professional group electronic devices
PGEM
    professional group engineering management
PGEWS
    professional group on engineering writing and speech
PGF
    presentation graphics feature
PGHFE
    professional group human factors in electronics
PGI
    perfo guide international
    peripherals general inc
    professional group instrumentation
PGIA
    pendulous gyro integrating accelerometer
    programmable gain instrumentation amplifier
PGLN
    page and line
PGM
    program
PGME
    professional group medical electronics
PGMILE
    professional group military electronics
PGMTT
    professional group microwave theory and techniques
PGNCS
    primary guidance and navigation control system

PGNS
    primary guidance and navigation system
PGP
    pulsed glide path
PGQC
    professional group quality control
PGR
    precision graphic recorder
    psychogalvanic response
PGRF
    pulse group repetition frequency
PGRFI
    professional group radio frequency interference
PGRQC
    professional group reliability and quality control
PGRTRC
    professional group radio telemetry and remote control
PGS
    power generator section
    program generator system
PGSET
    professional group space electronics and telemetry
PGT
    page table
PGTRC
    professional group telemetry and remote control
PGU
    power generator unit
    pressure gas umbilical
PGUE
    professional group ultrasonic engineering
PGVC
    professional group vehicular communications
PH
    per hour
    performance history
    period hours
    phantom
    phase
    phone
    phot
    picohenry
    power house
    power of hydrogen
    telephone
PHA
    prelaunch hazard area
    programmable host access
    pulse height analysis
    pulse height analyzer

PHCONST
  phase constant
PHD
  phase shift driver
  pulse height discrimination
PHENO
  precise hybrid element for nonlinear operation
PHENOS
  precise hybrid elements for nonlinear operations
PHERMEX
  pulsed high energy radiographic machine emitting x-rays
PHF
  plug handling fixture
PHI
  position and homing indicator
  position homing indicator
PHICT
  philips inventory control technique
PHIN
  position and homing inertial navigator
PHLAG
  philips petroleum load and go system
PHLODOT
  phase lock doppler tracking
PHM
  phase meter
  phase modulation
  phase modulator
PHOENIX
  plasma heating obtained by energetic neutral injection experiment
PHOFL
  photoflash
PHONE
  headphone
  telephone
PHONO
  phonograph
  phonographic
PHOTOM
  photometry
PHP
  parts, hybrids, and packaging
  performance, hermeticity and price
  planetary horizon platform
  pound per horsepower
PHR
  per hour
  ph-recorder
  pound force per hour
  process heat reactor program
  pulse height resolution

PHRC
  ph-recording controller
PHS
  pan head steel
  public health service
PHSPS
  preservation, handling, storage, packaging, and shipping
PHT
  phototube
PHTS
  primary heat transport system
PHV
  phase velocity
PHWR
  pressurized heavy water moderated and cooled reactor
  pressurized heavy water reactor
PHYSREV
  physical review
PI
  packaging institute
  paper insulated
  paper insulated point of intersection
  parallel input
  parametric industries
  penetration index
  performance index
  personal identification
  petrol injection
  photo international
  photointerpreter
  photogrammetric instrumentation
  pilotless interceptor
  plug in
  plug-in instruments
  point initiating
  point insulating
  point of intersection
  position indicator
  postal instructions
  power indicator
  power input
  power interlock
  precision instrument
  primary input
  print image
  priority interrupt
  problem input
  processor interface
  production index
  productivity index
  program indicator
  program interrupt

program of instrumentation
programmed instruction
proportional integral
proportional plus integral
PIA
   pakistan international airlines
   peripheral interface adapter
   place indicators in accumulators
   plastics institute of america
   preinstallation acceptance
PIACTIONELEMENT
   proportional integral action element
PIANC
   permanent international association of navigation congresses
PIAPACS
   psychophysiological information acquisition, processing and control system
PIB
   polar ionosphere beacon
   polar ionospheric beacon
   program information block
   publishing information bulletin
   pulse interference blanker
   pyrotechnic installation building
PIBD
   portable interface bond detector
PIBMRI
   polytechnic institute of brooklyn, microwave research institute
PIBS
   polar ionospheric beacon satellite
PIC
   particle in cell
   periodic inspection control
   peripherie controller
   photo interpretation console
   photographic interpretation center
   picture
   picture interactive computer system
   plastic igniter cord
   plastic insulated cable
   polyethylene insulated conductor
   position independend code
   positive impedance converter
   procurement information center
   program information center
   program interrupt control
   programmable interrupt controller
   pulse ionisation chamber
PICA
   power industry computer applications
PICAC
   power industry computer applications conference

PICAO
   provisional international civil aviation organisation
PICE
   programmable integrated control equipment
PICEL
   picture element
   picture elements per line
PICONTROL
   proportional integral action control
PICONTROLLER
   proportional integral action controller
PICS
   production information and control system
PICSEC
   picture per second
PICT
   philips inventory control technique
PID
   personnel identification device
   photo ionization detector
   program information department
   proportional integral and differential
   proportional integral derivation
   proportional integral differential
   proportional plus integral plus derivative
   public information division
PIDACTIONELEMENT
   proportional integral differential action element
PIDC
   pakistan industrial development corporation
PIDCONTROL
   proportional integral differential control
PIDEP
   preinterservice data exchange program
PIE
   peripheral interface element
   plug in electronics
   post irradiation examination
   program interrupt element
   pulse interference elimination
   pulse interference eliminator
   pulse interference emitting
PIEA
   petroleum industry electrotechnical association
PIEE
   proceedings of the institute of electrical engineers
PIELEMENT
   proportional integral action element
PIF
   package information form
   payload integration facility

PIFAL
　program instruction frequency analyzer
PIFR
　program interrupt flag register
PIG
　pendulous integrating gyro
　penning ionisation gage
　philips ionisation gage
PIGA
　pendulous integrating gyroscope accelerometer
PIGFET
　p channel insulated gate field effect transistor
PIGMI
　position indicating general measuring instrument
PII
　positive immittance inverter
PIL
　pitt interpretive language
　pittsburgh interactive language
　pittsburgh interpretive language
PILAC
　pulsed ion linear accelerator
PILC
　paper insulated lead covered cable
PILL
　programmed instruction language learning
PILMS
　precision insertion loss measurement set
PILOT
　permutation indexed literature of technology
　pilot data processor
　piloted low speed test
PILP
　pseudoinfinite, logarithmically periodic
PIM
　point indicating machine
　precision indicator of the meridian
　precision instrument mount
　pulse interval modulation
PIN
　personal identification number
　police information network
　position indicator
　positive intrinsic negative
　program identification number
PINO
　positive input, negative output
PINS
　portable inertial navigation system
PINSAC
　pins alignment console
PINT
　purdue interpretive programming and operating system
PIO
　peripheral input output
　physical input output
　precision iterative operation
　public information office
PIOC
　program input output cassette
PIOCS
　physical input output control system
PIOSA
　pan indian ocean science association
PIOTA
　post irradiation open test assembly or proximity instrumented open test assembly
PIOU
　parallel input output unit
PIOUS
　peripheral integrated off-line utility system
PIP
　peripheral interchange program
　persistent internal polarization
　position indicating probe
　predicted impact point
　primary indicating position
　probabilistic information processing
　problem input preparation
　program in progress
　program integrating plan
　programmable integrated process
　programmable integrated processor
　programmed individual presentation
　project on information processing
　prototypic inlet piping
　provabilistic information processing
　pulsed integrating pendulum
PIPA
　pulse integrating pendulum accelerometer
　pulsed integrating pendulum accelerometer
PIPER
　pulsed intense plasma for exploratory research
PIPS
　pattern information processing system
　production information processing system
　pulsed integrating pendulums
PIQSY
　probe for the international quiet solar year
PIR
　parallel injection readout
　petrolite irradiation reactor
　precision inspection request

program interrupt register
publication information register
PIRAD
    proximity information, range and disposition
PIRD
    program instrumentation requirements document
PIRE
    proceedings of the institute of radio engineers
PIRG
    public interest research group
PIRN
    preliminary interface revision notice
PIRS
    personal information retrieval system
PIRT
    precision infrared tracking
    precision infrared triangulation
    preliminary infrared triangulation
PIS
    postal inspection service
    prime implicant solution
PISAB
    pulse interference separation and blanking
PISH
    program instrumentation summary handbook
PISO
    parallel input with serial output
PISW
    process interrupt status word
PIT
    parameter input tape
    peripheral input tape
    print illegal and trace
    processing index terms
    program instruction tape
PITTC
    philips international telecommunications training center
PIU
    pilot indicator unit
    plug-in unit
PIUMP
    plug-in unit mounting panel
PIV
    peak inverse voltage
    positive infinitely variable
    post indicator valve
PIVGEAR
    positive infinitely variable gear
PIX
    parallel interface extender
PJM
    pennsylvania new jersey maryland interconnection

PJYLARC
    penn jersey young ladies amateur radio certificate
PK
    peak
    peck
PKA
    primary knocked-on atom
PKD
    packed
PKF
    polarity correlation function
PKG
    package
PKGD
    packaged
PKP
    preknock pulse
    or aid to navigation
    program language analyzer
PKPK
    peak to peak
PKV
    peak kilovolt
PL
    partial loss
    parts list
    peak loss
    phase line
    photoconductor lamp
    photoluminescence
    piping loads
    plain language
    plastic limit
    plate
    plateau length
    plated
    plug
    position line
    power line
    power loading
    pressurizer level
    private line
    production language
    profit and loss
    program level
    program library
    programming language
    proportional limit
    proportionality limit
    public law
    pulse length
PLA
    physiological learning aptitude
    power lever angle

programmable logic array
programmed logic array
proton linear accelerator
PLAAR
    packaged liquid air augmented rocket
PLACE
    postlandsat d advanced concept evaluation
    programming language for automatic checkout equipment
PLACO
    planning committee
PLAD
    plasma diode
PLAN
    positive locator aid to navigation
    program language analyzer
PLANET
    private local area network
PLANIT
    programming language for interactive teaching
PLANNET
    planning network
PLANS
    program logistics and network scheduling system
PLANT
    program for linguistic analysis of natural plants
    programming language for interactive teaching
PLAP
    prelaunch, launch, and ascent procedures
PLAR
    postal laws and regulations
PLARA
    packaged liquid air augmented
PLASTEC
    plastic evaluation center
PLAT
    pilot landing aid television
PLATO
    programmed logic for automated learning operation
    programmed logic for automatic teaching operations
PLATR
    pawling lattice test rig
PLB
    pull button
PLBR
    prototype large breeder reactor
PLC
    power line carrier
    power loading control

    private line carrier
    programmable logic controller
    programming language committee
    propellant loading console
PLCB
    pseudo line control block
PLCDR
    private line carrier divided ringing
PLCL
    phase locked control loop
PLCU
    propellant loading control unit
PLCURR
    plate current
PLD
    payload
    phase lock demodulator
    phase locked demodulator
    phase locked detector
    phase locked discriminator
    pogrammable logic device
    pulse length discriminator
PLDF
    pseudo load factor
PLDISS
    plate dissipation
PLDTS
    propellant loading data transmission system
PLE
    prudent limit of endurance
PLF
    parachute landing fall
    positive lock fastener
    proposition letter formula
PLI
    pilot location indicator
    preload indicator
    programming language i
PLIANT
    procedural language implementing analog techniques
PLIM
    postlaunch and instrumentation message
    postlaunch information message
PLIP
    preamplifier limited infrared photoconductor
PLL
    phase locked loop
PLLDF
    phase locked loop with decision feedback
PLM
    planetary rotation machine
    prelaunch monitor

    programming language for microcomputers
    pulse length modulation
PLMATH
    procedure library mathematics
PLN
    plan
    primary learning net
PLO
    pacific launch operation
    phase locked oscillator
PLOCAP
    post loss of coolant accident protection
PLOD
    planetary orbit determination
PLOO
    pacific launch operations office
PLOP
    pressure line of position
PLOT
    plotting
PLP
    passive low pass
    pattern learning parser
    photolithographic process
    preformed line products
PLPA
    permissive low pressure alarm
PLRACTA
    position, location, reporting, and control of tactical aircraft
PLRS
    position location reporting system
PLS
    plugging switch
    polynomial solution
    private line service
    programming language for system
PLSS
    portable life support system
    precision location strike system
    prelaunch status simulator
PLT
    pacific lighting
    planar tube
    power line transients
    princeton large torus
    private line teletype
    private line teletypewriter
    program library tape
    pulse light theodolite
PLTG
    plating
PLTT
    private teletypewriter service
PLTTY
    private line teletypewriter
PLU
    pert life cycle unified system
    phi lambda upsilon
    pluggable unit
    price look-up
PLUG
    propellant loading and utilization group
PLUGE
    picture line-up generating equipment
PLUOT
    parts listing used on line technique
    parts listing used on technique
PLUS
    pert life cycle unified system
    precision loading and utilization system
    program library update system
PM
    panel maintenance
    panel meter
    parts per million
    pattern maker
    per minute
    permanent magnet
    phase match
    phase modulation
    phase modulator
    photomultiplier
    post meridiem
    pounds per minute
    powder metal
    powder metallurgical
    powder metallurgy
    preventative maintenance
    preventive maintenance
    print matrix
    printing mechanism
    procedures manual
    processor module
    production monitoring
    programming method
    project manager
    pulse code modulation
    pulse modulation
    purpose made
PMA
    permanent magnet association
    phonograph manufacturers' association
    pole mounted amplifier
    precision measurement association
PMAF
    pharmaceutical manufacturers association foundation

PMAR
    preliminary maintenance analysis report
PMB
    physical metallurgy branch
    pilot make busy
PMBX
    private manual branch exchange
PMC
    plaster moulded cornice
    powdered metal cathode
    power mate corporation
    precise measurements company
    program marginal checking
    programmable machine control
    programmable machine controller
    programmable machine tool controller
    programmable matrix controller
    programmed marginal checking
    project management corporation
    pseudo machine code
PMCS
    pulse modulated communication system
PMD
    program monitoring and diagnosis
PMDOWNTIME
    preventive maintenance downtime
PME
    photomagnetoelectric
    photomagnetoelectric effect
    precision measuring equipment
    process and manufacturing engineering
    protective multiple earthing
PMEE
    prime mission electronic equipment
PMEL
    precision measurement equipment laboratory
PMEV
    panel mounting electronic voltmeter
PMF
    performance measurement facility
    polaris missile facility
    probability mass function
    probable maximum flood
    product measurement facility
    programmable matched filter
PMFPAC
    polaris missile facility, pacific
PMG
    permanent magnet generator
    prediction marker generator
PMH
    probable maximum hurricane
    production per man hour

PMI
    personnel managment information
    preventive maintenance inspection
    printed motors, incorporated
    programming methods, inc
PMIS
    personal management information system
    precision mechanisms in sodium
    program measurement information system
PMM
    partial matrix multiply
    permanent magnet motor
    post mast message
    pulse mode multiplex
PMMC
    permanent magnet movable coil
    permanent magnet moving coil
PMMEASUREMENT
    particulate matter measurement
PMMI
    packaging machinery manufacturers institute
PMNT
    permanent
PMOS
    p channel metal oxide semiconductor
    p type channel metal oxide semiconductor
    permanent manned orbital station
    physical movement of spacecraft
PMP
    planar metallization with polymer
    premodulation processor
    preventive maintenance plan
    probable maximum precipitation
    product and marketing planning
PMPE
    punch memory parity error
PMR
    pacific missile range
    paramagnetic resonance
    point of minimum radius
    projection microradiography
PMRF
    pacific missile range facility
PMRFAC
    pacific missile range facility
PMRM
    periodic maintenance requirements manual
PMRTF
    pacific missile range tracking facility
PMS
    performance management software
    permanent magnet speaker
    picturephone meeting service
    plant monitoring system
    plastic to metal seal

probability of mission success
probable maximum surge
processor memory switch
processor, memories and switches
programmed mode switch
project management system
public message service
PMSCHEDULE
    preventive maintenance schedule
PMSRP
    physical and mathematical sciences research paper
PMSS
    precision measuring subsystem
PMT
    page map table
    permanent magnet tester
    photomultiplier tube
    physical master tape
    precious metal tip
    prepare master tape
    production monitoring test
    program master tape
    pulse modulator tube
PMTE
    page map table entry
PMTS
    precision missile tracking system
    predetermined motion time standards
    predetermined motion time system
PMU
    performance monitor unit
    performance monitoring unit
    portable memory unit
    program management unit
PMUX
    programmable multiplexer
PMW
    private microwave
    pulse modulated wave
PMX
    private manual exchange
PN
    part number
    party notified
    perceived noise
    performance number
    phon
    pilot navigator
    position
    positive negative
    preliminary notification
    proportional navigation
    pseudonoise
    punch on

PNA
    project network analysis
PNBOUNDARY
    positive negative boundary
PNC
    programmed numerical control
PNCH
    punch
PND
    premodulation processor near earth data
    present next digit
PNDB
    perceived noise in decibels
    perceived noise level expressed in decibels
PNDC
    parallel network digital computer
PNDG
    pending
PNEC
    proceedings of the national electronics conference
PNGCS
    primary navigation, guidance, and control system
PNJUNCTION
    positive negative junction
PNL
    panel
    passenger name list
    perceived noise level
PNLBD
    panelboard
PNM
    pulse number modulation
PNMF
    pseudo noise matched filter
PNMT
    positive negative metal transistor
PNP
    positive negative positive
PNPF
    piqua nuclear power facility
PNPN
    positive negative positive negative
PNPS
    plant nitrogen purge system
    plant nuclear protection system
PNR
    passenger name record
    pittsburgh naval reactor office
PNSRC
    plant nuclear saftey review commission
PNTD
    personnel neutron threshold detector

PNTDMA
  pseudonoise time division multiple access
PNVS
  pilot night vision system
PNWL
  pacific northwest laboratories
PO
  parallel output
  parking orbit
  part of
  patent office
  phone order
  planetary office
  planetary orbit
  polarity
  pole
  post office
  postal order
  power oscillator
  power output
  pressure oscillation
  primary outlet
  primary output
  print out
  program objectives
  project office
  pull out
  pulse oscillator
  punch out
  purchase order
POA
  probability of acceptance
  provisional operating authorization
POAE
  port of aerial embarkation
POAR
  postal laws and regulations
POB
  post office box
  proceeding push out base
  push out base
POBA
  patent office board of appeals
POBOX
  post office box
POC
  patch output converter
  physics of control
  port of call
  process operator console
  production operational capability
POCO
  position computer
POCP
  program objectives change proposal
POCS
  patent office classification system
POD
  point of origin device
  port of debarkation
  post office department
PODAF
  power density exceeding a specified level over an area within an assigned frequency band
PODS
  post operative destruct system
POE
  port of embarkation
POED
  post office engineering department
POEE
  post office electrical engineer
POEEJ
  post office electrical engineers journal
POESID
  position of earth satellites in digital display
POET
  portable orders entry terminal
  program operation and environment transfer
POEU
  post office engineers' union
POF
  planned outage factor
POGO
  polar orbit geophysical observatory
  polar orbiting geophysical observatory
  program optimizer
  programmer oriented graphics operation
POH
  planned outage hours
POI
  parking orbit injection
  program of instruction
POIL
  power density imbalance limit
POINTER
  particle orientation interferometer
POIS
  prototype online instrument systems
POISE
  panel on in-flight scientific experiments
  preoperational inspection services engineering
POL
  patent office library
  petroleum, oil, and lubricants
  philips optical language
  polarity
  polarization

polarized
polling
problem oriented language
procedure oriented language
process oriented language
provisional operating license
POLAR
  polarity
  polarization
  polarize
POLREG
  polynomial regression
POLYGLASS
  polyester fibre fibre glass
POLYTRAN
  polytranslation analysis and programming
POM
  printer output microfilm
POMD
  program operation mode
POMI
  photochromic microimage
POMM
  preliminary operating and maintenance manual
POMS
  panel on operational meteorological satellites
POMSEE
  performance, operational and maintenance standards for electronic equipment
  preparation, operation, maintenance of shipboard electronics equipment
PON
  position
PONN
  positive on negative
POO
  post office order
POP
  particle oriented paper
  power on off protection
  pressurizer over pressure protection system
  printing out paper
  program operating plan
  programmed operators and primitives
  project optimization procedure
  proof of principle
POPD
  plans, operations and programs division
POPI
  post office position indicator
POPR
  prototype organic power reactor
POPS
  pantograph optical projection system

POR
  peak overshoot ratio
  power on reset
  problem oriented routine
PORACC
  principles of radiation and contamination control
PORIS
  post office radio interference service
PORT
  photooptical recorder tracker
  portable
PORV
  pilot operated relief valve
  power operated relief valve
POS
  permanent orbital station
  plant operating system
  point of sale
  position
  positive
  pressure operated switch
  primary operating system
  probability of survival
  products of sums
  programming optimizing system
POSA
  preliminary operating safety analysis
POSD
  program for optical system design
POSE
  program for optimizing sales effort
POSIP
  portable ship instrumentation package
POSS
  passive optical satellite surveillance
  photooptical surveillance system
  prototype optical surveillance system
POSTER
  post strike emergency reporting
POSV
  pilot operated solenoid valve
POT
  paint on tangent
  potential
  potentiometer
POTC
  pert orientation and training center
POTDIF
  potential difference
POTELECTROMET
  potentiometric electrometer
POTS
  photooptical terrain simulator

POV
  peak operating voltage
POW
  power
POWER
  priority output writers execution processor
  priority output writes execution process
POWS
  pyrotechnic outside warning system
POY
  partially oriented yarn
PP
  between perpendiculars
  panel point
  partial product
  patent pending
  peak power
  peak to peak
  peripheral processor
  pilot punch
  pilotless plane
  pipeline processor
  plate power
  plate pulse
  point to point
  postprocessor
  power package
  power plan
  preprocessor
  pressure proof
  print t punch
  product publication
  program package
  pseudoprogram
  pulse pair
  push pull
PPA
  photo peak analysis
  preliminary pile assembly
  princeton particle accelerator
  princeton pennsylvania proton accelerator
  professional programmers association
  program problem area
  protected partition area
  push pull amplifier
PPAL
  pacific power and light
  pennsylvania power and light
PPB
  parts per billion
PPBS
  planning programming budgeting system
PPC
  parallel path counter
  phased program construction
  planar positive column
  plane paper copier
  potentional performance capability
  power physics corporation
  preliminary phase correction
  program product center
  project planning and control system
  public power corporation
  pulsed power circuit
PPCC
  parts per cubic centimeter
PPCCD
  profiled peristaltic charge coupled device
PPD
  plans and program division
PPDD
  plan position data display
  preliminary project design description
PPE
  polyphenylether
  premodulation processing equipment
  premodulation processor equipment
  print punch editor
  problem program efficiency
  problem program evaluator
PPFP
  plane parallel fabry perot
PPG
  print pattern generator
  program pulse generator
  propulsion and power generation
  propulsion and power generator
PPH
  pound per hour
  pulse per hour
PPHM
  pulse phase modulation
PPI
  pictorial position indicator
  plan position indicator
  point per inch
  present position indicator
  programmable peripheral interface
  pulse per inch
PPIM
  programmable peripheral interface microcomputer
PPIU
  programmable peripheral interface unit
PPL
  pixel per line
  plasma physics laboratory
  power plant laboratory
  preferred parts list

preliminary power laboratory
process peripherals ltd
PPLB
postprocessor call library
PPM
parallel processing machine
parts per million
peak program meter
periodic permanent magnet
periodic permanent magnet focusing
periodic pulse metering
planned preventive maintenance
problem program monitor
pulse per minute
pulse phase modulation
pulse position modulation
PPMA
precision potentiometer manufacturers association
PPMS
program performance measurement system
PPMW
primary plant mineralized water
PPO
polyphenylene oxide
PPP
parallel push pull
peak pulse power
phased project planning
popular power package
push pull power
PPPA
push pull power amplifier
PPPCIRCUIT
parallel push pull circuit
PPPEE
pulsed pinch plasma electromagnetic engine
PPPI
precision plan position indicator
projection plan position indicator
PPPP
proposed partial package program
PPPPI
photographic projection plan position indicator
PPPS
pulse pairs per second
PPR
photoplastic recording
program planning report
PPRF
pulse pair repetition frequency
PPS
parallel processing system
partitioned priority system
period per second
periods per second
peripheral processor system
philadelphia programming society
phosphorous propellant system
picture per second
plant protection system
point to point system
pound per second
primary power system
primary propulsion system
program production supervisor
pulse per second
pulsed per second
PPSN
present position
PPT
precipitate
programmer productivity techniques
punched paper tape
PPTP
power proportioning temperature programmer
PPU
peripheral processing unit
PPV
pitch power valve
PQ
premium quality
PQA
protected queue area
PQAD
plant quality assurance director
PQC
procurement quality control
production quality control
PQGS
propellant quantity gauging systems
PQM
print quality monitor
PR
pattern recognition
pawling research reactor
pen record
periodic report
periodic reversal
photographic reconnaissance
physical record
picture ratio
plyrating
polarized relay
position register
position report
press release
primary radar
print register

print restore
printer
proceedings
processor
program register
program requirements
programming
pseudorandom
public relations
puerto rico
pulse rate
pulse ratio
pulse ration
PRA
page replacement algorithm
peak recording accelerographs
pendulous reference axis
pitch and roll attitude
polar regions award
precision axis
prime responsable authority
print alphanumerically
probabilistic risk assessment
production reader assembly
program reader assembly
PRAC
public relations advisory committee
PRAD
program research and development
PRADOR
pulse repetition frequency ranging doppler radar
PRAM
programmable random access memory
pseudorandom access memory
PRAN
production analyzer
PRB
panel review board
program request block
PRBS
pseudorandom binary sequence
PRBSG
pseudorandom binary sequence generator
PRC
partial response coding
parts release curve
periodic reverse current
periodic reverse current plating
planning research corporation
point of reverse curve
postal rate commission
pressure recording controller
production control

PRCCT
printed circuit
PRCF
plutonium recycle critical facility
PRCP
power remote control panel
PRCS
process
processing
PRD
power range detector
prime radar digitizer
printer dump
program requirement document
program requirements data
program requirements document
PRDC
power reactor development company
PRDF
page reference distribution function
PRDV
peak reading digital voltmeter
PRE
planetary rotation engine
preliminary amplifier
processing refabrication experiment
PREAMP
preamplifier
PREC
precision
PRECO
precision equipment company
PREDICT
prediction of radiation effects by digital computer techniques
PREF
prefix
propulsion research environmental facility
PREFAB
prefabricated
PRELORT
precision, long-range tracking
PREP
pacific range electromagnetic platform
programmed educational package
PREPARE
project for retraining of employable persons as relates to electronic data processing
PRES
proton resonance
PRESS
pacific range electromagnetic signature study
pressure
PRESSDUCTOR
pressure inductor

PRESSRA
   presentation equipment for show scan radar
   presentation equipment for slow scan radar
PRESTO
   program for rapid earth to space trajectory optimization
   program reporting and evaluation system for total operation
   project release status operation
PREV
   previous
PRF
   power radio frequency
   pulse rate frequency
   pulse recurrence frequency
   pulse repetition frequency
PRFCS
   pattern recognition feedback control system
PRFL
   pressure fed liquid
PRI
   photo radar intelligence
   plan repeater indicator
   pressure ratio indicator
   primary
   primary winding
   primer
   priority
   program interrupt
   pulse rate indicator
   pulse repetition interval
PRIC
   parts reliability information center
PRIDE
   personal responsability in daily effort
   programmed reliability in design
   programmed reliability in design engineering
PRIM
   plume radation intensity measurement
PRIME
   precision integrator for meteorological echoes
   precision recovery including maneuverable entry
   precision recovery maneuverable entry
   prime computer
   program independence, modularity, economy
   programmed instruction for management education
PRIMSCO
   pilot run item master schedule committee
PRINCE
   parts reliability information center
   programmed reinforced instruction necessary to continuing education
PRINCIR
   printed circuit
PRIND
   present indication
PRINSYS
   product information system
   production information system
PRINT
   preedited interpreter
   preedited interpretive
   preedited interpretive system
   printed
   printer
   printing
PRIOR
   program for in-orbit rendezvous
PRIP
   parts reliability improvement program
PRIS
   pacific range instrumentation satellite
   propeller revolution indicator system
PRISE
   program for integrated shipboard electronics
PRISECIMP
   primary secondary impedance
PRISM
   personnel record information system for management
   powerful resource for information and system management
   program reliability information system for management
   program reporting and information system for management
   programmed integrated system maintenance
PRITAC
   primary tactical radio circuit
PRIVX
   private exchange
PRK
   phase reversal keying
PRL
   postal reform league
   pulse reflection logic
PRLP
   planetary rocket launcher platform
PRM
   personal radiation monitor
   power range monitor
   process radiation monitor
   pulse rate modulation
PRN
   previous result negative
   print numerically

pseudorandom noise
pulse ranging navigation
PRNC
   puerto rico nuclear centers
PRO
   pen recorder output
   print octal
   propagation prediction
   report
PROB
   problem
PROC
   proceedings
   process
   programming computer
PROCAL
   programmable calculator
PROCIEE
   proceedings of the institute of electrical engineers
   proceedings of the institution of electrical engineers
PROCIRE
   proceedings of the institute of radio engineers
PROCNEC
   proceeding of the national electrical conference
PROCOL
   process control oriented language
PROCOM
   procedures committee
   prognose compiler
PROCOMP
   process compiler
   process computer
   program compiler
PROCTOR
   priority routing organizer for computer transfers and operations of registers
PROD
   product
   production
PRODAC
   programmed digital automatic control
PRODUCE
   production distribution using component evaluation
PROF
   profile
PROFAC
   propulsive fluid accumulator
PROFILE
   programmed functional indices for laboratory evaluation
PROFIT
   programmed reviewing, ordering and forecasting inventory technique
PROG
   program
PROGDEV
   program device
PROGRAM
   programming
PROJ
   project
   projection
   projector
PROLAMAT
   programming languages for numerically controlled machine tools
PROLAN
   processing language
PROLOG
   project logic planning
PROM
   pockels read out optical memory
   pockels read out optical modulator
   programmable read only memory
PROMIS
   project oriented management information system
PROMISS
   packaging requirements for optimum malfunction isolation by systematic substitution
PROMPT
   producting reviewing organizing and monitoring of performance techniques
   program monitoring and planning techniques
   program reporting, organization, and management planning technique
   project management and product team technique
PRONTO
   program for numerical tool operation
PROOF
   precision recording optical of fingerprints
PROP
   performance review for operating programs
   planetary rocket ocean platform
   primary operand unit
   production planning
PROSA
   programming system with symbolic addresses
PROSD
   performance records for optimizing system design
PROSEL
   process control and sequencing language

PROSIG
  procedure signal
PROSIN
  procedure sign
PROSPER
  profit simulation planning and evaluation of risk
PROSPRO
  process system program
PROT
  protected
PROTECT
  probabilities recall optimizing the employment of calibration time
PROV
  provide
  provision
  provisional
PROWL
  procedure work log system
PROXI
  projection by reflection optics of xerographic images
PRP
  placement route and patch
  position report print out
  print out
  pseudorandom pulse
  pulse recurrence period
  pulse repetition period
PRPF
  planar radial peaking factor
PRPQ
  programming request for price quotation
PRR
  pawling research reactor
  pulse recurrence rate
  pulse repetition rate
PRS
  pacific rocket society
  pattern recognition society
  pattern recognition system
  process radiation sampler
  programmed radar simulator
PRSA
  public relations society of america
PRT
  parachute radio transmitter
  pattern recognition technique
  permanent recording traffic
  portable remote terminal
  printer
  program reference table
  prompt relief trip
  protable remote terminal
  pulse recurrence time
  pulse repetition time
PRTOT
  prototype real time optical tracker
PRTP
  prototype
PRTR
  plutonium recycle test reactor
PRU
  photo reconnaissance unit
  physical research unit
  programs research unit
PRUNIT
  photo roentgen unit
PRV
  peak rated voltage
  peak reserve voltage
  pressure reducing valve
  pressure regulating valve
  pressure relief valve
PRVS
  penetration room ventilation system
PRW
  per cent rated wattage
PRWRA
  puerto rico water resources authority
PS
  pack switch
  pakistan standard
  parallel to serial
  parity switch
  part surface
  phase shift
  phase shifter
  physical sequential
  picosecond
  pisosecond
  planetary surface
  planning and scheduling
  plotter system
  point of switch
  polarity selector
  postal service
  potentiometer synchron
  power series
  power source
  power supply
  presentation services
  pressure sensitive
  pressure switch
  print scan
  problem statement
  process storage
  product service

program section
program source
program stateword
program status
program store
programmable switch
programming system
proof stress
proton synchrotron
pull switch
pulse per second
pulse shaper
pulse stretcher
PSA
philippine standards association
polysilicon self aligned
power and servo assembly
power servo amplifier
pressure sensitive adhesive
problem statement analyzer
process service area
product saftey association
PSAA
plasma sciences and applications
PSAC
president's science advisory committee
PSAD
prediction, simulation, adaption and decision
PSAL
permanent supplementary artificial lighting
PSALI
permanent supplementary artificial lighting of interiors
PSAR
preliminary safety analysis report
pressure system automatic regulator
process storage address register
programmable synchronous asynchronous receiver
PSAT
programmable synchronous asynchronous transmitter
PSB
process specification block
program specification block
PSBLS
permanent, space based logistics system
PSBMA
professional services business management association
PSC
per standard compass
permanent split capacitor
phase sensitive converter
photosensitive cell

power system communications
preferred semiconductor circuits
production scheduling and control
program switching center
public service commission
pulse shape control circuit
PSCAN
p scanner
PSCC
power system communications
power system computation conference
PSCRT
passive satellite communications research terminal
PSCS
pacific scatter communication system
PSD
phase sensitive demodulator
phase sensitive detector
polysilicon diode
position sensitve light detector
power spectral density
pulse shape discrimination
pulse shape discriminator
PSDD
preliminary system design description
PSDF
propulsion systems development facility
PSDR
process storage data register
PSDU
power switching distribution unit
PSE
packet switched exchange
packet switching exchange
power system engineering
pressurized subcritical experiment
pulse sense
PSEAG
public service electric and gas
PSEAGCO
public service electric and gas company
PSEC
picosecond
PSEP
passive seismic experiment package
PSF
personal silicon foundry
pound force per square foot
pound per square foot
power separation filter
process signal former
PSG
phosphosilicate glass
phrase structure grammar

planning system generator
prask structive grammar
pulse signal generator
pulsed strain gauge
PSI
  pacific semiconductors, incorporated
  pakistan standards institution
  peripheral subsystm interface
  plas speed indicator
  pound force per square inch
  pound per square inch
  pounds per square inch absolute
  power semiconductors, incorporated
  preprogrammed self instruction
  problem solving and inference machine
  proctorial system of instruction
PSIA
  pound per square inch absolute
PSICOMP
  predicted speech intelligibility computer
PSID
  pounds per square inch differential
  preliminary safety information document
PSIEP
  project on scientific information exchange in psychology
PSIG
  pounds per square inch gauge
PSIM
  power system instrumentation and measurement
PSJ
  point spread junction
PSK
  phase shift keyed
  phase shift keying
PSL
  phase sequence logic
  photographic science laboratory
  problem statement language
PSLT
  port side light
PSM
  peak selector memory
  production systems management
  progressive series modulator
  pulse slope modulation
PSMA
  professional services management association
PSMR
  parts specification management for reliability
PSMS
  permanent section of microbiological standardization
PSMT
  pedestal sight manipulation test
PSN
  position
  public switched network
PSNR
  power signal to noise ratio
PSO
  pilot systems operator
PSP
  packet switching processor
  peak sideband power
  performance shaping parameter
  planet scan platform
  portable service processor
  power system planning
  primary sodium pump
  pseudostatic spontaneous potential
PSPS
  planar silicon photoswitch
PSPSK
  previous signaling element phase shift keying
PSQ
  personnel security questionnaire
PSR
  parts and supply requisition
  pennsylvania state university research reactor
  peripheral shim rods
  point source range
  power system relaying
  pressure
  processor state register
  procurement status report
  program study request
  programming support representative
PSREGISTER
  processor status register
PSRP
  physical sciences research paper
PSRR
  power supply rejection ratio
PSS
  packet switch service
  packet switched service
  personal signaling system
  power supply subsystem
  premature separation switch
  process switching services
  proprietary support system
  propulsion support system
PSSCC
  private sector standards coordinating center
PSSP
  partition with self substitution property

PST
  pacific standard time
  paired selected ternary
  point of spiral tangent
  polished surface technique
  pressure sensitive tape
  primary surge tank
  production sampling test
  program status table
  program synchronization table
PSTC
  pressure sensitive tape council
PSTCODE
  paired selected ternary code
PSTF
  pressure suppression test facility
  proximity sensor test facility
  pump seal test facility
PSTL
  pressure model static and transient launch configuration
PSTN
  public switched telephone network
PSU
  microprocessor program storage unit
  port sharing unit
  power supply unit
PSUR
  pennsylvania state university reactor
PSV
  portable sensor verifier
PSVM
  phase sensitive voltmeter
PSW
  philosophical society of washington
  potential switch
  potentiometer slide wire
  processor status word
  program status word
PSWP
  plant service water pump
PSWR
  power standing wave ratio
PT
  pacific time
  page table
  paper tape
  part
  partially tested
  partition table
  pencil tube
  per ton
  perforated tape
  performance test
  period tapering
  picture telegraphy
  pint
  pipe thread
  plasma thermocouple reactor
  point
  point of tangency
  port
  portable
  post and telegraph
  potential transformer
  pressure transducer
  print
  printer terminal
  priority telegram
  product test
  program time
  propellant transfer
  prototype
  pulse time
  pulse timer
  pulse train
  pulse transformer
  punch through
  punched tape
PTA
  page table address
  phototransistor amplifier
  pitch trim adjustment
  planar turbulence amplifier
  power transfer assembly
  program time analyzer
  proposed technical approach
  pulse torquing assembly
PTAB
  program status table
PTAT
  proportional to absolute temperature
PTB
  page table base
  production and test branch
PTBL
  portable
PTBR
  punched tape block reader
PTC
  pacific telecommunications council
  passive thermal control
  periscope television camera
  pitch trim compensator
  portable temperature controller
  positive temperature coefficient
  postal and telegraphic censorhsip
  postal telegraph cable
  power transfer coefficient
  power transmission council

programmed transmission control
pulse time code
PTCCS
   polaris target card computing system
PTCP
   parameter test control program
PTCR
   pad terminal connection room
   positive temperature coefficient of resistance
   positive temperature coefficient resistor
PTCRESISTOR
   positive temperature coefficient resistor
PTCS
   propellant tanking computer system
PTCTHERMISTOR
   positive temperature coefficient thermistor
PTCV
   pilot operated temperature control valve
PTD
   patented
PTDB
   point target data base
PTDTL
   pumped tunnel diode transistor logic
PTDTLCIRCUIT
   pumped tunnel diode transistor logic circuit
PTE
   page table entry
   peculiar test equipment
   pressure tolerant electronics
PTF
   program temporary fix
   programmable transversal filter
PTFE
   polytetrafluoroethylene
PTG
   place to go
   printing
   program test group
PTH
   plated-through hole
PTI
   pipe test insert
   plugging temperature indicator
   power tool institute
PTIB
   program testing information bulletin
PTJ
   pulse train jitter
PTL
   phase tracking loop
   pittsburgh testing lab
   pulse transmission logic
PTM
   mancake torque motor

phase time modulation
photomultiplier
pretuned module
program timing and maintenance
program timing and miscellaneous
program trouble memorandum
programmable terminal multiplexer
proof test model
pulse time modulation
pulse time multiplex
punch through modulation
PTML
   pnpn transistor magnetic logic
PTO
   patent and trademark office
   permeability tuned oscillator
   permeability tuned oscillator power take off
   please turn over
   power take off
PTP
   paper tape punch
   parameter test program
   point to point
   preferred target point
   print to point
   programmed text processing
   proximity test plug
PTPCONTROL
   point to point control
PTPS
   parallel tuned parallel stabilized
PTPSK
   pilot tone phase shift keyed
   pilot tone phase shift keying
PTR
   paper tape reader
   photoelectric tape reader
   polar to rectangular
   pool test reactor
   position track radar
   power transformers
   pressure tube reactor
   pretransmit receiving
   printer
   processor tape read
   programmer trouble report
   proof test reactor
   punch tape reader
PTRP
   paper tape reader punch
PTRV
   peak transient reverse voltage
PTS
   paper tape system
   parameter test set up

perforated tape subsystem
permanent threshold shift
photogrammetric triangulation system
phototransmission system
pneumatic test set
points
post and telecommunications service
power transient suppressor
program test system
programmable terminal system
propellant transfer system
pure time sharing
PTSP
pressure tube to spool piece
PTSR
presssure tube superheat reactor
PTSS
parallel tuned series stabilized
photon target scoring system
PTT
post, telephone, and telegraph
program test tape
push to talk
PTTC
perforated tape and transmission code
PTTI
post, telegraph and telephone international
PTTOPT
point to point
PTTS
private telegraph and telephone service
PTU
parallel transmission unit
pice terminal unit
pice transfer unit
PTV
predetermined time value
punch through varactor
PTVA
propulsion test vehicle assembly
PTVAA
portland television appliance association
PTW
page table word
PTWAM
page table word associative memory
PTX
polythermalex
PTY
parity
PU
peripheral unit
physical unit
pick up

pluggable unit
power unit
processing unit
propulsion unit
PUAC
propellant utilization acoustical checkout
PUAS
postal union for america and spain
PUB
physical unit block
PUC
processing unit cabinet
public utilities commission
punctured uniform code
PUCK
propellant utilization checkout kit
PUCP
process unit control panel
PUCS
propellant utilization control system
PUD
pick-up and delivery
public utility district
PUDL
push down list
PUDT
propellant utilization data translator
PUFFS
passive underwater fire control sonar feasibility study
PUFFT
purdue university fast fortran translator
PUGS
propellant utilization gauging system
PUI
physical unit of information
PUJT
programmable unijunction transistor
PULLBSW
pull button switch
PULSE
public urban locator service
PUMP
parts usage maintenance program
PUNC
program unit counter
PUP
peripheral unit processor
plutonium utilization program
program unit punch
PUR
procurement request
program utility routine
purdue university reactor
purdue university research

PUSS
　pilots' universal sighting system
PUT
　programmable unijunction transistor
PUTWS
　put word in string
PV
　peak to valley
　photovoltaic
　positive volume
　pressure control valve
　primary valve
　process variable
PVA
　polyvinyl acetate
　polyvinyl alcohol
PVAC
　present value of annual charges
PVB
　potentiometric voltmeter bridge
PVC
　photovoltaic cell
　polyvinyl carbazol
　polyvinyl chloride
　position and velocity computer
　potential volume change
　puch validity check
PVD
　paravisual director
　plasma vapour deposition
PVDC
　polyvinylidene chloride
PVF
　polyvinyl fluoride
PVM
　projection video monitor
PVOR
　precision vhf omnidirectional range
　precision vhf omnirange
PVR
　precision voltage reference
PVRC
　pressure vessel research committee
PVS
　performance verification system
　program validation services
PVSPSENSINGELEMENT
　process variable set point sensing element
PVST
　premate verification system test
PVT
　polyvinyl toluene
　pressure volume temperature
PVTI
　piping and valve test insert

PVX
　phosphorous doped vapor deposited oxide
PW
　parallel with
　picowatt
　pilot wire
　plated wire
　postal wire
　printed wiring
　program word
　pulse width
PWB
　printed wiring board
　programmer's workbench
PWC
　pulse width coded
PWD
　post write disturb
　power distribution
　power distributor
　powered
　process word
　pulse width discriminator
PWE
　pulse width encoder
PWF
　present worth factor
PWI
　piecewise linear
　pilot warning indicator
　pilot warning instrument
　proximity warning indicator
PWL
　piecewise linear
　power level
PWM
　pulse width modulated
　pulse width modulation
　pulse width multiplier
PWMAF
　pulse width modulated audio frequency
PWMFM
　pulse width modulation frequency modulation
PWP
　picowatt psophometric
PWR
　power
　pressurized water reactor
PWRAMPL
　power amplifier
PWRDEVELENGR
　power development engineer
PWRMON
　power monitor

PWRPLT
  power plant
PWRPNL
  power panel
PWRS
  programmable weapons release system
PWRSEMICOND
  power semiconductor
PWRSUP
  power supply
PWRU
  power unit
PWS
  plasma waveguide switch
  predicted wave signaling
  private wire service
  private wire system
  programmer's workstation
  proximity warning system
PWSS
  port war signal station
PWT
  pacific war time
  pacific winter time
  pennyweight
  propulsion wind tunnel
PWTO
  principal wireless telegraphy officer
PX
  private exchange
PXA
  place index in address
  pulsed xenon arc
PXD
  place index in decreement
PY
  pyrometer
PZ
  peak to zero
  plus zero
PZM
  piezoelectric mount
PZR
  pressurizer
PZT
  lead zirconate titanate
  photographic zenith time
  piezoelectric transducer

Q

quad
quadded
quality
quality factor
quantity
quantity of electric charge
quenched
quenching
query
queue
quiescent
quotient
QA
  quality analysis
  quality assurance
  query analyzer
  question answer
  quick acting
  quickly acting
  quiescent antenna
QAA
  question and answer
QAAO
  quality assurance and operations
QAAS
  quality assurance acceptance standard
QAC
  quality assurance checklist
QAD
  quick attach detach
QADS
  quality assurance data system
QAGC
  quiet automatic gain control
QAI
  quality assurance instruction
QAIP
  quality assurance inspection procedure
QAL
  quality assurance laboratory
QAM
  quadrature amplitude modulation
QAP
  quality assurance program
QAPI
  quality assurance program index
QAPP
  quality assurance program plan
QAR
  quality and reliability
  quality assurance representative at vendor plant
  quality assurance requirements
QASK
  quadrature amplitude shift keying
QASL
  quality assurance systems list

QATP
  quality assurance technical publication
QATS
  quality assurance and test service
QAVC
  quiet automatic volume control
QB
  query buffer
  quick break
QBG
  qualified binary grouping
QC
  quality control
  quartz crystal
  quick change
QCAT
  quality control and techniques
QCB
  queue control block
QCC
  quality control chain
  quartz crystal company
QCCR
  quality control change request
QCD
  quality control division
QCE
  quality control engineering
QCENGR
  quality control engineer
QCESC
  quad cities engineering and science council
QCF
  quality control form
QCL
  quality control level
QCM
  quality control manual
  quantitative computer management
QCNC
  charge controlled negative capacitance
QCO
  quality control officer
QCPLL
  quadrature channel phase locked loop
QCPM
  quality control procedures manual
QCPP
  quality control planning procedure
QCR
  quality control reliability
  quality control representative
QCRI
  quality control reliability investigator

QCS
  quality control standard
  quality control survey
QCSR
  quality control service request
QCVTI
  quality control verification test inspection
QCW
  quadrature continuous wave
QCWA
  quarter century wireless association
QD
  quick disconnect
  quick disconnection
QDC
  quick dependable communications
QDM
  quick disconnector, miniature
QDRI
  qualitative development requirements information
QE
  quantum efficiency
QEA
  quantum electronics and applications
  queue element area
QEBR
  quasi equilibrium boltzmann relations
QEC
  quick engine change
QF
  quality factor
  quality form
  quick firing
QFIRC
  quick fix interference reduction capability
QFM
  quantized frequency modulation
QG
  quadrature grid
QGV
  quantized gate video
QH
  quadrature hybrid
QHC
  quarter half circle
QHM
  quartz horizontal magnetometer
QHR
  quality history records
QI
  quiet ionosphere
QIAM
  queued indexed access memory
  queued indexed access method

QIC
  quality insurance chain
  quarter inch compatibility
QIL
  quad in line
QIP
  quad in line package
  query interpretation program
QIS
  quality information system
  quality insurance system
QISAM
  queued indexed sequential access method
QIT
  quality information and test
QL
  query language
QLAP
  quick look analysis program
QLDS
  quick look data station
QLIT
  quick look intermediate tape
QLP
  query language processor
QLR
  quick look report
QLTY
  quality
QM
  quadrature modulation
  quick make
QMB
  quick make and break
QMDO
  qualitative material development objective
QMETER
  quality meter
QMI
  qualification maintainability inspection
QMOSCIRCUIT
  quick metal oxide semiconductor circuit
QMR
  qualitative material requirement
QNT
  quantizer
QO
  quick opening
QOD
  quick opening device
QOR
  qualitive operational requirement

QP
  quality product
  quartered partition
  quasi-peak
QPAC
  qualified productivity aid for computing
QPC
  quality performance chart
QPD
  quadrature phase detector
  quantized probability design
QPL
  qualified parts list
  qualified producers list
  qualified products list
QPM
  quality program manager
QPP
  quantized pulse position
  quiescent push-pull
QPPM
  quantized pulse position modulation
QPSK
  quadrature phase shift keying
QPT
  quadrant power tilt
  quartz pressure transducer
QQPRI
  qualitative and quantitative planning requirements information
QR
  quality and reliability
  quality requirement
  quart
  quarter
  quick reaction
  quick response
QRA
  quality and reliability assurance
  quality reliability assurance
QRBM
  quasi-random band model
QRC
  quick ram change
  quick reaction capability
QRCD
  qualitative reliability consumption data
QRI
  qualitative requirements information
  quick reaction interceptor
QRTR
  quarter
QS
  quadruple screw
  quick sweep

QSAM
  queued sequential access method
QSE
  qualified scientist and engineer
QSG
  quasi-stellar galaxy
QSO
  quasar
  quasi-stellar object
QSP
  quench spray pump
QSR
  quarterly statistical report
QSRS
  quasi-stellar radio source
QSS
  quasi-steady state
  quasi-stellar source
  quench spray subsystem
QSSP
  quasi-solid state panel
QSTOL
  quiet short take off and landing
QT
  quadruple thermoplastic
  qualification test
  quantity
  quart
  quasi-transverse
  quiet
QTAM
  queued telecommuniction access method
QTOL
  quiet take off and landing
QTP
  qualification test procedure
  qualification test program
  quality test plan
QTPT
  qualification test and proof test
QTR
  quarter
QTY
  quantity
QTZ
  quartz
QTZN
  quantization
QTZR
  quantizer
QU
  quarter
QUAD
  quadrat

  quadrilateral
  quadruple
QUAL
  qualification
  quality
QUALENGR
  quality engineer
QUAM
  quadrature amplitude modulation
QUANT
  quantity
QUAP
  questionnaire analysis program
QUASER
  quantum amplification by stimulated emission of radiation
QUASIEFF
  quasi-effective
QUED
  quick editor
QUEST
  quality electrical system test
QUIL
  quad in line
  quadruple in line
QUIP
  quad in line package
  query interactive processor
  questionnaire interpreter program
QUISAM
  queued index sequential access method
QUOT
  quotient
QUP
  quantity per unit pack
QVT
  quality verification test
  quality verification testing
QWA
  quarter wave aerial
  quarter wave antenna
QWSSUS
  quasi-wide sense stationary uncorrelated scattering

R

R
  radical
  radio
  radiolocation

radius
range
rankine
ratio
read
readability
reading
received
receiver
receiving
reception
reconnaissance
record
recovery
rectifier
red primary
redundancy
register
registered
regulator
relative signal strength
relay
reliability
reluctance
render
repeater
reset
resistance
resistor
reverse
revolute joint
right
rocket
roentgen
rotary
routine
rubber
rydberg constant

RA
    radar altimeter
    radioactive
    radius of action
    random access
    range
    read amplifier
    receiver attenuation
    receiver auxiliary
    record address
    recording annunciator
    rectifier
    reduction of area
    register allocator
    register allotter
    remote access
    rental agreement
    right accumulator
    right angle right ascension

RAA
    radar aircraft altitude
    research and analysis

RAB
    radio advertising bureau

RABAL
    radiosonde balloon

RABAR
    raytheon advanced battery acquisition radar

RABH
    reported altitude block height

RAC
    racemic
    radio service code
    random access controller
    reactor accident calculations
    read address counter
    rectified alternating current
    remote access
    remote access computing system
    remote access execution
    research advisory committee
    research analysis corporation
    rework after completion
    rules of the air traffic control

RACC
    radiation and contamination control

RACE
    radiation adaptive compression equipment
    random access computer equipment
    random access control equipment
    rapid automatic checkout equipment
    regional automatic circuit exchange
    response analysis for call evaluation
    routing and cost estimate

RACEP
    random access and correlation for extended performance

RACES
    radio amateur civil emergency service

RACFO
    radio and communications facilities operative

RACI
    reported altitude change indicator

RACMD
    radio countermeasures and detection

RACOM
    random communication

RACON
    radar beacon

RACS
  remote access calibration system
  remote access computing system
  remote automatic calibration system
RACT
  reasonably available control technique
  remote access computer technique
RAD
  radial
  radian
  radiation
  radiation absorbed dose
  radiation detector
  radiation dosage
  radiator
  radical
  radio
  radius
  radix
  random access data
  random access disk
  rapid access data
  rapid access device
  rapid access disk
  rapid access drum
  relative air density
  research and development
RADA
  radioactive
  random access discrete address
RADAC
  radar anaolg digital data and control
  rapid digital automatic computing system
RADAL
  radio detection and location
RADAN
  radar doppler automatic navigation
  radar doppler automatic navigator
RADANT
  radome antenna
RADAR
  radio detecting and ranging
  radio detection and ranging
  random access dump and reload
RADARC
  radially distributed annular rocket combustion chamber
RADAS
  random access discrete address system
RADAT
  radar data transmission
  radio direction and track
  radiosonde observation data

RADATA
  radar automatic data transmission assembly
  radar data transmission and assembly
RADATAC
  radiation data acquisition chart
RADC
  radar countermeasure
  rome air development center
RADCM
  radar countermeasure
RADCO
  radio company
RADCON
  radar data converter
RADELECTENG
  radio and electronic engineer
RADEM
  random access delta modulation
RADFAC
  radiating facility
RADHAZ
  radiation hazard
  radio frequency hazard
RADI
  radiographic inspection
RADIAC
  radio activity detection, identification and computation
RADIC
  radio interior communication
  redifon analog digital computing system
RADICS
  research and development in computer system
RADINT
  radar intelligence
  radio intelligence
RADIOCOM
  radio communication
RADIOG
  radiography
RADIOSCI
  radio science
RADIOTELENGR
  radiotelephone engineer
RADIQUAD
  radio quadrangle
RADIR
  random access document indexing and retrieval
RADIST
  radar distance
  radar distance indicator
RADIT
  radio teletype

RADLMON
 radiological monitor
RADNOTE
 radio note
RADOC
 remote automatic detection of contingencies
RADOME
 radar dome
RADOP
 radar doppler
 radar operator
 radar optical
RADOPR
 radio operator
RADOPWEAP
 radar optical weapons
RADOT
 realtime automatic digital optical tracker
RADPLANBD
 radio planning board
RADPROPCAST
 radio propagation forecast
RADREL
 radio relay
RADRELRON
 radio relay squadron
RADREPMN
 radio repairman
RADRON
 radar squadron
 radio squadron
RADRONMOB
 radio squadron, mobile
RADRONSAGE
 radar squadron sage
RADS
 radiation and dosimetry services
RADSEC
 radio section
RADSIM
 random access discrete address system simulator
RADSO
 radiological survey officer
RADSONDE
 radiosonde
RADSTA
 radio station
RADTT
 radioteletype
RADVS
 radar altimeter and doppler velocity sensor
RADWASTE
 radioactive waste

RAE
 radio astronomoy explorer satellite
 radio astronomy explorer
 range azimuth elevation
 research and engineering
 right arithmetic element
 royal aeronautical establishment
 royal aircraft establishment
RAEN
 radio amateur emergency network
RAES
 remote access editing system
 royal aeronautical society
RAF
 royal air force
 royal aircraft factory
RAFAC
 radio aids and facilities
RAFAR
 radioautomated facsimile and reproduction
RAFIAM
 royal air force institute of aviation medicine
RAFT
 radially adjustable facility tube
 recomp algebraic formula translator
 reentry advanced fusing test
RAI
 random access and inquiry
 receiving and inspection
 removal and installation
RAIC
 redstone arsenal information center
RAID
 random access image device
 random access interactive debugger
RAIDS
 rapid availability of information and data for safety
RAILS
 remote area instrument landing system
RAINBO
 research and instrumentation for national bioscience operations
RAINDX
 random access index edit
RAINIT
 random access initializer
RAIP
 requester's approval in principle
RAIR
 random access information retrieval
 remote access immediate response
RAL
 redifonastrodata, limited
 riverbend acoustical laboratory

RALLOC
   random access allocation
RALT
   radio altitude
   reported altitude
RALU
   register and arithmetic logic unit
RALW
   radioactive liquid waste
RAM
   radar absorbing material
   radiation attenuation measurement
   radio attenuation measurement
   random access memory
   rapid access memory
   recovery aids material
   relaxing avalanche mode
   reliability and maintainability
   repair and maintenance
   resident access method
RAMA
   rome air material area
RAMAC
   radio marine associated companies
   random access memory accounting computer
   random access memory accounting machine
   random access method of accounting and control
RAMARK
   radar marker
RAMD
   random access memory device
   receiving agency material division
RAMIS
   rapid access management information system
   receive, assemble, maintain, inspect, and store
   repair, assemble, maintain, issue, and supply
RAMM
   random access metal oxide semiconductor memory
RAMNAC
   radio aids to marine navigation application committee
RAMP
   radar masking parameter
   radiation airborne measurement program
   raytheon airborne microwave platform
   reliability, availability, maintainability program
RAMPART
   radar advanced measurement program for analysis of reentry techniques

RAMPS
   resources allocation and multiproject scheduling
RAMS
   radio amateur megacycle society
   random access memorix storage
   repair, assembly and maintenance shop
   right ascension of mean sun
   right ascension of the mean sun
RAN
   read around number
RANCID
   real and not corrected input data
RANCOM
   random communication
RAND
   research and development
RANDAM
   random access nondestructive advanced memory
RANDANAL
   randomization analyser
RANDID
   rapid alpha numeric digital indicating device
RANDIS
   random disc file
RANDO
   radiotherapy analog dosimetry
RANN
   research applied to national needs
RANSAD
   random access noiselike signal address
RAO
   radio astronomical observatory
RAOB
   radiosonde observation
RAOBS
   radiometeorograph observation
RAP
   random access projector
   redundancy adjustment of probability
   relational associative processor
   reliability assurance program
   reset after punch
   resident assembler program
   rocket assisted projectile
RAPC
   radio administrative plenipotentiary conference
RAPCAP
   radar picket combat air patrol
RAPCOE
   random access programming and checkout equipment

RAPCON
  radar approach control
RAPD
  reach avalanche photodiode
  reach through avalanche photodiode
RAPEC
  rocket assisted personnel ejection catapult
RAPI
  rubber and plastics industry
RAPID
  rapid accurate polynomial interpolation device
  reactor and plant integrated dynamics
  relative address programming implementation device
  remote automatic parts input for dealers
  research in automatic photocomposition and information dissemination
  rocketdyne automatic processing of integrated data
  ryan automatic plot indicator device
RAPIT
  record and process input tables
RAPLOT
  radar plotting
RAPO
  resident apollo project office
RAPPI
  random access plan position indicator
RAPR
  radar processor
RAPS
  radioactive argon processing system
  rate and position sensor
  retrieval analysis and presentation system
  risk appraisal of program
  risk appraisal of programs system
  rotary phase shifter
RAPT
  reusable aerospace passenger transport
RAPTAP
  rapid access parallel tape
RAQA
  reliability and quality assurance
RAR
  radio acoustic ranging
  rapid access recording
  record and report
  reliability action report
  remove and replace
  remove audible ring
  repair and retrofix
  revenue and retrieval
  routing and recording
  rules and regulations

RARA
  random access to random access
RARAD
  radar advisory
RARC
  reactors and reactor controls
RARE
  radiation and repair engineering
RAREF
  radiation and repair engineering facility
RAREP
  radar weather report
RARG
  regulatory analysis review group
RARR
  range and range rate
RAS
  radar advisory service
  radar assembly spare
  radio astronomy satellite
  random access storage
  reactor alarm system
  reactor analysis and safety
  rectified air speed
  reference address for small core memory
  reliability, availability, security
  reliability, availability, serviceability
  research and statistics
  reset and start
  row address select
  row address strobe
  royal aeronautical society
RASC
  rear area security controller
  rome air service command
RASCC
  rear area security control center
RASE
  rapid acquisition by sequential estimation
RASER
  radio frequency amplification by stimulated emission of radiation
  range and sensitivity extending resonator
RASPO
  resident apollo spacecraft project office
RASS
  radar attitude sensing system
  register, address, skip and special chip
RASSR
  reliable advanced solidstate radar
RASTA
  radiation augmented special test apparatus
  radiation special test apparatus
RASTAC
  random access storage and control

RASTAD
  random access storage and display
RASUI
  reliability, availability, serviceability, useability, installability
RAT
  rate
  rating
  ration
  reliability assurance test
  reserve auxiliary transformer
  rocket assisted torpedo
  rocket launched antisubmarine torpedo
RATAC
  radar analog target acquisition computer
  radar coverage via tactical air navigation
  radar target acquisition
RATAN
  radar and television aid to navigation
RATC
  rate aided tracking computer
RATCC
  radar airtraffic control center
RATE
  remote automatic telemetry equipment
RATEL
  radiotelephone
RATER
  response analysis tester
RATG
  radiotelegraph
  radiotelegraphy
RATIO
  radiotelescope in orbit
  ratiometer
RATO
  rocket assisted takeoff
RATOG
  rocket assisted takeoff gear
RATRAN
  radar triangle navigation
RATS
  rate and track subsystem
  raytheon automatic test system
RATSC
  rome air technical service command
RATSCAT
  radar target scatter
RATT
  radio teletypewriter
  radioteletype
RAU
  remote acquisition unit
  repairs and utilities
RAVE
  radar acquisition visual tracking equipment
RAVEN
  ranging and velocity navigation
RAVIR
  radar video recording
RAVU
  radiosonde ananlysis and verfication unit
RAWARC
  radar report and warning coordination system
RAWIN
  radar wind sounding
RAWINDS
  radar wind sounding
RAWINDSONDE
  radar wind sounding and radiosonde
RAX
  rural automatic exchange
RAYCI
  raytheon controlled inventory
RAYCOM
  raytheon communications
RAYDAC
  raytheon digital automatic computer
RAYSISTOR
  raytheon resistor
RAYSPAN
  raytheon spectrum analyzer
RAYTEL
  raytheon telephone
RAZEL
  range, azimuth, and elevation
RB
  radar beacon
  radio beacon
  rate beacon
  reactor building
  read back
  read backward
  read buffer
  reclamation bureau
  relative bearing
  remote batch
  renegotiation board
  reserve blocked
  return to bias
  ring back
  road buffer
  roller bearing
RBA
  radar beacon antenna
  recovery beacon antenna
  relative byte address
RBAF
  royal belgian air force

RBAM
  remote batch access method
RBC
  remote balance control
RBCCW
  reactor building closed coolant water
RBCH
  rod bank coil unit
RBCWS
  reactor building cooling water system
RBD
  reliability block diagram
RBDE
  radar bright display equipment
RBDP
  rocket booster development program
RBDT
  reverse blocking diode thyristor
RBE
  radiation biological effectiveness
  relative biological effectiveness
  relative biological efficiency
  remote batch entry
RBGS
  radio beacon guidance system
RBHPF
  reactor building hydrogen purge fan
RBI
  radio berlin international
  rippleblanking input
  rocketborne instrumentation
RBM
  real time batch monitor
  rod block monitor
RBN
  radio beacon
RBO
  rippleblanking output
RBOD
  required beneficial occupancy data
RBOF
  receiving basins for offsite fuels
RBP
  registered business programmer
  remote batch processing
RBQ
  request block queue
RBR
  radar boresight range
  refracted bottom reflected
RBS
  radar beacon station
  radar beacon system
  radar bomb scoring
  radio beacon station
  random barrage system
  rated breaking strength
  reactor building sump
  recoverable booster system
RBSS
  recoverable booster space system
  recoverable booster support
  recoverable booster support system
RBT
  radio beam tube
  remote batch terminal
  resistance bulb thermometer
RBU
  remote buffer unit
RBV
  reactor building vent
  return beam vidicon
RBWO
  resonant backward wave oscillator
RC
  radar control
  radio command
  radio compass
  radio components
  radio control
  range clearance
  range control
  range correction
  rapid curing
  rate of climb
  ray control
  ray control electrode
  reaction coupling
  reactor
  read and compute
  read clock
  reader code
  receiver
  record carriers
  record code
  recording controller
  redundancy check
  regional center
  register containing
  reinforced concrete
  reinstate card
  relay computer
  remote computer
  remote computing
  remote control
  research center
  resistance capacitance
  resistance coupled
  resistor capacitor
  resistor capacitor circuit

RCA

resolver control
reverse current
right center
ringing circuit
robot controller
rockwell c
rotary compression
rotary converter
rubber covered
RCA
   radio corporation of america
   radio correspondents association
RCAC
   radio corporation of america,
   communications
RCADXC
   radio club amsterdam dx certificate
RCAENG
   rca engineer
RCAF
   returned customer assginment form
   royal canadian air force
RCAG
   remote controlled airground communication site
RCAREV
   radio corporation of america review
RCAT
   radiocode aptitude test
   radiocontrolled aerial target
RCAVIC
   radio corporation of america victor
RCB
   radiation control board
   radio-controlled boat
   reactor containment building
   rubber covered braided
RCBC
   rapid cycling bubble chamber
RCBWP
   rubber covered, braided, and weatherproof
RCC
   radio common carrier
   radio common channel
   radio communication company
   radiochemical center
   range commanders council
   range control center
   read channel continue
   reader common contact
   recovery control center
   regional computer center
   remote communications center
   remote communications complex
   remote communications console

remote control complex
resistance capacitance coupling
ring closed circuit
rod cluster control
routing control center
RCCA
   remote control rod cluster assembly
RCCPLD
   resistance capacitance coupled
RCCS
   royal canadian corps of signals
RCCTL
   resistor capacitor coupled transistor logic
RCD
   received
   record
   reverse current device
RCDB
   rubber covered double braided
RCDC
   radar course directing center
   radar course directing control
RCDG
   recording
RCDR
   recorder
RCDTECHNIQUE
   resistor capacitor diode technique
RCDTTECHNIQUE
   resistor capacitor diode transistor technique
RCE
   rapid circuit etch
   ray control electrode
RCEEA
   radio communication and electronic engineering association
RCEI
   range communications electronics instructions
RCEIVER
   receiver
RCEME
   royal canadian electrical and mechanical engineers
RCF
   radio correction factor
   recall finder
   remote cluster facility
RCFC
   reactor core fan cooling
RCFCU
   reactor core fan cooling unit
RCG
   radio command guidance
   radioactivity concentration guide

radioelectrocardiograph
receiving
recovery control group
reverberation control of gain
RCGM
reactor cover gas monitor
RCGS
radio command guidance system
RCGUGA
resistance capacitance grounded unity gain amplifier
RCI
radar coverage indicator
range communications instructions
read channel initialize
request for change information
RCIA
remote control interface adapter
RCIC
reactor core isolation cooling
RCICS
reactor core isolation cooling system
RCL
radiation counting laboratory
radio command linkage
read clock
receive clock
RCLC
reactor coolant leakage calculation
RCLM
reclaim
reclamation
RCM
radar countermeasure
radio-controlled mine
radio countermeasure
reaction control motor
read clutch magnet
reliability corporate memory
revenue cost model
rod cell memory
RCMF
radio component manufacturers' federation
RCO
radar control officer
reactor core
remote communications outlet
remote control office
remote control oscillator
rendezvous compatibility orbit
RCOA
radio club of america
RCOM
remote communication message
RCP
radiological control program used in reference to states
reactor coolant pump
recovery command post
relative corrector program
remote communication processor
righthanded circular polarization
RCPA
rural cooperative power association
RCPB
reactor coolant pressure boundary
RCPT
receipt
RCPTN
reception
RCR
reactor control room
reader control relay
reverse contactor
reverse current relay
RCRC
reinforced concrete research council
RCRDC
radio components research and development committee
RCRDTC
radio components research and development technical committee
RCS
radar cross section
radio command system
reaction control subsystem
reaction control system
reactor coolant system
rearward communications system
recurrent change of station
reentry control system
reloadable control storage
remote control system
reversing color sequence
royal corps of signals
RCSC
radio components standardization committee
RCSDE
reactor coolant system dose equivalent
RCSG
restarting computer and symbol generator
RCSS
random communication satellite system
RCSTN
radio compass station
RCT
radar control trailer
radiation chemical technician

receipt
reference clock trigger
region control task
regional control task
resistor capacitor transistor
resolver control transformer
reverseconducting thyristor
rework completion tag
RCTL
resistor capacitor transistor logic
RCTSR
radio code test speed on response
RCU
receiver control unit
relay control unit
remote control unit
resistor capacitor unit
rocket countermeasure unit
RCV
receive
receiving
relative conductor volume
RCVD
received
RCVG
receiving
RCVR
receiver
RCW
read, compute, write
register containing word
RCWP
rubber covered, weather proof
RCWV
rated continuous working voltage
RCX
remote cluster executive
RCZ
radiation control zone
RD
radar data
radar display
radiation absorbed dose
radiation detection
random drift
rate of descent
rated
rated duty
ratio detector
read
read delay
readiness date
reading
received data
recognition differential

recorder
recording demand meter
rectifier diode
register drive
register driver
relay driver
relocatable directory
reply delay
required date
requirements document
research and development
resistor diode
resolver differential
restricted data
ring down
roof diameter
root diameter
round
rutherford
RDA
reliability design analysis
rod drop accident
runtime debugging aid
RDAA
range doppler angle angle
RDARA
regional and domestic air route area
RDB
radar decoy balloon
reference data and bias
research and development board
RDBL
readable
RDC
rand development corporation
reliability data center
reliability data control
remote data collection
remote data concentrator
reply delay compensation
rotating disk contactor
RDCHK
read check
RDCLP
response document capability list positive
RDCO
reliability data control office
RDD
read disconnect delay
RDDM
reactor deck development mock-up
RDDP
response document discard positive
RDE
radial defect examination

RDEP
  response document end positive
RDF
  radio direction finder
  radio direction finding
  record definition field
  recursive digital filter
  reflection direction finding
  repeater distribution frame
RDG
  radio directors' guild
  resolver differential generator
RDGL
  radiological
RDGR
  response document general reject
RDGY
  radiology
RDH
  remote device handler
RDHD
  round head
RDI
  radio, doppler inertial
RDL
  replaceable lamp display light
  resistance diode logic
  rocket development laboratory
RDM
  radarman
  recording demand meter
RDME
  range and distance measuring equipment
RDMU
  rangedrift measuring unit
RDN
  redundancy
RDO
  radio
  readout
RDOINT
  radio intelligence
RDOS
  real time disc operating system
RDOSEC
  radio section
RDOSERV
  radio service
RDOSTN
  radio station
RDOTRANS
  radio transmitter
RDP
  radar data processing
  radio distribution point
  radiosonde data processor
  reactor development program
  remote data processing
RDPBN
  response document page boundary negative
RDPBP
  response document page boundary positive
RDPC
  radar data processing center
RDPS
  radar data processing system
RDR
  radar
  radar diagnostic report
  radar display room
  read drum
  reliability diagnostic report
  research and development report
  restart delay relay
RDRINT
  radar intermittent
RDRP
  response document resynchronisation positive
RDRREL
  radar relay
RDRSMTR
  radar transmitter
RDS
  radar distribution switchboard
  raytheon data systems
  read select
  rendezvous and docking simulator
  resolver differential transmitter
  rural development service
RDT
  reactor development and technology
  remote data transmitter
  research, development, and test
  resistor diodetransistor technique
  resource definition table
  rotational direction transmission
RDTE
  research, development, test and evaluation
RDTL
  resistor diodetransistor logic
RDU
  remote display unit
RDVU
  rendezvous
RDY
  ready
RDZ
  radiation danger zone

RE
  radiated emission
  radiation effects
  radio equipment
  radio exposure
  rare earth
  rate effect
  read emitter
  read error
  real
  recovery equipment
  recursively enumerable
  reentry
  regular expression
  relay engineer
  reset
  restriction of extension
  reynolds number
REA
  reentry angle
  request for engineering action
  rural electric association
  rural electrification administration
REAC
  reactance
  reactive
  reactivity
  reactor
  reeves electronic analog computer
REACT
  radio emergency associated citizens teams
  reactance
  record evaluate and control time system
  register enforced automated control technique
  rese engineering automatic core tester
READ
  radar echo augmentation device
  real time electronic access and display
  remote electronic alphanumeric display
READI
  rocket engine analysis and decision instrumentation
  rocket engine analyzer and decision instrument
READJ
  readjust
  readjustment
READYMAIDS
  ready multipurpose automatic inspection and diagnostic system
REAL
  recourse allocation
REALCOM
  real time communications
  real time communications computer

REAP
  remote entry acquisition package
REAR
  reliability engineering analysis report
REASM
  reassemble
REASTAN
  reton electrical analog for solution of thermal analog networks
REB
  radar evaluation branch
  rebecca eureka beacon
  reentry body building
REBA
  relativistic electron beam accelerator
REBE
  recovery beacon evaluation
REBECCA
  remote electrical block energization clock control arrangement
REC
  radiant energy conversion
  receipt
  receive
  received
  receiver
  receiving
  receptacle
  record
  recorder
  recreation
  rectifier
  registration of engineers committee
  request for engineering change
  rural electrification conference
RECAU
  receipt acknowledged and understood
RECCEN
  reception center
RECD
  received
RECG
  radio electro cardiograph
RECM
  recommend
RECMARK
  record mark
RECMF
  radio and electronic component manufacturers' federation
RECMFA
  radio and electronic component manufacturers association

RECOG
  recognition
  recognize
RECOGSIG
  recognition signal
RECON
  reconnaissance
RECONST
  reconstruct
RECOV
  recover
  recovery
RECP
  receptacle
RECR
  receiver
RECSTA
  receiving station
RECT
  receipt
  reception
  rectification
  rectified
  rectifier
  rectify
RED
  radio equipment department
  reduce
  reducing
  reduction
REDAC
  racal electronic design and analyses by computer
REDAP
  reentrant data processing
REDC
  reinsertion of direct current
REDIT
  random access edit
REDSOD
  repetitive explosive device for soil displacement
REE
  reliance electric and engineering
REEA
  radio and electronics engineering association
REEG
  radioelectroencephalograph
REEP
  range estimating and evaluation procedure
  regression estimation of event probabilities
REF
  range error function
  refer
  reference
  refraction
REFFREQ
  reference frequency
REFL
  reference line
  reflectance
  reflection
  reflector
REFLEC
  reflection
REFLECT
  reflectometer
REFORS
  replacement forecasting system
REFTEMP
  reference temperature
REG
  radioisotope electrogenerator
  regeneration
  region
  register
  regression analysis
  regular
  regulate
  regulation
  regulator
REGAL
  range and evaluation guidance for approach and landing
REGD
  registered
REGEN
  regenerate
  regeneration
  regenerative
REGIS
  remote graphic instruction set
REGRA
  regression analysis
REI
  recognition equipment
  request for engineering investigation
  research engineering interaction
REIC
  radiation effects information center
REIG
  rare earth iron garnet
REIL
  runway end identifier lights
  runway end identification light
REINS
  radio equipped inertial navigation system
  requirements electronic input system

REJ
  reject
  rejection
REL
  radiation evaluation loops
  radio engineering laboratories
  rapidly extensible language
  rate of energy loss
  relais
  relation
  relay
  release
  relief
  relieve
  reluctance
RELCODE
  relative coding
RELE
  radio electrician
RELIA
  regional european long lines agency
REM
  radiation equivalent man
  rapid eye movement
  reliability engineering model
  removable
  roentgen equivalent man
REMAD
  remote magnetic anomaly detection
REMAS
  radiation effects machine analysis system
REMC
  radio and electronics measurements committee
  resinencapsulated mica capacitor
REMG
  radioelectromyograph
REMIS
  real estate management information system
REMON
  real-time event monitor
REMS
  registered equipment management system
REMSA
  railway electrical and mechanical supply association
REN
  remote enable
RENE
  rocket engine nozzle ejector
RENM
  ready for next message
REO
  regenerated electrical output

REON
  rocket engine operations nuclear
REOURRAD
  reference our radio
REOURTEL
  reference our telegram
REOURTT
  reference our teletype
REP
  radar evaluation pod
  range error probable
  recovery and evacuation program
  reliability evaluation program
  rendesvouz evaluation pod
  repair
  roentgen equivalent physical
REPERF
  reperforator
REPET
  repetition
REPL
  replace
REPLAB
  responsive environmental programmed laboratory
REPM
  representative of electronic products manufacturers
REPOP
  repetitive operation
REPPAC
  repetitive pulses plasma accelerator
REPRO
  reproduce
  reproducer
  reproduction
REPROM
  reprogrammable read only memory
REPT
  report
REQ
  request
REQU
  require
REQVER
  requirements verification
RER
  radar effects reactor
  radiation effects reactor
  redundant element removal
RERL
  residual equivalent return loss
RES
  rapid evaluation system
  remote access editing system

remote entry services
remote entry subsystem
reserve
reservoir
reset
resistance
resistant
resistor
resolver
restore
RESA
  research society of america
RESAR
  reference safety analysis report
RESC
  royal engineers and signal corps
RESCAN
  reflecting satellite communications antenna
RESCAP
  resistor capacitor
RESCU
  rocket ejection seat catapult, upward
RESD
  research and engineering support division
RESER
  reentry system evaluation radar
  reentry system and evaluation radar
RESEX
  resource executive
RESFLD
  residual field
RESG
  research engineering standing group
RESIS
  resistance
  resistor
RESMA
  railway electric supply manufacturers' association
RESNA
  rehabilitation engineering society of north america
RESOJET
  resonant pulse jet
RESOLUT
  resolution
RESP
  regulated electrical supply package
RESS
  radar echo simulation subsystem
  radar echo simulation system
  radio echo simulation study
REST
  radar electronic scan test
  reentry environment and systems technology

restrict
restricted
restricted radar electronic scan technique
restriction
RESUP
  resupply
RET
  return
RETMA
  radio electronics television manufacturing association
  radio, electronics, and television manufacturers' association
RETP
  retape
RETS
  radio electronics television school
RETSPL
  reference equivalent threshold sound pressure level
REU
  radio engineering unit
  remote entry unit
REV
  reentry vehicle
  reverse
  review
  revise
  revolution
REVCUR
  reverse current
REVMIN
  revolutions per minute
REVOCON
  remote volume control
REVOP
  random evolutionary operation
REVR
  receiver
REVS
  revolutions
  rotor entry vehicle system
REW
  read, execute, write
  rewind
REWSONIP
  reconnaissance electronic warfare special operation and naval intelligence processing
REX
  reed electronic exchange
REYRTWX
  reference your telegraph wire exchange
RF
  radar frequency
  radio facilities

radio frequency
radio france
range finder
rating factor
reactive factor
reactive factor meter
read forward
register finder
representative fraction
resistance factor
running forward
RFA
  radio frequency amplifier
  radio frequency authorization
  register field address
RFC
  radio facility chart
  radio frequency chart
  radio frequency choke
  radio frequency coil
RFCO
  radio facilities control officer
  radio frequency checkout
  range facilities control officer
RFCP
  radio frequency compatibility program
RFD
  reactor flight demonstration
  read for data
  ready for data
  reentry flight demonstration
RFEI
  request for engineering information
RFFD
  radio frequency fault detection
RFG
  radar field gradient
  ramp function generator
  receive format generator
  register finder grid
RFH
  radio frequency head
RFI
  radio frequency interchange
  radio frequency interference
  remote file inquiry
  request for information
  request for inspection
RFICAPACITOR
  radio frequency interference capacitor
RFIT
  radio frequency interference test
RFL
  radio frequency laboratories
  rotating field logic

RFM
  reactive factor meter
  reliability figure of merit
RFMO
  radio frequency management office
RFNA
  radio frequency noise analyzer
  red fuming nitric acid
RFO
  radio frequency oscillator
RFP
  request for proposal
RFQ
  request for quotation
  request for quote
RFR
  reject failure rate
RFS
  radio frequency shift
  ready for sending
  remote file sharing
RFSAT
  radio frequency saturation
RFSP
  rigid frame selection program
RFSS
  reliability failure summary support
RFST
  rapid frequency settling time
RFSTF
  rf systems test facility
RFT
  radio frequency transformer
  recursive function theory
RFTD
  radial flow torr deposition
RFW
  reversible full wave
RFWAC
  reversible full wave alternating current
RFWDC
  reversible full wave direct current
RG
  radar guidance
  radiation guidance
  radio guidance
  radio guide
  range
  rate gyroscope
  rategrown
  reception good
  recording
  reduction gear
  register
  regulatory guide

report generator
report program generator
reset gate
residue generator
resticulated grating
reverse gate
RGA
rate gyro assembly
remote gain amplifier
routing accumulator
RGB
red, green, blue
RGBY
red, green, blue, yellow
RGE
range
RGL
radar gunlaying
report generator language
RGM
recorder group monitor
RGN
ranging
RGNG
ranging
RGO
royal greenwich observatory
RGP
radar glider positioning
rate gyro package
RGR
routing register
RGS
radio guidance system
rate gyro system
rocket guidance system
RGSAT
radio guidance surveillance and automatic tracking
RGSC
ramp generator and signal converter
RGT
resonant gate transistor
RGZ
recommended ground zero
RH
radiation homing
radiological health
receiver hopping mode
red heat
relative humidity
reserve shutdown hours
response header
rheostat
right half

right hand
right-hand
rockwell hardness
roentgen per hour
RHAW
radar homing and warning
RHB
radar homing beacon
radar homing bomb
RHC
range height converter
right-hand circular
right-hand components
rotation hand controller
RHCP
right-hand circular polarization
RHCTL
right-hand controls
RHD
radar horizon distance
right-hand drive
RHE
radiation hazard effects
RHEED
reflected high energy electron diffraction
RHEL
rutherford high energy laboratory
RHEO
rheostat
RHFEB
right-hand forward equipment bay
RHI
radar height indicator
range height indicator
RHJ
rubber hose jacket
RHM
right-hand polarized mode
roentgen per hour atonemeter
RHOGI
radar homing guidance
RHP
rated horsepower
reduced hard pressure
RHR
rejectable hazard rate
residual heat removal
roentgen per hour
RHRP
residual heat removal pump
RHRSW
residual heat removal service water
RHS
right-hand side

RHTS
reactor heat transport system
RHU
reserved for hardware use
RHW
righthalf word
RHWAC
reversible half wave alternating current
RHWDC
reversible half wave direct current
RI
radar input
radar intercept
radio industries
radio inertial
radio influence
radio inspector
radio interference
range instrumentation
read in
receiving inspection
reflective insulation
reliability index
repulsion induction
resistance inductance
reverberation index
rhode island
ring indicator
rubber insulation
RIA
reactivity initiated accident
removable instrument assembly
research institute of america
reset indicators from accumulator
robot institute of america
RIAA
record industry association of america
recording industry association of america
RIAEC
rhode island atomic energy commission
RIAL
runway identifiers and approach lighting
RIAS
research institute for advanced studies
RIC
radar input control
radar intercept calculator
radio industry council
range instrumentation conference
range instrumentation coordination
receiver impulse characteristic
RICASIP
research information center and advisory service on information processing
RICMO
radar input countermeasures officer
RICMT
radar input countermeasures technician
RICNETWORK
resistance inductance capacitance network
RICS
range instrumentation control system
RID
radar input drum
radio intelligence division
range instrumentation division
reset inhibit driver
reset inhibit drum
review item disposition
RIDD
range instrumentation development division
RIDS
receiving inspection data status report
RIE
radio interference elimination
radio interference eliminator
reactive ion etching
rhodesian institution of engineers
RIF
radar identification set
radio influence field
reduction in force
reliability improvement factor
RIFI
radio interference free instrument
ratio interference filed intensity
RIFIM
radio interference field intensity meter
RIFT
reactor inflight test
RIFTS
reactor inflight test system
RIG
radio inertial guidance
RIGFET
resistive insulated gate field effect transistor
RIGS
radio inertial guidance system
runway identifier with glide slope
RII
receiving inspection instructions
RIL
radio influence level
radio interference level
radiolocation
red indicating lamp
red indicator lamp
reset indicators of the left half
rivlin instruments, limited

RIM
   radar input mapper
   radar input monitor
   reaction injection molded
   receiving inspection and maintenance
   request initialization mode
   research instrument module
   resource interface module
RIME
   radio inertial missile equipment
   radio inertial monitoring equipment
RIN
   record identification number
   register in
   register in instruction
   regular inertial navigator
RINAL
   radar inertial altimeter
RING
   ringing
RINS
   rotorace inertial navigation system
RINT
   radar intermittent
RIOMETER
   relative ionospheric opacity meter
RIOPR
   rhode island open pool reactor
RIOS
   remote input output system
RIOT
   real time input output transducer
   resolution of initial operational techniques
RIP
   reactor instrument penetration
   receiving inspection plan
   recoverable item program
   regulatory and information policy
RIPPLE
   radio active isotope powered pulse light equipment
   radio isotope powered prolonged life equipment
RIPS
   radar impact prediction system
   radio isotope power supply
   radio isotope power system
   range instrumentation planning study
RIPV
   reactor isolation pressure valve
RIQAP
   reduced inspection quality assurance program
RIR
   read only memory instruction register
   receiving inspection report
   reliability investigation requests
   reporting interface record
   reset indicators of the right half
   ribbon to ribbon
RIRTI
   recording infrared tracking instrument
RIS
   range instrumentation ship
   record input subroutine
   reporting identification symbol
   requirements planning and inventory control system
   reset indicators form storage
   resistor insulator semiconductor
   revolution indicating system
RISC
   reduced instruction set computer
RISE
   relative integral squared error
   research in supersonic
   research in supersonic environment
RISP
   recoverable interplanetary space probe
RISS
   range instrumentation and support system
RISSWITCH
   resistor insulator semiconductor switch rise of resistance method
RIST
   radar installed system tester
RIT
   radar inputs test
   radio information test
   radio network for inter-american telecommunications
   receiving and inspection test
   rocket interferometer tracking
RITA
   recoverable interplanetary transport approach
   reusable interplanetary transport approach
   reusable interplanetary transport approach vehicle
RITE
   rapid information technique for evaluation
   rapidata interactive editor
RITS
   remote input terminal system
RITU
   research institute of temple university
RIV
   radio influence voltage
RIVA
   recreational industry vehicle association

RIVAL
  rapid insurance valuation language
RIW
  reliability improvement warranty
RJE
  remote job entry
RJP
  remote job processing
RKHS
  reproducing kernel hilbert space
RKO
  radio keith orpheum
  range keeper operator
RKT
  rocket
RKVA
  reactive kilovolt ampere
RL
  radiation laboratory
  radiolocation
  reactive loss
  reactor licensing
  record length
  red lamp
  reel
  reference library
  reference line
  relay logic
  relocatable library
  relocation
  remote location
  research laboratory
  resistance inductance
  resistor logic
  return loss
  right line
  rocket launcher
RLBG
  relative bearing
RLBI
  right left bearing indicator
RLBM
  rearward launched ballistic missiles
RLC
  radio launch control system
  remote line concentrator
  resistance inductance capacitance
  resistor inductor capacitor
  run length coding
RLCS
  radio launch control system
RLD
  relocation dictionary
RLE
  rate of loss of energy
  research laboratory of electronics
RLF
  reactive load factor
RLHTE
  research laboratory of heat transfer in electronics
RLI
  red line instrumentation
RLL
  right lower lobe
RLM
  rearward launched missile
  reflector and lighting equipment manufacturers
  reflector lamps manufacturers
  right middle lobe
RLOP
  reactor licensing operating procedure
RLQ
  right lower quadrant
RLR
  reverse locking relay
RLS
  radar line of sight
  reels
  reusable launch system
  rotating lighthouse system
RLTS
  radio linked telemetry system
RLU
  relay logic unit
RLV
  roving lunar vehicle
RLWL
  reactor low water level
RLY
  relay
RM
  rack mounted
  rack mounting
  radar mapper
  radar missile
  radarman
  radiation measurement
  radio marker
  radio message
  radio monitor
  radio monitoring
  radioman
  range mark
  range marker
  reactance meter
  reaction mass

read out material
read out matrix
receiver, mobile
record mark
rectangular module
reference manual
reference mark
reflection modulation
remote
residue manipulator
resolution multiplier
rod memory
rotating machinery
routing matrix
RMA
radio maufacturers' association
rubber manufacturers' association
RMAAS
reactivity monitoring and alarm system
RMAC
reactor monitoring and control
RMAG
rocky mountain association of geologists
RMAI
radio maufacturers' association of india
RMAX
range maximum
RMBAA
rocky mountain business aircraft association
RMC
radio materials company
radio modifications committee
radio monte carlo
relay mode control
remote control
remote message concentrator
rendezvous mercury capsule
rod memory computer
RMCS
reactor manual control system
RMDI
radio magnetic deviation indicator
RMF
reactivity measurement facility
RMG
radar mapper gap filler
RMI
radio magnetic indicator
reliability maturity index
RMICBM
road mobile intercontinental ballistic missile
RMIN
roentgen per minute
RML
radar mapper long range

radar microwave link
read major line
rescue motor launch
RMM
radar map matching
read mostly memory
RMO
radio material officer
RMOS
real memory operating system
refractory metal oxide semiconductor
RMP
reentry measurements program
RMS
radar maintenance spare
radiation monitoring satellite
radiation monitoring system
radio merchandise sales
reactor monitor system
recovery management support
reentry measurement system
regulatory manpower system
remote manipulator system
remote multiplexer system
resource management system
root mean square
royal marine signaller
RMSD
root mean square deviation
RMSE
root mean square error
RMSI
royal marine signalling instructor
RMSV
root mean square value
RMT
radioman telegrapher
remote
RMTE
remote
RMU
remote maneuvering unit
remote measuring unit
RMV
reentry measurement vehicle
RMWS
reactor make-up water system
RN
radio navigation
radio noise
random number
reference noise
reynolds number
RNA
ribonucleic acid

RNAS
    royal naval air station
RND
    round
RNDM
    random
RNF
    radio noise figure
    receiver noise figure
RNFP
    radar not functioning properly
RNG
    radio range
    range
RNGCOMP
    range computer
RNIT
    radio noise interference test
RNM
    radio navigation mobile
RNR
    receive not ready
    ring number read
RNTWPA
    radio newsreel television working press association
RNV
    radio noise voltage
RNVPR
    royal naval volunteer postal reserve
RNVWR
    royal naval volunteer wireless reserve
RNW
    ring number write
RNWAR
    royal naval wireless auxiliary reserve
RNY
    radio new york
RO
    radar observer
    radar operator
    radio operator
    radio orient
    range only
    range operation
    range operator
    read only
    read out
    receive only
    reportable occurrence
    round off
    route order
ROA
    radio operator's aptitude
    rules of the air

ROAMA
    rome air material area
ROAP
    reorganization, office of assistant to president
ROAR
    read only storage address register
    royal optimizing assembly routine
ROAT
    radio operator's aptitude test
ROB
    radar order of battle
    radar out of battle
ROBIN
    remote on-line business information network
    rocket balloon instrument
ROBO
    rocket orbital bomber project
ROBOMB
    robot bomb
ROCP
    radar out of commission for parts
    remote operator control panel
ROCR
    remote optical character
    remote optical character recognition
ROD
    release order directive
RODATA
    registered organization data bank
RODIAC
    rotary dual input for analog computation
ROE
    reflector orbital equipment
ROFT
    radar off target
ROI
    range operational instructions
    range operations instructions
    resources objectives inc
    return on investment
ROINST
    range operations instructions
ROIS
    radio operational intercom system
ROJ
    range of jamming
ROLF
    remotely operated longwall face
ROLR
    receiving objective loudness rating
ROLS
    recoverable orbital launch system
ROLTS
    remote on-line testing system

ROM
  read only memory
  read out memory
  reciprocal ohmmeter
  rotating piston machine
  rough order magnitude
  run of mine
ROMAC
  range operations monitor analysis center
ROMACC
  range operational monitoring and control center
ROMBUS
  reusable orbital module booster and utility shuttle
ROMON
  receiving only monitor
ROMOTAR
  range only measurement of trajectory automatic recording
ROMR
  read only memory register
ROMS
  read only memory storage
RON
  research octane number
  run occurence number
ROOST
  reusable one stage orbital space truck
ROOT
  relaxation oscillator optically tuned
ROP
  record of purchase
  recovery operating plan
  rotating observation platform
ROPIMA
  rotary piston machine
ROPP
  receive only page printer
ROR
  range only radar
  rate of return
  rocket on rotor
RORO
  roll on roll off
ROS
  range operation station
  read only storage
  read only store
  remote operating system
  resident operating system
  review of systems
ROSA
  recording optical spectrum analyzer
ROSCOE
  remote operating system conventional operating environment
ROSE
  remotely operated special equipment
  retrieval by on-line search
  rising observational sounding equipment
ROSIE
  reconnaissance by orbiting ship identification equipment
ROT
  rate of turn
  reusable orbital transport
  rotary
  rotary switch
  rotate
  rotating
  rotation
  rotor
  rule of thumb
ROTAB
  rotary table
ROTAC
  rotary oscillating torque actuators
ROTE
  range optical tracking equipment
ROTEL
  rotational telemeter
ROTI
  recording optical tracking instrument
ROTR
  receive only tape reperforator
  receive only typing reperforator
ROTRSP
  receive only typing reperforator series to parallel
ROTS
  range on target signal
  reusable orbital transport system
  rotary out trunk switch
ROUT
  register out
  routine
ROVD
  relay operated voltage divider
ROW
  right of way
RP
  radio paris
  reception poor
  recommended practice
  recorded program
  recovery phase
  reference point
  relative pressure

release point
reliability program
remote pick up
repeater
reply paid
reply prepaid
reprint
reprinting
reproducer
reproducing punch
revertive pulsing
rocket projectile
round punch
RPA
radar performance analyzer
RPAO
radium plague adaptometer operator
RPC
radar planning chart
radar processing center
remote position control
resistance products company
row parity check
RPCRS
reactor protection control rod system
RPCS
reactor plant control system
RPD
radar planning device
radar prediction device
reactor plant designer
recorded program department
resistance pressure detector
retarding potential difference
RPDG
rand program development group
RPE
radial probable error
radio production executive
registered professional engineer
related production equipment
remote peripheral equipment
resource planning and evaluation
rocket propulsion establishment
rotating platinum electrode
RPEP
register of planning emergency producers
RPF
radar performance figure
radio position finding
radiometer performance factor
RPFC
recurrent peak forward current

RPG
radiation protection guide
report program generator
RPH
revolutions per hour
RPI
radar precipitation integrator
reversals per inch
revolutions per inch
rework print image
rod position indicator
RPIE
real property installed equipment
RPIS
rod position indicator system
RPL
radar processing language
radiation physics laboratory
radio physics laboratory
replacement
rocket propulsion laboratory
running program language
RPLS
reactor protection logic system
RPM
radar performance monitor
random phase modulator
rate per minute
regulated power module
reliability performance measure
remote performance monitoring
resupply provisions module
revolutions per minute
rotations per minute
runs per minute
RPMC
remote performance monitoring and control
RPMI
revolutions per minute indicator
RPN
reverse polish notation
reversed polish notation
RPO
rotor power output
RPOA
recognized private operating agency
RPP
radar power programmer
RPPA
repetitively pulsed plasma accelerator
RPPI
remote plan position indicator
RPQ
request for price quotation

RPR
   rapid power reduction
   read printer
   reverse phase relay
   reverse power relay
RPROM
   reprogrammable rom
RPRS
   random pulse radar system
RPRT
   report
RPRWP
   reactor plant river water pump
RPS
   radar plotting sheet
   reactor protection systems
   realtime programming system
   regulated power supply
   regulatory performance summary report
   remote processing service
   remote processing system
   revolution per second
RPV
   remotely piloted vehicle
RPVT
   relative position velocity technique
RPW
   running process word
RPY
   roll, pitch, and yaw
RQ
   request
   resolver quantizer
RQA
   recursive queue analyzer
RQC
   radar quality control
   receiving quality control
RQD
   rock quality designation
RQL
   reference quality level
RQR
   require
   requirement
RQS
   rate quoting system
RR
   radiation response
   radio receptor
   radio regulations
   radio relay
   radio research
   railroad
   range recorder

   readout and relay
   receive ready
   receiving report
   record retransmit
   recorder
   recurrence rate
   register to register instruction
   register to register operation
   remington rand
   remove and replace
   rendezvous radar
   repetition rate
   research reactor
   residue register
   retro rockets
   return rate
   round robin
   running reverse
RRAC
   radio regulations atlantic city
RRB
   radio range beacon
   radio research board
   railroad retirement board
RRC
   radio receptor company
   radio research company
   receiving report change
RRCN
   receiving report change notice
RRDA
   rendezvous retrieval docking and assembly
RRDE
   radar research and development establishment
RRDR
   raw radar data recorder
RRDTRL
   resistor resistor diode transistor logic
RRE
   radar research establishment
   royal radar establishment
RREAC
   royal radar establishment automatic computer
RREG
   r register
RRF
   resonant ring filter
RRHICMD
   remote reading high intensity constant monitoring device
RRI
   radio republic indonesia
   range rate indicator
   rocket research institute

RRIC
    radar radio industries of chicage
    radar repeater indicator console
RRIS
    remote radar integrating station
    remote radar integration station
RRL
    radar research laboratory
    radio research laboratory
RRLG
    rocket, radio, longitudinal, generator powered
RRMG
    reactor recirculation motor generator
RRNS
    redundant residue number system
RRO
    radio research organization
RRP
    radio ripple proximity
    reactor refueling plug
RRPG
    regular right part grammar
RRPI
    relative rod position indication
    rotary relative position indicator
RRPM
    representatives of radio parts manufacturers
RRR
    raleigh research reactor
    range and range rate
RRRV
    rate of rise of restriking voltage
RRS
    radiation research society
    radio research station
    reaction research society
    reactor recirculation system
    reactor refueling system
    reactor regulation system
    required response spectrum
    restraint release system
    retrograde rocket system
RRSA
    radio republic south africa
RRSQ
    radio relay squadron
RRTTL
    resistor resistor diode transistor logic
    resistor resistor transistor transistor logic
RRU
    radiobiological research unit
    remington rand univac
RRV
    rate of rise of voltage

RS
    radar simulator
    radar start
    radiated susceptibility
    radio communication supervisor
    radio school
    radio set
    radio station
    radiospares
    radiotelegram service
    range safety
    range selector
    range setter
    reader stop
    ready service
    record separator
    register storage
    regulating station
    relay selector
    relay set
    remote station
    render and set
    reset
    reset set
    reset steering
    resetting
    resistor
    resonator
    resynchronizing state
    return to saturation
    reverberation strength
    reverse signal
    rocket station
    root mean square average
    rotary selector
    rotary switch
    rotary system
    route signal
    routing selector
    run stop
RSA
    random sequential automaton
    rate subsystem analyst
    redstone arsenal
    remote station alarm
    rotary switch art
RSAC
    radiological safety analysis computer
    reactor safety advisory committee
    reliability surveillance and control
RSAF
    royal swedish air force
RSAI
    rules, standards and instructions

RSAP
response session abort positive
RSARR
republic of south africa research reactor
RSB
reactor service building
RSCCP
response session change control positive
RSCH
research
RSCIE
remote station communication interface equipment
RSCS
rate stabilization and control system
rod sequence control system
RSCSS
range safety command shutdown system
RSCW
research reactor, state college of washington pullman
RSD
radar system development
refueling shutdown
reliability status document
responsible system designer
runcible system duplexer
RSDC
range safety data coordinator
RSDP
remote shutdown panel
remote site data processing
remote site data processor
RSDS
range safety destruct system
RSDU
radar storm detector unit
RSE
request for self enhancement
RSEP
response session end positive
RSERV
relocatable library service function
RSET
receiver signal element timing
RSF
remote service facility
RSG
rate signal generator
RSGB
radio society of great britain
RSGS
range and space ground support
RSH
radar status history

RSI
radar scope interpretation
rationalization, standardization and integration
reactor siting index
reflected signal indicator
replacement stream input
research studies institute
ring state indicator
RSIC
radiation shielding information center
redstone scientific information center
RSID
resource indentification table
RSIGG
rocket signal, green
RSIGR
rocket signal, red
RSJ
rolled steel joist
RSL
radio standards laboratory
received signal level
RSLA
range safety launch approval
RSLS
reply path side lobe suppression
RSLT
result
RSM
radio squadron, mobile
readability, strength, modulation
resident sector management
resource management system
RSMPS
romanian society for mathematics and physical sciences
RSN
radiation surveillance network
record sequence number
RSO
radiosonde observation
revenue sharing office
RSOPN
resumed operation
RSP
radio switch panel
reactivity surveillance procedures
record select program
resource sharing protocol
responder beacon
rotating shield plug
route selection program
RSPP
radio simulation patch panel

RSPT
  real storage page table
RSR
  rapid solidification rate
  reactor safety research
  relay set receiver
  remote start relay
  reverse switching rectifier
  right element shift right
  rod select relay
RSRI
  regional science research institute
RSROM
  row select read only memory
RSRS
  radio and space research station
RSS
  range safety switch
  range safety system
  reactive system sensitivity
  reactor safety study
  reactor shutdown system
RST
  reset
  reset set trigger
  residential subsurface transformer
  routine sequence table
RSTN
  relay station
RSTR
  restricted
RSTRT
  restart
RSTS
  recovery systems track site
  resource sharing time sharing
RSU
  relay storage unit
  reserved for software use
RSV
  ready storage vessel
  reserve
RSVP
  radiation spectral visual data distribution
  radiation spectral visual photometer
RSVR
  reservoir
RSW
  retarded surface wave
RT
  radar tracking radiotelegraphy
  radio technician
  radio tower
  radiographic test
  radiotelegraph
  radiotelegraphy
  radiotelephone
  radiotelephony
  range time
  range tracking
  rapid transit
  rated time
  ratio transfer
  ratio transformer
  reaction time
  reactor trip
  real time
  receive transmit
  receiver transmitter
  receiving terminal
  receiving tube
  record transfer
  recovery time
  reduction table
  register ton
  register translator
  registered transmitter
  regression testing
  rejection tag
  relocatable term
  remote terminal
  rendezvous transponder
  reperforator transmitter
  research and technology
  reset trigger
  resistance thermometer
  resistor tolerance
  resistor transistor
  resolver transformer
  resolving transmitter
  resonant transfer
  revolving transmitter
  right
  ring time
  ringing tone
  room temperature
  rotary transformer
  routine tag
  routine test
RTA
  reliability test assembly
  resident transient area
  rise time analyzer
RTAC
  roads and transportation association of canada
RTAM
  remote telecommunication access method
RTAS
  rapid telephone access system

RTASCV
   radio tv association of santa clara valley
RTB
   radial time base
   range and true bearing
   read tape binary
   resistance temperature bridge
   return to base
   return to bias
   rocket test base
   rural telephone bank
RTBCN
   rate beacon
RTC
   radar tracking center
   radar tracking control
   radiotransmission control
   radiotelegraph communication
   radiotelephone communication
   reader tape contact
   real time clock
   real time command
   real time computation
   real time computer
   real time control
   removable top closure
RTCA
   radio technical commission for aeronautics
   radio television correspondents' association
RTCC
   real time computer complex
RTCF
   real time computer facility
RTCM
   radio technical commission for marine service
RTCMS
   radio technical commission for marine services
RTCOMN
   radio telephone communication
RTCP
   radio transmission control panel
   real time control program
RTCS
   real time computation system
   real time computer system
RTCU
   real time control unit
RTDD
   remote timing and data distribution
RTDDC
   real time digital data correction
RTDG
   radio and tv directors' guild

RTDHS
   real time data handling system
RTDS
   real time data system
RTDTL
   resistor tunnel diode transistor logic
RTE
   real time event
   real time executive
   residual total elongation
   route
RTEB
   radio trades examination board
RTEM
   radar tracking error measurement
RTES
   radio and television executives society
RTEX
   real time executive
   real time telecommunications executive
RTF
   radiotelephone
   radiotelephony
   resistance transfer factor
   rocket test facility
RTG
   radioisotope thermoelectric generator
RTGB
   reactor turbine general board
RTI
   real time interface
   real time interference
   referred to input
   research triangle institute
RTIOC
   real time input output control
   real time input output controller
RTIRS
   real time information retrieval system
RTK
   range tracker
RTL
   radioisotope transport loop
   real time language
   resistor transistor logic
RTLOC
   root locus
RTM
   rapid tuning magnetron
   real time monitor
   receiver transmitter modulator
   recording tachometer
   register transfer module
   registered trade mark

RTMA
    radio television manufacturers association
RTMON
    real time executive monitor
RTMOS
    real time multiprogramming operating system
RTNET
    radiotelephone network
RTO
    referred to output
RTOS
    real time operating system
RTP
    reactor thermal power
    real time peripheral
    real time processing
    real time program
    real time programming
    requirement and test procedure
RTPH
    round trips per hour
RTQC
    real time quality control
RTR
    repeater test rack
    return and restore status
RTRP
    remote terminal routine package
RTS
    radar tracking station
    reactive terminal service
    reactor trip system
    real time system
    remote terminal scanning system
    remote terminal supervisor
    remote terminal system
    remote test system
    request to send
RTSD
    resources and technical services division
RTSS
    real time scientific system
RTST
    radio technician selection test
RTT
    radioteletype
    radioteletypewriter
    real time telemetry
RTTDS
    real time telemetry data system
RTTV
    real time television
RTTW
    radioteletypewriter
RTTY
    radioteletype
RTU
    rate of a transfer unit
    remote terminal unit
RTV
    rocket test vehicle
    room temperature vulcanizing
RTVM
    real time virtual memory
RTWS
    raw tape write submodule
RTX
    real time executive
RU
    reproducing unit
    request unit
    response unit
    roentgen unit
RUDH
    reserve shutdown unplanned derated hours
RUG
    recomp users group
    report and update program generator
RUL
    refractoriness under load
RUM
    remote underwater manipulator
RURALELECT
    rural electric
RUSH
    remote use of shared hardware
RV
    rated voltage
    reactor vessel
    reduced voltage
RVA
    reactive volt ampere
    reactive volt amperemeter
    recorded voice announcement
    reliability variation analysis
RVI
    reverse interrupt
RVIS
    reactor and vessel instrumentation system
RVLIS
    reactor vessel water level indication system
RVLR
    road vehicles lighting regulations
RVM
    reactive voltmeter
RVR
    runway visual range
RVSS
    reactor vessel support system

RW
  read write
  resistance welding
RWCS
  reactor water clean up system
  report writer control system
RWCU
  reactor water clean up unit
RWD
  rewind
RWG
  roebling wire gage
RWM
  read write memory
  rectangular wave modulation
  rod worth minimizer
RWMA
  resistance welder manufacturers' association
RWO
  right wrong omit
RWP
  radiation work permit
  rain water pipe
  reactor work permit
RWRAMREFRESH
  read write random access memory refresh
RWRAMVOLATILITY
  read write random access memory volatility
RWS
  radwaste system
RWSS
  river water supply system
RWST
  refueling water storage tank
RWVRC
  read write vertical redundancy check
RWW
  read while write
RX
  register to indexed storage operation
RXOS
  rank xerox operating system
RXS
  real time executive system
RY
  railway
  relay
  rydberg
RZ
  return to zero
RZL
  return to zero level
RZM
  return to zero mark

S

S
  satellite
  saybolt second
  screen
  search
  second
  secondary
  secondary winding
  sector
  security
  side
  side signal
  siemens
  sign
  signal strength
  signed
  silk
  single
  single silk
  soft
  solid
  source
  south
  space
  speech
  spherical joint
  spool
  stoke
  storage
  store
  substrate
  sulphur
  supervisor
  switch
  symbol
  symmetrically substituted
  syn
SA
  safe arrival
  safety analysis
  sales aid
  saturn apollo
  security agency
  select address
  semi automatic
  sense amplifier
  sequential access
  service aids
  service area
  single amplitude
  single armored
  slow acting

slow acting relay
spacecraft adapter
spectrum analyzer
speech amplifier
spin axis
stress anneal
string analysis
subassembly
successive approximation
surface air
switching assembly
symbolic assembler
SAA
signal appliance association
south african airways
standards association of australia
system application architecture
SAAC
swiss american aircraft corporation
SAAD
san antonio air depot
SAAEB
south african atomic energy board
SAAMA
san antonio air materiel area
SAB
scientific advisory board
sync address bus
system advisory board
systems analysis branch
SABC
south african broadcasting corporation
SABE
society for automating better education
society for automation in business education
SABL
serialized assembly breakdown list
SABM
set asynchronous balanced mode
SABMIS
sea based antiballistic missile
SABO
sense amplifier blocking oscillator
SABOC
sage bomarc
SABRAC
sabra computer
SABRE
sage battery routing equipment
sales and business reservations done electronically
secure airborne radar bombing equipment
secure airborne radar equipment
self aligning boost and reentry

SABS
south african bureau of standards
SABU
sage back-up
semi automatic back-up
SAC
scientific advisory committee
semi automatic coding
service area computer
single address code
society for analytical chemistry
storage access channel
storage access counter
storage address counter
store access control
store and clear
strategic air command
synchronous astro compass
SACCOMNET
strategic air command communications network
SACCS
strategic air command communications system
strategic air command control system
SACDIN
stategic air command digital information
SACDU
switch and cable distribution unit
SACE
shore based acceptance checkout equipment
SACM
strategic air command manual
strategic air command missiles
SACMAPS
selective automatic computational matching and positioning system
SACNET
secure automatic communications network
SACO
select address and contract operate
SACOM
senior aircraft communicator
ship's advanced communications
SAD
safety assurance diagrams
silicon alloy diffused
single administration document
situation attention display
special adapter device
sympathetic aerial detonation
SADA
seismic array data analyzer
SADAP
simplified automatic data plotter

SADAS
  sperry airborne data acquisition system
SADC
  sequential analog digital computer
  sequential analog to digital computer
SADE
  superheat advanced demonstration experiment
SADGE
  sage data generator
SADIC
  solid state analog digital computer
  solid state analog to digital computer
SADIE
  scanning analog digital input equipment
  scanning analog to digital input equipment
  secure automatic data information exchange
  semi automatic decentralized intercept environment
  sterling and decimal invoicing electronically
SADOPS
  ships angle tracking and doppler system
SADR
  six hundred megacycle air defense radar
SADSAC
  sample data simulator and computer
  sampled data simulator and computer
  seiler algol digitally simulated analog computer
SADT
  structural analysis and design technique
  surface alloy diffused base transistor
SADTAC
  selective automatic decade turnover, absolute control
SAE
  self aligned emitter
  semiconductor assembly equipment
  shaft angle encoder
  society of automotive engineers
  standard of automotive engineers
SAEH
  society for automation in english and the humanities
SAES
  sputter auger electron spectroscopy
SAESIP
  saturn apollo electrical systems integration panel
SAF
  sound and flash
  spacecraft assembly facility
  strategic air force
SAFCS
  steam and feedwater rupture control system
SAFCUTTER
  side and face milling cutter
SAFE
  security audit and field evaluation
SAFEA
  space and flight equipment association
SAFER
  special aviation fire and explosion reduction
SAFI
  semiautomatic flight inspection
SAFMS
  secretary of the air force for missile and satellite systems
SAFO
  safe altitude fuzing option
SAFOC
  semiautomatic flight operations center
SAFRD
  secretary of the air force for research and development
SAG
  self-aligned gate
  senior advisory group
  standard address generator
  system application group
SAGA
  studies analysis and gaming agency
  system for automatic generation and analysis
SAGCI
  semi-automatic ground control of interceptors
SAGE
  semi-automatic ground environment
SAGMOS
  self-aligning gate metal oxide semiconductor
  self-aligning gate mos
SAGP
  saturable absorber giant pulsing
SAGW
  surface to air guided weapon
SAH
  sample and hold
  semiactive homer
  semiactive homing
SAHARA
  synthetic aperture high altitude radar
SAHF
  semiautomatic height finder
  semiautomatic height finding
SAHYB
  simulation of analog and hybrid computers
SAI
  storage and inspection
  sudden auroral intensity
  surveillance and inspection

SAIC
switch action interrupt count
SAID
safety analysis input data
semi automatic integrated documentation
space and information systems division
speech autoinstruction device
SAIEE
south africa institute of electrical engineers
SAIFER
safe arm initiation from electromagnetic radiation
SAILS
simplified aircraft instrument landing system
SAIM
semiautomatic inserting machine
system analysis and integration model
SAIMS
selected acquisitions information management system
SAINT
satellite inspection technique
satellite inspector
satellite intercept
satellite interceptor
SAIP
submarine antenna improvement program
SAIS
south african interplanetary society
SAL
savings and loan
south american program library
supersonic aerophysics laboratory
supply and logistics
symbolic assembly language
systems and logistics
systems assembly language
SALDRI
semiautomatic low data rate input
SALE
safeguards analytical laboratory evaluation
simple algebraic language for engineers
society for airline meteorologists
SALON
satellite balloon
SALOON
satellite launched from a balloon
SALS
solid state acoustoelectric light scanner
SALT
signal and homing light
stand alone terminal
strategic arms limitation talks
symbolic algebraic language translator

SALUT
sea, air, land, underwater target
SAM
safety activation monitor
scanning auger microprobe
school of aerospace medicine
selective access memory
selective automonitoring
selective automonitoring tracing routine
selective automatic monitoring
self addressing memory
semantic analyzing machine
semiautomatic mathematics
semiconductor active memory
semiconductor advanced memory
sequence and monitor
sequencer and monitor
sequential access memory
sequential access method
serial access memory
simulation of analog methods
society for advancement of management
sort and merge
space age microcircuits
substitute alloy material
surface to air missile
symbolic and algebraic manipulation
synchronous amplitude modulation
system activity monitor
system analysis machine
systems analysis module
SAMA
scientific apparatus makers' association
SAMD
surface to air missile development
SAME
society of american military engineers
SAMI
socially acceptable monitoring instrument
SAMIS
structural analysis and matrix interpretive system
SAMMIE
scheduling analysis model for mission integrated experiments
SAMOS
satellite and missile observation system
silicon aluminium metal oxide semiconductor
stacked gate avalanche injection type mos
surveillance and missile observation system
SAMP
sample
sampling

SAMPE
  society for the advancement of material and process engineering
  society of aerospace material and process engineers
SAMS
  satellite automatic monitoring system
  satellite automonitor system
SAMSO
  space and missile systems office
  space and missile systems organization
SAMSON
  system analysis of manned space operations
SAMTEC
  space and missile test center
SAMUX
  serial addressable multiplexer
SAN
  strong acid number
  styrene acrylo nitrile resin
SANDS
  speech and simplex
SANDW
  sandwiched
SANFM
  source range neutron flux monitor
SANOVA
  simultaneous analysis of variance
SANZ
  standards association of new zealand
SAO
  select address and operate
  smithsonian astrophysical observatory
SAOC
  space and aeronautics orientation class
SAOS
  select address and provide output signal
SAP
  semiarmour piercing
  share assembly program
  single axis platform
  sintered aluminium powder
  society for applied spectroscopy
  symbolic address program
  symbolic assembly program
  systems assurance program
SAPC
  suspended, acoustical plaster ceiling
SAPCH
  semiautomatic program checkout
SAPCHE
  semiautomatic program checkout equipment
SAPIR
  system for the automatic processing and indexing of reports
  system of automatic processing and indexing of reports
SAPOAD
  systems applications project operation action detail
SAR
  safety analysis report
  save address register
  search and rescue
  solar aircraft company
  special apparatus rack
  stack address register
  starting air receiver
  storage address register
  successive approximation register
  synthetic aperture radar
  system analysis report
SARA
  sequential automatic recorder and annunciator
  student admission records administration
SARAC
  steereable array for radar and communications
SARAH
  search and rescue and homing
  semiautomatic range, azimuth, and height
SARBE
  search and rescue beacon
SARCAP
  search and rescue combat air patrol
SARCC
  search and rescue control center
SARD
  simulated aircraft radar data
SAREF
  safety research experiment facilities
SARI
  small airport runway indicator
SARM
  set asynchronous response mode
SARO
  special applications routine
SARP
  scrample and recovery procedure
SARPS
  standards and recommended practices
SARS
  single axis reference system
SARUC
  southeastern association of regulatory utility commissioners
SARUS
  search and rescue using satellites

SAS
- secondary alarm station
- secondary alerting system
- security agency study
- shift accumulator left, including sign
- small astronomy satellite
- society for applied spectroscopy
- stability augmentation system
- staff activity system
- statistical analysis system
- storage address switch
- super accuracy simplex
- surface active substances

SASC
- semiautomated stock control

SASCS
- scintillation and semiconductor counter symposium

SASI
- south african standards institution

SASR
- subaqueous sound ranging

SASRDI
- subaqueous sound ranging development installation

SASS
- sage atabe simulation system
- special army signal service
- strategic alert sound system

SASSC
- senate aeronautical and space sciences committee

SASSE
- synchronous altitude spin stabilized experiment

SASTRO
- sage strobe training operator

SASTU
- signal amplitude sampler and totalizing unit

SAT
- satellite
- saturation
- society of acoustic technology
- solar atmospheric tide
- south african time
- stabilization assurance test
- stepped atomic time
- subscriber access terminal
- surface alloy transistor
- system acceptance test
- system access technique
- system analysis table

SATAN
- satellite automatic tracking antenna
- self contained automatic tactical air navigation
- sensor for airborne terrain analysis

SATANAS
- semiautomatic analog setting

SATAR
- satellite for aerospace research

SATC
- suspended, acoustical tile ceiling

SATCC
- southern air traffic control center

SATCO
- semiautomatic air-traffic control
- signal automatic air-traffic control

SATCOM
- satellite communication
- satellite communication agency
- satellite communication center
- scientific and technical communication committee

SATD
- saturated

SATE
- semiautomatic test equipment

SATEL
- satellite

SATELLAB
- satellite laboratory

SATF
- shortest access time first

SATIC
- scientific and technical information center

SATIF
- scientific and technical information facility

SATIN
- sage air-traffic integration
- satellite inspection

SATIRE
- semiautomatic technical information retrieval

SATL
- satellite

SATO
- self-aligned thick oxide

SATOBS
- satellite observation

SATP
- single aircraft tracking program

SATPATT
- satellite paper tape transfer

SATRAC
- satellite automatic terminal rendezvous and coupling

SATRACK
- satellite tracking

SATSA
   signal aviation test and support activity
SATT
   semiautomatic transistor tester
   strowger automatic toll ticketing
SATUR
   saturate
SAU
   search attack unit
SAUG
   swiss apl user group
SAV
   space air vehicle
SAVA
   society for accelerator and velocity apparatus
SAVDAT
   save data
SAVE
   society of american value engineers
   system analysis of vulnerability and effectiveness
   system for automatic value exchange
SAVITAR
   sanders associates video input output terminal access resource
SAVOR
   signal actuated voice recorder
SAVS
   safeguards for area ventilation system
   status and verification system
SAW
   signal air warning
   software analysis workstation
   surface acoustic wave
SAWDEVICE
   surface acoustic wave device
SAWE
   society of aeronautical weight engineers
   society of allied weight engineers
SAWG
   schedule and allocations working group
SAWP
   society of american wood preservers
SAYE
   save as you earn
SAZO
   seeker azimuth orientation
SB
   secondary battery
   selection board
   sense byte
   serial binary
   serial block
   service bureau
   sideband
   simultaneous brodcasting
   single braid
   sleeve bearing
   slow blowing
   slow burning
   space booster
   space branch
   stabilized breakdown
   standby
   stilb
   straight binary
   substitute blank
   sustained breakdown
   switchboard
   synchronization bit
SBA
   small business administration
   standard beam approach
   system for business automation
SBAC
   society of british aerospace companies
SBAEDS
   satellite based atomic energy detection system
SBAFWP
   standby aux feed water pump
SBASI
   single bridge apollo standard initiator
SBB
   silicon borne bond
SBC
   service bureau cooperation
   service bureau corporation
   single board computer
   small bayonet cap
   small business computer
   standard buried collector
SBCC
   separate bias, common control
SBCIC
   standard buried collector integrated circuit
SBCODE
   straight binary code
SBCT
   schottky barrier collector transistor
SBD
   schottky barrier diode
   service bureau division
SBDET
   switchboard detachment
SBDO
   space business development operation
SBDT
   schottky barrier diode transistor
   surface barrier diffused transistor

SBDTTL
  schottky barrier diode transistor transistor logic
SBE
  simple boolean expression
  society of broadcast engineers
  sub bit encoder
SBF
  short backfire antenna
SBFET
  schottky barrier gate fet
SBFM
  silver band frequency modulation
SBFU
  standby filter unit
SBGTS
  standby gas treatment system
SBH
  switch busy hour
SBI
  satellite borne instrumentation
SBIS
  satellite based interceptor system
SBK
  single beam klystron
SBLC
  standby liquid control
SBM
  subtract magnitude
  system balance measure
SBN
  strong base number
SBO
  sidebands only
SBOPERDET
  switchboard operation detachment
SBOR
  successive block overrelaxation
SBOS
  silicon borne oxygen system
SBP
  simulated bomarc program
  society of biological psychiatry
  special boiling point
SBR
  storage buffer register
  styrene butadiene rubber
SBRC
  single braided rubber covered
SBS
  satellite business system
  silicon bidirectional switch
  small business system
  stimulated brillouin scattering
  straight binary second

SBSC
  separate bias single control
SBSTA
  sound bearing station
SBT
  safe break terminator
  sheet, bar, tubing
  six bit transcode
  surface barrier transistor
SBTM
  s band telemetry modification
SBTTL
  schottky barrier transistor transistor logic
SBUE
  switch back up entry
SBUV
  solar backscatter ultraviolet experiment
SBV
  shield building vent
SBVS
  shield building vent system
SBW
  space bandwidth product
SC
  satellite carriers
  saturable core
  scanner
  screwed
  search control
  secondary containment
  sectional center
  selectional center
  selector channel
  self check
  semiconductor
  sequence counter
  shaped charge
  shift control
  short circuit
  short-circuited
  shunt capacitor
  side contact
  signal conditioner
  signal conditioning
  signal corps
  silk covered
  silvered copper
  simulation council
  sine cosine
  single contact
  single current
  situation console
  slave clock
  slow curing
  solar cell

solar coil
solar constant
south carolina
space capsule
spacecraft
spacecraft communicator
speach comminication
specification control
speech communication
speed control
squirrel cage
stabilization and control
standard channel
standard conditions
statistical control
steel cored
step counter
storage circuit
stromberg carlson
subcarrier
subchannel
subcommittee
super calandered
super computer
super current
superimposed current
supervisor call
supervisory control
suppressed carrier
switched capacitor
switching cell
symbolic code
synchroncyclotron
system controller

SCA
secondary communications authorization
selectively clear accumalator
sequence chart analyzer
sequence control area
servo corporation of america
short circuit ampere
simulated core assembly
sneak circuit analysis
spacecraft adapter
subsidiary channel authorization
subsidiary communications allocation
subsidiary communications authorization
synchronous communication adapter
system control adapter

SCAD
subsonic cruise armed decoy

SCADA
supervisory control and data
supervisory control and data acquisition

SCADAR
scatter detection and ranging

SCADS
scanning celestial attitude determination system
simulation of combined analog digital systems

SCAL
steel cored aluminium

SCALAMP
scale and lamp

SCALC
steel cored aluminium conductor

SCALE
space checkout and launch equipment

SCALO
scanning local oscillator

SCALP
small card automated layout program

SCAM
subcarrier amplitude modulation

SCAMA, SCAMA
switching conference and monitoring arrangement

SCAMP
signal conditioning amplifier
single channel amplitude monopulse processing
standard configuration and modification program

SCAMPS
small computer analytical and mathematical programming system

SCAN
scanning
selected current aerospace notices
self-containing automatic navigation
self-correcting automatic navigation
self-correcting automatic navigator
stock market computer answering network
switched circuit automatic network

SCAND
single crystal automatic neutron diffractometer

SCANIT
scan only intelligent terminal

SCANS
scheduling and control by automated network system
scheduling control and automation by network systems
systems checkout automatic network simulator

SCAP
  silent compact auxiliary power
  star computer assembly program
SCAPE
  self-contained atmospheric protective ensemble
SCAR
  satellite capture and retrieval
  scientific committee on antartic research
  subcaliber aircraft rocket
  submarin celestial altitude recorder
  submerged celestial altitude recorder
SCARF
  santa cruz acoustic range facility
  side locking coherent all range focused
  strategic cislunar advanced retaliatory force
SCAS
  spacecraft adapter simulator
SCASP
  sequence of coverage and speed
SCASS
  senate committee on aeronautics and space sciences
SCAT
  security control of air trafffic
  share compiler assembler translator
  space communication and tracking
  speed command of attitude and thrust
  speed control approach take off
  supersonic commercial air transport
  surface controlled avalanche transistor
  system configuration acceptance test
SCATE
  stromberg carlson automatic test equipment
SCATER
  security control of air traffic and electromagnetic radiation
SCATHA
  spacecraft charging at high altitudes
SCATS
  sequentially controlled automatic transmitter start
  simulation, checkout, and training system
SCATSD
  signal corps aviation test and support detachment
SCB
  shallow cathode barrier
  supervisory circuit breaker
SCBD
  scan conversion and bright display
  signal corps base depot
SCBR
  steam cooled breeder reactor

SCC
  sage control center
  satellite control center
  scientific control corporation
  secondary containment cooling
  security control center
  sequence control chart
  sequence controlled calculator
  simulation control center
  single conductor cable
  single cotton covered
  slice control central
  small center contact
  society of cosmetic chemists
  specialized common carrier
  standards coordinating committee
  standards council of canada
  status change character
  storage connecting circuit
  supply control center
SCCC
  space surveillance command and control center
SCCD
  surface channel charge coupled device
  surface charge coupled device
SCCHLL
  standards coordinating committee on high level language
SCCO
  steel cored copper
SCCOC
  steel cored copper conductor
SCCS
  sodium chemistry control system
  software controlled communication service
  source code control system
  straight cut control system
SCD
  satellite control department
  scheduled
  scintillation detector
  screwed
  semiconductor device
  source control drawings
  space control document
  specification control drawing
  stored charge diode
  subcarrier discriminator
SCDA
  small card design automation
SCDC
  single commutation direct current
  single commutation direct current signaling
  source code and data collection

SCDP
society of certified data processors
SCDSB
suppressed carrier double sideband
SCDSBTRANSMISSION
suppressed carrier double sideband transmission
SCE
saturated calomel electrode
selection control element
service checkout equipment
short channel effect
signal conditioning equipment
single cotton covering enamel insulated wire
single cotton enamelled
single cycle execute
situation caused error
society of cuban engineers north eastern regional division
solder circuit etch
stage calibration equipment
station cable equalizer
SCEA
side cutting edge angle
signal conditioning electronic assembly
SCEAR
scientific committee on the effects of atomic radiation
SCEC
south central electric companies
SCED
schedule
SCEI
serial carry enable input
SCEL
signal corps engineering laboratory
small components evaluation loop
SCEO
satellite control engineering office
SCEPTRE
system for circuit evaluation and prediction of transient radiation effects
SCERT
system and computer evaluation and review technique
SCETV
south carolina educational television
SCF
satellite control facility
self consistent field
sequence compatibility firing
snap critical facility
sodium cleaning facility
standard cascade form
standard cubic foot
SCFBR
steam cooled fast breeder reactor
SCFH
standard cubic foot per hour
SCFM
standard cubic foot per minute
subcarrier frequency modulation
SCFS
subcarrier frequency shift
SCG
specification control group
SCGA
sodium cooled graphite assembly
SCGD
specification control group directive
SCGDL
signal corps general development laboratory
SCGP
self-contained guidance package
SCGRL
signal corps general research laboratory
SCGSA
signal corps ground signal agency
SCGSS
signal corps ground signal service
SCH
store channel
supporting checkout
SCHED
schedule
SCHG
supercharge
supercharged
supercharger
SCHO
standard controlled heterodyne oscillator
SCHS
small component handling system
SCI
scientific computers, incorporated
ship controlled intercept
ship controlled interception
simulation councils, incorporated
society of chemical industry
society of computer intelligence
soft cast iron
supervisor call instruction
SCIC
semiconductor integrated circuit
SCIM
speech communication index meter
SCINT
scintillator

SCIP
   scanning for information parameter
   self-contained instrument package
SCIPP
   silicon computing instrument, patch programmed
SCIS
   safety containment isolation system
SCISD
   signal corps intermediate supply depot
SCK
   set clock
SCL
   scale
   selectivity cross linked
   space charge limited
   standard classification list
   symbolic correction loader
   symmetric clipper
   system control language
SCLC
   space charge limited current
SCLD
   space charge limited diode
SCLWR
   scientific computing laboratory work request
SCM
   scratch pad memory
   segment control module
   service command module
   signal conditioning module
   simulated command module
   simulated core mock-up
   simulated core model
   sine cosine multiplier
   small capacity memory
   small core memory
   smith corona marchant
   software configuration management
   standard cubic meter
   superconducting magnet
   symmetrically cyclically magnetized
SCMA
   southern california meter association
   systems communications management association
SCMP
   software configuration management plan
SCMTVS
   signal corps mobile television system
SCN
   satellite control network
   scan
   sensitive command network
   specification change notice

SCNA
   sudden cosmic noise absorption
SCNS
   self-contained navigation system
SCO
   signal company
   spacecraft observer
   staff communication office
   subcarrier oscillator
SCOCOM
   scope communications
SCODA
   scan coherent doppler attachment
SCOMO
   satellite collection of meteorological observations
SCOOP
   scientific computation of optimal programs
   scientific computation of optimum procurement
SCOPE
   schedule cost performance
   sequential customer order processing, electronically
   special committee on paperless entries
   stromberg carlson operations panel electrical
   subsystem for the control of operations and plan evaluation
   supervisor control of program execution
   system to coordinate the operation of peripherial equipment
SCOPLT
   scope plot
SCOPT
   subcommittee on programming terminology
SCOR
   scientific committee on oceanographic research
   self-calibration omnirange
SCORE
   satellite computer operated readiness equipment
   selection copy and reporting
   signal communications by orbital relay equipment
   signal communications by orbiting relay equipment
   systematic control of range effectiveness
SCOSS
   senior chief officer, shore signal service
SCOST
   special committee on space technology
SCOSWS
   senior chief officer, shore wireless service

SCOT
  stand-by compatible one tape
SCOTICE
  scotland iceland
SCOTTR
  super critical, once thru, tube reactor experiment
SCP
  sage change proposal
  sage computer program
  sage computer project
  semiconductor products
  spherical candlepower
  support control program
  symbolic conversion program
  system communication pamphlet
  system control panel
  system control processor
  system control program
SCPA
  stabilization and control system control panel
SCPC
  single channel per carrier
SCPT
  sage computer programming training
SCR
  scanning control register
  scruple
  selective chopper radiometer
  semiconductor controlled rectifier
  series control relay
  short circuit ratio
  signal conditioner
  signal corps radio
  silicon controlled rectifier
  single card reader
  sodium cooled reactor
  solar corpuscular radiation
  space charge recombination
  strip chart recorder
  styrene chloroprene rubber
SCRA
  standford center for radar astronomy
SCRAM
  safety control rod axe man
  several compilers reworked and modified
  space capsule regulator and monitor
SCRAP
  super caliber rocket assisted projectile
SCRID
  silicon controlled rectifier indicator driver
SCRIPT
  scientific and commercial interpreter and program translator
  scientific and commercial subroutine interpreter and program translator
SCRL
  signal corps radio laboratory
SCRN
  screen
SCROPT
  scientific and commercial subroutine interpreter and program translator
SCRP
  small card release processing
SCRR
  supercircular reentry research
SCS
  scientific control system
  secret control station
  section control station
  security control system
  sequence coding system
  short circuit stable
  signal communication system
  silicon control switch
  silicon controlled switch
  simulation control subsystem
  simultaneous color system
  single change of station
  single channel simplex
  society for computer simulation
  sodium characterization system
  soil conservation service
  southern computer service
  space cabin simulator
  space command station
  space communication system
  spacecraft control system
  special computer service
  stabilization and control system
  standard coordinate system
SCSCP
  system coordination for sage computer programming
SCSE
  south carolina society of engineers
SCSG
  sage computer support group
SCSI
  small computer systems interface
SCSL
  scientific continuous simulation language
SCSM
  spacecraft systems monitor
SCSPLS
  south caroline society of professional land surveyors

SCT
  scanning telescope
  signal corps training
  subroutine call table
  surface charge transistor
  surface controlled transistor
  system component test
  system configuration table
  systems and computer technology
SCTC
  signal corps training center
  spacecraft test conductor
SCTCA
  strategic air command channel and traffic control agency
SCTD
  scattered
SCTE
  society of cable television engineers
  society of carbide and tool engineers
SCTF
  santa cruz test facility
  sodium chemical technology facility
SCTL
  schottky coupled transistor logic
  schottky transistor logic
  short-circuited terminating line
  short-circuited transmission line
  small components test loop
SCTP
  straight channel tape print
SCTPP
  straight channel tape print program
SCTR
  sector
  standing conference on telecommunications research
SCTY
  security
SCU
  s-band cassegrain ultra
  scan control unit
  scanner control unit
  secondary control unit
  sequencing and control unit
  signal conditioning unit
  static checkout unit
  statistical control unit
  subscriber channel unit
  suit cooling unit
  system control unit
SCUBA
  self-contained underwater breathing apparatus
SCUDS
  simplification, clarification, unification, decimalization, standardization
SCUP
  school computer use plan
SCUTG
  signal corps unit training group
SCV
  subclutter visibility
SCVE
  spacecraft vicinity equipment
SCW
  silk covered wire
  slow cyclotron wave
  subchannel control word
SCWS
  space combat weapon system
SD
  sample delay
  scan data
  schottky diode
  seasonal derating
  section definition
  selenium diode
  selenoid driver
  self dual
  semiconductor devices
  sequential disk
  signal digit
  signal to distortion
  signed digit
  simplex drop out
  single density
  situation display
  slow down
  sof drawn
  solenoid driver
  south dakota
  spaced doublet
  spectral distribution
  speech plus duplex
  standard deviation
  standardization directory
  standards development
  stereo directional
  sweep driver
  system director
SDA
  satellite data area
  shaft drive axis
  shut down amplifier
  source data acquisition
  source data automation
  supplier data approval
  supporting data analysis

symbols digits alphabetics
systems dynamic analyzer
SDAD
satellite digital and analog display
SDAP
systems development analysis program
SDAS
scientific data automation system
SDAT
spacecraft data analysis team
SDB
standard device byte
SDC
sage direction center
sage division commander
self-destruct circuit
semiconductor devices council of jedec
setpoint digital control
shut-down cooling
signal data converter
situation display converter
space data corporation
stabilization data computer
station directory control
structural design criteria
system development corporation
SDCC
san diego computer center
small diameter component cask
SDCE
society of die casting engineers
SDCS
space borne data conditioning system
station digital command system
SDD
software design description
system design description
SDDRL
self loading disk dump and reload
SDDS
single dimensional deflection system
SDDU
simplex data distribution unit
SDE
society of data educators
state difference equation
students for data education
SDF
seasonal derating factor
single degree of freedom
spectral density function
static direction finder
supergroup distributing frame
SDFC
space disturbance forecast center

SDFD
system data flow diagram
SDFL
schottky diode fet logic
SDG
simulated data generation
simulated data generator
situation display generator
SDGA
single conductor, degaussing, armored
SDGE
situation display generator element
SDH
seasonal derated hours
SDHE
spacecraft data handling equipment
SDI
selective dissemination of information
source data information
system diagram index
SDL
system descriptive language
systems design laboratory
SDLC
single level data link control
synchronous data link control
SDLE
transparent dle
SDM
sequency division multiplexing
shut-down mode
situation display matrix
software development methodology
space division multiplexing
standardization design memoranda
statistical delta modulation
system data module
system decision manager
SDMA
space division multiple access
SDMD
sequential decision making device
SDO
scan data out
signal distributing office
software distribution operation
SDOP
ship doppler
SDOS
source data operating system
SDP
selective data processing
signal data processor
single dry plate
site data processor

standards development program
station directory program
SDPL
  servomechanism and data processing laboratory
SDR
  self-decoding read-out
  sender
  significant deficiency report
  single drift region
  snap developmental reactor
  statistical data recorder
  storage data register
  system design review
  system development requirement
SDRNG
  sound ranging
SDRP
  simulated data reduction program
SDS
  safety data sheet
  sales distribution system
  scientific data system
  signals dispatch service
  significant digit scanner
  simulation data subsystem
  software development specifications
  system data synthesizer
  systems and data services
SDSI
  scientific data systems israel
SDSS
  self deploying space station
SDT
  simulated data tape
  space detection and tracking
  start data traffic
  step down transformer
SDTDL
  saturated drift transistor diode logic
SDTK
  supported drift tube klystron
SDTL
  schottky clamped dtl
  schottky diode transistor logic
SDTP
  startover data transfer and processing
SDU
  signal distribution unit
  spectrum display unit
  subcarrier delay unit
SDV
  scram discharge volume
  slowed down video
  start device

SDW
  segment descriptor word
  standing detonation wave
SE
  safety evaluation
  secondary electron
  secondary electron multiplier
  secondary emission
  self-extinguishing
  service entrance
  service equipment
  shielding effectiveness
  sign extended
  single end
  single-ended
  single-engined
  single entry
  software engineer
  sounding equipment
  space equivalent
  space exploration
  starter electrode
  sterling electronics
  storage element
  subcritical experiment
  superior electric
  switching element
  system engineer
  system engineering
  system equalizer
SEA
  science and education administration
  society of electronics and automation
  standard electronic assembly
  sudden enhancement of atmospherics
  systems effectiveness analyzer
SEAC
  single-engined aircraft
  south-east architects collaboration
  standard eastern automatic computer
  standard electronic automatic computer
  structural engineers association of colorado
SEACOM
  southeast asia extension of the commonwealth cable
SEADAC
  seakeeping data analysis center
SEAIMP
  solar eclipse atmospheric and ionospheric measurements project
SEAL
  standard electronic accounting language
SEALS
  severe environmental air launch study
  stored energy actuated lift system

SEAM
  subset extraction and association measurement
  symposium on the engineering aspects of magnetohydrodynamics
SEAO
  structural engineers association of oregon
SEAOC
  structural engineers association of california
SEAOCC
  structural engineers association of central california
SEAONC
  structural engineers association of northern california
SEAOSD
  structural engineers association of san diego
SEASC
  structural engineers association of southern california
SEASCO
  southeast asia science cooperation office
SEB
  source evaluation board
  southern electricity board
SEBECC
  scanning electron beam excited charge collection
SEBIC
  sustained electron bombardment induced conductivity
SEC
  sacramento engineers club
  safeguards equipment cabinet
  sanitary engineering center
  scientific and engineering computation
  secant
  second
  secondary
  secondary electron conduction
  secondary emission conductivity
  securities and exchange commission
  simple electronic computer
  southern electronics corporation
SECAM
  sequential color and memory
SECAP
  system experience correlation and analysis program
SECAR
  secondary radar
SECD
  secondary
SECDED
  single error correction double error detection
SECFT
  second foot
SECIMP
  secondary impedance
SECL
  sequential emitter coupled logic
  symmetrical emitter coupled logic
  symmetrically operated emitter coupled logic
SECO
  self-regulating error correcting coder decoder
  sequential coding
  sequential control
  sequential encoder decoder
SECOM
  system engineering communication
SECOMACS
  sector commander, alascan communications system
SECOR
  sequential collation of range
  sequential correlation of range
SECORD
  secure voice cord board
SECPS
  secondary propulsion system
SECS
  sequential events control system
  space environmental control system
SECT
  section
  skin electric tracing
SECTAM
  southern conference on theoretical and applied mechanics
SECV
  state electricity commission of victoria
SECWND
  secondary winding
SED
  sanitary engineering division
  signal equipment depot
  society for electrical development
  space engineering document
  space environment division
  special expanded display
  spectral energy distribution
  suppressed electrical discharge
SEDD
  systems evaluation and development division
SEDGE
  sage experimental display generator
  special experimental display generation
SEDPC
  scientific and engineering data processing center

SEDR
  systems engineering department report
SEDS
  society for educational data systems
  space electronic detection system
SEE
  sage evaluation exercise
  senior electronic engineer
  signal experimental establishment
  society of environmental engineers
  society of explosives engineers
  southeastern electric exchange
  space environmental experiment
SEEB
  southeastern electricity board
SEEK
  systems evaluation and exchange of knowledge
SEEL
  singapore electronic and engineering, limited
SEENY
  society of engineers of eastern new york
SEER
  steam electric evaluating and recording
  systems engineering, evaluation and research
SEF
  shielding effectiveness factor
  shock excited filter
  simple environment factor
  single equivalent formant
  space education foundation
  space environmental facility
SEFAR
  sonic end fire for azimuth and range
SEFR
  shielding experiment facility reactor
SEG
  society of economic geologists
  society of exploration geophysics
  solartron electronic group
  standardization evaluation group
  system engineering group
SEGMOS
  service goods movement system
SEGS
  solar energy generating system
SEGSYS
  segmentation system
SEI
  safety equipment institute
  salford electrical instruments
  systems engineering and integration
SEIA
  solar energy industries assn

SEIC
  solar energy information center
SEIP
  strategy for exploration of the inner planets
  system engineering implementation plan
SEIT
  satellite educational and information television
SEL
  select
  selection
  selective
  selector
  selector channel
  signal engineering laboratories
  standard electric lorenz
  stanford electronics laboratory
  system engineering laboratories
SELCAL
  selective calling
SELECT
  selectivity
SELFCL
  self-closing
SELN
  selection
SELR
  saturn engineering liaison request
  selector
SELRECT
  selenium rectifier
SELRIP
  selected release improvement program
SELS
  self-synchronizing
  self-synchronous
SELSYN
  self-synchronizing
  self-synchronous
SELT
  sage evaluation library tape
SELV
  safety extra low voltage
SEM
  satellite to earth missile
  scanning electron micrograph
  scanning electron microscope
  secondary emission material
  semiconductor electronic memories
  singularity expansion method
  standard estimating module
  system effectiveness measure
  system engineering management

SEMI
   semiconductor equipment and materials institute
SEMIAUT
   semiautomatic
SEMICON
   semiconductor
SEMIRAD
   secondary electron mixed radiation dosimeter
SEMLAM
   semiconductor laser amplifier
SEMLAT
   semiconductor laser array technique
SEMM
   scanning electron mirror microscope
SEMS
   severe environment memory system
SEN
   sense
   steam emulsion number
SENL
   standard equipment nomenclature list
SENS
   sensitivity
SENTOS
   sentinel operating system
SEOP
   secondary operand unit
SEOS
   strategic earth orbit system
   synchronous earth observation satellite
SEP
   separate
   separation
   separation control character
   society of engineering psychologists
   space electronic package
   sperical error probable
   standard electronic package
   star epitaxial planar
   symbolic equations program
   systematic evaluation program
SEPA
   southeastern power administration
SEPOL
   settlement problem oriented language
   soil engineering problem oriented language
SEPS
   secondary electric power system
   service environment power system
   service module electrical power system
   severe environment power system
   sevice module electrical power system
SEPT
   silicon epitaxial planar transistor

SEQ
   sequence
   sequencer
   sequencing
SEQL
   sequential
SEQUIN
   sequential quadrature inband system
SER
   safety evaluation report
   sandia engineering reactor
   serial
   significant event report
   snap experimental reactor
   system environment recording
SERAPE
   simulator equipment requirements for accelerating procedural evolution
SERB
   study of enhanced radiation belt
SERC
   southeastern electric reliability council
SEREP
   system environment recording and edit program
SERF
   sandia engineering reactor facility
SERI
   solar energy research institute
SERL
   service electronics research laboratory
SERME
   sign error root modulus error
SERPS
   service propulsion system
SERT
   single electron rise time
   society of electronic and radio technicians
   space electrical rocket test
   space electronic rocket test
SERVO
   servomechanism
SES
   shorted emitter switch
   small edison screw
   society of engineering science
   solar energy society
   space environment simulator
   standards engineering society
   standards engineers society
   strategic engineering survey
   sylvania electronic systems
SESA
   signal equipment support agency

society for environmental stress analysis
society for experimental stress analysis
SESAME
service sort and merge
SESCO
secure submarine communications
SESE
secure echo sounding equipment
SESIP
systems engineering summary of installation and program planning
SESL
self erecting space laboratory
space environment[al] simulation laboratory
SESLINK
sesam linked in data base handler
SESOME
service, sort and merge
SESUPP
safety evaluation supplement
SET
selective employment tax
self-extending translator
service evaluation telemetry
software engineering terminology
solar energy thermionic
space electronics and telemetry
special engineering test
system evaluation technique
SETA
simplified electronic tracking
simplified electronic tracking apparatus
SETAB
sets tabular material
SETAC
sector and tactical air navigation
SETAR
serial event timer and recorder
SETC
solid electrolyte tantalum capacitor
SETE
secretariat for electronic test equipment
SETF
snap experimental test facility
SETI
search for extraterrestrial intelligence
SETS
solar energy thermionic conversion system
SETTECHNIQUE
stepped electrode transistor technique
SEU
small end-up
source entry utility

SEURE
systems evaluation code under radiation environment
SEV
scout evolution vehicle
SEVA
system evaluation program
SEVAS
secure voice access system
SEW
sonar early warning
SF
safety
safety factor
sampled filter
scale factor
selective filter
service factor
shift forward
side frequency
signal frequency
significant figure
single feed
single frequency
standard frequency
store and forward
subframe
SFA
scientific film association
sequential functional analysis
single failure analysis
sun finder assembly
SFAR
sound fixing and ranging
special federal aviation regulation
system failure analysis report
SFAS
safety feature activation system
SFB
semiconductor functional block
solid state functional block
SFBAEC
san francisco bay area engineering council
SFC
sectored file controller
short form catalog
solar forecast
space flight center
specific fuel consumption
SFCS
secondary flow control system
SFD
sudden frequency deviation
system function description

SFDALGOL
   system function description algorithmic language
SFE
   secondary feedback element
   society of fire engineers
   solar flare effect
SFEA
   space and flight equipment association
SFEC
   standard facility equipment card
SFERT
   spinning satellite for electric rocket test
SFET
   surface fet
SFF
   silicone rubber insulated fixture wire, flexible stranding
   solar forecast facility
SFG
   screen format generator
SFI
   space flight instrumentation
SFID
   supplementary flight information documentation
SFIT
   swiss federal institute of technology
SFL
   sequenced flashing light
   slip full load
   substrate fed logic
   symbolic flowchart language
SFLIPFLOP
   sign flip flop
SFM
   signal flow matrix
   simulated flow method
   split field motor
   surface feet per minute
   swept frequency modulation
SFMR
   transformer
SFN
   strategic facsimile network
SFOC
   space flight operations center
SFPM
   surface feet per minute
SFTS
   standard frequency and time signal
SFU
   signals flying unit
SG
   safety guide
   sawtooth generator
   screen grid
   segment
   set gate
   signal generator
   single groove
   solar generator
   specific gravity
   standing group
   steam generator
   symbol generator
   synchronous generator
SGA
   self-gating and
   shipboard, general use, armored
SGC
   simulation generation control
   space general corporation
   sperry gyroscope company
SGCA
   spacecraft ground controlled approach
SGCF
   snap generalized critical facility
SGCS
   silicon gate controlled switch
SGD
   silicon grown diffused
SGDE
   system ground data equipment
SGDF
   supergroup distribution frame
SGE
   slow glass etch
SGF
   smoke generating fuel
SGFP
   steam generator feed pump
SGHWR
   steam generating heavy water reactor
SGIS
   steam generator isolation signal
SGITS
   spacecraft ground operational support system interface test system
SGJP
   satellite graphic job processor
SGL
   signal
   system generation language
SGLS
   space ground link system
   space to ground link subsystem
SGM
   spark gap modulation
   spark gap modulator

stationary gaussian markov
strict good middling
SGN
   scan gate number
SGO
   strict good ordinary
SGP
   subscriber group plant
SGT
   segment table
   silicon gate transistor
SGTR
   steam generator test rig
SGTS
   standby gas treatment
SGV
   screen grid voltage
SH
   sample and hold
   service hours
   shield
   shunt
   single heterostructure
   subharmonic
SHA
   sideral hour angle
   sodium hydroxide addition
   software houses association
   solid homogeneous assembly
SHACO
   shorthand coding
SHAOB
   strategic high altitude orbital bomber
SHAPE
   sage high altitude prototype environemt
SHARE
   society for help to avoid redundant efforts
   systems for heat and radiation energy
SHAS
   shared hospital accounting system
SHCIRCUIT
   sample and hold circuit
SHCON
   shore connection
SHCRT
   short circuit
SHD
   second harmonic distortion
SHE
   semihomogeneous experiment
   signal handling equipment
   standard hydrogen electrode
   substrate hot electron
SHED
   solar heat exchanger drive

SHEMP
   system hydraulic, electrical, mechanical, pneumatic
SHEP
   solar high energy particles
SHF
   sensible heat factor
   storage handling facility
   superhigh frequency
   supra high frequency
SHFA
   single conductor, heat and flame resitant, armored
SHFUNCTION
   source handshake function
SHG
   second harmonic generation
   self-holding gate
SHI
   sheet iron
SHIEF
   shared information elicitation facility
SHIELD
   sylvania high intelligence electronic defense
SHIP
   search height integration program
SHIRAN
   s band high accuracy ranging and navigation
   s band high presicion short range navigation
   s band of high precision short range electronic navigation
SHLD
   shield
SHM
   ship head marker
   ship heading marker
   simple harmonic motion
   society for hybrid microelectronics
SHO
   super high output
SHOAP
   symbolic horribly optimizing assembly program
SHODOP
   short-range doppler
SHOF
   shipboard cable, heat and oil resistant, flexible
SHORAN
   short-range aid to navigation
   short-range navigation
SHOT
   society for the history of technology

SHP
 shaft horsepower
 standard hardware program
SHPE
 society of hispanic professional engineers
SHPO
 subharmonic parametric oscillator
SHPS
 sodium hydroxide purge system
SHR
 semi homogeneous fuel reactor
 shift register
SHREWD
 system for holding and retrieving wanted data
SHRS
 supplementary heat removal system
SHS
 sheet steel
SHSMB
 safety and health standards management board
SHSS
 superhigh speed steel
SHT
 society for the history of technology
SHTC
 short time constant
SHTH
 sheath
SHTHG
 sheathing
SHTL
 small heat transfer loop
SHTONF
 short ton force
SHW
 short wave
SI
 safety injection
 safety inspection
 sample interval
 scientific instrumentation
 screen grid input
 seal in
 self-induction
 semiinsulating
 sense indicator
 shift in
 sign code
 signal to interference
 signal to intermodulation
 simulation routine
 smithsonian institute
 spark ignition
 special instruction
 specific impulse
 spectrum index
 speed indicator
 status indicator
 storage immediate
 switch interpretation
 system international
 systems integration
SIA
 semiconductor industry association
 service in informatics and analysis
 software industry association
 subminiature integrated antenna
 swiss society of engineers and architects
 system integration
 system integration area
SIAM
 signal information and monitoring
 society for industrial and applied mathematics
 system integrated access method
SIAP
 standard instrument approach
SIAPT
 siemens interactive apt
SIAS
 safety injection actuation signal
 single integrated attack team
SIB
 satellite ionospheric beacons
 snake in the box
SIBOL
 sweden integrated banking on-line
SIBS
 stellar inertial bombing system
SIC
 satellite information center
 science information council
 scientific information center
 security intelligence corps
 semicconductor integrated circuit
 semiconductor
 semiconductor integrated circuit
 silicon coated
 silicon integrated circuit
 social implications of computers
 special interest committee
 specific inductive capacity
 standard industrial classification
 survey information center
 systems integration contractor
SICBM
 small intercontinental ballistic missile
 super intercontinental ballistic missile

SICEJ
   society of instrument and control engineers of japan
SICO
   switched in for checkout
SICOM
   securities industry communications
SICS
   safety injection system
   semiconductor integrated circuits
SICSAM
   special interest committee on symbolic algebraic manipulation
SICSIC
   special interest committee on social impications of computing
SICSOC
   special interest committee on social science computing
SICT
   scientific inventory control technique
SID
   silicon imaging device
   situation display
   society for information display
   sodium ionization detector
   space and information systems division
   speech input device
   standard instrument departure
   sudden ionospheric disturbance
   syntax improving device
   system integrational diagnostic
SIDASE
   significant data selection
SIDEB
   sideband
SIDIODE
   shift-in diode
SIDS
   speech identification system
   stellar inertial doppler system
SIDT
   silicon integrated device technology
SIE
   science information exchange
   southwestern industrial electronics company
SIF
   selective identification feature
   sound intermediate frequency
   standard interface
SIFCS
   sideband intermediate frequency communications system
SIFR
   simulated instrument flight rules

SIFT
   share internal fortran translator
   simplified input for tiros operational satellite system
SIG
   signal
   signaling
   significance testing
   significant testing
   special interest group
SIGACT
   special interest group on automata and computability theory
SIGAGCY
   signal agency
SIGAIRDEFENGRAGCY
   signal air defense engineering agency
SIGARCH
   special interest group on computer architecture
SIGART
   special interest group on artifical intelligence
SIGAVNCO
   signal aviation company
SIGBDP
   special interest group on business data processing
SIGBIO
   special interest group on biomedical computing
SIGBN
   signal battalion
SIGC
   signal corps
SIGCAPH
   special interest group on computers and the physically handicapped
SIGCAS
   special interest group on computers and society
SIGCO
   signal company
SIGCOA
   signal company, airline
SIGCOC
   signal company, cable
SIGCOM
   signal communication
SIGCOMM
   special interest group on data communication
SIGCOMMAGCY
   signal communication agency
SIGCOMMSECAGCY
   signal communication security agency

SIGCONDNET
 signal conditioning network
SIGCONSBN
 signal construction battalion
SIGCOSIM
 special interest group on computer system installation management
SIGCOW
 signal company, wireless
SIGCOWG
 signal company, wing
SIGCOY
 signal company
SIGCP
 signal group
SIGCPR
 special interest group on computer personnel research
SIGCSE
 special interest group on computer science education
SIGCUE
 special interest group on computer use-in education
SIGDA
 special interest group on design automation
SIGDEP
 signal depot
SIGDEPCO
 signal depot company
SIGDIV
 signal divsion
SIGDOC
 special interest group on documentation
SIGENGRAGCY
 signal engineering agency
SIGEQUIP
 signal equipment
SIGFET
 silicon gate fet
SIGGEN
 signal generator
SIGGND
 signal ground
SIGGRAPH
 special interest group on computer graphics
SIGHVCONSTBN
 singal heavy construction battalion
SIGINT
 signal intelligence
SIGINTELAGCY
 signal intelligence agency
SIGIR
 special interest group on information retrieval
SIGL
 signal lieutenant
SIGLASH
 special interest group on language analysis and studies in the humanities
SIGMA
 shielded inert gas metal arc
SIGMAP
 special interest group on mathematical programming
SIGMICRO
 special interest group on microprogramming
SIGMINI
 special interest group on minicomputers
SIGMN
 signalman
SIGMOD
 special interest group on management of data
SIGMR
 signal master
SIGMSGCEN
 signal message center
SIGMSLSPTAGCY
 signal missile support agency
SIGN
 signal
SIGNO
 signal officer
SIGNUM
 special interest group on numerical mathematics
SIGO
 signal officer
SIGOFFR
 signal officer
SIGOPNBN
 signal operation battalion
SIGOPS
 special interest group on operating systems
SIGPC
 special interest group on personal computing
SIGPLAN
 special interest group on programming languages
SIGPROCOFC
 signal procurement office
SIGRADIOMAINTTEAMAVN
 signal radio maintenance team, aviation
SIGREPCO
 signal repair company
SIGRES
 signal corps reserve
SIGS
 signals

simplified inertial guidance system
stellar inertial guidance system
SIGSAM
special interest group on symbolic and algebraic manipulation
SIGSCH
signal school
SIGSEC
signal section
SIGSERVCO
signal service company
SIGSIM
special interest group on simulation
SIGSMALL
special interest group on small computing systems and applications
SIGSOC
special interest group on social and behavioral science computing
SIGSOFT
special interest group on software engineering
SIGSPTBN
signal support battalion
SIGSTR
signal strength
SIGSUPAGNCY
signal supply agency
SIGSUPBN
signal supply battalion
SIGSVCBN
signal service battalion
SIGTC
signal training center
SIGTNG
signal training
SIGTNGCEN
signal training center
SIGTNGDET
signal training detachment
SIGUCC
special interest group on university computing centers
SIIL
schottky integrated injection logic
SIM
simulated approach
simulation
simulator
simulator program
single rotation machine
symbolic integrated maintenance
SIMA
scientific instrument manufacturers' association

SIMAC
sonic instrument measurement and control
SIMAJ
scientific instrument manufacturers' association of japan
SIMANNE
simulation of analogical network
SIMBOL
simulated boolean oriented language
SIMCHE
simulation and checkout equipment
SIMCOM
simulator compiler
SIMCON
scientific inventory management and control
scientific inventory management control
simplified control
SIMD
single input multiple data stream
single instruction multiple data
single instruction stream multiple data stream
SIMDCOMPUTER
single instruction multiple data computer
SIMDIS
discrete simulation
SIMDS
single instruction multiple data stream
SIMDSYSTEM
single instruction flow multiple data flow system
SIMFAC
simulation facility
SIMICORE
simultaneous multiple image correlation
SIMILE
simulator of immediate memory in learning experiments
SIMM
symbolic integrated maintenance manual
SIMMR
simulation monitor recorder
SIMOS
stacked gate injection mos
SIMP
satellite interface message processor
simulation program
simulator program
specific impulse
SIMPAC
simplified programming for acquisition and control
simulation package
SIMPLER
system for information management and program logic for education and research

SIMPU
  simulation punch
SIMR
  simulator
SIMS
  secondary ion mass spectrometer
  secondary ion mass spectroscopy
  single item, multisource
  symbolic integrated maintenance system
SIMU
  simulation
  stellar inertial measuring unit
SIMULA
  simulation language
SIN
  security information network
  sensitive information network
  sine
  symbolic integrator
SINAD
  signal to noise ratio and distortion
SINB
  southern interstate nuclear board
SINS
  ship's inertial marine navigational system
  ship's inertial naviation system
  stellar inertial navigation system
  submarine inertial navigation system
SIO
  serial input and output
  staged in orbit
  start input output
SIOC
  serial input output channel
SIOP
  selector input output processors
  single integrated operations
  single integrated operations plan
SIOUX
  sequential iterative operation unit x
SIOV
  siemens metal oxide varistor
SIP
  safety injection pump
  sage improvement program
  satellite inspection program
  satellite interceptor program
  scheduling and control by automated
  network systems implementation plan
  self-interpreting program
  short irregular pulse
  simulated input processor
  single in line package
  single in line plastic
  solar instrument probe
  sonar instrumentation probe
  symbolic input program
  system initialize program
SIPB
  safety injection permissive blocks
SIPE
  system internal performance evaluator
SIPI
  scientist's institute for public information
SIPO
  serial in parallel out
SIPOP
  satellite information processor operational program
SIPROS
  simultaneous processing operating system
  simultaneous processing operation system
SIPS
  sac intelligence data processing system
  simulated input preparation system
SIPT
  simulating part
SIR
  selective information retrieval
  semantic information retrieval
  signal to interference ratio
  simultaneous impact rate
  speaker independent recognition
  statistical information retrieval
  submarine intermediate reactor
  symbolic input routine
SIRA
  safety investigation regulations
  scientific instrument research association
SIRC
  spares integrated recording and control
SIRE
  satellite infrared experiment
  symbolic information retrieval
SIRO
  service in random order
SIRS
  salary information retrieval system
  satellite infrared spectrometer
SIRSA
  special industrial radio service association
SIRU
  strapdown inertial reference unit
SIRWT
  safety injection reserve water tank
SIS
  safety injection signal
  safety injection system
  sage interceptor simulator
  satellite intercept system

satellite interceptor system
semiconductor insulator semiconductor
short interval scheduling
shorter interval scheduling
simulation interface subsystem
simulation interface system
sound in sync
stage interface simulator
stored information system
successor instruction set
swedish interplanetary society
SISCO
singer information services co
SISD
single instruction single data
single instruction single data stream
single instruction stream single data stream
space and information systems division
SISDSYSTEM
single instruction flow single data flow system
SISI
surveillance and in service inspection
SISO
shift in, shift out
SISP
sudden increase of solar particle
SISS
single item, single source
submarine integrated sonar system
SIT
safety injection tank
self-ignition temperature
self-induced transparency
silicon intensifier target tube
silicon intesifier target
simulation input tape
situation
society of instrument technology
software integration test
spontaneous ignition temperature
static induction transistor
stepped impedance transformer
SITAR
system for interactive text editing, analysis and retrieval
SITC
standard international trades classification
SITE
search information tape equipment
spacecraft instrumentation test equipment
SITP
system integration test plan
SITRANSISTOR
shift in transistor

SITS
sage intercept target simulation
SITT
system integration of triad technology
SITVC
secondary injection thrust vector control
SIU
system integration unit
SIUNIT
standard international unit
SIXPAC
system for inertial experiment priority and attitude control
SJ
simultation journal
SJAE
steam jet air ejector
SJCC
spring joint computer conference
SJCM
standing joint committee on metrication
SJP
stacked job processing
SJQ
selected job queue
SK
seek command
SKILL
satellite kill
SKU
stock keeping unit
SL
safety limit
saturated logic
sea level
searchlight
separate lead
single lead
soft landing
solid logic
sound locator
source language
source library
space laboratory
surface launch
symmetrizing line
SLA
sound line alarm
spacecraft lunar module adapter
special libraries association
SLAC
stanford linear accelerator center
stanford linear accelerator computer
SLAE
standard lightweight avionics equipment

supplementary leak collection and release system
SLAET
society of licensed aircraft engineers and technicians
SLAM
single layer metallization
space launched air missile
stored logic adaptable metal oxide semiconductor transistor
supersonic low altitude missile
SLAMS
simplified language for abstract mathematical structures
SLANG
systems language
SLANT
simulator landing attachment for night landing training
SLAP
subscriber line access protocol
symbolic language assembler program
SLAR
side looking aerial radar
side looking airborne radar
SLASH
seiler laboratory algol simulated hybrid
SLATE
small lightweight altitude transmission equipment
stimulated learning by automated typewriter environment
SLB
side lobe blanking
SLBM
sea launched ballistic missile
space launched ballistic missile
submarine launched ballistic missile
SLBMDWS
submarine launched ballistic missile detection and warning system
SLBMW
submarine launched ballistic missile warning
SLC
searchlight control
shift left and count
side lobe cancellation
simulated linguistic computer
simultaneous lobe comparison
single lead covered
single line controller
slicer
specific line capacitance
straight line capacitance

straight line capacitor
subscriber line circuit
SLCB
single line color bar
SLCC
saturn launch control computer
saturn launcher computer complex
soft launch control center
SLCM
submarine launched cruise missile
SLCRS
supplementary leak collection and release system
SLCS
standby liquid control system
SLD
sealed
simplified logic diagram
simulated launch demonstration
solder
solid
solid logic dense
source language debug
straight line depreciation
symmetrized logarithmic derivative
SLDA
solid logic design automation
SLDCIRCUIT
solid logic dense circuit
SLDTECHNOLOGY
solid logic dense technology
SLE
society of logistics engineers
SLEAT
society of laundry engineers and allied trades
SLEEP
scanning low energy electron probe
SLEP
service life extension program
SLET
self-loading tape edit
SLEW
static load error washout
static load error washout system
SLF
straight line frequency
SLG
single line to ground
SLGM
surface launched guided missile
SLI
sea level indicator
steam line isolation
SLIC
subscriber line interface circuits

SLICBM
   sea launched intercontinental ballistic missile
SLIDPROCESS
   solid liquid interdiffusion process
SLIM
   special language interpreting matrix
   standards laboratory information manual
   submarine launched inertial missile
SLION
   solution ion
SLIP
   symmetric list processor
SLIRBM
   sea launched intermediate range ballistic missile
SLIS
   shared laboratory information system
SLIV
   steam line isolation valve
SLM
   ship launched missile
   sound level meter
   space laboratory module
   spatial light modulation
   statistical learning model
SLMP
   self-loading memory print
SLMS
   saturn launched meteoroid satellite
   ship launched missile system
SLN
   subsidiary learning net
SLO
   swept local oscillator
SLOMAR
   space logistics maintenance and repair
   space logistics maintenance and rescue
SLOP
   standard listen output program
SLOR
   successive line overrelaxation
SLOSYN
   slow synchronization
SLOT
   slotted
SLP
   segmented level programming
   selective line printing
   single langmuir probe
   single layer polysilicon
   source language processor
SLPA
   solid logic process automation
SLPC
   signal lines pair combinations

SLR
   short length record
   side look radar
   side looking radar
   sideways looking radar
   single lens reflex camera
   solar
   storage limits register
SLRAP
   standard low frequency range approach
SLRD
   searchlight radar
SLRE
   self-loading random access edit
SLREG
   stepwise linear regression
SLRI
   shipboard long-range input
SLRN
   select read numerically
SLRV
   surveyor lunar roving vehicle
SLS
   side lobe suppression
   side looking sonar
   strategic lunar system
SLSF
   sodium loop safety facility
SLSS
   system library subscription service
SLT
   searchlight
   simulated launch test
   solid logic technique
   solid logic technology
   standard light source
SLTE
   self-loading tape edit
SLTR
   service life test report
SLU
   subsriber's line unit
SLUC
   standard level user charge
SLUG
   superconducting low inductance undulatory galvanometer
SLV
   satellite launching vehicle
   soft landing vehicle
   space launch vehicle
   space launcher vehicle
   standard launch vehicle
   standardized laucher vehicle

SLW
   store logical word
   straight line wavelength
SLWL
   straight line wavelength
SLX
   self-lubricating exterior
SM
   scheduled maintenance
   scheduling maintenance
   screw motorship
   segment mark
   semimat
   sequence and monitor
   service module
   servomotor
   set mode
   shared memory
   simpson multiplier
   simulator
   stack mark
   storage mark
   strategic missile
   strict middling
   submarine
   subminiature
   sulphurized mineral oil
   superimpose
   synchronous modem
   system manual
SMA
   science masters association
   screen manufacturers association
   simulated machine analysis
   switch, modular, attenuator
SMAC
   simulation model of automobile collisions
   space maintenance analysis center
   special mission attack computer
SMACH
   sounding machine
SMACNA
   sheet metal and air conditioning contractors national association
SMACS
   simulated message analysis and conversion subsystem
SMALGOL
   small computer algorithmic language
SMALL
   selenium diode matrix alloy logic
SMART
   satellite maintenance and repair techniques
   space maintenance and repair techniques
   supermarket allocation and recorder technique
   supersonic military aircraft research track
   supersonic missile and rocket track
   system malfunction analysis reinforcement trainer
   systems management analysis, research, and test
SMARTII
   simple minded artificial intelligence
SMBL
   semimobile
SMC
   sage maintenance control
   scientific manpower commission
   segmented maintenance cask
   shunt mounted chip
   systems management committee
   systems, man, and cybernetics
SMCC
   simulation monitor and control console
SMD
   scheduling management display
   semiconductor magnetic field detector
   storage module device
   storage module drive
   surface mounted device
   systems measuring device
SMDC
   superconductive materials data center
SMDCODE
   sign magnitude binary coded decimal code
SMDL
   spares master data log
SMDR
   station message detailed recording
SME
   society of manufacturing engineers
   society of military engineers
   static mission equivalent
SMEAIME
   society of mining engineers of aime
SMEK
   summary message enable keyboard
SMET
   simulated mission endurance testing
SMETER
   signal strength meter
SMF
   solar magnetic field
   system management facility
   system measurement facility
SMG
   sort merge generator
   spacecraft meteorology group

SMGP
  strategic missile group
SMH
  simple harmonic motion
SMI
  standard measuring instrument
  start manual input
SMICBM
  semimobile intercontinental ballistic missile
SMILE
  significant milestone integration lateral evaluation
SMIS
  section of medical information science
  society for management information systems
SMIT
  spin motor interruption technique
SMITE
  simulation model of interceptor terminal effectiveness
SML
  symbolic machine language
SMLM
  simple-minded learning machine
SMM
  standard method of measurement
  system maintenance management
  system maintenance monitor
SMMP
  standard methods of measuring performance
SMMT
  society of motor manufactureres and traders
SMO
  small machine organizer
  stabilized master oscillator
SMODOS
  self-modulating derivative optical spectrometer
SMOG
  special monitor output generator
SMOOTH
  spectra mode of operation through hardware
SMOS
  submicrometer mos
SMP
  sampler
  servo meter panel
  synthesis measurement plan
  system maintenance program
  systems management processor
SMPS
  simplified message processing simulation
SMPTE
  society of motion picture and television engineers

SMR
  series mode rejection
  shield mock-up reactor
  signal master
  solid moderated reactor
SMRD
  spin motor rate detector
  spin motor rotation detector
SMRE
  safety in mines research establishment
  surface mining reclamation and enforcement office
SMRF
  series mode rejection factor
SMS
  semiconductor metal semiconductor
  service module simulator
  small mass store
  standard modular system
  standard modular system structure
  storage management system
  strategic missile squadron
  surface missile system
  synchronous altitude meteorological satellite
  synchronous meteorological satellite
  system measurement software
SMSA
  signal missile support agency
  standard metropolitan statistical area
SMSAE
  surface missile system availability evaluation
SMSQ
  strategic missile squadron
SMSSTRUCTURE
  semiconductor metal semiconductor structure
SMSTR
  signal master
SMT
  service module technician
  ship mean time
  small missile telecamera
  square mesh tracking
  system master tape
SMTE
  segment map table entry
SMTI
  selective moving target indication
  selective moving target indicator
  sodium mechanism test installation
SMTS
  special machine tool standard
SMU
  self-maneuvering unit
  statement match unit

SMUD
   sacramento municipal utility district
SMW
   strategic missile wing
SMWG
   strategic missile wing
SMX
   submultiplex
SN
   saponification number
   saturn nuclear
   sector number
   semiconductor network
   sequence number
   serial number
   sign
   signal to noise
   signal to noise ratio
   sine of the amplitude
   siren
   sound negative
   speech to noise
   stress number
   sub nanosecond
   subnetwork
   synchronizer
SNA
   standard numerical attributes
   system network architecture
   system numerical attributes
   systems network architecture
SNAC
   sonar automatic controller
SNACS
   share news on automatic coding systems
SNAE
   society of norwegian american engineers
SNAFU
   situation normal, all fouled up
SNAME
   society of naval architects and marine engineers
SNAP
   sharp numeric assembler program
   simplified numerical automatic programmer
   space nuclear auxiliary power
   system for nuclear auxiliary power
SNATCH
   sna and transdata coupling of hosts
SNC
   stored program numeric control
SND
   sound
SNDPLG
   sandwich plug

SNDPRF
   soundpoof
SNDR
   signal to noise density ratio
SNDT
   society of nondestructive testing
SNE
   single nylon enamelled
   suppress normal end
SNEMSA
   southern new england marine sciences association
SNF
   system noise figure
SNG
   synthetic natural gas
SNGL
   single
SNI
   sequence number indicator
   signal to noise improvement
SNIF
   signal to noise improvement factor
SNIRD
   supposedly noiseless infrared detector
SNL
   standard nomenclature list
SNM
   society of nuclear medicine
   special nuclear material
   spent nuclear material
SNMP
   spent nuclear material pool
SNN
   signal plus noise to noise
SNOS
   silicon nitride oxide silicon
SNP
   synchro null pulse
SNPM
   standard and nuclear propulsion module
SNPO
   space nuclear propulsion office
SNR
   signal to noise ratio
   supplier nonconformance report
SNRCN
   signal to noise ratio due to channel noise
SNS
   simulated network simulations
SNSR
   sensor
SNT
   sign on table
   society for nondestructive testing

SNUPPS
   standardized nuclear unit power plant system
SO
   select order
   send only
   sending only
   shift out
   signal office
   signal officer
   slow operate
   socket
   sorting office
   spin orbital
   spring opening
   stationary orbit
   substitution oscillator
   switch over
SOA
   safe operating area
   start of address
SOALM
   scanned optically addressed light modulators
SOAP
   self-optimizing automatic pilot
   symbolic optimum assembly program
   symbolic optimum assembly programming
SOAR
   safe operating area
SOARAD
   solar radiation
SOAV
   solenoid operated air valve
SOB
   space orbital bomber
   start of block
SOC
   satellite orbit control
   self-organizing control
   separated orbit cyclotron
   set overrides clear
   signal officer
   signal operations center
   simulation operation computer
   simulation operations center
   single orbit computation
   socket
   space operations center
   spacecraft orientation control
   specific optimal controller
SOCCS
   summary of component control status
SOCF
   spacecraft operations and checkout facility
SOCO
   switched out for checkout
SOCOM
   solar communications
   solar optical communications system
   solar orbital communications
SOCP
   satellite orbit control program
SOCR
   sustained operations control
   sustained operations control room
SOCRATES
   simulation of closure and rendezvous approach techniques for early spacecraft
   simulator or creative reasoning applied to eduction systems
SOCS
   space operation command system
   spacecraft orientation control system
SOD
   small object detector
   small oriented diode
   sum of the years digit
   surface oriented diode
SODA
   source oriented data acquisition
   system optimization and design algorithm
SODAR
   sound detecting and ranging
   sound detection and ranging
SODAS
   structure oriented description and simulation
SOE
   significant operating experience
   silicon overlay epitaxial
   special operations executive
   stripline opposed emitter
SOEP
   solar oriented experiment package
SOER
   significant operating event report
SOERO
   small orbiting earth resources observatory
SOF
   signal officer
   sound on film
SOFAR
   sound fixing and ranging
SOFCS
   self-organizing flight control system
SOFF
   sign off
SOFI
   software information
SOFNET
   solar observing and forecasting network

SOFT
    simple output format translator
SOFTMARK
    software marketing
SOG
    same output gate
SOH
    sonarman harbor defense
    start of header
    start of heading
    start of heading character
SOI
    signal operation instruction
    specific operating instruction
    standard operation instruction
SOICS
    summary of installation control status
SOINC
    signal officer in chief
SOL
    sequence operated lock
    short octal load
    simulation oriented language
    solar
    solenoid
    solid
    system oriented language
SOLAR
    serialized on-line automatic recording
    shared on-line airline
    shared on-line airline reservations
SOLARIS
    submerged object locating and retrieving identification system
SOLAS
    safety of life at sea
SOLDIER
    solution of ordinary differential equations routine
SOLE
    society of logistics engineers
SOLID
    self-organizing large information dissemination system
SOLION
    solution ion
SOLIONDEVICE
    solution of ions device
SOLIS
    symbionics on-line information system
SOLLAR
    sof lunar landing and return
SOLO
    selective optical lock-on
    system for ordinary life operations
SOLOMON
    simultaneous operation linked ordinal modular network
SOLR
    sidetone objective loudness rating
SOLRAD
    solar radiation
SOLV
    solenoid valve
    super open frame low voltage
SOM
    self-organizing machine
    shift operation manager
    soundman
    start of message
    suborbital mission
SOMADA
    self-organizing multiple access discrete address
SOMP
    start of message priority
SON
    sign on
SONAC
    sonar nacelle
SONAD
    sound operated noise attenuation device
SONAR
    sound navigation and ranging
SONCM
    sonar countermeasures and deception
SONCR
    sonar control room
SONG
    satellite for orientation, navigation and geodesy
SONI
    staff officer navigation instructor
SONIC
    system wide on-line network for informational control
SONOAN
    sonic noise analyzer
SOP
    sea of peace
    signal operating procedure
    simulated output program
    simulation operations plan
    special operating procedure
    standard operating procedure
    station operations plan
    strategic orbit point
    study organization plan

sum of products
symbolic optimum program
system operator's manual
SOPM
standard orbital parameter message
SOPR
spanish open pool reactor
SOR
single operation responsibility
single operator responsibility
slow operating relay
society of rheology
start of record
successive overrelaxation
SORAP
standard omnirange approach
SORC
site operations review committee
sound ranging control
systems objectives and requirements committee
SORD
society of record dealers of america
SORI
staff officer radio instructor
SORNG
sound ranging
SORO
sort program, sort routine
SORTE
summary of radiation tolerant electronics
SORTI
satellite orbital track and intercept
star oriented real time tracking instrument
SORTIE
suborbital reentry test integrated environment
supercircular orbital reentry test integrated environment
SOS
safety observation station
service off the shelf
share operating system
silicon on sapphire
silicon on spinel
simulator operating system
sophisticated operating system
sophisticated operation system
sound on sync
speed of service
start of significance
station operating supervisor
strategic orbital system
symbolic operating system
synchronous orbit satellite

SOSCMOS
silicon on sapphire complementary metal oxide semiconductor
SOSFET
silicon on sapphire field effect transistor
SOSRAM
silicon on sapphire random access memory
SOSS
strategic orbital system study
SOSTECHNIQUE
silicon on sapphire technique
SOSUS
sound surveillance system
SOT
scanning oscillator technique
sky wave observation timer
syntax oriented translator
SOTAS
standoff target acquisition system
SOTIM
sonic observation of the trajectory and impact of missiles
SOTUS
sequentially operated teletypewriter universal selector
SOV
solenoid operated valve
sound on vision
SOVAS
scanning optical vibration analysis system
SOW
start of word
SP
self-powered
self-propelled
separate element pricing
sequence programmer
sequential phase
serial to parallel
series parallel
service processor
set pattern
set point
setting point
shift pulse
short persistance
signal procesor
silver plate
silver plated
singing point
single-peaked
single-phase
single pole
single precision
softening point

soil pipe
sort program
sound positive
sound-powered
sound-powered telephone
source program
space
space character
space probe
spare
spare part
sparking plug
special projects
special purpose
specific
splash proof
spontaneous potential
spool
square punch
stack pointer
standard peripherals
standard pressure
storage protection
structure programming
structured programming
subliminal perception
subprogram
summary punch
surveillance procedure
symbol programmer
system processor

SPA
s band power amplifier
servo power assembly
single position automatic tester
southwestern power administration
southwestern psychological association
spectrum analyzer
state power authority
subject to particular average
sudden phase anomaly
systems and procedures association

SPAA
systems and procedures association of america

SPAC
spatial computer

SPACCS
space command and control system

SPACE
self-programming automatic circuit evaluator
sequential position and covariance estimation
sidereal polar axis celestial equipment
space program american citizens' effort
spacecraft prelaunch automatic checkout equipment
symbolic programming anyone can enjoy

SPACECOM
space communication

SPACELOOP
speech analog compression and editing loop

SPACETRACK
space tracking

SPACON
space control

SPACS
sodium purification and characterization system

SPAD
satellite position prediction and display
satellite position predictor and display
satellite protection for air defense
space patrol for air defense

SPADATS
space detection and tracking system

SPADATSIMP
space detection and tracking system improved

SPADE
single channel per carrier pcm multiple access demand assignment equipment
spare parts analysis, documentation and evaluation
sparta acquisition digital equipment
sperry air data equipment

SPADESYSTEM
single channel per carrier pcm multiple access demand system

SPAM
scratch pad memory address multiplexer
ship position and attitude measurement

SPAMS
ship position and attitude measurement system

SPAN
solar particle alert network
space communications network
space navigation
statistical processing and analysis
stored program alphanumerics

SPANDRA
space and range

SPANNET
space navigation network

SPANRAD
superimposed panoramic radar display
superposed panoramic radar display

SPAQUA
sealed package quality assurance

SPAR
    satellite position adjusting rocket
    sea-going platform for acoustic research
    superprecision approach radar
    symbolic program assembly routine
    synchronous position attitude recorder
SPARC
    space air relay communications
    space program, analysis and review council
    standards planning and requirements committee
SPARR
    self-contained perspective approach rotor blade radar
    steerable paraboloid azimuth radio reflector
SPARS
    space precision attitude reference system
SPARSA
    sferics pulse, azimuth, rate, and spectrum analyzer
SPARTA
    sequential programmed automatic recording transistor analyzer
    spatial antimissile research tests in australia
SPAS
    serial poll active state
SPASEC
    space track sensor computer
SPASM
    system performance and activity software monitor
SPASUR
    space surveillance
SPAT
    silicon precision alloy transistor
SPAU
    signal processing arithmetic unit
SPC
    set point control
    silver plated copper
    single paper covered
    single paper covered wire
    small peripheral controller
    starting point counter
    static power converters
    storage programmed computer
    stored program command
    stored program control
    stored program controlled
    stored program controller
    stored program controller system
    stored programmed command
    summary punch control
    systems planning and computing board
SPCB
    single pole circuit breaker
SPCC
    sample polarity coincidence correlator
    staggered phase carrier cancellation
SPCHGR
    supercharger
SPCHR
    supercharger
SPCN
    silver plated copperweld conductor
SPCR
    scratch pad control register
    silicon planar controlled rectifier
SPD
    serial poll disable
    sigma phi delta fraternity
    single path doppler
    special products division
    surge protective devices
SPDC
    stored program data compressor
SPDP
    society of professional data processors
SPDS
    safe practice data sheet
    self-power density spectrum
SPDT
    single pole double throw
SPDTCONTACT
    single pole double throw contact
SPDTDB
    single pole, double throw, double break
SPDTNCDB
    single pole, double throw, normally closed, double break
SPDTNO
    single pole, double throw, normally open
SPDTNODB
    single pole, double throw, normally open, double break
SPDTSWITCH
    single pole double throw switch
SPE
    serial poll enable
    service propulsion engine
    silicon planar epitaxial
    society of petroleum engineers
    society of plastics engineers
    systems performance effectiveness
SPEA
    panamanian society of engineers and architects
SPEAR
    stanford positron electron asymmetric ring

SPEARS
   satellite photo electric analog rectification system
   satellite photo electronic analog rectification system
SPEC
   specification
   specifications
   speech predictive encoding system
   stored program educational computer
SPECON
   system performance effectiveness conference
SPECS
   specifications
SPECTOR
   single path error correcting teleprinter over radio
SPECTROL
   scheduling, planning, evaluation, cost control
SPECVER
   specification verification
SPED
   supersonic planetary entry decelerator
SPEDAC
   solid state parallel expandable differential analyzer computer
SPEDE
   state system for processing educational data electronically
SPEDTAC
   stored program educational transistorized automatic computer
SPEDTAR
   stored program eductional transistorized automatic computer
SPEE
   society for the promotion of engineering education
   society of petroleum evaluation engineers
SPEECS
   speech parameter extraction experimental comparison system
SPEED
   self-programmed electronic equation delineator
   signal processing in evacuated electronic device
   subsistance preparation by electronic energy diffusion
SPEEL
   shore plant electronic equipment list
SPERT
   schedule performance evaluation and review technique
   schedule program evaluation and review technique
   special power excursion reactor test
SPES
   stored program element system
SPET
   solid propellant electric[al] thruster
SPF
   site population factor
   standard project flood
SPFP
   single point failure potential
SPFW
   single-phase full wave
SPG
   self-propelled gun
   single point ground
   sort program generator
   synchronization pulse generator
SPGR
   specific gravity
SPGS
   secondary power generating subsystem
SPH
   single-phase
   space heater
   superphantom
SPHD
   spool piece head
SPHE
   society of packaging and handling engineers
SPHL
   sun present horizon lost
SPHT
   specific heat
SPHW
   single-phase half wave
SPI
   self-paced instruction
   shared peripheral interface
   signal point identification
   single program initiated
   single program initiator
   site population index
   society of the plastics industry specific productivity index
   special position identification pulse
   specific productivity index
   sun position indicator
SPIA
   solid propellent information agency
SPIAM
   sodium purity in line analytical module
SPIC
   ship's position interpolation computer

society of the plastics industry of canada
spare parts inventory package
SPICBM
solid propellant intercontinental ballistic missile
SPICE
sales point information computing equipment
SPIDER
sonic pulse echo instrument designed for extreme resolution
SPIE
scavenging precipitation ion exchange
self-programmed individualized education
simulated problem input evaluation
society of photooptical instrumentation engineers
SPINDEX
selective permuting index
SPIRAL
sperry inertial radar altimeter
SPIRBM
solid propellant intermediate range ballistic missile
SPIRE
spatial inertial reference equipment
SPIS
serial poll idle state
SPIT
selective printing of items from tape
SPK
spark
storage protection key
SPKPRF
spark proof
SPKR
speaker
SPL
software programming language
sound pressure level
south pacific program library
spaceborne programming language
special
special priority logic
speed phase lock
standard programming logic
symbolic programming language
systems programming language
systems programming ltd
SPLICE
shorthand programming language in cobol environment
SPLIT
sundstrand processing language internally translated

SPM
scratch pad memory
scratch pad module
self-propelled mount
sequential processing machine
serial parallel multiplier
source program maintenance
strokes per minute
symbol processing machine
symbolic processing machine
SPMD
silicon planar multiple diode
SPMS
serial poll mode state
solar particle monitoring system
system program management survey
SPN
series parallel network
switched public network
SPNS
switched private network service
SPO
saturn project office
short period oscillation
special project office
system and programming organization
SPOC
single point orbit calculator
splicing of cross correlation function
SPOCK
simulated procedure for obtaining common knowledge
SPOLE
south seeking pole
SPOOK
supervisory program over other kinds
SPOOL
simultaneous peripheral operations on line
spontaneous peripheral operations on line
spooling
SPOS
system program offices
SPOT
satellite positioning and tracking
speed, position, and track
steel plate ordering technique
SPOUT
system peripheral output utility
SPP
solar photometry probe
sound powered phone
speed power product
structured programming processor
system package plan
system package program

SPPH
  split phase
SPPM
  serial parallel pipeline multiplier
SPQ
  special product quotation
SPR
  sandia pulsed reactor
  serial printer
  short pulse radar
  silicon power rectifier
  simplified practice recommendation
  specific resistance
  sudden pressure relay
SPRA
  space probe radar altimeter
SPRAT
  small protable radar torch
SPRC
  self-propelled robot craft
SPRDS
  steam pipe rupture detector system
SPRF
  space propulsion research facility
SPRI
  scott polar research institute
SPRINT
  selective printer
  selective printing
  simultaneous print
  solid propellant rocket intercept missile
SPRITE
  solid propellant rocket ignition and evaluation
SPRT
  sequential probability ratio test
SPRTAP
  specially preparad tape program
SPS
  sample per second
  secondary propulsion system
  sequential partitioned system
  serial parallel serial
  service propulsion system
  society of physics students
  solar power system or satellite
  solar probe spacecraft
  solar proton stream
  space power system
  spacecraft propulsion system
  speed switch
  standard pipe size
  supplementary power supply
  symbolic programming system
  synchronous program supervisor

SPSC
  shield plug support cylinder
SPSE
  society of photographic scientics and engineers
SPSP
  small power system program
SPSS
  shield plug storage station
  single pole snap switch
  statistical package for social sciences
  supplementary power supply set
SPSSTRUCTURE
  serial parallel serial structure
SPST
  single pole single throw
  spent resin storage tank
SPSTCONTACT
  single pole single throw contact
SPSTNC
  single pole single throw, normally closed
SPSTNO
  single pole, single throw, normally open
SPSTRELAY
  single pole single throw relay
SPSTSWITCH
  single pole single throw switch
SPSW
  single pole switch
SPT
  silicon planar thyristor
  silicon planar transistor
  special purpose tests
  symbolic program tape
  symbolic program translator
SPTC
  specified period of time contact
SPTF
  sodium pump test facility
SPTV
  supersonic parachute test vehicle
SPU
  self-propelled underwater missile
  sense punch
  small peripheral unit
  standard propulsion unit
SPUD
  solar power unit demonstrator
  stored program universal demonstrator
SPUR
  single precision unpacked rounded
  single precision unpacked rounded floating point package

single precision unpacked rounding floating point package
space power unit reactor
SPURM
special purpose unilateral repetitive modulation
SPURT
spinning unguided rocket trajectory
SPVOL
specific volume
SPWG
spare parts working group
SPWLA
society of professional well log analysts
SPX
superheat power experiment
SQ
sequential
square
squint quoin
superquick
SQA
software quality assurance
squaring amplifier
supplier quality assurance
SQC
station quality control
statistical quality control
SQCG
squirrel cage
SQCM
square centimeter
SQFT
square foot
SQIN
sequential quadrature inband
square inch
SQKM
square kilometer
SQL
structured query language
SQM
square meter
SQMI
square mile
SQPCM
slope quantized pulse code modulation
SQR
sequence relay
square root
supplier quality representative
SQRT
square root

SQUID
sperry quick updating of internal documentation
superconducting quantum interference device
superconducting quantum interferometer device
SQUP
software quality assurance plan
SQW
square wave
SQYD
square yard
SR
safety rods
saturable reactor
scientific report
secondary radar
sedimentation rate
segment root
selective ringing
selenium rectifier
self rectifying
send receive
sending receiving
sensitivity ratio
series relay
shift register
shift reverse
shifting register
short range
shunt reactor
signal regulation
silicon rectifier
slip ring
slow release
slow release relay
slow running
society of rheology
soft radiation
solar radiation
solid rocket
sound ranging
sound rating
specific resistance
speed recorder
speed regulator
split ring
standard resistor
starting relay
starting resistor
steradian
stirred tank reactor
storage and retrieval
subroutine
supervisor

support reaction load
surveillance radar
system requirement
SRA
science research associates
shop replaceable assembly
SRAM
short-range attack missile
static random access memory
SRAMS
short-range attack missile system
SRAP
standard-range approach
SRB
send receive bomb
SRBM
shortrange ballistic missile
SRBP
synthetic resin bonded paper
SRBR
surface reflected bottom reflected
SRBTRY
sound ranging battery
SRC
science research council
single reflex camera
sound ranging central
sound ranging control
source range channel
space research corporation
spares receiving checklist
special release card
sperry rand corporation
standard requirements code
SRCC
shift, rotate, check, control
simplex remote communications center
SRCH
search
SRCRA
shipowners refrigeration cargo research association
SRCS
special reverse charge service
SRD
secret restricted data
self-reading dosimeter
shift register drive
shipper receiver difference
spacecraft research division
standard rate and data
standard reference data
step recovery diode
swing rate discriminator
SRDA
sodium removal development apparatus
SRDAS
service recording and data analysis system
SRDE
signal research and development establishment
SRDL
semiconductor research and development laboratory
signal research and development laboratory
SRDS
standard reference data system
SRE
scientific research and experiments department
search radar element
senio reliability engineer
series relay
single rotation engine
society of relay engineers
sodium reactor experiment
surveillance radar element
surveillance radar equipment
SRETL
screened resistor evaporated transistor logic
SRF
self-resonant frequency
signal strength radio frequency
spectral redistribution function
strength of radio frequency
SRFB
space research facilities branch
SRFLIPFLOP
set reset flipflop
SRFUNCTION
service request function
SRG
shift register generator
short range
sound ranging
statistical research group
SRGR
short-range guided rocket
sound ranging group
SRH
shockley read hall
SRI
servo repeater indicator
southern research institute
spalling resistance index
stanford research institute
SRIAER
scientific research institute for atomic energy reactors

SRILTA
   stanford research institute lead time analysis
SRIMO
   senior radio installation and manufacture officer
SRIVISIONMODULE
   stanford research institute vision module
SRJE
   sna remote job entry
SRL
   system reference library
SRLY
   series relay
SRM
   sealectro reliable miniature
   short-range missile
   single register machine
   source range monitors
   subarea routing manager
SRME
   submerged repeater monitoring equipment
SRML
   short-range missile launcher
SRMU
   signal radar maintenance unit
SRN
   simulation reference number
SRNFC
   source range neutron flux channel
SRNH
   service request not honoured
SRO
   senior reactor operator
   singly resonant oscillator
   society of radio operators
   specification release order
   station routine order
   supervisor of range operations
SROB
   short-range omnidirectional beacon
SROS
   special run operations sheet
SRP
   seat reference point
   shift register partition
   standard relative power
   standard review plan
   status response field
SRPM
   single reversal permanent magnet
SRQ
   service request
   status request field
SRQS
   service request state

SRR
   search and range radar
   security rules and regulations
   shift register recognizer
   short-range radar
   short-range recovery
   software requirements review
   sound recorder reproducer
SRRB
   search and rescue radio beacon
SRS
   seat reservation system
   selenium rectifier stack
   send receive switch
   short-range search
   simulated raman scattering
   simulated remote sites
   simulated remote station
   sodium removal station
   software requirements specification
   solar radiation satellite
   sound ranging section
   space and reentry system
   stimulated raman scattering
   stimulated rayleigh scattering
   submarine reactor small
   subscriber response system
SRSA
   scientific research society of america
SRSAGM
   short-range surface to air guided weapon
SRSCC
   simulated remote station control console
SRSK
   short-range station keeping
SRSS
   simulated remote sites subsystem
SRST
   system resource and status table
SRT
   search radar terminal
   single requestor terminal
   single run time
   society of radiological technologists
   supporting research and technology
   surface craft radio transmitter
   synchro and resolver transmission
   system reaction time
   system recovery table
   systems readiness test
SRTC
   signal replacement training center
SRTF
   short-range task force

SRTFLIPFLOP
  set reset trigger flipflop
SRU
  self-recording unit
  shop replaceable unit
  signal responder unit
  subscriber response unit
  subscriber's response unit
SRV
  space rescue vehicle
SRVAMPL
  servo amplifier
SRVIN
  servo inlet
SRVRET
  servo return
SS
  sample per second
  satellite space system
  satellite system
  secret service
  selective signaling
  selector switch
  sequence switch
  shift supervisor
  signal strength
  single shot
  single sideband
  single signal
  single silk
  sliding scale
  small signal
  solar system
  solid state
  space sciences
  space simulator
  space station
  space system
  special source materials
  spin stabilized
  spin stabilizer
  stabilization system
  stabilized screen
  stainless steel
  stationary satellite
  statistical standards
  steady state
  steamship
  step size
  stop scan
  storage to storage
  storage to storage operation
  submarine scout
  subsystem
  summary sheet
  summing selector
  summing selector supersonic
  supersonic
  surveillance station
  synchro standard
SSA
  seismological society of america
  slave service area
  small search aera
  social security administration
  space suit assembly
  synchro signal amplifier
SSAC
  suspended sprayed acoustical ceiling
SSAM
  society for science and mathematics
SSAR
  site safety analysis report
  special save register
  spin stabilized aircraft rocket
  standard safety analysis report
SSARC
  standards screeing and review committee
SSAV
  self sealing aerospace vehicle
SSB
  signal sight back
  single sideband
  soft service building
  source selection baord
  space science board
  supersonic balloon
SSBAM
  single sideband amplitude modulation
SSBARA
  single sideband amateur radio association
SSBD
  single sideboard
SSBFM
  single sideband frequency modulation
SSBM
  single sideband amplitude modulation
  single sideband angle modulation
  single sideband modulation
SSBN
  ship submersible nuclear
SSBO
  single swing blocking oscillator
SSBS
  surface to surface strategic ballistic missile system
SSBSC
  single sideband with suppressed carrier
  single switched suppressed carrier

SSBSCAM
single sideband with suppressed carrier, amplitude modulated
SSBSCOM
single sideband suppressed carrier optical modulator
SSBTRANSMISSION
single sideband transmission
SSBWC
single sideband with carrier
SSBWPHENOMENON
surface skimming bulk wave phenomenon
SSC
second search character
sensor signal conditioner
short segmented cask
single silk covered
small scientific computer
solid state circuit
space systems center
stepping switch counter
systems science and cybernetics
SSCBM
shipping and storage container ballistic missile
SSCC
solid state circuits conference
space surveillance control center
spin scan cloud cover
SSCDS
small ship combat data system
SSCF
signal strength center frequency
SSCP
system services control point
SSCR
spectral shift control reactor
SSCT
solid state circuits
SSCU
special signal conditioning unit
SSCW
single silk covered wire
slow space charge wave
SSD
satellite system development
seize signal detector
sequence switch driver
silicon single diffused
single station doppler
solid state detector
solid state devices
source skin distance
space systems division
stabilized ship detector
steady state distribution
surveillance situation display
SSDC
signal source distribution center
system safety development center
SSDF
space science development facility
SSDFNAA
space science development facility north american aviation
SSDR
satellite situation display room
SSDS
small ship data system
SSE
safe shutdown earthquake
self-sustained emission
single silk enamel covered
single silk enamelled
solid state electrolyte
solid state electronics
spacecraft simulation equipment
special support equipment
switching system engineer
SSEB
south of scotland electricity board
SSEC
selective sequence electronic calculator
static source error correction
subsystem executive control program
SSEP
system safety engineering plan
SSESM
spent stage experimental support module
SSF
safe shutdown facility
seconds saybolt furol
service storage facility
space simulation facility
supersonic frequency
SSFC
sequential single frequency code
SSFL
steady state fermi level
SSFM
single sideband frequency modulation
SSG
search signal generator
small signal gain
standard signal generator
subscriber switching grid
surface discharge spark gap
SSGA
single conductor, shipboard general use, armored

SSGP
  signal service group
SSGS
  standard space guidance system
  standardized space guidance system
SSGW
  surface to surface guided weapon
SSHP
  single shot hit probability
SSI
  sector scan indicator
  semiconductor specialists, incorporated
  single scale integration
  small scale integration
  standing signal instructions
  steady state irradiation
  storage to storage instruction
  strategic systems, incorporated
  supplemental security income
  system status index
SSIG
  single signal
SSIP
  subsystem integration plan
SSL
  secondary standards laboratory
  self-aligned superintegration logic
  shift and select
  solid state lamp
  source statement library
  super speed logic
  support system language
SSLC
  synchronous single line controller
SSLORAN
  sky wave synchronized long-range navigation
SSLS
  standard space launch system
  standardized space launch system
SSLV
  standard space launch vehicle
  standardized space lauch vehicle
SSM
  scientific survey module
  single sideband modulation
  solid state materials
  spacecraft system monitor
  spread spectrum modulation
  stochastic sequential machine
  surface to surface missile
  system software message
SSMA
  spread spectrum multiple access
SSMD
  silicon stud mounted diode

SSMS
  solid state mass spectrometer
SSMTG
  solid state and molecular theory group
SSMUX
  spread spectrum multiplexing
SSN
  social security number
  switched services network
SSNM
  strategic special nuclear materials
SSNPP
  small size nuclear power plant
SSO
  saturn systems office
  space sciences office
  space station office
  spacecraft systems officer
  spindle speed override
  steady state oscillation
SSOC
  space surveillance operations center
SSP
  scientific subroutine package
  signal processing peripheral
  sodium sampling package
  solid state preamplifier
  solid state pneumatics
  solid state products
  stand by status panel
  static sodium pots
  static spontaneous potential
  steady state pulse
  system service program
  system status panel
SSPC
  steel structures painting council
SSPI
  solid state products, incorporated
SSPL
  steady state power level
SSPM
  single sideband phase modulation
SSPS
  satellite solar power station
  satellite solar power system
  solid state protection system
  sunflower space power system
SSPWR
  small size pressurized water reactor
SSQ
  sum of the squares
SSR
  satellite situation report
  scratched surface recording

secondary surveillance radar
separate superheater reactor
single signal receiver
site suitability report
solid state relay
specification status report
standby supply relay
static oscillation squelch range
subsynchronous resonance
synchronous stable relaying
SSRC
  single sideband reduced
  single sideband reduced carrier
  structural stability research council
SSRD
  secondary surveillance radar digitizer
SSRL
  systems simulation research laboratory
SSRM
  sealectro small reliable miniature
SSRS
  start stop restart system
SSRT
  subsystem readiness test
SSS
  scientific subroutine system
  sequential scheduling system
  simulation study series
  single signal superhet
  single signal superheterodyne receiver
  small scientific satellite
  software specification sheet
  solid state scientific
  solid state systems
  special safety safeguards
  strategic satellite system
SSSA
  soil science society of america
SSSB
  system source selection board
SSSC
  single sideband suppressed carrier
  special spectrum study committee
  surface subsurface surveillance
  surface subsurface surveillance center
SSSG
  scientific system support group
SSSR
  sage system status report
SST
  scroll symbolic tracer
  secondary surge tank
  simulated structural test
  single step
  solid state technology

  solid state transmitter
  spacecraft systems test
  step by step test
  subsystem test
  super surface treatment
  supersonic telegraphy
  supersonic transport
SSTC
  single sideband transmitted carrier
SSTDMA
  satellite switched time division multiple access
SSTF
  shortest seek time first
  space simulation test facility
SSTM
  sage system training mission
SSTO
  second stage tail off
SSTP
  subsystem test procedure
SSTV
  slow scan television
  slow scanning television
SSU
  saybolt seconds universal
SSV
  ship to surface vessel
  ship to surface vessel radar
  space shuttle vehicle
  static self verification
SSW
  safety switch
  synchro switch
SSWAM
  single-sided wideband analog modulation
SSWF
  sudden short wave fading
SSX
  small systems executive
ST
  sawtooth
  schmitt trigger
  scientific and technical
  screw terminal
  segment table
  serial tasking
  service tabulating
  service test
  set trigger
  shop telegraph
  short time
  short ton
  single throw
  sonic telegraphy

sound telegraphy
sounding tube
special test
standard temperature
standard time
star tracker
start
start signal
start timing
stoke
stone
store
stores transfer
straight time
studio to transmitter
sub tuner
surface tension
symbol table
synchronization table
system table
systems technology
STA
  satellite test annex
  segment table address
  shuttle training aircraft
  single tape armored
  spark thrust augmentor
  station
  stationary
  stator
  steel tape armored
  store accumulator
  store address
  swedish telecommunications adminstration
STAA
  signal training, all arms
STAB
  stabilization
  stabilize
  stabilized
  stabilizer
STABAMP
  stabilizing amplifier
STAC
  science and technology advisory committee for msf
STACO
  standing committee for the scientific principles of standardization
STADAN
  satellite tracking and data acquisition network
  space tracking and data acquisition network
STADAR
  servo tester with automatic data acquisition and reduction

STAE
  second time around echo
STAF
  scientific and technological applications forecast
  statistical analysis of files
STAFF
  stellar acquisition feasibility flight
STAI
  simulation tape alarm indicator
STAIR
  structural analysis interpretive routine
STAIRS
  storage and information retrieval system
STALO
  stabilized local oscillator
  stable local oscillator
STAMIC
  set theoretic analysis and measure of information characteristics
STAMO
  stabilized master oscillator
STAMOS
  sortie turn around maintenance operations simulation
STAMP
  systems tape addition and maintenance program
STAMPS
  stabilized translation and maneuvering propulsion system
STAN
  standard
STANAG
  standardization agreement
STANDAN
  space tracking and data acquisition network
STAPP
  simulation tape print program
  single thread all purpose program
  standard tape print program
STAQ
  security trader automated quotations
STAR
  scientific and technical aerospace report
  segment table address register
  self-steering array repeater
  self testing and repairing
  shield test air reactor
  ship tended acoustic relay
  simultaneous temperature alarm read-out
  space technology and advanced research
  space terminal auxiliary reactor
  space thermionic auxiliary reactor
  special tube analyzing recorder

standard routine
statistical analysis routine
storage address register
system for telephone administrative response
STARADC
star analog digital computer
STARCOM
strategic army communications
STARE
steerable telemetry antenna receiving equipment
STARFIRE
system to accumulate and retrieve financial information with random extraction
STARS
satellite telemetry automatic reduction system
services and techniques for advanced real time systems
simplified three axes reference system
system test and astronaut requirements simulation
START
selection to activate random testing
spacecraft technology and advanced reentry test
spacecraft technology and reentry test
system of transportation applying rendezvous technology
systematic tabular analysis of requirements technique
STASS
special tactical air surveillance system
surveillance target acquisition support system
STAT
static
status
STATAMPERE
static ampere
STATCOULOMB
static coulomb
STATE
simplified tactical approach and terminal equipment
STATFARAD
static farad
STATHENRY
static henry
STATMHO
static mho
STATOHM
static ohm
STATPAC
statistic package
STATVOLT
static volt

STB
segment table base
subsystem test bed
STBY
standby
STC
satellite test center
scientific and technical committee
sensitivity time control
serving test center
short time constant
signal training center
silicon transistor corporation
simulation tape conversion
society for technical communication
spacecraft test conductor
standard telephones and cables
standard transmission code
storage technology corporation
symbol table counter
system test complex
STCL
source term control loop
STCW
system time code word
STD
salinity temperature depth
semiconductor on thermoplastic on dielectric
set driver
short time duty
silicon triple diffused
spacecraft technology division
spectral theory of diffraction
standard
subscriber trunk dialing
superconductive tunneling device
system technology division
STDA
selenium tellurium development association
STDBY
stand by
STDM
statistical time division multiplexing
synchronous time division multiplexing
STDN
spaceflight tracking and data network
standardization
STDP
service technicians development program
single throw double pole
single throw double pole switch
STDS
set theoretic data structure

STDTECHNOLOGY
   semiconductor on thermoplastic on dielectric technology
STE
   segment table entry
   service technique externe
   shield test experiment
   single threshold element
   society of tractor engineers
STEC
   solar to thermal energy conversion
STED
   solar turboelectric drive
STEDI
   space thrust evaluation and disposal investigation
STEIN
   system test environment input
STEM
   scanning transmission electron microscope
   self storing tubular extensionable member
   shaped tube electrolytic machining
   stay time extension module
   stellar tracker evaluation missile
STEP
   safety test engineering program
   scientific and technical exploitation program
   simple transition electronic processing
   simple transition to economical processing
   simple transition to electronic processing
   simulated tracking evaluation program
   space terminal evaluation program
   space thermoelectric power
   standard equipment practice
   standard tape executive program
   standard terminal program
   supervisory tape executive program
   system for test and plug
STEPS
   solar thermionic electric[al] power system
   solar thermionic electrical propulsion system
STEREO
   stereophonic
   stereophony
STET
   specialized technique for efficient type setting
STEXIS
   south texas interconnected system
STF
   safety test facility
   snap shield test facility reactor
   space track facility
STG
   space task group
   stage
   starting
   steering task group
   storage
STGT
   secondary target
STI
   scientific technical information
STIAP
   standard instrument approach
STIC
   scientific technical intelligence center
STICO
   standard interpretation and compiling system
   standard system for interpretation and compilation
STID
   scientific and technical information division
STIF
   short-term irradiation facility
STINFO
   scientific and technical information
STINGS
   stellar inertial guidance system
STIO
   store input output
STIR
   snap shield test irradiation reactor
STISP
   science and technology information service for parliament
STL
   schottky transistor logic
   selective tape listing
   simulated table load
   space technology laboratory
   standard telecommunications laboratories
   standard telegraph level
   station transmission link
   studio transmitter link
   support table load
   symmetrizing and transformation line
   synchronous transistor logic
   system test loop
STLE
   senior test laboratory engineer
STLO
   scientific and technical liaison office
STLT
   studio transmitter link television
STM
   send test message
   service test model
   structural test model
   supersonic tactical missile

STME
stellar television monitor equipment
STMGR
station manager
STMIS
system test manufacturing information system
STMNT
statement
STMP
single track master operational recording tape processing
STMT
statement
STMU
special test and maintenance unit
STN
satellite tracking network
station
strategic air command telephone network
STO
system test objectives
system test operator
STOCS
small terminal oriented computer system
STOKPAC
stock control package
STOL
short takeoff and landing
slow takeoff and landing
STOP
storage protector
STOR
segment table origin register
storage
STORET
storage and retrieval
STORLAB
space technology operations and research laboratory
STORM
statistically oriented matrix program
statistically oriented matrix programming
STP
selective tape print
simultaneous track processor
space technology products
space test program
standard program
standard temperature and pressure
system test plan
STPF
shield test pool facility
STPO
science and technology policy office

STPS
series tuned parallel stabilized
STR
sea test range
start address register
store
strobe
submarine thermal reactor
synchronous transmit receive
synchronous transmitter receiver
STRAD
signal transmission reception and distribution
switching, transmitting, receiving and distributing
STRADAP
storm radar data processor
STRAM
synchronous transmit receive access method
STRANGE
sage tracking and guidance evaluation
STRAP
stretch assembly program
STRATCOM
strategic communications
STRATCOMMEX
strategic communications military exchange
STRATOSCOPE
stratosphere telescope
STRAW
simultaneous tape read and write
STRESS
structural engineering system solver
STRINGS
stellar inertial guidance system
STRIP
standard taped routines for image processing
STRIVE
standard techniques for reporting information on value engineering
STRL
schottky transistor resistor logic
STROBE
satellite tracking of balloons and emergencies
STROBES
shared time repair of big electronic systems
STRUDL
structural design language
STS
satellite to satellite
satellite tracking station
scientific terminal system
ship to shore
space technology satellite

standard technical specification
stockpile to target sequence
structural transition section
STSS
  series tuned series stabilized
STSV
  satellite to space vehicle
STSWITCH
  single throw switch
STT
  short time test
  single transition time
  start timing
  store tag
STTL
  schottky clamped transistor transistor logic
  schottky transistor transistor logic
STU
  short ton unit
  systems test unit
STV
  satellite test vehicle
  separation test vehicle
  small test vessel
  space test vehicle
  special test vehicle
  standard test vehicle
  standardized test vehicle
  structural test vehicle
  subpersonic test vehicle
  surveillance television
STVC
  space thermal vacuum chamber
STW
  store word
  system tape writer
STWP
  society of technical writers and publishers
STX
  start of text
  start of text character
STY
  space time yield
STZ
  store zero
SU
  selectable unit
  sensation unit
  service unit
  set up
  signal unit
  sonics and ultrasonics
  start up
  storage unit
  support

  suppressor
  surface to underwater
  symbolic unit
SUA
  state universities association
SUAS
  system for upper atmospheric sounding
SUB
  subdrift
  sublevel
  submarine
  substations
  substitute
  substitute character
  substitution
SUBASSY
  subassembly
SUBDIZ
  submarine defense identification zone
SUBIC
  submarine integrated control
SUBL
  sublimation point
SUBROCK
  submarine rocket
SUBSTA
  substation
SUBSYS
  subsystem
SUBTEL
  submarine telegraph
  submarine telephone
SUDT
  silicon unilateral diffused transistor
SUE
  strontium unit equivalent
  sudden expansion
  system user engineered
SUHL
  sylvania ultra high level logic
  sylvania universal high level logic
SUI
  stanford university institute for plasma research
SUIAP
  simplified unit invoice accounting plan system
SUM
  shallow underwater mobile
  summator
  surface to underwater missile
  system utilisation monitor
SUMC
  space ultrareliable modular computer

SUMM
   summarize
SUMMIT
   supervisor of multiprogramming multiprocessing, interactive time sharing
SUMS
   sperry universal automatic computer material system
SUMT
   sequential unconstrained maximization technique
   sequential unconstrained minimization technique
SUN
   symbols, units and nomenclature
SUP
   supervisor
   supply
   suppress
   suppressor
SUPARCO
   space and upper atmosphere research committee
SUPER
   superheterodyne
SUPERB
   supervisor b
SUPERPH
   superphantom
SUPO
   superpower water boiler
   superpower
SUPP
   support
SUPROX
   successive approximation
SUPRVRADSTA
   supervisory radio station
SUR
   surface
SURANO
   surface radar and navigation operation
SURCAL
   surveillance calibration
SURE
   symbolic utility revenue environment
SURGE
   sorting, updating, report generating
SURV
   surveillance
   survey
SURVAL
   simulator universal radar variability library
SURWAC
   surface water automatic computer

SUS
   saybolt universal second
   silicon unidirectional switch
   silicon unilateral switch
   single underwater sound
SUSAN
   system utilizing signal processing for automatic navigation
SUSIE
   sequential unmanned scanning and indicating equipment
   stock updating sales invoicing electronically
SUSY
   subsystem
   survey system
SUT
   system under test
SUTHEASTCON
   south eastern convention
SUV
   saybolt universal viscosity
SV
   safety valve
   saponification value
   self-verification
   set value
   simulated video
   single silk varnish
   slowed down video
   sluice valve
   space vehicle
   stop valve
   supervisor
SVA
   shared virtual area
SVBT
   space vehicle booster test
SVC
   service
   serviced
   servicing
   supervisor call
   supervisor call instruction
   switched virtual circuit
SVCE
   service
SVCG
   servicing
SVCT
   supervisor call address table
SVD
   simultaneous voice data
SVE
   space vehicle electronics

sum of vector elements
swept volume efficiency
SVEA
  scioto valley electric association
  slowly varying envelope
    approximation
SVF
  set vertical format
  standard volume flow
SVIC
  shock and vibration information center
SVLL
  short vertical lower left
SVLR
  short vertical lower right
SVO
  servo
SVP
  saturated vapor pressure
  service processor
  software verification plan
  surge voltage protection
SVR
  software verification report
  supply voltage rejection
SVS
  space vehicle simulator
  space vehicle system
  stationary control variable speed
  supervisory signal
SVTL
  services valve test laboratory
SVTP
  sound velocity, temperature and pressure
SVTSV
  space vehicle to space vehicle
SVUL
  short vertical upper left
SVUR
  short vertical upper right
SW
  sandwich wound
  shock wave
  short wave
  single weight
  software
  specific weight
  specification of wiring
  standing wave
  stationary wave
  status word
  steel wire
  struthers wells
  switch
  switchband wound

SWA
  scheduler work area
  single wire armored
SWAC
  standard western automatic
    computer
SWACS
  space warning and control system
SWADS
  scheduler work area data set
SWALM
  switch alarm
SWAMI
  software aided multipoint input
  standing wave area motion indicator
SWAT
  sidewinder ic acquisition track
  stress wave analysis technique
SWB
  short wheelbase
SWBD
  switchboard
SWBDOP
  switchboard operator
SWBP
  service water booster pumpe
SWC
  soft wired control
  solid waste cask
  surge withstand capability
  switching control
SWCL
  sea water conversion laboratory
SWCS
  space warning and control
    system
SWD
  smaller word
SWE
  shift word, extracting
  society of women engineers
SWEB
  south wales electricity board
SWEE
  southwest electronic exhibition
SWET
  simulated water entry test
SWF
  short wave fade out
  sudden wave fade out
SWFR
  slow write fast read
SWG
  standard wire gage
  stubs wire gage

SWGR
  switchgear
SWI
  short-wave interference
  standing wave indicator
  switch
SWIEEECO
  southwestern ieee conference
SWIFR
  slow writing fast reading
SWIFT
  sequential weight increasing factor technique
  society for worldwide interbank financial telecommunication
  software implemented friden translator
SWINC
  soft wired integrated numerical controller
SWINGR
  sweep integrator
SWITT
  surface wave independent tap transducer
SWL
  safe working load
  short wave listener
  short wavelength limit
SWM
  stewart warner microcircuits
SWO
  signal wireless officer
  squadron wireless officer
SWOP
  stereo wave observation project
  structural weight optimization program
SWP
  safe working pressure
  service water pump
SWPA
  southwestern power administration
SWPP
  service water pressurization pump
  southwest power pool
SWR
  service water reservoir
  standing wave ratio
  steel wire rope
  stepwise refinement
SWRI
  southwestern research institute
SWS
  service water system
  shift word substituting
  single white silk
  space weapon system
  stripline with stud

SWSR
  standing wave signal ratio
SWST
  service water storage tank
SWT
  supersonic wind tunnel
SWTL
  surface wave transmission line
SWU
  separative work unit
SWVR
  standing wave voltage ratio
SWW
  severe weather warning
SX
  simplex
  simplex signaling
SXA
  store index in address
SXAP
  soft x-ray appearance potential
SXD
  store index in decrement
SXT
  space sextant
SY
  symbol
  synchronized
  synchronoscope
SYC
  symbolic corrector
SYCATE
  symptom cause test
SYCOM
  synchronous communications
SYCR
  synchronize
SYDAS
  system data acquisition system
SYFA
  system for access
SYM
  symmetrical
SYMBOL
  system for mass balancing in off line
SYMPAC
  symbolic program for automatic control
  symbolic programming for automatic control
SYN
  synchronism
  synchronization
  synchronize
  synchronizing
  synchronous

synchronous idle character
synchronous working
SYNAC
synthesis of aircraft
SYNC
synchronism
synchronization
synchronize
synchronizer
synchronizing
synchronizing character
synchronous
SYNCCODE
synchronization code
SYNCD
synchronized
SYNCG
synchronizing
SYNCH
synchronization
SYNCIN
synchronization input
SYNCOM
synchronous communication satellite
synchronous orbiting communications
synchronous orbiting communications satellite
SYNCOUT
synchronization output
SYNCR
synchronizer
SYNCS
synchronous
SYNCSCP
synchronoscope
SYNSCP
synchronoscope
SYNSEM
syntax and semantics
SYNTOL
syntagmatic organization of language
SYS
system
SYSCMA
system core image library maintenance program
SYSDEV
systems developement
SYSEC
system synthesizer and evaluation center
SYSGEN
system generation
system generator
SYSIN
system input

SYSIPT
system input unit
SYSLOG
system log
SYSOUT
system output
SYSPOP
system programmed operator
SYST
system
SYSTRAN
systems analysis translator
SYSTSW
system software
SYSVER
system specification verification
SYSX
systems exchange
SZ
size
SZI
service zone indication
SZVR
silicon zener voltage regulator

T

T
tank
target
tee
telemetering
telephone
teletype
temperature
tera
terminal
tesla
thrust
time
time constant
timer
toggle
toggle flipflop
ton
tone
torque
track
transfer
transformer
transistor
transmission
transmitter

transmitting
transponder
trigger
trimmer
triode
tritium
troy system
true
trunk
tube
tubular
TA
  tape
  tape adapter
  target
  target area
  technical advisor
  telegraphic address
  telephone apparatus
  terminal address
  test announcer
  track address
  trank amplifier
  transactional analysis
  true attitude
  trunk amplifier
  turbulence amplifier
  type approval
TAA
  television appliance association
  time and attendance
  trans australia airlines
  transfer and accounability
  transportation association of america
TAAM
  theoretical and applied mechanics
  tomahawk airfield attack missile
TAAR
  target area analysis radar
TAAS
  three axes attitude sensor
TAB
  tabular language
  tabulate
  tabulating
  tabulating machine
  tabulation
  tabulator
  tape automated bonding
  technical abstract bulletin
  technical activities board
  technical advisory board
  title abstract bulletin
  tone answer back
  top and bottom
  totalizer agency boards
  turned and bored
TABC
  tabulator character
TABS
  terminal access to batch service
TABSIM
  table simulation
  table simulator
  tabulating equipment simulation
  tabulating equipment simulator
  tabulator simulator
TABSOL
  tabular systems oriented language
TABSTONE
  target and background signal to noise evaluation
  target and background signal to noise experiment
TABTECHNIQUE
  tape automated bonding technique
TAC
  tactical air command
  tape adapter cabinet
  targeting and control
  technical activities commitee
  technical area coordinator
  technical assistance committee
  technical assistance contract
  television advisory committee
  time and charges
  time to amplitude converter
  tokyo automatic computer
  total automatic color
  tracking accuracy control
  transac assembler compiler
  transformer analog computer
  transistorized automatic computer
  transistorized automatic control
  translator assembler compiler
  transmitter assembler compiler
  trapped air cushion
TACALS
  tactical air control and landing system
TACAN
  tactical air navigation
TACANDME
  tactical air navigation distance measuring equipment
TACC
  tactical air control center
TACCAR
  time averaged clutter coherent airborne radar

TACCS
  tactical air command control system
TACDEN
  tactical data entry
  tactical data entry device
TACE
  turbine automatic control equipment
TACF
  temporary alteration control form
TACFIRE
  tactical fire direction system
TACH
  tachometer generator
TACL
  time and cycle log
TACMAR
  tactical memory address register
  tactical multifunction array radar
TACO
  technical appliance corporation
TACODA
  target coordinate data
TACOL
  thinned aperture computed lens
TACPOL
  tactical procedure oriented language
TACR
  time and cycle record
TACS
  tactical air control system
  talker active state
  technical assignment control system
  test assembly conditioning station
  theater area communications system
TACSAT
  tactical satellite
TACSATCOM
  tactical satellite communication
TACT
  transistor and component
  tester
TACV
  tracked air cushion vehicle
TAD
  tactical air direction
  target acquisition data
  technical acceptance demonstration
  telemetry analog to digital
  telemetry and data
  test acceptance document
  thrust augmented delta
  top assembly drawing
  transmissio and distribution
TADA
  tracking and data acquisition

TADC
  tactical air direction center
TADIC
  telemetry analog to digital information converter
TADIL
  tactical digital information link
TADOR
  table data organization and reduction
TADP
  terminal area distribution processing
TADS
  tactical automatic digital switching talker addressed state
  teletypewriter automatic dispatch system
TADSS
  tactical automatic digital switching system
TAE
  technician aeronautical engineering
  test and evaluation
  time and event
TAEC
  thailand atomic energy commission for peace
TEAM
  terminal area energy management
TAEREC
  time and events recorder
TAF
  top of active fuel
  transaxle fluid
TAFCSD
  total active ferderal commissioned service to date
TAFMSD
  total active ferderal military service to date
TAFOR
  terminal airport forecast
TAFUBAR
  things are fouled up beyond all recognition
TAFX
  tapping fixture
TAG
  technical advisory group
  technical assitance group
  terminating and grounding
  test automation growth
  the acronym generator
  time automated grid
  tongued and grooved
  transient analysis generator
TAGIS
  tracking and ground instrumentation system
TAGRET
  thermal advanced gas cooled reactor exploiting thorium

TAGS
    tactical aircraft guidance system
TAHA
    tapered aperture horn antenna
TAI
    time to autoignition
TAIC
    tokyo atomic industrial consortium
TAL
    terminal application language
TALAR
    tactical approach and landing radar
    tactical landing approach radar
TALT
    tracking altitude
TAM
    tactical air missile
    telecommunication access method
    telephone answering machine
    teleprocessing access method
    television audience measurement
    terminal access method
    the access method
    time and materials
TAMCO
    training aid for mobidic console operators
TAMIS
    telemetric automated microbial dentification system
TAMP
    tactical antimissile measurement program
TAMPNL
    triggering and monitoring panel
TAN
    tangent
    total acid number
TANDEL
    temperature autostabilizing nonlinear dielectric element
TAO
    technical assistance order
    test and operation
TAOC
    tactical air operations center
TAP
    target and penetration
    term availability plan
    terminal applications package
    terrestrial auxiliary power
    thermal analyzer program
    time sharing accounting package
    time sharing assembler program
    time sharing assembly program
TAPAC
    tape automatic positioning and control

TAPAK
    tape pack
TAPAT
    tape programmed automatic tester
TAPE
    tape automatic preparation equipment
    technical advisory panel for electronics
TAPEX
    tape executive program
TAPP
    two axes pneumatic pickup
TAPPI
    technical association of the pulp and paper industry
TAPRE
    tracking in an active and passive radar environment
TAPS
    tactical area positioning system
    turboalternator power system
TAQA
    test and quality assurance
TAR
    technical assistance request
    temporary accumulator
    temporary accumulator register
    terrain avoidance radar
    test and return
    thrust augmented rocket
    track address register
    trajectory analysis room
    transaction area
    transmit and receive
    transmitter and receiver
TARA
    terrain avoidance radar
TARAC
    terminology, aids, references, applications and coordination
TARAN
    tactical attack radar and navigation
    tactical attack radar and navigator
TARBIT
    three axes rout byro inertial tracker
TARC
    television allocation research committee
    television allocation research council
    through axis rotational control
TARE
    telegraph automatic relay equipment
    telemetry automatic reduction equipment
    transistor analysis recording equipment
TAREF
    the acronym generator reference

TARFU
   things are really fouled up
TARGET
   thermal advanced reactor gas coumpled exploiting thorium
TARIT
   telegraph automatic routing in the field
TARMAC
   terminal area radar moving aircraft
TARP
   tactical airborne recording package
   technical assistance research programs
   test and repair processor
TARPS
   tactical aerial reconnaisance pod system
   true and relative motion plotting system
TARS
   technical aircraft reliability statistics
   terrain analog radar simulator
   terrain and radar simulator
   three axes reference system
TART
   twin accelerator ring transfer
TAS
   target acquisition system
   telegraphy with automatic switching
   telephone answering service
   teleprogrammer assembly system
   temperature actuated switch
   terminal address selector
   true air speed
TASC
   terminal area sequence and control
   terminal area sequencing and control
TASCON
   television automatic sequence control
TASD
   tactical action situation display
TASI
   time assignment by speech interpolation
   time assignment speech interpolation
TASK
   temporary assembled skeleton
TASO
   television allocation study organization
TASR
   temperature auto stabilizing regime
   terminal area surveillance radar
TASS
   tactical avionics system simulator
   tech assembly system
   theater army signal system
TASSA
   the army signal supply agency

TAST
   thermoacoustic sensing technique
TAT
   target aircraft transmitter
   telephone and telegraph
   thrust augmented thor
   transatlantictelephone
   tuned aperiodic tuned
   turbine trip and trottle valve
   type approval test
TATAN
   radar and television aid to navigation
TATC
   terminal air traffic control
   transatlantic telephone cable
TATCS
   terminal air traffic control system
TATS
   tactical transmission system
TATSA
   transportation aircraft test and support activity
TAU
   tape adapter unit
TAVE
   thor agena vibration experiment
TAVET
   temperature, acceleration, vibration environmental tester
TAVG
   temperature average
TAWAR
   tactical all weather attack requirements
TAWCS
   tactical air weapons control system
TAWDS
   target acquisition weapon delivery system
TAWS
   thomasville aircraft and warning station
TB
   talk back
   technical bulletin
   terminal block
   terminal board
   time bandwidth
   time base
   transmitter blocker
   transmitter buffer
   triple braided
   true bearing
   trunk barrier
TBA
   tires, batteries, and accessories
   to be activated
   to be added

TBAX
  tube axial
TBC
  tie line bias control
  toss bomb computer
  trunk block connector
TBCF
  time buffered coarse fine
TBD
  target bearing designator
TBE
  time base error
  total binding energy
TBED
  time base error difference
TBF
  tail bomb fuse
TBI
  target bearing indicator
  threaded blind inserts
  to be initiated
TBIG
  terbium iron garnet
TBIRD
  terrestrial ballistic infrared development
TBL
  table
  terminal ballistics laboratory
TBLE
  trouble
TBM
  tactical ballistic missile
  terabit memory
  terabit memory system
TBMX
  tactical ballistic missile experiment
TBO
  time between overhauls
TBP
  twisted bonded pair
TBR
  temporary base register
  to be released
TBS
  tape and buffer system
  taut band suspension
  terminal business system
  tokyo broadcasting system
  tanslator bail switch
TBT
  transitional butterworth
  thomson
TBW
  time bandwidth

TBWP
  triple braid weatherproof
  triple braided weatherproof
TBX
  tactical ballistic missile experiment
  tactical range ballistic missile
TC
  tactical computer
  tank circuit
  tantalum capacitor
  tape core
  technical characteristics
  technical committee
  technical control
  telecommunication
  telecomputing corporation
  telefunken computer ag
  telephone center
  telephone central office
  television receiver
  temperature coefficient
  temperature compensated
  temperature compensation
  temperature controller
  temperature in degrees centigrade
  teracycle
  terminal computer
  terminal concentrator
  termocouple
  test case specification
  test conductor
  test console
  test controller
  thermal conductivity
  thermocouple
  thermocurrent
  thrust chamber
  time and charge
  time closing
  time compression
  time constant
  time controlled
  time to computation
  timed closing
  timing channel
  tinned copper
  toll center
  tone control
  top cap
  towed cable
  tracking camera
  traffic control
  transaction code
  transfer control
  transistorized carrier

transitional control
transmission control
transmission controller
transmitter tuning circuit
trim coil
trip cell
trip coil
true course
tuned circuit
tungsten carbide
type certificate
TCA
  tactical combat aircraft
  task control area
  telecommunication association
  telemetering control assembly
  temperature control amplifier
  terminal control area
  thermal critical assembly
  transcanada airlines
  transfer control a register
TCAI
  tutorial computer assisted instruction
Telecommunications access method
  telegraph construction and maintenance
TCAP
  tactical channel assignment panel
TCAT
  test coverage analysis tool
TCB
  tape control block
  task control block
  technical coordinator bulletin
  the computer bulletin
TCBM
  transcontinental ballistic missile
TCBV
  temperature coefficient of breakdown voltage
TCC
  tactical control computer
  tactical control console
  technical computing center
  telecommunications coordinating committee
  telegraph condenser company
  television control center
  temperature coefficient of capacitance
  test conductor console
  test control center
  test controller console
  thermofor catalytic cracking
  time compression coding
  total cobol capability
  tracking and control center
  tractor computing corporation

traffic control center
traffic control complex
transfer channel control
transportable cassette converter
triple cotton covered
TCCELL
  thermal conductivity cell
TCCN
  transcanada computer communications network
TCCO
  temperature comensated crystal oscillator
  temperature controlled crystal oscillator
TCD
  telemetry and command data
  temperature control device for crystal units
  test communications division
  thermal conductivity detector
  thyratron core driver
  transistor controlled delay
TCDETECTOR
  thermal conductivity detector
TCE
  talker communication error
  telemetry checkout equipment
  test connection equipment
  total composite error
  trichloroethylene
TCEA
  training center for experimental aerodynamics
TCED
  thrust control exploratory development
TCEWA
  twin cities electronic wholesalers association
TCF
  technical control facility
TCG
  time controlled gain
  transponder control group
  tune controlled gain
TCH
  transfer on channel
TCI
  technology for communications, international
  telemetry components information
  terrain clearance indicator
  theoretical chemistry institute
  trunk cut in
TCINPUT
  thermocouple input
TCK
  track
  tracking

TCL
  tansistor coupled logic
  telephone cables, limited
  time and cycle log
  tranfer chemcial laser
  transistor coupled logic
  transmit clock
TCLD
  thermocouple lead detector
TCM
  telemetry code modulation
  temperature compensated mask
  temperature compensation
  terminal to computer multiplexer
  time compression multiplex
  tone code modulation
  translator cam magnet
TCMA
  tabulating card manufacturers' association
TCMF
  touch calling multifrequency
TCN
  track channel number
  transfer on channel not in operation
TCO
  temperature coefficient of offset
  time and charges, operate
  transfer on channel in operation
  trunk cut off
TCOA
  translational control a
TCOB
  translational control b
TCOCD
  thermocouple open circuit detection
TCP
  telecommunications processor
  test checkout procedure
  the acronym generator converter program
  thermofor continuous percolation
  thrust chamber pressure
  traffic control post
TCPC
  tab card punch control
TCPLD
  tunable compound phase locked demodulator
TCR
  tactical control radar
  telemetry compression routine
  television cathode ray
  temperature coefficient of resistance
  temperature coefficient of resistivity
  temperature coefficient of resistor
  thermal coefficient of resistance

  transfer control register
  transistorized car radio
  transportable cassette recorder
TCS
  tactical call sign
  telecommunication control system
  telecommunication system
  telephone conference summary
  temperature coefficient of sensitivity
  temporary change of station
  terminal control system
  terminal count sequence
  terminal countdown sequencer
  thermal conditioning service
  tone call squelch
  traffic control station
  transmission control system
  transportable communications system
  transportation and communication service
TCSC
  the computer software company
  trainer control and simulation computer
TCSL
  transistor current switching logic
TCT
  terminal control table
  toll connecting trunk
  tool change time
  translator and code treatment frame
TCU
  tape control unit
  target control unit
  teletypewriter control unit
  terminal control unit
  test control unit
  time controlling unit
  torpedo control unit
  transmission control unit
TCV
  temperature control valve
  terminal configured vehicle
TCVC
  tape control via console
TCVR
  transceiver
TCW
  time code word
  tinned copper weld
TCWG
  telecommunications working group
TD
  dabular data
  tape distributor
  tape drive

technical data
telecommunications department
telemetry data
telephone depot
temperature differential
terminal distributor
test and diagnostics
test data
tst design specification
testing device
thoria dispersed
threshold decoding
time delay
time device
time difference
time division
timing device
track data
track display
transaction driven
transducer
transfer dolly
transmit data
transmitted data
transmitter distributor
trapped domain
tunnel diode

TDA
target docking adapter
tracking and data acquisition
tunnel diode amplifier

TDAFP
turbine driven auxiliary feed pump

TDAS
tactical data automation system
tracking and data acquisition system

TDC
technical data center
technical development capital
technical development center
technical development corporation
thermal diffusion coefficient
time data card
time delay closing
time of day clock
tone digital command
top dead center
top desk computer
totally decentralized control
track data central
training device center
transistor digital circuit
transistor digital control
transistorized digital circuit
transmission distribution center
two-dimensional finite cylinder
transport code for computer

TDCC
tactical data communications center
transportation data coordinating committee

TDCM
transistor driver core memory

TDCO
torpedo data computer operator

TDCT
track data central tables
tunnel diode charge transformer

TDCTL
tunnel diode charge transformer logic

TDCU
target data control unit

TDD
target detecting device
target detection device
technical data digest
telemetry data digitizer
test design description
timing data distributor

TDDL
time division data link

TDDLPO
time division data link print-out

TDDO
time delay drop-out

TDDR
technical data department report

TDDS
two-dimensional deflection system

TDE
time displacement error

TDEC
technical development evaluation center
technical division and engineering center
telephone line digital error checking

TDEFWP
turbine driven emergency feed water pump

TDF
task deletion form
transformer differential
trunk distributing frame
trunk distribution frame
two degree of freedom

TDFL
tunnel diode fet logic

TDG
test data generator

TDH
total dynamic head

TDI
    target data inventory
    target doppler indicator
    time delay and integration
    tool and die institute
TDIA
    transient data input area
TDIC
    target data input computer
TDIO
    timing data input output
TDIPRE
    target data inventory master tape preparation
TDL
    tapped delay line
    telemetry data link
    tunnel diode logic
TDM
    tandem
    telecommunications data link monitor
    telemetric data monitor
    time division multiplex
    time division multiplexer
    time division multiplexing
    torpedo detection modification
    trouble detection and monitoring
TDMA
    time distributed multiple access
    time division multiple access
TDMG
    telegraph and data message generator
TDMS
    telegraph distortion measurement set
    telegraph distortion measuring set
    telegraph distortion measuring system
    telegraphic distortion measuring set
    time shared data management system
    transmission distortion measuring set
TDN
    target doppler nullifier
TDO
    time delay opening
    transistor dip oscillator
TDOS
    tape disk operating system
TDP
    technical development plan
    teledata processing
    tracking data processor
    traffic data processing
    traffic data processor
TDPM
    time domain prony method
TDPS
    tactical data processing system

TDPSK
    time differential phase shift keying
TDR
    target discrimination radar
    technical data relay
    technical data report
    test data report
    test disable reset
    time delay relay
    time domain reflectometer
    time domain reflectometry
    torque differential receiver
TDRS
    tracking and data relay satellite
TDRSS
    tracking and data relay satellite system
TDS
    tactical data system
    tactical display system
    tape data selector
    target designation system
    technical data system
    telemetering decommutation system
    test data sheet
    test data specification
    time division switch
    time division switching
    time domain spectroscopy
    total dissolved solids
    track data simulation
    track data storage
    tracking and data system
    transaction driven system
    transistor display and data handling system
    translation and docking simulator
    tuned laser differential spectrometry
TDSA
    telegraph and data signal analyser
TDSSC
    tone dial switching system control
TDT
    target designation transmitter
    tunnel diode transducer
TDTL
    tunnel diode transistor logic
TDV
    technology development vehicle
    touch-down velocity
TDW
    tons deadweight
TDX
    time division exchange
TDZ
    touch-down zone

TE
  tape error
  telemetry event
  tension equalizer
  terminal
  test equipment
  testing equipment
  thermal element
  thermal expansion load
  thermoelectric
  threshold element
  time equipment
  totally enclosed
  trailing edge
  transistor equivalent
  transverse electric
  transverse electrostatic
  tuning eye

TEA
  technical engineers association
  tetraethylammonium
  texas electronic association
  transferred electron amplifier
  transversely excited atomospheric
  transversely excited atomospheric lasers
  transcersely excited atmospheric pressure
  tunnel emission amplifier
  tyrethylaluminum

TEADDA
  teledyne electrically alterable digital differential analyzer

TEAM
  technique for evaluation and analysis of maintainability
  terminology evaluation and acquisition method
  the electronic association of missouri

TEAMS
  test evaluation and monitoring system

TEASE
  tracking errors and simulation evaluation

TEAWC
  totally enclosed air water cooled

TEB
  tape error block

TEC
  tactical electromagnetic coordinator
  test evaluation and control
  thermionic energy conversion
  thermoelectric cooler
  thermonic energy conversion
  tokyo electric co
  total estimated cost
  transistor electronics corporation

TECHCLUB
  technical club of dallas

TECHSPECS
  technical specifications

TED
  technological equipment department
  teleprinter error detector
  test engineering division
  transferred electron device
  translation error detector

TEDAR
  telemetered data reduction

TEDIT
  tape edit

TEE
  telecommunications engineering establishment

TEEAR
  test equipment error analysis report

TEF
  total energy feasibility
  transfer on end of file

TEFC
  totally enclosed fan cooled
  totally enclosed force cooled

TEG
  test element group
  thermoelectric generator

TEI
  transearth injection
  transfer on error indication

TEIC
  tissue equivalent ionization chamber

TEKTEST
  tektronik test programming language

TEL
  telecommunication
  telegraph
  telegraphic
  telegraphy
  telephone
  telephone group
  telephone station
  telephony
  teletype
  teletype writer
  television
  tetraethyl lead
  tokyo electron ltd
  total energy loss
  transporter, erector, launcher

TELATEL
  telephone and telegraph

TELCO
   telephone communications
   telephone company
TELCON
   telephone conversation
TELD
   transferred electron logic device
TELDEC
   telefunken decca
TELE
   telegraph
TELECOM
   telecommunications
TELECON
   telephone conversation
   teletypewriter conference
TELEDAC
   telemetric data converter
TELEG
   telegraph
TELENET
   telenet communications
TELENGR
   telephone engineer
TELEPH
   telephone
TELERAN
   television and radar navigation
TELESAT
   telecommunications satellite
TELESIM
   teletypewriter simulator
TELEVISIONSOCJ
   television society journal
TELEX
   telegraph exchange
   teleprinter exchange
   teleprocessing executive
   telex computer products
TELN
   telephone
TELNO
   telephone number
TELOPS
   telemetry on-line processing system
TELRY
   telegraph reply
TELS
   turbine engine load simulator
TELSCOM
   telemetry surveillance communications
TELSEC
   telegraph section
   telephone section

TELSIM
   teletype simulator
   teletypewriter simulator
TELSTAR
   television star
TELSTATS
   telemetry station system
TELTAP
   telephone tape
TELTRAC
   telemetry tracking
TELUS
   telemetric universal sensor
TEM
   telex extended memory
   template
   test excursion module
   transmission electron microscope
   transmission electron microscopy
   transmission engineering memorandum
   transverse electric and magnetic
   transverse electromagnetic
   transverse electromagnetic mode
TEMA
   tank equipment manufacturers' association
   telecommunication engineering and manufacturing association
   test macro
   tubular exchanger manufacturers' association
TEMMODE
   transversal electromagnetic mode
TEMP
   temporary
   texas educational microwave project
   the expanded memory print program
TEMPO
   total evaluation of management and production output
TENS
   tension
TENV
   totally enclosed, non ventilated
TEO
   transferred electron oscillator
TEOM
   transformer environment overcurrent monitor
TEP
   tetraethyl pyrophosphate
TEPG
   thermionic electrical power generator
TEPP
   tetraethyl pyrophosphate
TEPRSSC
   technical electronic product radiation safety standards committee

TER
   teleprinter retransmitting
   transmission engineering recommendations
   transmission equivalent resistance
   triple ejection rack
TERAC
   tactical electromagnetic readiness advisory council
TERCOM
   terrain contour matching
TEREC
   tactical electromagnetic reconnaissance
TERM
   terminal
TERP
   terrain elevation retrieval program
TERRA
   terrain evaluation and retrieval for road alignment
TERS
   tactical electronic reconnaissance system
TES
   telemetering evaluation station
   texas electric service
   text editing system
   tidal electric station
TESA
   television and electronics service association
TESCO
   texas electric service company
TESS
   tactical electromagnetic systems study
   total energy systems service
TEST
   thesaurus of engineering and scientific terms
TESTAS
   turkish electronics and trade assocication
TESTRAN
   test translator
TET
   total elapsed time
   traveling wave tube
TETRA
   terminal tracking telescope
   terminal trajectory telescope
TETRAC
   tension truss antenna concept
TETROON
   tethered meteorological balloon
TEU
   telemetering equipment unit
   transducer excitation unit
TEVROC
   tailored exhaust velocity rocket

TEW
   tactical early warning
TEWC
   totally enclosed, water cooled
TEWR
   thrust to earth weight ratio
TEX
   teleprinter exchange service
   telex
TEXTINDY
   textile industry
TEXTIR
   text indexing and retrieval
TF
   talker function
   test facility
   test fixture
   test to failure
   thermionic field
   thin film
   threshold function
   time frequency
   time of fall
   total float
   transfer
   transfer function
   transformer
   transversal filter
TFA
   technology forecasting and assessment
   telex file adapter
   transfer function analyzer
TFAD
   thin film active device
TFC
   thin film circuit
   traffic
   transfer function computer
TFCS
   torpedo fire control system
TFD
   television feasibility demonstration
   time frequency dissemination
   total frequency deviation
TFE
   tetrafluorethylene
   thermal field emission
   thermionic fuel element
TFET
   thin layer field effect transistor
TFF
   thermoplastic covered fixture wire, flexible stranding
   thin film fet
   toggle flip-flop

TFFET
  thin film fet
  thin film field effect transistor
TFFT
  thin film field effect
TFG
  transmit format generator
TFHC
  thick film hybrid circuit
TFI
  the fertilizer institute
TFIC
  thin film integrated circuit
TFL
  transfomerless
  transient fault locator
TFLIPFLOP
  trigger flip-flop
TFM
  time quantized frequency modulation
TFO
  tuning fork oscillator
TFOL
  tape file octal load
TFR
  television film recorder
  terrain following radar
  trouble and failure report
TFSK
  time frequency shift keying
TFT
  thin film field effect transistor
  thin film technique
  thin film technology
  thin film transistor
  threshold failure temperatures
  time to frequency transformation
TFTC
  thin film thermocouple
TFU
  telecommunications flying unit
TG
  tangent
  tape gage
  telegram
  telegraph
  terminal guidance
  terminator group
  timing gage
  torque generator
  tracking and guidance
  transfer gate
  transmission group
  trigger
  tuned grid

TGA
  thermogravimetric analysis
TGAM
  training guided aircraft missile
TGC
  terminator group controller
  time gain control
  transmit gain control
  transmitter gain control
TGCA
  transportable ground control approach
TGCS
  transportable ground communictions station
TGF
  through group filter
TGI
  target intensifier
TGL
  toggle
  triangular guideline
TGLORAN
  traffic guidance loran
TGM
  telegram
TGS
  telemetry ground station
  transfer generator system
  translator generator system
  triglycine sulfate
  true ground speed
TGSE
  telemetry ground support equipment
TGT
  target
  transformational grammar tester
TGTP
  tuned grid tuned plate
  tuned grid tuned plate circuit
TGZMP
  temperature gradient zone melting process
TH
  tape handler
  telegraph
  tracking head
  transmission header
  true heading
THC
  thermal converter
  third harmonic distortion
  thrust hand controller
  total harmonic distortion

THD
  third harmonic distortion
  thread
  total harmonic distortion
THE
  technical help to exporters
THERM
  thermal
  thermometer
  thermostat
  thermostatic
THERMISTOR
  thermal resistor
THERMOC
  thermocouple
THERP
  technique for human error rate prediction
THFA
  three conductor, heat and flame resistant, armored
THFR
  three conductor, heat and flame resistant radio
THI
  temperature humidity index
THIR
  temperature humidity infrared radiometer
THL
  tuned hybrid lattice
THOF
  triple conductor, heat, oil and flame resistant
THOMIS
  total hospital operating and medical information system
THOMOTROL
  thyratron motor control
THOPS
  tape handling operational system
THOR
  tape handling option routines
  transistorized high speed operations recorder
THORAD
  thor advanced
THP
  thousands position
  thrust horsepower
THR
  total heat rejection
THRES
  threshold
THS
  thermostat switch
THTF
  thermal hydraulic test facility
THTR
  thorium high temperature reactor
THTRA
  thorium high temperature reactor association
THY
  thyratron
THYMOTRO
  thyratron motor control
THZ
  terahertz
TI
  tape indicator
  tape inverter
  target identification
  target identifier
  target indication
  target indicator
  technicl inspection
  technology innovation
  temperature indicator
  terminator interrupt
  test instruction
  texas instruments
  thallium
  time index
  time interval
  track initiation
  traffic identification
  transfer impedance
  tuning indicator
  tuning inductance
TIAM
  terminal interactive access method
TIARA
  target illumination and recovery aid
TIAS
  target identification and acquisition system
TIB
  technical information bureau
TIBOE
  transmitting information by optical electronics
TIC
  tape identification card
  tape intersystem connection
  target intercept computer
  technical information center
  telemetry instruction conference
  temperature indicating controller
  thermionic integrated circuit
  transducer information center
  transfer in channel
TICA
  technical information center administration

TICCIT
    time shared interactive computer controlled information television
TICE
    time integral cost effectiveness
TICOL
    texas instruments cassette operating language
TICS
    teacher interactive computer system
TICTAC
    time compression tactical communications
TID
    technical information department
    technical information division
    test instrument division
    touch input device
    traveling ionospheric disturbance
TIDAR
    time delay array radar
TIDDAC
    time in deadband digital attitude control
TIDE
    tactical international data exchange
    transponder interrogation and decoding equipment
TIDES
    time division electronics switching system
TIDG
    taper isolated dynamic gain
TIDOS
    table and item documentation system
TIDS
    talker idle state
TIE
    technical information exchange
    technical integration and evaluation
TIES
    transmission and information exchange system
TIF
    tape inventory file
    task initiation form
    technical information file
    telephone influence factor
    telephone interference factor
TIFET
    thin layer field effect transistor
TIFS
    total in-flight simulator
TIG
    teletype input generator
    transearth injection geometry
    tungsten inert gas
TIGOR
    time interval gage of relays

TIH
    time in hold
TII
    table and item inventory
TIIF
    tactical image interpretation facility
TIIPS
    technically improved interference prediction system
TIL
    technical information and library services
TILL
    total initial lamp lumen
TILS
    tactical instrument landing system
TILT
    texas instruments language translator
TIM
    technical information on microfilm
    temperature independent material
    temperature indicator monitor
    terminal interface monitor
    time indicator
    time interval measurement
    time interval meter
    time meter
    tracking instrument mount
TIMIX
    texas instruments minicomupter information exchange
TIMM
    thermionic integrated micromodule
TIMNET
    timeshare international data communications network
TIMS
    telecommunications instruction module system
    the institute of management sciences
    transmission impairment measuring set
    transmission impairment measuring system
TIN
    temperature independent
TINS
    trains inertial navigation system
TIO
    television information office
    test input output
    time interval optimization
    transistorized image orthicon camera
TIOT
    task input output table
TIP
    technical information processing
    technical information program

technical information project
technology in production
teletype input processing
terminal interface message processor
terminal interface processor
total information processing
track initiation and prediction
transaction interface package
transient in core probe
traveling in core probe
TIPACS
  texas instrument planning and control system
TIPI
  tactical information processing and interpretation
TIPL
  teach information processing language
TIPP
  time phasing program
TIPS
  technical information processing system
  technology integration project
  telemetry impact prediction system
TIPSY
  task input parameter synthesizer
TIPTOP
  tape input, tape output
TIR
  target illuminating radar
  technical information release
  technical information report
  test incident report
  tolerance of inside radius
  total indicator reading
  total internal reflecting
  total internal reflection
  true indicated radius
TIREC
  tiros ice reconnaissance
TIRKS
  trunks integrated records keeping system
TIROS
  television and infrared observation satellite
  television infrared observation satellite
TIRP
  total internal reflection prism
TIRPF
  total integrated radial peaking factor
TIS
  target information sheet
  target information system
  technical information series
  technical information service
  temperature indicating switch
  test instrumentation system
  total information system
  transit injection station
TISE
  technical information service
TIT
  ternary digit
  trouble indicator trunk
TITAN
  teamster's international terminal and accounting network
TITF
  test item transmittal form
TITTL
  texas instruments transistor transistor logic
TIU
  tape identification unit
TIV
  total indicator variation
TIVICON
  texas instruments vidicon
TIX
  transfer on index
TJ
  trunk junctor
TJC
  trajectory chart
TJD
  trajectory diagram
TJF
  test jack field
TJID
  terminal job identification
TJS
  transverse junction stripe
TJT
  tri junction transistor
TK
  trunk
TKOF
  take-off
TKW
  thermal kilowatt
TL
  talk listen
  tape library
  target language
  task leader
  terminology library
  test link
  test load
  test log
  thermo luminescence
  tie line
  time limit
  total load

total loss
  transformation line
  transistor logic
  transmission level
  transmission line
  transmission loss
  trunk load
TLA
  telex line adapter
TLAB
  translation look aside buffer
TLC
  temperature level control
  thin layer chromatography
TLCC
  training launch control center
TLD
  telephone line doubler
  thermoluminescent dosimeter
TLE
  test laboratory engineer
  thin layer electrophoresis
  tracking light electronics
TLF
  terminal launch facility
  trunk line frame
TLG
  telegram
  telegraph
  telegraphy
  timing level generator
TLI
  telephone line interface
  translunar injection
TLIR
  time limited impulse response
TLK
  test link
TLLE
  twin linear loop exciter
TLLM
  temperature and liquid level monitor
TLM
  telemeter
  telemetry
  transformer load management
TLMCTLPNL
  telemetry control panel
TLO
  total loss only
TLP
  term lease plan
  threshold learning process
  top load pad

  total language processor
  transmission level point
TLR
  toll line release
TLRP
  trace last reference position
TLS
  telecommunication liaison staf
  telemetry listing submodule
  terminal landing system
  test line signal
  total logic solution
TLT
  telecommunications translator
  terminal list table
  transportable link terminal
TLTA
  two loop test apparatus
TLTP
  teletype
TLTR
  translator
TLU
  table look-up
  threshold logic unit
TLV
  television
  threshold limit value
TLX
  telex
TLZ
  transfer on less than zero
TM
  tactical missile
  tactical monitor
  tangent mechanism
  tape mark
  task memory
  technical manual
  telemetering
  telemetry
  temperature meter
  temperature monitor
  terminal multiplexer
  time and materials
  time modulation
  tone modulation
  top management
  track monitor
  traffic manager
  transaction mode
  translator code magnet
  transmission matrix
  transverse magnetic
  trunk mark

tuning meter
turing machine
twisting moment
TMA
   telecommunications manager association
   telemetry manufacturers' association
TMAMA
   textile machinery and accessory manufacturers' association
TMC
   talker per megacycle
   telephone manufacturing company
   three mode control
TMCC
   time multiplexed communication channel
TMCOMP
   telemetry computation
TMCP
   technical manual control panel
TMD
   tactical munitions dispenser
   telemetered data
   timed
   total mean downtime
   transient mass distribution code
TMDT
   total mean downtime
TME
   temperature measuring equipment
TMES
   tactical missile electrical simulator
TMF
   time marker frequency
TMG
   tactical missile group
   thermal meteoroid garment
   thermometeroid garnet
TMGE
   thermomagnetogalvanic effect
TMGS
   terrestrial magnetic guidance system
TMHF
   transit missile hold facility
TMHP
   turing machine halting problem
TMI
   telemeter magnetics, incorporated
   texas microelectronics, incorporated
   transfer on minus
   tuning meter indicator
TMIS
   television management information system
   transmission impairment measuring set
TML
   tetramethyl lead
   two mixed layer
TMLP
   thermal margin low pressure
TMM
   telex main memories
   test message monitor
TMMD
   tactical moving map display
TMN
   technical and management note transmission
TMO
   thermomagnetic optic
TMOS
   two main orbiting spacecraft
TMP
   temperature
   terminal monitor program
   theodolite measuring point
TMPRLY
   temporarily
TMPROC
   telemetry processing
TMR
   teledyne materials research
   timer
   triple modular redundancy
   triple modular redundant
   true motion radar
TMRBM
   transportable medium range ballistic missile
TMS
   tactical missile squadron
   tape management software
   telegraphy with manual switching
   telemeter transmitter
   temperature measurements society
   the metallurgy society
   time shared monitor system
   track monitor supervisor
   transmission measuring set
   turing machine system
   type, model, and series
TMT
   transmit
TMTC
   tri mode tape converter
TMTR
   thermistor
   transmitter
TMU
   time measurement unit

time multiplexer unit
twin and multiply
TMV
   triplicated majority voting
TMW
   tactical missile wing
   thermal megawatt
TMX
   tactical missile, experimental
TN
   task number
   technical note
   tennessee
   thermonuclear
   ton
   track number
   tuning
   tuning unit
TNA
   time of nearest approach
   transient network analyzer
   transistor noise analyzer
TNC
   threaded nut connector
   total numerical control
   track navigation computer
   track number conversion
   twisted nematic liquid
TNDS
   tactical navigational display system
TNFE
   twisted nemetic field effect
TNL
   technical newsletter
   terminal net loss
TNLCD
   twisted nematic liquid crystal display
TNM
   twisted nematic mode
TNO
   transfer on no overflow
TNT
   theater network television
   trinitrotoluene
   tuned not tuned
TNX
   transfer on no index
TNZ
   transfer on no zero
TNk
   thanks
TO
   take-off
   tandem outlet
   technical order
   telegraph office
   telemetry oscillator
   telephone office
   terminal office
   time of launch
   time open
   time over
   tool offset
   tracking officer
   transistor outline
   turn over
TOA
   total obligational authority
   transferred on assembly
TOB
   take-off boost
   test one bit
TOC
   television operating center
TOCS
   technological aides to creative thoughts
TOD
   technical objective directive
   time of day
   time of delivery
TODA
   third octave digital analyzer
TODC
   triple overriding dual control
TODS
   test oriented disk system
TOE
   tables of organization and equipment
   tape overlap emulator
TOF
   time of filing
   time of flight
TOGO
   time to go
TOHM
   terohmmeter
TOI
   technical operation instruction
TOJ
   time on jamming
   track on jamming
TOL
   test oriented language
TOLIP
   trajectory optimization and linearized pitch
TOLO
   time of lock-out
TOLR
   transmitting objective loudness rating

TOLTEP
   teleprocessing on-line test executive program
TOLTS
   total on-line test system
TOM
   teleprinter on multiplex
   teleprinting on multiplex
   test set, overall, missile
   toolmanager
   transistor oscillator multiplier
   translator octal mnemonic
   transparent office manager
TOMCAT
   telemetry on-line monitoring compression and transmission
TOMCIS
   test of multiple corridor identification system
TOMS
   torus oxygen monitoring system
TON
   talk only
TONLAR
   tone operated net loss adjuster receiving
TOO
   time of origin
TOOS
   transaction oriented operating system
TOP
   technical office protocol
   tool package
   top of potentiometer
   training for opportunities in programming
   transient over power
TOPACKAGE
   transistor outlines package
TOPI
   tons of paper in
TOPICS
   total on-line program information and control system
TOPL
   terminal operated production language
TOPO
   tons of paper out
TOPR
   taiwan open pool reactor
   thermoplastic optical phase recorder
TOPS
   teletype optical projection system
   the operational pert system
   thermoelectric outer plant spacecraft
   total operations processing system
TOPSI
   topside sounder, ionosphere
TOPSY
   on-line order processing system
   test operations and planning system
TOPTS
   test oriented paper tape system
TOR
   technical operations research
   teleprinter on radio
   teleprinting over radio
   teletype on radio
   thermal overload relay
   time of receipt
TORTOS
   terminal oriented real time operating system
TOS
   tactical operating system
   tape operating system
   temporarily out of service
   terminal oriented softwae
   terminal oriented system
   time ordered system
   time sharing operating system
   tiros operational satellite
   tiros operational system
TOSS
   tiros operational satellite system
   transient or steady state
TOT
   time of transmission
   time on tape
   time on target
   total
   total outage time
TOU
   trace operate unit
TOV
   transfer on overflow
TOW
   tube launched optically tracked wire guided antitank missile
TOWA
   terrain and obstacle warning and avoidance
TOWT
   take off weight
TP
   tape
   technical publication
   telemetry processor
   telephone
   teleprinter
   teleprocessing
   test panel
   test point
   test pressure
   test procedure

text processing
time pulse
timing pulse
tin plate
tracking program
train printer
transaction processing
trigger pulse
triple play
triple pole
true position
tunable pulsed
tuned plate
turbo prop engine
turn point
turning point
TPA
  tape pulse amplifier
  technical publications announcements
  test point access
  ton per annum
  transient program area
TPAD
  teleprocessing analysis and design program
TPBVP
  two point boundary value problem
TPC
  telecommunications planning committee
  telecommunications program and control
  television policy committee
  time polarity control
  total print control
  triple paper covered
TPCOMP
  tape compare
TPCU
  test power control unit
TPD
  television and promotion department
  total program diagnostic
  trivial problem discriminator
TPDT
  triple pole double throw
TPDUP
  tape duplicate
TPE
  total publishing environment
  transmission parity error
TPF
  terminal phase finalization
  two photon fluorescence
TPFW
  three phase full wave

TPG
  timing pulse generator
  transmission project group
TPH
  through plated hole
  ton per hour
TPHC
  time to pulse height converter
TPHW
  three phase half wave
TPI
  tape phase inverter
  target position indicator
  task parameter interpretation
  teeth per inch
  terminal phase initiate
  test parts list
  thread per inch
  track per inch
  trim position indicator
  turn per inch
TPINIT
  tape initializer
TPINT
  tape initial label
TPL
  telecommunications programming language
  terminal per line
  terminal programming language
  test parts list
  total peak loss
  transfer on plus
  transistorized portable laboratory
  turn per layer
TPLAB
  tape label
  tape label information
TPLAN
  test plan
TPM
  tape preventive maintenance
  telemetry processor module
  ton per month
  transmission and processing model
  tri plate module
TPMA
  thermodynamics properties of metals and alloys
TPOS
  track position
TPP
  test point pace
  test point prelaunch automatic checkout equipment

TPPD
   technical program planning division
   technical program planning document
TPPR
   tape to printer
TPR
   t pulse response
   tape programmed raw
   technical proposal requirements
   teleprinter
   telescopic photographic recorder
   terrain profile recorder
   transmitter power rating
TPRA
   tape to random access
TPRC
   thermophysical properties research center
TPROC
   test procedure specification
TPRV
   transient peak reverse voltage
TPS
   task parameter synthesizer
   teleprocessing system
   terminal per station
   thermal protection system
   thyristor power supply
   tracking antenna pedestal system
   training package system planning
   translunar propulsion stage
   tree pruning system
TPSI
   torque pressure in pounds per square inch
TPST
   triple pole single throw
TPTC
   temperature pressure test chamber
TPTG
   tuned plate tuned grid
TPTP
   tape to tape
TPU
   tape preparation unit
   task processor unit
   terminal processing unit
TPV
   thermophotovoltaic
TPY
   ton per year
TQC
   time quality cost
TQE
   technical quality evaluation
TR
   tactical radar
   tape reader
   tape recorder
   tape resident
   technical report
   technical review
   telegraph repeater
   temperature recorder
   test request
   tetrode
   thermal resistance
   time delay relay
   time of rise
   time routine
   tons registered
   torque receiver
   torque repeater
   torque synchro receiver
   trace
   tracking radar
   transfer reset
   transformation ratio
   transformer
   transformer rectifier
   transient response
   transistor
   translate
   translation
   transmit receive
   transmit receiving
   transmitter
   transmitter receiver
   transponder radar
   tunnel rectifier
TRA
   the razorback award
   triaxial recording accelerometer
TRAAC
   transit research and attitude control
TRAACS
   transit research and attitude control satellite
TRAC
   text reckoning and compiling
   tracking and communications
   transient radiation analysis by computer
TRACAL
   traffic control and landing
TRACALS
   traffic control and landing system
TRACE
   tactical readiness and checkout equipment
   tape controlled reckoning and checkout equipment
   tape controlled recording automatic checkout equipment

taxiing and routing of aircraft coordinating equipment
teleprocessing recording for analysis by the customer
test equipment for rapid automatic checkout and evaluation
time shared routines for analysis classification and evaluation
tolls recording and computing equipment
total remote assistance center
tracking and communication extraterrestrial
transaction accounting control and endorsing
transaction control and encoding
transistor radio automatic circuit evaluator
transportable automated control environment
TRACIS
traffic records and criminal justice information system
TRACOMP
tracking comparison
TRACON
terminal radar approach control facility
terminal radar control
TRACS
traffic reporting and control system
TRAD
traductrice
training research and development
TRADAT
transit data transmission system
TRADEX
target resolution and discrimination experiment
tracking radar experiment
TRADIC
transistor digital computer
transistorized airborne digital
transistorized digital computer
TRADIS
tape repeating automatic data integrating system
TRAFO
transformer
TRAIN
telerail automated information network
TRAJ
trajectory
TRAM
target recognition and attack multisensor
TRAMP
target radiation measurement program
test retrieval and memory print
time shared relational associative memory program

TRAMPS
temperature regulator and missile power supply
text information retrieval and management program system
transportation movement planing system
TRAN
computer transmission
transient
transmit
TRANDIR
translation director
TRANET
tracking network
transit network
TRANEX
transaction executive
TRANS
transaction
transfer
transformer
transistorized
transmission
transmitter
TRANSAC
transistorized automatic computer
TRANSAIEE
transactions of the aiee
TRANSC
transcribe
transcription
TRANSCEIVER
transmitter receiver
TRANSEC
transmission security
TRANSF
transformer
TRANSFDESENGR
transformer design engineer
TRANSM
transmission
TRANSRA
transistorized radar
TRANSV
transverse
TRAP
terminal radiation airborne program
tracker analysis program
TRAPATT
trapped plasma avalanche transit time
trapped plasma avalanche triggered transit
TRAPATTMODE
trapped plasma avalanche triggered transit mode

TRASER
 transformer laser
TRASTA
 training station
TRAVERSE
 traverse analysis
TRAWL
 tape read and write library
TRAX
 three axis
TRB
 transportation research board
TRC
 tape record coordinator
 technical review committee
 telemetry and remote control
 temperature recording controller
 time ratio control
 tracking, radar inputs and correlation
 transmitter circuit
 transportation research command
 type requisition code
TRCCC
 tracking radar central control console
TRCVR
 transceiver transmitter receiver
TRDET
 trouble detection
TRDTO
 tracking radar data take off
 tracking radar take off
TRE
 telecommunication research establishment
 timing read error
TREAT
 transient radiation effects automated tabulation
 transient reactor test facility
TREE
 transient radiation effects on electronics
TRF
 thermal radiation at microwave frequencies
 tuned radio frequency
TRG
 trailing
TRI
 technical research institute
 torsion reaction integrating
 transmission interface converter
 trichloroethylene
 triclinic
 triode
TRIA
 temperature removable instrument assembly

TRIAC
 triode alternating current
TRIAL
 technique for retrieving information from abstracts of literature
TRIASS
 triumph adler assembler
TRIB
 transfer rate of information bits
 tributary
TRIC
 tri camera
TRICE
 transistorized real time incremental computer, expandable
TRICON
 tri coincidence navigation
 triple coincidence navigation
TRID
 track identity
TRIDOP
 tri doppler
TRIG
 trigger
TRIGA
 training reactor, isotopes production, general atomic
TRIM
 test rules for inventory management
 thin region integral method
 tri mask process
 trimmer
TRIMCELL
 tri mask cell
TRIMIS
 tri service medical information systems
TRIMOSCIRCUIT
 triode alternating current switch and metal oxide semiconductor circuit
TRIMPROCESS
 tri mask process
TRIP
 truck routing improvement program
TRIPOLD
 transit injector polaris derived
TRITET
 triode-tetrode
TRIUMF
 tri university meson facility
TRIXIE
 transistors and nixie tube
TRK
 track
 trunk

TRL
  thermodynamics research laboratory
  transistor resistor logic
TRLR
  trailer
TRM
  thermal remanent magnetization
  time ratio modulation
TRML
  terminal
TRMPS
  temperature regulator and missile power supply
TRMT
  terminate
TRMUP
  target radiation measurement program
TRN
  technical research note
  track reference number
  transfer
  translation
TRODI
  touchdown rate of descent indicator
TROO
  transponder on off
TROS
  tape resident operating system
  timesharing realtime operating system
  transducer read only storage
  transformer read only storage
TRP
  television remote pick-up
TRPO
  track reference print out
TRPSK
  transmitted reference phase shift keying
TRQ
  torque
TRR
  tactial range recorder
  take real result
  target ranging radar
  topical report request
  topical reports review
  trouble recorder
TRS
  telephone repeater station
  test response spectrum
  tetrahedra research satellite
  ticket reservation system
  time reference system
  tough rubber sheathed
  track and store
  transmit receive switch
  transmitter
TRSA
  terminal radar service area
TRSB
  time reference scanning beam
TRSR
  taxi and runway surveillance radar
TRSSGM
  tactical range surface to surface guided missile
TRSSM
  tactical range ship to shore missile
  tactical range surface to surface missile
TRT
  translate and test
  tropical radio telegraph
TRTL
  transistor resistor transistor logic
TRU
  transformer rectifier unit
  transmit receive unit
  transportable radio unit
  transuranic
  transuranium processing plant
TRUD
  time remaining until dive
TRUMP
  target radiometric ultraviolet measuring program
  teller register unit monitoring program
TRUST
  television relay using small terminals
TRUT
  time remaining until transition
TRV
  transient recovery voltage
TRVM
  transistorized voltmeter
TRW
  thompson ramo woolridge
TRX
  two region physics critical experiment
TRZON
  three ton range and azimuth only
TS
  taper shank
  target seeker
  target strength
  telephone set
  television society
  temperature switch
  tensile strength
  terminal series
  terminal station

terminal strip
terminating system
test and set
test set
test solution
test specification
threaded studs
three state
time service
time shared
time sharing
time switch
timing selector
toll station
tool steel
torque synchro transmitter
torque transmitter
track store unit
tracking system
transfer set
transit storage
transmit
transmitter
transmitting station
transverse staggering
tristate
turboshaft engine
turbosynchro transmitter
turn per second
type of shift

TSA
   time series analysis
   total system analyzer
   transfer system a
   tree structured attribute
   tube support assemblies
   two step antenna

TSAD
   trajectory sensitive arming device

TSADV
   television service association of the delaware valley

TSAM
   television service association of michigan

TSAR
   telemetry systems application requirements
   time sharing acitivity report system

TSAZ
   target selector azimuth

TSB
   transfer system b
   twin sideband

TSBP
   time sharing business package

TSC
   tactical support center
   technical subcommittee
   telecommunications systems corporation
   test shipping cask
   time sharing control task
   transfer system c
   transmitter start code
   transmitting switch control
   transportable communications system
   transportation system center
   tristate control
   two subcarrier

TSCA
   target satellite controlled approach

TSCC
   telemetry standards coordination committee

TSCF
   task schedule change form

TSCLT
   transportable satellite communication link terminal

TSD
   track situation display

TSDA
   television service dealers' association

TSDAGR
   television service dealers' association of grand rapids

TSDD
   temperature salinity density depth

TSDF
   target system data file

TSDI
   tactical situation display indicator

TSDM
   time shared data management

TSDMS
   time shared data management system

TSDOS
   time shared disk operating system

TSDU
   target system data update

TSE
   telecommunications systems engineering
   telemetry support equipment
   total shielding effectiveness
   transmission secondary emission

TSEB
   twin sideband

TSEL
   target selector elevation

TSEM
  transmission secondary electron multiplication
  transmission secondary emission multiplier
TSEQ
  time sequence
TSET
  transmitter signal element timing
TSEXEC
  time shared executive
TSF
  ten statement fortran
  thin solid films
  through supergroup filter
  tower shielding facility
TSFR
  transfer
TSG
  technical service group
  technical specialty group
  time signal generator
  transversely adjusted gap
  triggered spark gap
TSGA
  three conductor, shipboard, general use, armored
TSI
  technical standardization inspection
  technical systems, incorporated
  threshold signal to interference ratio
  time slot input
  tons per square inch
  transistor specialities, incorporated
  true speed indicator
TSIC
  time slot interchange circuit
TSIMS
  telemetry simulation submodule
TSK
  the technical society of knoxville triangle fraternity
  time shift keying
TSL
  three state logic
  total signal lines
  translator
  tri state logic
TSM
  time shared monitor system
  time sharing monitor
  time sharing monitor system
  time sharing multiplex
  trouble shooting manual
TSMDA
  test section melt down accident
TSMT
  transmit
TSMU
  time share multiplex unit
TSN
  task sequence number
TSNC
  time sharing and multiplexing numerical control
TSO
  technical standard order
  time sharing option
  time slot zero
TSOC
  tape system output converter
  time sharing operating control system
TSOS
  time sharing operating system
TSOSC
  test set operational signal converter
TSOUTPUT
  three state output
TSP
  time share peripherals
  transponder
TSPS
  time sharing programming system
TSR
  test summary report
  thermal shock rig
  total stress range
  transistor saturable reactor
TSRC
  transmitter receiver
TSRE
  tropospheric scatter radio equipment
TSRELAY
  thermal synthesis relay
TSS
  tangential signal sensitivity
  target selector switch
  technical specification sheet
  technical staff surveillance
  telecommunications switching system
  time shared supervisory system
  time sharing system
  total suspended solids
  track store switch
  train supervisory system
TSSC
  the technical and scientific societies' council of cincinnati
TSSTAGE
  three state stage

TST
  test
  time shared terminal
  trouble shooting time
TSTA
  transmission, signaling, and test access
TSTICT
  test incoming trunk
TSTR
  transistor
TSTS
  tracking system test stand
TSU
  technical service unit
  technical support unit
  time standard unit
  transfer switch unit
  twin and subtract
TSUP
  trunk supervision
TSV
  turbine stop valve
TSW
  test switch
  transfer switch
  transmitting slide wire
TSX
  telephone satellite, experimental
  time sharing execution
  time sharing executive system
  transfer and set index
TT
  technical test
  telegraphic transfer
  teletype
  teletypewrite
  teletypewriter
  temporarily transferred
  tension transformer
  terminal timing
  test temperature
  test terminator
  test time
  test tube
  testing time
  thermally tuned
  thermomagnetic treatment
  thermostat switch
  timing and telemetry
  top to top
  touch tone
  tracking telescope
  transfer trip
  transmitting typewriter
  triple thermoplastic
  trunk test
  turbine trip
TTA
  television technicians' association
  turbine alternator assembly
TTAC
  telemetry tracking and control
  tracking, telemetry and command
  tracking, telemetry and control
TTAT
  television technicians' association of tillamook
TTB
  test two bits
TTBWR
  twisted tape boiling water reactor
TTC
  tactical telephone central
  tape to card
  telecommunications training center
  telephone toll call
  teletype center
  teletype message converter
  terminating toll center
  test transfer cask
  tight tape contact
  tracking, telemetry and command
  trans telecommunication corporation
  translunar trajectory characteristics
TTCE
  tooth to tooth composite error
TTCP
  transmitting typewriter with card punch
TTD
  temporary text delay
TTDR
  tracking telemetry data receiver
TTDT
  tactical test data translator
TTE
  telephone terminal equipment
  thermal transfer equipment
  time to event
TTF
  test to failure
  thoriated tungsten filament
  transient time flowmeter
TTG
  technical translation group
  time to go
TTGR
  time to go rating
TTHE
  thermal transient histogram equivalent

TTHFA
    twisted pairs, telephone, heat and flame resistant, armored
TTI
    telecommunication technology, inc
    teletype input
    teletype test instruction
    time temperature indicator
TTL
    teletype telling
    transistor transistor logic
    transistor transistor micrologic
TTLS
    team training launch station
TTM
    two tone modulation
TTML
    transistor transistor micrologic
TTMS
    telephoto transmission measuring set
TTO
    teletype output
    transmitter turn-off
TTP
    tabular tape processor
    tabulator tape processor
    tape to print
    test transfer port
TTR
    take triplex result
    tape reading typing relay
    target track radar
    target tracking radar
    teletype translator
    terminal radiation airborne program translator
    thermal test reactor
    time temperature recorder
    time to repair
    toshiba training reactor
    touch tone receiver
    transistor telegraph relay
TTRC
    transistorized thyratron ring counter
TTRSA
    twisted pair, telephone radio, shielded and armored
TTS
    teletype setter
    teletype setting
    temporary threshold shift
    testing and training satellite
    three state transceiver
    transistor transistor logic schottky barrier
    transmission test set
    triple transit suppression

TTT
    time temperature transformation
    tyne tees television
TTTL
    transistor transistor transistor logic
TTTP
    transmitting typewriter with tape punch
TTU
    terminal time unit
    tracer test unit
TTVM
    thermal transfer voltmeter
TTX
    teletex
TTY
    teletype
    teletypewriter
    teletypewriter equipment
TTYQRSS
    teletype query reply subsystem
TU
    take-up
    tape unit
    terminal unit
    thermal unit
    timing unit
    toxic unit
    traffic unit
    transmission unit
    transmitting unit
    tuning unit
TUCC
    triangle universities computation center
    triangle university computer center
TUD
    technology utilization division
    total underground distribution
TUDAT
    tunnel diode arithmetic tester
TUFCDF
    thorium uranium fuel cycle development facility
TUFX
    turning fixture
TUG
    transac users group
    transistorized automatic computer users' group
TUM
    tuning unit member
TUN
    transfer unconditionally
    tune
    tuning

TUNNET
  tunnel transit time
TUR
  traffic usage recorder
TURBOALT
  turbo alternator
TURBOGEN
  turbo generator
TURPS
  terrestrial unattended reactor power system
TUT
  transistor under test
TV
  television
  test vehicle
  test voltage
  testing voltage
  thermal vacuum
  traverse
  traffic unit
  tube voltmeter
TVA
  tennessee valley authority
  thrust vector alignment
  time variant automation
TVB
  television bureau
TVC
  technical valve committee
  thrust vector control
TVCS
  thrust vector control system
TVDP
  terminal vector display point
TVEL
  track velocity
TVG
  time varied gain
  triggered vacuum gap
TVI
  television interference
TVIC
  television interference committee
TVIG
  television and inertial guidance
TVIS
  television information storage
TVIST
  television information storage tube
TVL
  time variation of loss
  transmit varilosser
  transmitting varilosser
TVM
  tachometer voltmeter
  track via missile
  transistor voltmeter
  transistorized voltmeter
TVOC
  television operations center
TVOM
  transistorized volt ohm milliammeter
TVOR
  terminal very high frequency omnidirectional range
  terminal very high frequency omnirange
TVP
  time variable parameter
TVPPA
  tennessee valley public power association
TVR
  television recording
  tennessee valley region
  time variable reflection
TVRO
  television receive only
TVSM
  time varying signal measurement
TVSO
  television space observatory
TVSYS
  television systems
TVTVA
  tualatin valley television association
TW
  tail warning
  thermal wire
  thermoplastic wire
  time word
  travelling wave
  true watt
  twin
  twisted
  typewriter
TWA
  transworld airlines
  travelling wave amplifier
  two way alternate
TWC
  teletype service without voice communication
TWCRT
  travelling wave cathode ray tube
TWG
  telemetry working group
TWIS
  technical writing improvement society
TWK
  traveling wave klystron

TWM
 traveling wave magnetron
 traveling wave maser
TWMBK
 traveling wave multiple beam klystron
TWMR
 tungsten water moderated reactor
TWO
 travelling wave oscillator
TWOM
 travelling wave optical maser
TWPA
 travelling wave parametric amplifier
TWR
 tower
 transworld radio
 travelling wave resonator
TWRI
 texas water resources institute
TWS
 tactical warning system
 tail warning set
 teletypewriter exchange service
 track while scan
 track while scanning
 translator writing system
 two way simultaneous
TWSB
 twin sideband
TWSO
 tactical weapon system operation
TWST
 torus water storage tank
TWT
 time wire transmission
 travelling wave tube
TWTA
 travelling wave tube amplifier
TWTS
 travelling wave tubes
TWU
 trace watch unit
TWW
 television wales and the west of england
TWX
 teletypewriter exchange
TX
 telex
 texas
TXA
 task extension area
TXE
 telephone exchange electronics
TXH
 transfer on index high

TXI
 transfer with index incremented
TXL
 transfer on index low
TXT
 text
 text section
TY
 teletypwriter
TYDAC
 typical digital automatic computer
TYPOUT
 typewriter output
TYPW
 typewriter
TZE
 transfer on zero
TZM
 tantalum zirconium molybdenum alloy

U

U
 undefined
 unified atomic mass unit
 unit
 unit of measure
 unnumbered acknowledge
 up
 uranium
 utility
UA
 ultra audible
 unassigned
 unnumbered acknowledge
UAA
 united arab airlines
 university aviation association
UAC
 upper air control
 utility assemble communication pool
UAD
 user attribute definition
UADPS
 uniform automatic data processing system
UADS
 user attribute data set
UAF
 universal active filter
UAIDE
 users of automatic information display equipment

UAL
  universal assembly language
  upper acceptance limit
UAM
  underwater to air missile
UAMC
  utility assemble master communication
UAP
  upper atmosphere phenomena
UAR
  upper atmosphere research
UARBC
  united arab republic broadcasting corporation
UARI
  university of alabama research institute
UARL
  united aircraft research laboratories
UART
  universal asynchronous receiver transmitter
UASCS
  united states army signal center and school
UAT
  uniform asymptotic theory
UAU
  underwater to air to underwater
UAUM
  underwater to air to underwater missile
UAW
  united auto workers
UAX
  unit automatic exchange
UBAEC
  union of burma atomic energy center
UBC
  uniform building code
  universal buffer controller
UBD
  utility binary dump
UBHR
  user block handling routine
UBI
  unibus interface
UBITRON
  undulating beam interaction electron tube
UC
  unclassified
  under construction
  undercarriage
  undercurrent
  unichannel
  unit call
  unloader coil
  upper case
  upper characters
  upper control

UCA
  unitized component assembly
UCAL
  universal cable adapter
UCB
  unit control block
UCC
  uniform commercial code
  university computing center
  university computing company
  utility control console
UCCRS
  underwater coded command release system
UCCS
  universal camera control system
UCDP
  uncorrected data processor
  uncorrelated data processor
UCE
  union carbide electronics
  unit checkout equipment
  unit control error
  unit correction entry
UCG
  ultrasound cardiogram
UCI
  utility card input
UCL
  unclamp
  upper confidence limit
  upper control limit
UCLA
  university of california at los angeles
UCLR
  university of california lawrence radiation laboratory
UCML
  university of california, microwave laboratory
UCORC
  university of california operations research center
UCOS
  uprange computer output system
UCP
  uninterruptable computer power
  unit construction practice
  unit construction principle
  utility control program
UCPTE
  union for the coordination of the production and transport of electric power
UCRI
  union carbide research institute

UCRL
university of california research laboratory
UCS
universal camera site
universal character set
universal connector strip
UCSD
universal communications switching device
UCSEL
university of california structural engineering laboratory
UCSSL
university of california space sciences laboratory
UCW
unit control word
UCWE
underwater countermeasures and weapons establishment
UD
underground distribution
universal dipole
unplanning derating
up down
UDACS
underwater detection and classification system
UDAR
universal digital adaptive recognizer
UDAS
unified direct access standards
UDB
universal data base
universal data base access system
up data buffer
UDC
unidirectional current
universal decimal classification
universal decimal code
universal digital controller
upper dead center
UDCD
unit data and control diagram
UDEC
unitized digital electronic calculation
unitized digital electronic calculator
UDF
uhf direction finding station
unit derating factor
UDG
unit derated generation
UDH
unplanned derated hours
UDI
universal digital instrument

UDL
uniform data link
up data link
UDOFT
universal digital operational flight trainer
UDOFTT
universal digital operational flight trainer tool
UDOP
ultrahigh frequency doppler
UDP
united data processing
UDR
universal document reader
utility data reduction
UDRC
utility data reduction control
UDRO
utility data reduction output
UDS
ultronic data systems
universal data set
UDT
unidirectional transducer
united detector technology
universal data transcriber
UDTI
universal digital transducer indicator
UE
unit entry
UEC
united engineering center
utah engineers council
UECL
ultra electronics components, limited
UED
united electro dynamics
UEL
ultra electronics, limited
UEP
underwater elecric potential
unequal error protection
UEPG
united european power grid
UER
unsatisfactory equipment report
UERMWA
united electrical radio machine workers of america
UERN
utilities emergency radio network
UES
united engineering societies
UET
underground explosion test

united engineering trustees
universal engineer tractor
UEW
united electrical workers
UF
ultrasonic frequency
unavailability factor
urea formaldehyde
UFD
unit functional diagram
UFET
unipolar field effect transistor
UFO
unidentified flying object
users files on line
UFR
underfrequency relay
UFTR
university of florida teaching reactor
UG
union guide
UGLIAC
united gas laboratory internally programmed automatic computer
UGS
upper guide structure
UH
unavailable hours
unit heater
UHF
ultra high frequency
UHFDF
ultra high frequency direction finding
UHI
upper head injection
UHL
user header label
UHMW
ultrahigh molecular weight
UHR
ultrahigh resistance
ultra high resolution
UHS
ultimate heat sink
UHT
ultra heat treated
ultrahigh temperature
umbilical handling technician
UHTREX
ultrahigh temperature reactor experiment
UHV
ultrahigh vacuum
ultrahigh voltage
UI
unnumbered information

UIC
united insulator company
UID
universal identifier
universal individual identifier
UIEO
union of international engineering organizations
UIL
univac interactive language
UIM
union of international motor boating
UIR
upper information region
UIS
united inventors and scientists
UITA
union of international technical associations
UJT
unijunction transistor
UJTO
unijunction transistor oscillator
UKAACREG
united kingdom airways and communication region
UKAC
united kingdom automation council
UKAEA
united kingdom atomic energy authority
UKNR
university of kansas nuclear reactor
UKR
united kingdom atomic energy authority office at risley
UL
ultralinear
underwriters laboratories
ULA
uncommitted logic array
universal logic array
ULB
universal logic block
ULC
universal logic circuit
ULCC
university of london computer center
ULD
unit logic device
universal language description
ULF
ultralow frequency
ULI
universal logic implementer

ULICP
universal log interpretation computer program
ULLLC
upper limit, lower limit comparator
ULLV
unmanned lunar logistics vehicle
ULM
ultrasonic light modulator
universal logic module
ULMA
upper level management advisor
ULMS
undersea long-range missile system
ULN
unlaunchable
ULO
unmanned launch operation
ULOW
unmanned launch operations western test range
ULPA
united lightning protection association
ULPR
ultralow pressure rocket
ULRGW
ultralong range guided weapon
ULSI
ultralarge scale integration
ULSV
unmanned launch space vehicle
ULT
uniform low frequency technique
unique last term
ULTRA
universal language for typographic reproduction applications
ULTRASON
ultrasonic
ULV
ultralow volume
UM
unit of measure
unscheduled maintenance
UMA
ultrasonic manufacturers association
universal measuring amplifier
UMAP
university of michigan assembly program
UMASS
unlimited machine access from scattered sites
UMF
users master file
UMLC
universal multiline controller

UMOST
u groove power mosfet
UMP
uniformly most powerful
UMRR
university of missouri research reactor
UMTA
urban mass transportation administration
UMVPP
upper mississippi valley power pool
UMW
ultramicrowaves
UNACOM
universal army communication system
UNADS
univac automated documentation system
UNAMACE
universal automatic map compilation equipment
UNB
universal navigation beacon
UNBAL
unbalanced
UNC
unconditional
undercurrent
unified coarse thread
unified national coarse
UNCAST
united nations conference on the applications of science and technology
UNCDRP
universal card read-in program
UNCL
unified numerical control language
UNCOL
universal computer oriented language
UNCOPOUS
united nations committee on the peaceful uses of outer space
UNCTAD
un conference on trade and development
UND
unit derating
UNDEF
undefined
UNDH
unit derated hours
UNDP
united nations development program
UNE
universal nonlinear element
UNEF
unified national extra fine

UNESCO
  united nations educational, scientific, and cultural organization
UNF
  unified fine thread
  unified national fine
UNH
  uranyl nitrate hexahydrate
UNIBI
  unipolar bipolar
UNIBORS
  univac bill of material processor random system
UNIBOSS
  univac bill of material processor sequential system
UNIC
  voltage negative impedance converter
UNICOH
  unidensity coherent light recording
UNICOL
  universal computer oriented language
UNICOM
  universal integrated communications
UNICOMP
  universal compiler
UNIDO
  united nations industrial development organization
UNIFET
  unipolar field effect transistor
UNIPOL
  universal procedure oriented language
UNISAP
  universal automatic computer share assembly program
UNITAR
  united nations institute for training and research
UNITRAC
  universal trajector compiler
UNIV
  universal
UNIVAC
  universal automatic computer
UNL
  unlisten
UNLD
  unload
UNMON
  unable to monitor
UNOPAR
  universal operator performance analyzer and recorder

UNPKD
  unpacked
UNPS
  universal power supply
UNSCC
  united nations standards coordinating committee
UNSCEAR
  united nations scientific committee on the effects of atomic radiation
UNSERV
  unserviceable
UNSVC
  unserviceable
UOC
  ultimate operating capability
  ultimate operational capability
UOF
  unplanned outage factor
UOH
  unplanned outage hours
UOHM
  microohm
UOL
  underwater object locator
  utility octal load
UOR
  unplanned outage rate
UORS
  unusual occurence report
UP
  units position
  user programmer
  utility path
  utility program
UPA
  united power association
UPC
  united power company
  universal postal congress
  universal postal convention
  universal product code
UPD
  universally programmable digitizer
  update
UPDATE
  unlimited potential data through automation technology in education
UPF
  ultrapherical polynomial filter
UPL
  universal programming language
  user programming language
UPM
  universal permissive modules

UPO
undistorted power output
UPOS
utility program operating system
UPP
utility print punch
UPR
ultrasonic parametric resonance
uranium production reactor
UPS
microprocessor power supply
uninterrupted power supply
uninterruptible ac electric power system
uninterruptible power supply
uninterruptible power system
UPSI
user program switch indicator
UPTS
undergraduate pilot training system
UPU
universal postal union
UPW
union of post office workers
UR
ultrared
underrange
unit real
unit record
up range
URA
universities research association
URAEP
university of rochester atomic energy project
URBM
ultimate range ballistic missile
URC
uniform resistance capacitance
unit record control
unit record controller
URD
underground residential distribution
underground rural distribution
URG
universal radio group
URI
unintentional radar interference
utility read-in program
URIPS
undersea radioisotope power supply
URIR
unified radioactive isodromic regulator
URISA
urban and regional information systems association

URM
unlimited register machine
URPA
university of rochester, department of physics and astronomy
URRI
urban regional research institute
URS
unate ringe sum
uniform reporting system
uniformly reflexive structure
united research service
unviseral regulating system
US
unconditional stop
underwater to surface
undistorted signal
unit separator
utility satellite
USA
united states of america
USAABNAELCTBD
united states army airborne and electronics board
USAAESWBD
united states army airborne, electronics and special warfare board
USAAFO
united states army avionics field office
USAAS
united states army armor signals
USAASO
us army aeronautical services office
USAAVLABS
us army aviation material laboratories
USAAVNTA
us army aviation test activity
USAB
united states activities board
USABAAR
us army board for aviation accident research
USABESRL
us army behavioral science research laboratory
USABRL
us army ballistics research laboratories
USAC
united states activities committee
USACDA
us army control and disarmament agency
USACECDA
united states army communications electronics combat developments agency

USACOMZEUR
   united states army communications zone in europe
USACSC
   us army computer systems command
USACSSC
   us army computer systems support and evaluation command
USADSC
   united states army data service and administrative systems command
USAEC
   united states army electronics command
   us army energy commission
USAECOM
   us army electronics command
USAEL
   united states army electronics laboratories
USAELCTPG
   united states army electronic proving ground
USAELRDL
   united states army electronics research and development laboratory
USAEMA
   united states army electronics material agency
USAEPG
   united states army electronics proving ground
USAERDA
   united states army electronics research and development agency
USAERDL
   united states army electronics research and development laboratory
   us army engineer research and development laboratories
USAF
   united states air force
USAFCED
   united states air force communications electronics doctrine
USAFETAC
   usaf environmental technical applications center
USAIC
   united states army intelligence center
USAIDSCOM
   united states army information and data systems command
USAINTC
   united states army intelligence center
USAM
   unified space applications mission
USAMERDC
   united states army mobility equipment research and development center
USAMICOM
   united states army missile command
USANWSG
   us army nuclear weapons surety group
USAOGMS
   united states army ordnance guided missile school
USAOMC
   us army ordnance missile command
USAREPG
   us army electronic proving ground
USARIEM
   us army research institute of environmental medicine
USARP
   us army research program
USARPA
   united states army radio propagation agency
USART
   universal synchronous asynchronous receiver transmitter
USAS
   united states of america standard
USASC
   united states army signal corps
USASCA
   united states army satellite communications agency
USASCC
   united states army strategic communications command
USASCII
   united states american standard code for information interchange
USASCS
   united states army signal corps school
USASCSOCR
   united states of america standard character set for optical characters
   usa standard character set for optical character recognition
USASEL
   united states army signal engineering laboratory
USASI
   united states of america standards institute
USASIGENGLAB
   united states army signal engineering laboratory
USASIGRSCHUNIT
   united states army signal research unit

USASIGTC
united states army signal training center
USASII
united states of america standard code for information interchange
USASMSA
us army signal missile support agency
USASRDL
united states army signal research and development laboratory
USASSA
united states army signal supply agency
USASTCEN
united states army signal training center
USATEA
us army transportation engineering agency
USB
unified s-band
upper sideband
USBCODE
unipolar straight binary code
USBE
unified s-band equipment
USBS
unified s-band system
united states bureau of standards
USBUC
upper sideband upconverter
USC
united states code
united states components
USCAGS
us coast and geodetic survey
USCAL
university of southern california aeronautical laboratory
USCC
united states capacitor corporation
USCEC
university of southern california engineering center
USCIB
united states communications intelligence board
USCOLD
us committee on large dams
USD
ultimate strength design
USDA
us department of agriculture
USDOD
us department of defense
USE
unit support equipment
univac scientific exchange
universal automatic computer scientific exchange
USEC
united system of electronic computers
USERC
united states environment and resources council
USERID
user identification
USEUCOM
united states european communications
USF
university sciences forum
USFS
united states frequency standard
us frequency standard
USFSS
us fleet sonar school
USFTHREAD
united states form thread
USG
united states gage
united states gallon
USGAL
united states gallon
USGPO
us government printing office
USGR
us government report
USGRDR
us government research and development report
USGS
us geological survey
USGSG
united states government standard gage
USIA
us information agency
USIC
us information center
USIS
us information service
USITA
united states independent telephone association
USJCB
united states joint communication board
USJPRS
us joint publication research service
USL
underwater sound laboratory
upper square law limit
USM
underwater to surface missile

USMID
  ultrasensitive microwave infrared detector
USNAS
  united states naval air station
USNBS
  united states national bureau of standards
USNC
  united states national committee
USNCWEC
  us national committee world energy conference
USNEL
  united states naval electronics laboratory
  us navy electronics laboratory
USNMTC
  united states naval missile test center
USNO
  united states naval observatory
USNRDL
  us naval radiological defense laboratory
USNSCF
  united states naval shore communications facility
USNSISSMFE
  us national society for the international society of soil mechanics and foundation engineers
USNUSL
  united states navy underwater sound laboratory
  us navy undersea laboratory
USP
  united states patent
USPO
  united states patent office
  united states post office
USR
  universal series regulator
USRL
  underwater sound reference laboratory
USRS
  united states rocket society
USRT
  universal synchronous receiver transmitter
USS
  united states standard
  universal scheduling system
USSA
  united states space administration
USSG
  united states standard gage
USTAG
  united states technical advisory group
USTOL
  ultrashort take-off and landing

USTS
  united states travel service
USW
  ultrashort wave
  undersea warfare
USWB
  us weather bureau
USWP
  ultrashort wave propagation panel
UT
  ultrasonic test
  umbilical tower
  unitech
  units tens
  universal time
  universal tube
  utah
  utility
UTA
  upper testing area
UTC
  unit time coding
  united technology center
  united technology corporation
  united telephone cables
  united transformer corporation
  universal time coordinated
  utilities telecommunications council
UTCE
  utah university college of engineering
UTCS
  urban traffic control system
UTD
  uniform theory of diffraction
  universal transfer device
UTE
  underwater tracking equipment
UTEC
  university of toronto electronic computer
UTELRAD
  utilization of enemy electromagnetic radiations
UTIA
  university of toronto institute of aerophysics
UTIAS
  university of toronto institute for aerospace studies
UTL
  user trailer label
UTLM
  up telemetry
UTM
  universal test message
  universal transverse mercator

UTP
  utility tape processor
UTR
  unprogrammed transfer register
  up time ratio
UTRIP
  universal triangulation program
UTRR
  university of teheran research reactor
UTS
  ultimate tensile strength
  underwater telephone system
  unified transfer system
  universal test station
  universal time sharing system
UTTAS
  utility tactical transport aircraft system
UTTC
  universal tape to tape converter
UU
  ultimate user
UUA
  universal automatic computer users' association
UUM
  underwater to underwater missile
UUM, UUM
  underwater to underwater missile
UUT
  unit under test
UV
  ultraviolet
  under voltage
UVAR
  university of virginia reactor
UVASER
  ultraviolet amplification by stimulated emission of radiation
UVD
  undervoltage device
UVL
  ultraviolett light
UVR
  university of virginia reactor
UVROM
  ultraviolet read only memory
UVS
  universal versaplot software
UVVIS
  ultraviolet visible
UW
  ultrasonic wave
  unique word
UWAL
  university of washington aeronautical laboratory
UWBIC
  university of washington basic interpretative compiler
UWRR
  university of wyoming research reactor
UWTR
  university of washington training reactor
UXB
  unexploded bomb
UY
  unit years

V

V
  vacuum
  vacuum tube
  valve
  vanadium
  variable
  variant
  vartan
  vector
  velocity
  vertical
  video
  viscosity
  visual telegraphy
  voice
  volt
  voltage
  voltmeter
  volume
VA
  value analysis
  variometer
  vertical amplifier
  veterans administration
  video amplifier
  virginia
  vital area
  voltammeter
  volt-ampere
  voltage amplifier
VAAT
  vibration and acoustic testing
VAB
  van allen belt
  vehicle assembly building
  vertical assembly building
  voice answer back

VAC
vacant
vacuum
value added carrier
vector analog computer
video amplifier chain
volt alternating current
VAD
voltmeter analog to digital
VADE
versatile auto data exchange
VAEP
variable, attributes, error propagation
VAI
video assisted instruction
VALSAS
variable length word symbolic assembly system
VAM
vector airborne magnetometer
virtual access method
voltammeter
VAMFO
variable angle monochromatic fringe observation
VAMP
vector arithmetic multiprocessor
visual acoustic magnetic pressure
visual anamorphic motion picture
VAN
value added network
VANT
vibration and noise tester
VAOR
very high frequency aural omnirange
VAP
versatile automatic test equipment assembly program
VAPOX
vapor deposite oxide
VAPPRF
vapor proof
VAR
vaccum arc
variable
variation
variometer
varistor
varying
video audio range
visual aural radio range
visual aural range
volt-ampere reactive

VARACTOR
variable reactor
voltage variable capacitor
VARAD
varying radiation
VARCAP
variable capacitor
VARCOND
variable condenser
VARH
volt-ampere reactive hour
VARHM
var hour meter
VARHMETER
var hour meter
VARIAC
variable capacitor
VARICAP
variable capacitor
VARIMU
variable mu tube
VARISTOR
variable resistor
VARITRAN
variable voltage transformer
VARM
varmeter
VARR
variable range reflector
visual aural radio range
VARS
vertical azimuth reference system
VASCA
electronic valve and semiconductor manufacturers' association
VASCAR
visual average speed computer and recorder
VASI
visual approach slope indicator
VASIS
visual approach slope indicator system
VAST
versatile automatic specification tester
versatile avionics ship test
versatile avionics shop test
VAT
value added tax
variable autotransformer
vernier auto track
virtual address translator
VATE
versatile automatic test equipment
vertical anisotropic etch
VATEPROCESS
versatile automatic test equipment process

VATLS
  visual airborne target location system
  visual airborne target locator system
VB
  vacancy bit
  valence bond
  valve box
  vibrating
  vibration
  vibrator
  voltage board
VBA
  vibrating beam accelerometer
VBD
  voice band data
VBI
  vital bus inverter
VBL
  voyager biological laboratory
VBO
  voltage, breadover
  voltage, breakover
VBOMP
  virtual bomp
VBWR
  vallecitos boiling water reactor
VC
  validity check
  variable capacitor
  varnished cambric
  varnished cambric insulated wire
  varnished cambric tape
  vector control
  versatility code
  vertical circle
  vertical spacing
  video correlation
  video correlator
  virtual circuit
  voice ciphony
  voice coil
  voice coil of speaker
  volt coulomb
  voltage changer
  voltage comparator
VCA
  valve control amplifier
  voice connecting arrangements
  voice controlled carrier
  voltage controlled amplifier
  voltage current adapter
VCASS
  visually coupled airborne systems simulator
VCC
  variable cycle controller
  visual communications congress
  voice control center
  voice controlled carrier
  voltage controlled clock
VCCO
  voltage controlled crystal oscillator
VCCS
  voltage controlled current source
VCD
  variable capacitance diode
VCF
  variable crystal filter
  voltage controlled filter
  voltage controlled frequency
VCG
  vectorcardiogram
  vectorcardiography
  vertical line through center of gravity
  vertical location of the center of gravity
VCI
  volatile corrosion inhibitor
VCL
  vertical center line
VCLF
  vertical cask lifting fixture
VCM
  vibrating coil magnetometer
  vinyl chloride monomer
  voltage controlled multivibrator
VCNC
  voltage controlled negative capacitance
VCNR
  voltage controlled negative resistance
VCO
  variable crystal oscillator
  voice coder
  voltage controlled oscillator
VCOS
  voltage controlled oscillators
VCP
  vacuum condensing point
  vector collecting program
  virtual communication path
  virtual control processor
  visual comfort probability
VCR
  video cartridge recorder
  video cassette recorder
  voltage control resistor
  voltage controlled resistor
VCS
  vacuum actuated control switch
  variable correlation synchronization

ventilation control system
victorian computer society
visually coupled system
voltage calibration set
VCSL
 voice call sign list
VCSR
 voltage controlled shift register
VCT
 voltage clock trigger
 voltage control transfer
 voltage controlled transfer
 volume control tank
VCU
 variable correction unit
 video combiner unit
 voltage control unit
VCVS
 voltage controlled voltage source
VD
 vapor density
 video display
 voice data
 voltage detector
 voltage drop
VDA
 variable data area
 video distribution amplifier
 vision distribution amplifier
VDAC
 video display controller
VDAM
 virtual data access method
VDAS
 vibration data acquisition system
VDC
 volt direct current
 voltage to digital converter
 volume description card
VDCT
 volt direct current test
VDCW
 volt direct current working
VDD
 visual display data
 voice digital display
VDDI
 voyager data detailed index
VDDL
 voyager data distribution list
VDDS
 voyager data description standards
VDE
 variable display equipment

VDES
 voice data encoding system
VDET
 voltage detector
VDF
 very high frequency direction finder
 very high frequency direction finding
 video frequency
VDFG
 variable diode function generator
VDH
 variable length divide or halt
 vickers diamond hardness
VDI
 variable duration impulse
 virtual device interface
 visual doppler indicator
VDL
 variable delay line
 vienna definition language
 voice direct line
VDM
 varian data machines
VDMOST
 vertical dmost
VDOS
 vacuum distillation overflow sampler
VDP
 variable length divide or proceed
 vertical data processing
VDPI
 vehicle direction and position indicator
VDR
 video disk recorder
 video tape recorder
 voltage dependent resistor
VDRA
 voice and data recording auxiliary
VDS
 variable depth sonar
 visual docking simulator
VDT
 vertical deflection terminal
 video display terminal
 visual display terminal
VDU
 variable delay unit
 video display unit
 visual display unit
VE
 value engineering
 vernier engine
 voltage
VEA
 value engineering association

VEB
   variable elevation beam
VEC
   variable energy cyclotrons
   vector
VECI
   vehicular equipment complement index
VECO
   vernier engine cutoff
VECOS
   vehicle checkout set
VECP
   value engineering change proposal
VEDAR
   visible energy detection and ranging
VEDS
   vehicle emergency detection system
VEFCO
   vertient functional checkout
VEL
   velocity
VELF
   velocity filter
VEM
   vasoexcitor material
   vector element by element multiply
VENUS
   variable and efficient network utility service
VEPCO
   virginia electric power company
VEPIS
   vocational education program information system
VERA
   versatile experimental reactor assembly
   vision electronic recording apparatus
VERDAN
   versatile differential analyzer
   vertical digital analyzer
VERLORT
   very long range tracking
VERNITRAC
   vernier tracking by automatic correlation
VERS
   versed sine
VERT
   vertical
VES
   vector element by element sum
VESR
   vallecitos experimental superheat reactor
VEST
   vertical earth scanning test
   volunteer engineers, scientists and technicians

VEV
   voice excited voice coder
VEWS
   very early warning system
VF
   variable frequency
   vector field
   verification of function
   video frequency
   viscosity factor
   voice frequency
   voltage to frequency
VFB
   vertical format buffer
VFC
   variable frequency clock
   variable frequency crystal
   video frequency carrier
   video frequency channel
   voice frequency carrier
   voice frequency channel
   voltage frequency converter
   voltage to frequency converter
VFCONVERTER
   voltage to frequency converter
VFCT
   voice frequency carrier telegraphy
   voice frequency carrier teletype
VFD
   vacuum fluorescence display
   voltage fault detector
VFDIAL
   voice frequency dialing
VFEQT
   voice frequency equipment
VFET
   v groove fet
   vertical field effect transistor
VFL
   variable field length
VFLA
   volume folding and limiting amplifier
VFO
   variable frequency oscillator
VFR
   visual flight rules
VFT
   voice frequency carrier telegraph
   voice frequency telegraphy
VFTG
   voice frequency telegraph
VFU
   vertical format unit
VFX
   variable frequency crystal oscillator

variable frequency mixer
vector float to fix
VG
   viscosity grade
   voltage gain
VGA
   variable gain amplifier
VGC
   viscosity gravity constant
VGI
   vertical gyro indicator
VGPI
   visual glide path indicator
VGSI
   visual glide slope indicator
VH
   vertical hook
   very hard
VHAA
   very high altitude abort
VHES
   vitro hanford engineering service
VHF
   very high frequency
VHFDF
   very high frequency direction finding
VHM
   vibrating head magnetometer
   virtual hardware monitor
VHN
   vickers pyramid hardness number
VHO
   very high output
VHP
   very high performance
VHPIC
   very high performance integrated circuit
VHRR
   very high resolution radiometer
VHSI
   very high speed integration
VHSIC
   very high speed integrated circuit
VHTR
   very high temperature gas cooled reactor
VHV
   very high voltage
VHVI
   very high viscosity index
VI
   vertical incidence
   virgin islands
   viscosity index
   volume indicator

VIA
   versatile interface adapter
VIAS
   voice interference analysis set
   voice interference analysis system
VIB
   vertical integration building
   vibrate
   vibration
   vibrator
VIBR
   vibrate
   vibration
VIBRECON
   vibration recording console
VIBT
   vibrator
VIC
   variable instruction computer
   vicinal
VICC
   visual information control console
VICI
   velocity indicating coherent integrator
VID
   variable intermittent duty
   video
VIDAC
   visual information display and control
VIDAMP
   video amplifier
VIDAT
   visual data acquisition
VIDF
   vertical intermediate distributing frame
   video frequency
VIDIAC
   visual information display and control
VIE
   visual indicating equipment
   visual indication equipment
   visual indicator equipment
VIF
   voice interface frame
VIFCS
   vtol integrated flight control system
VIFI
   voyager information flow instructions
VIG
   video integrating group
VIL
   vertical injection logic
VILP
   vector impedance locus plotter

VIM
vehicle identification number
VIMS
virginia institute of marine science
VINIC
voltage inverting negative impedance converter
VINS
velocity inertial navigation system
VINT
video integration
VIOC
variable input output code
VIP
v groove isolation with polysilicon backfill
v isolation with polysilicon backfill
v shaped isolation regions filled with polycrystalline silicon
variable information processing
vector inner product
verifying interpreting punch
virtual instruction package
visual image processor
visual indicator panel
visual information projection
visual integrated presentation
voltage impulse protection
VIPER
video processing and electronic reduction
VIPP
variable information processing package
VIPS
verbal instruction programmed system
voice interruption priority system
VIPTECHNIQUE
v isolation with polysilicon backfill technique
VIR
vertical interval retrace
VIRNS
velocity inertia radar navigation system
VIRS
vertical interval reference signal
VIS
visual instrumentation subsystem
voltage inverter switch
VISAM
virtual index sequential access method
VISSR
visible infrared spin scan radiometer
VISTA
visual talking
VIT
vertical interval test
very intelligent terminal

VITA
volunteers for international technical assistance
VITAL
variably initialized translator for algorithmic languages
VITEAC
video transmission engineering advisory committee
VITROLAIN
vitreous enamel porcelain
VITS
vertical interval test signal
VIV
variable inlet vanes
VJSW
voice jamming simulator, weapons
VKIFD
von karman institute for fluid dynamcis
VL
vapour to liquid ratio
variolosser
video logic
voltage logic
VLA
very large antenna
very large array
very low altitude
VLBI
very long baseline interferometry
VLC
video logic corporation
VLCC
very large crude carrier
VLCD
very low cost display
VLCR
variable length cavity resonance
VLCS
voltage logic, current switching
VLDB
very large data base
VLDS
variable length distinguishing sequence
VLE
voice line expansion
VLED
visible light emitting diode
VLF
vertical launch facility
very low frequency
VLFD
very low frequency direct
VLFS
variable low frequency standard

VLI
  very low impedance
VLN
  variable length
VLP
  video long play
VLR
  very long range
  very low range
VLS
  vapor liquid solid
  volume loadability speed
VLSI
  very large scale integration
VLV
  valve
VM
  vector message
  velocity modulation
  vertical magnet
  virtual machine
  virtual memory
  virtual multi access
  volatile matter
  volt per meter
  voltage meter
  voltmeter
VMA
  valid memory address
  valve manufacturers association
VMC
  variable message cycle
  vector move convert
  visual meteorological condition
  vitramon microwave corporation
VMD
  vertical magnetic dipole
  vertical main distribution
VMDF
  vertical main distribution frame
VMGSE
  vehicle measuring gse
VMIL
  volt per mil
VMM
  virtual machine monitor
VMOS
  v groove mos
  vertical metal oxide semiconductor
  virtual memory operating system
VMS
  valve monitoring system
  variable magnetic shunt
  variable memory system
  vibration measuring system
  virtual memory system
  voice mail system
VMT
  video matrix terminal
VMTSS
  virtual machine time sharing system
VMU
  variable match unit
VNAV
  vertical navigation mode
VNL
  via net loss
VNSP
  vacant nozzle shield plug
VO
  vacuum tube oscillator
  valve oscillator
  vertical output
VOA
  voice of america
VOB
  vacuum optical bench
VOC
  voice operated coder
  voice operated control
  voice operated relay circuit
VOCA
  voltmeter calibrator
VOCODER
  voice coder
VOCOM
  voice communication
VOCSU
  voice operated carrier switching unit
VOD
  velocity of detonation
  vertical on board delivery
VODACOM
  voice data communication
VODAS
  voice operated device antisinging
VODAT
  voice operated device for automatic transmission
VODER
  voice coder
  voice operation demonstrator
VOGAD
  voice operated gain adjusting device
VOIR
  venus orbiting imaging radar
VOIS
  visual observation instrumentation subsystem

VOL
  volume
  volume label
VOLCAT
  volume catalog
VOLERE
  voluntary legal regulatory
VOLLIM
  voltage limiter
VOLTAN
  voltage amperage normalizer
VOM
  volt ohm milliammeter
  volt ohm milliamperemeter
  volt ohmmeter
VOMA
  volt ohm milliampere
VOR
  very high frequency omnidirectional radio range
  very high frequency omnidirectional range
  visual omnidirectional radiator
  visual omnirange
VORDAC
  very high frequency omnidirectional range and distance measuring equipment for average coverage
VORDME
  very high frequency omnidirectional range distance measuring equipment
VORTAC
  very high frequency omnidirectional range tactical air navigation
  vor co located with tacan
VORTEX
  varian omnitaste real time executive
VOS
  vertical obstactle sonar
  virtual operating system
  vision on sound
  voice operated switch
  voice operated transmission
  voice operated transmitter
  voice operated transmitter keyer
VOSE
  vacuum operation of spacecraft equipment
VOT
  very high frequency omnidirectional range test
  visual omnirange test
VOTA
  vibration open test assembly
VOTEM
  voice operated typewriter employing morse

VOY
  voice operated relay
VP
  vapor pressure
  variable pitch propeller
  velocity of propagation
  vent pipe
  verifying punch
  vertical polarization
  virtual machine control program
  volume pressure
  vulnerable point
VPB
  virtually pivoted beam laser
VPC
  virginia panel corporation
  voltage to pulse converter
VPCDS
  video prelaunch command data system
VPCF
  vapor pressure correction factor
VPE
  vapor growth epitaxy
  vapor phase epitaxial
  vapor phase epitaxy
  vulcanized polyethylene
VPI
  vacuum pressure impregnation
  vapor phase inhibitor
  virginia polytechnic institute
VPLAN
  veitch plan
VPLCC
  vehicle propellant loading control center
VPM
  vehicles per mile
  volt per meter
  volt per mil
  volts per mile
VPN
  vickers pyramid hardness number
VPOC
  variable performance optimizing controller
VPP
  variable pitch propeller
  volt peak peak
  volt peak to peak
VPR
  variable parameter record
  virtual plan position indicator reflectoscope
VPRF
  variable pulse repetition frequency
VPS
  vibrations per second

VPSS
vector processing support subsystem
VR
variable resistance
variable resistor
vertical redundancy
virtual equals real
voltage regulated
voltage regulation
voltage regulator
voltage relay
VRAM
video random access memory
VRB
very high frequency recovery beacon
VRC
vertical redundancy check
visible record computer
visual record computer
VRCI
variable resistive components institute
VRCR
vertical redundancy check register
VRD
vacuum tube relay driver
variable ratio devider
VRE
vibrating reed electrometer
VRL
vibration research laboratory
VRM
virtual resource manager
viscus remanent magnetization
VRMS
volt root mean square
voltage root mean square
VRPF
voltage regulated plate filament
VRPS
voltage regulated power supply
VRR
visual radio range
VRSA
voice reporting signal assembly
VRU
voice read out unit
voice response unit
VS
vacuum switch
variable speed
variable sweep
vector scan
versus
very soft
video and synchronization

virtual storage
virtual storage system
virtual system
visual signalling equipment
voice switching
voltage switching
voltmeter switch
volumetric solution
VSAM
variable, spanned and undefined mode
virtual storage access method
VSB
vestigial sideband
VSBS
voluntary standards bodies
VSC
vibration safety cutoff
voltage saturable capacitor
voltage saturated capacitor
VSCF
variable speed constant frequency
VSD
vertical situation display
VSE
vermont society of engineers
vessel steam explosion
virtual storage extension
VSFR
vertical seismic floor response
VSFS
voice store and forward messaging system
VSI
vertical side-band
vertical signal indicator
vertical speed indicator
VSL
ventilation sampling line
VSM
vestigial sideband modulation
vibrating sample magnetometer
VSMF
visual search microfilm file
VSMS
video switching matrix system
VSN
volume serial number
VSO
very stable oscillator
voltage sensitive oscillator
VSP
vehicle scheduling program
voltage stabilized polyethylene
VSPC
virtual storage personal computing

VSPX
   vehicle scheduling program extended
VSR
   vallecitos experimental superheat reactor
   variable length shift register
   very short range
VSRBM
   very short range ballistic missile
VSS
   vapor suppression system
   variable stability system
   voice storage system
VST
   visible speech translator
VSTOL
   vertical and/or short take -ff and landing
VSTP
   visual satellite tracking program
VSTR
   volt second transfer ratio
VSW
   very short wave
VSWR
   voltage standing wave ratio
VSYNC
   vertical synchronization
VSYNCH
   vertical synchronization
VT
   vacuum telegraphy
   vacuum tube
   variable threshold
   variable threshold velocity time
   variable time
   vehicular technology
   velocity time
   vermont
   vertical tabulation
   vertical tail
   video tape
   video terminal
   virtual terminal
   visual telegraphy
   voice tube
   voltage transformer
VTA
   vacuum tube amplifier
   variable transfer address
VTAM
   varian telecommunication access method
   virtual telecommunications access method
VTB
   voltage time to break down
VTC
   vaccum thermal chamber
   vehicular traffic control
   vertical redundancy check
   viscosity temperature coefficient
VTCS
   vehicular traffic control system
   video telemetering camera system
VTD
   vacuum tube detector
   vertical tape display
VTDC
   vacuum tube development committee
VTE
   vertical tube evaporator
VTF
   vertical test fixture
VTL
   variable threshold logic
   vertical turret lathe
VTM
   vacuum tube modulator
   voltage tunable magnetron
VTO
   vacuum tube oscillator
   vertical take-off
   voltage tunable oscillator
   voltage tuned oscillator
VTOC
   volume table of contents
VTOF
   voltage to frequency
VTOFCONVERTER
   voltage to frequency converter
VTOHL
   vertical take-off, horizontal landing
VTOL
   vertical take-off and landing
VTOVL
   vertical take-off, vertical landing
VTR
   video tape recorder
   video tape recording
VTS
   vandenberg tracking station
   vertical test site
   vertical test stand
   vessel traffic service
   virtual terminal system
VTSU
   virtual terminal support
VTVM
   vacuum tube voltmeter
VU
   vehicle unit
   voice unit
   volume

VUCDT
 ventilation unit condensate drain tank
VUTS
 verification unit test set
VUVM
 voluntary universal marking program
VV
 valve voltmeter
VVC
 variable voltage capacitor
 voltage variable capacitance
 voltage variable capacitor
VVHR
 vibration velocity per hour
VVM
 valve voltmeter
VVR
 variable voltage rectifier
VWL
 variable word length
VWOA
 veteran wireless operators' association
VWP
 variable width pulse
VWSS
 vertical wire sky screen

W

W
 microwatt
 watt
 wattage
 wattmeter
 waveguide
 weber
 west
 width
 wire
 wireless
 word
 write
 writing
WA
 washington
 wave analyzer
 waveform analyzer
 wide angle
 wire armored
 wire association
 word add
 word after

WAA
 worked all america
WAAC
 west african airways corporation
 working ampere alternating current
WAACS
 western airways and air communications service
WAAM
 wide area antiarmor munitions
WAC
 wagner computer
 worked all continents
 worked all countries
 world aeronautical chart
 write address counter
WACB
 world association for christian broadcasting
WACES
 wyoming association of consulting engineers and surveyors
WACK
 wait and acknowledge
 wait before transmit positive acknowledgement
WACM
 western association of circuit manufacturers
WACS
 workshop attitude control system
WADC
 wright air development center
WADD
 wright air development division
WADS
 wide area data service
WAE
 worked all europe
WAF
 wiring around frame
WAFFLE
 wide angle fixed field locating equipment
WAG
 worked all goose
WAGR
 windscale advanced gas cooled reactor
WAH
 wage and hour division
WAI
 worked all italy
WAIP
 worked all italian provinces
WAL
 westinghouse astronuclear laboratory
WALDO
 wichita automatic linear data output

WALP
weapons assignment linear programming
WAM
washburn and moen
worth analysis model
WAMCE
western association of minority consulting engineers
WAMGA
washburn and moen gage
WAMI
world association for medical informatics
WAML
wright aero medical laboratory
WAMOSCOPE
wave modulated oscilloscope
WAN
wide area network
WAND
westinghouse alphanumeric display
WANG
wang laboratories
WAOE
worked all oe
WAP
work assignment procedure
WAPA
western area power administration
WAPDA
water and power development authority
WARC
world administrative radio conference
WARLA
wide aperture radio location array
WAS
worked all states
WASP
weather atmosphere sounding projectile
workshop analysis and scheduling program
WAST
western alaska standard time
WASTN
wireless auxiliary station
WAT
weight, altitude and temperature
WATA
wisconsin automatic test apparatus
WATS
wide area telecommunications service
wide area telephone service
wide area transmission system
WAVE
westinghouse audio visual electronics
WAVEGD
waveguide standards

WAY
worked all yokosuka
WAZ
worked all zones
WB
weather bureau
weber
wheel base
wide band
wire bundle
word before
word buffer register
write buffer
WBAN
weather bureau, air force navy
WBCO
waveguide below cut ott
WBCT
wide band current transformer
WBCV
wide band coherent video
WBD
wide band data
wire bound
WBDL
wide band data link
WBFM
wide band frequency modulation
WBGT
wet bulb globe thermometer
WBIF
wide band intermediate frequency
WBL
wide band limiting
WBNL
wide band noise limiting
WBNS
water boiler neutron source
WBNWRC
weather bureau national weather records center
WBP
weather and boilproof
WBPB
wide band patch bay
WBR
word buffer register
WBRR
weather bureau radar remote system
WBRS
wide band remote switch
WBS
wide body stol
work breakdown structure

WBSIGSTA
  weather bureau signal station
WBSM
  weber per square meter
WBTEMPERATURE
  wet bulb temperature
WBTS
  wide band transmission system
WC
  water cooled
  watt per candle
  wave change
  wire cable
  wireless communication
  without charge
  word count
  word counting
  working current
  write and compute
WCAP
  westinghouse commercial atomic
  power
WCATT
  worcester county association of television
  technicians
WCC
  world computer conference
  write control character
WCCPPS
  waste channel and containment
  pressurization and penetration
  system
WCD
  weather card data
  worse case difference
WCDPC
  war control data processing center
WCEMA
  west coast electronic manufacturers'
  association
WCF
  waste calcinating facility
  white cathode follower
WCM
  wire core matrix
  word combine and multiplexer
  writable control memory
WCP
  waste collector pump
WCPSC
  western conference of public services
  commissioners
WCR
  wire contact relay
  word control register

WCS
  writable control storage
  writable control store
WCT
  water cooled tube
WD
  watt demand meter
  waveform digitizer
  waveform distortion
  williams domain
  wiring diagram
  word
  wound
WDAG
  word driver and gate
WDB
  wide band
WDC
  waste disposal cask
  world data center
WDG
  winding
WDM
  wavelength division multiplexing
WDPC
  western data processing center
WDR
  write drum
WDT
  weight data transmitter
WDTRS
  westinghouse development test requirement
  specification
WE
  weir electronics
  western electric
  write enable
WEADES
  western electric air defense engineering
  service
WEC
  western electric company
  westinghouse electric company
  world energy conference
WECO
  western electric company
WEDAC
  westinghouse digital airborne computer
WEDGE
  waterless electrical data generating effortless
WEERC
  western electric engineering research
  center
WEF
  write end of file

WEFC
  weather facsimile
WEICO
  westinghouse electric international company
WEMA
  western electronic manufacturers' association
WEN
  write enable
WEP
  water extended polyester
WEPA
  welded electronic packaging association
WERC
  world environment and resources council
WERS
  war emergency radio service
WES
  women's engineering society
  wyoming engineering society
WESAR
  westhinghouse safety analysis report
WESCON
  western electronics show and convention
WESRAC
  western research application center
WEST
  western electric system test
  western energy supply and transmission
WESTAR
  waterways experiment station terrain analyzer radar
WET
  weighted effective temperature
  western european time
WETAC
  westinghouse electronic tubeless analog computer
WEW
  western electronic week
WF
  wave frequency
  waveform
  wind finding radar
  write forward
  wrong fount
WFAEQ
  wave filters and equalizers
WFCMV
  wheeled fuel consuming motor vehicle
WFD
  work function difference
WFEO
  world federation of engineering organizations
WFF
  well formed formula

WFM
  waveform monitor
WFMC
  welding filter material control
WG
  water gage
  waveguide
  wire gage
  working group
WGBC
  waveguide below cutoff
  waveguide operating below cutoff
WGDT
  waste gas decay tank
WGEEIA
  western ground electronics engineering installation agency
WGL
  weapon guidance laboratory
  wire glass
WGN
  white gaussian noise
WGS
  waveguide glide slop
WGST
  waste gas storage tank
WGT
  weapons guiding and tracking
WGTC
  waveguide to coaxial
WH
  watt hour
  we have
WHC
  watt hour meter with contact device
WHDM
  watt hour demand meter
WHIP
  wide band high intercept probality
WHL
  watt hour meter with loss compensator
WHM
  watt hour meter
WHO
  world health organization
WHOI
  woods hole oceanographic institution
WHR
  watt hour
WHRM
  watt hour meter
WHT
  watt hour demand meter, thermal type

WI
wisconsin
wrought iron
WIAA
women's international association of aeronautics
WIDE
wiring integration design
WIF
water immersion facility
WIL
white indicating lamp
white indicator lamp
WILCO
will comply
WIN
weapons interception
WINA
witton network analyzer
WINB
western interstate nuclear board
WINCON
winter convention
WIND
weather information network and display
WINDS
weather information network and display system
WINPUT
write input
WIP
work in process
work in progress
WIPO
world intellectual property organization
WIRECOM
wire communication
WIRELESSENGR
wireless engineer
WIRELWLD
wireless world
WIRES
women in radio and electrical service
WISC
wisconsin integrally synchronized computer
WISE
wang intersystem exchange
wheaton information system for education
woman in space earliest
WISHA
washington industrial safety and health act
WISL
westinghouse information systems laboratory
WISTCI
wisconsin university, theoretical chemistry institute
WIT
wire in tube
WITNESS
wire installation tester for negating errors by sequencing and stationization
WITS
weather information telemetry system
WJ
watkins johnson
WJCC
western joint computer conference
WKB
wentzel kramers brillouin
WKGV
working voltage
WKNL
walter kidde nuclear laboratories
WL
warning light
water line
wavelength
width to length
word length
work light
WLD
warning light driver
WLDG
welding
WLESS
wireless
WLM
working level month
WLN
wiswesser line notation
WLRD
warning light relay driver
WM
watermark
wattmeter
wavemeter
western microwave
wire mesh
word mark
working memory
WMEC
western military electronics center
WMIN
words per minute
WMK
watt per meter kelvin
WMLI
western microwave laboratories, incorporated

WMO
world meteorological organization
WMS
waste management system
world magnetic survey
WMSC
weather message switching center
WMSI
western management science institute
WMSO
wichita mountains seismological observatory
WMSS
westinghouse microscan system
WMT
waste monitor tank
WND
winding
wound
WNG
wiring
WNRC
washington national records center
WNT
waste neutralizer tank
WNYNRC
western new york nuclear research center reactor
WO
water in oil
wireless officer
wireless operator
write out
WOM
wireless operator mechanic
write optional memory
WOP
wireless operator
WOPAG
wirless operator and air gunner
WOPTR
wirless operator
WORC
washington operations research council
WORCONNECTED
wired-or connected
WORDCOM
word computer
WORM
write once read multiple
WORMTECHNIQUE
write once read many time technique
WORTAC
westinghouse overall radar tester and calibrator

WOSAC
worldwide synchronizing of atomic clocks
WOTCU
waveoff and transition control unit
WOW
without whiskers
word on way
WOWM
write once write mostly
WOWS
wire obstacle warning system
WP
waterproof
weather permitting
weatherproof
word processing
word processor
work station
working pressure
workspace register pointer
worst pattern
WPA
with particular average
WPAFB
wright patterson air force base
WPC
watt per candle
watt per channel
world petroleum congress
world power conference
WPCF
water pollution control federation
WPDES
waste pollution discharge elimination system
WPI
worcester polytechnic institute
WPIS
wafer parameter identification system
WPL
wisconsin power and light
WPM
word per minute
WPPC
warning point photocell
WPPSS
washington public power supply system
WPRL
water pollution research laboratory
WPS
wisconsin public service
word per second
WPSL
western primary standard laboratory

WPU
  wet pick up
  write punch
WPWM
  wide pulse width modulator
WPX
  worked all prefixes
WQ
  water quenched
WQC
  water quality certification
WQM
  write circuit for queuing messages
WR
  weather relay broadcast center
WR
  wall receptacle
  warehouse receipt
  waveguide, rectangular
  wilson repeater
  workshop reporting
  write
WRA
  water research association
  wrinkle recovery angle
WRAC
  willow run aeronautical center
WRAIR
  walter reed army institute of research
WRBC
  weather relay broadcast center
WRC
  water resources committee
  welding research council
WRCHK
  write check
WREDAC
  weapons research establishment digital automatic computer
WRH
  world radio handbook
WRIA
  worked republic of india award
WRK
  work
WRL
  westinghouse research laboratories
  willow run laboratory
WRLS
  wireless
WRNI
  wide-range neutron indicator
  wide-range nuclear instrument
WRQ
  westinghouse resolver quantizer
WRRC
  willow run research center
WRRS
  wire relay radio system
WRU
  western reserve university
  who are you
WRUSS
  western reserve university relay searching selector
WRV
  water retention value
  winged reentry vehicle
WS
  watt second
  waveform synthesizer
  winding specification
  wireless set
  wireless station
  workshop control
WSCC
  western systems coordinating council
WSCS
  weapon system communications system
WSD
  working stress design
WSDD
  weapon status digital display
WSE
  washington society of engineers
  western society of engineers
WSEC
  washington state electronics council
  watt second
  winston salem engineers club
WSEIAC
  weapon system effectiveness industry advisory committee
WSEM
  weapon system evaluator missile
WSFC
  wallops space flight center
WSHT
  wave superheater hypersonic tunnel
WSI
  wafer scale integration
WSIT
  washington state institute of technology
WSL
  warren spring laboratory
WSMR
  white sands missile range
WSO
  weapons systems officer

WSP
water supply point
WSPACS
weapon system programming and control system
WSPG
white sands proving ground
WSR
watt per steradian
weather search radar
WSS
wide sense stationary
WSSA
white sands signal agency
WSSCA
white sands signal corps agency
WSSL
western secondary standards laboratories
WSSUS
wide sense stationary uncorrelated scattering
WST
world satellite terminal
write symbol table
WSTF
white sands test facility
WSUOPR
washington state university, open pool reactor
WT
differential temperature
watertight
watt
weight
will talk
wireless telegraphy
wireless telephony
world trade
WTB
wireless telegraphy board
write tape binary
WTC
world trade corporation
WTCI
western telecommunications inc
WTD
write tape decimal
WTDF
wireless telegraph direction finder
wireless telegraphy direction finding
WTDR
wireless telegraphy direction
WTF
waste treatment facility
WTM
wind tunnel model

WTMGE
wireless telegraphy message
WTO
wireless telegraphy officer
worked three oceans
WTP
world tape pals
WTPFT
weight per foot
WTR
western test range
westinghouse test reactor
WTRPR
waterproof
WTRPRF
waterproof
WTS
wireless telegraphy station
word terminal synchronous
world terminal synchronous
WTTELE
world trade telegraph
WU
western union
WUCU
western union computer utilities
WUF
where used file
WUI
western union international
WUIS
work unit information system
WUPS
westinghouse uninterruptible power system
WUS
western union message
WUT
western union telegraph
WUTC
western union telegraph company
WV
wave
weight in volume concentration
west virginia
wind velocity
working voltage
WVAC
working volt, alternating current
WVDC
working volt, direct current
working voltage, direct current
WVL
wavelength
WVT
water vapor transmission

WW
  weight in weight concentration
  winding to winding
  wire wound
  wire wound resistor
  wire wrap
  wire way
WWG
  warrington wire gage
WWM
  welded wire matrix
WWMCCS
  world wide military command and control system
  world wide military command control computer system
WWP
  washington water power
WWRMEMORY
  write while read memory
WWW
  world weather watch
WXTRN
  weak external reference
WY
  wyoming
WYSIWYG
  what you see is what you get

X

X
  cross
  ex
  exchange
  expansion
  experimental
  extra
  hexadecimal
  index
  index register
  x-ray tube
  x skip
  xenon
XA
  auxiliary amplifier
XAAM
  experimental air to air missile
XACT
  x automatic code translation
XCT
  execute

XDP
  x-ray density probe
XE
  experimental engine
XECF
  experimental engine cold flow
XEG
  x-ray emission gage
XGAM
  experimental guided air missile
  experimental guided aircraft missile
XHV
  extreme high vacuum
XID
  exchange identification
  xerox intel dec
XIO
  execute input output
XLL
  extra lightly loaded
XLR
  experimental liquid rocket
XM
  experimental missile
XMAS
  expandable machine accounting system
  extended mission apollo simulation
XOP
  extended operation
XOR
  exclusive or
XPL
  explain
  explanation
  explosive
XPS
  expert system
  x-ray photoelectron spectroscopy
XPT
  external page table
XR
  index register
XRA
  x-ray assistant
XRCD
  x-ray crystal density
XRM
  external rom mode
  x-ray microanalyzer
XRPM
  x-ray projection microscope
XSAM
  experimental surface to air missile
XSL
  experimental space laboratory

XSM
  experimental strategic missile
  experimental surface missile
XSONAD
  experimental sonic azimuth detector
XSPV
  experimental solid propellant vehicle
XSSM
  experimental surface to surface missile
XTASI
  exchange of technical apollo simulation information
XTC
  excess three code
XTEN
  xerox telecommunications network
XTRAN
  experimental translation language
XU
  x-unit
XUV
  extreme ultraviolet
XWAVE
  extraordinary wave

Y

Y
  y-punch
  yard
  yellow
  yoke
  yttrium
YAG
  yttrium alumina garnet
  yttrium aluminium garnet
YAIG
  yttrium aluminum iron garnet
YCTA
  yamhill county television association
YD
  yard
YEA
  yale engineering association
YEB
  yorkshire electricity board
YIG
  yttrium iron garnet
  yttrium iron garnet plate
YIL
  yellow indicating lamp
  yellow indicating light
YL
  yellow glow lamp
  yellow lamp
YLCC
  yl century certificate
YM
  your message
YP
  yield point
YR
  year
YRGB
  yellow red green blue
YS
  yield strength
YSAT
  youngston sheet and tube
YSB
  yacht safety bureau
YSF
  yield safety factor
YSLF
  yield strength load factor
YST
  yukon standard time
YTD
  year to date
YV
  yield value
YVC
  yellow varnish cambric

Z

Z
  zero
  zone
  zone code
  zuse
ZA
  zero adjusted
  zero and add
ZAR
  zeus acquisition radar
ZB
  zero based
  zero beat
ZBB
  zero base budgeting
ZCCFAR
  zero crossing constant false alarm rate
ZCD
  zone controlled deposition

ZCR
　zone of correct reading
ZCT
　zero count table
ZD
　zener diode
　zenith distance
　zero defect
ZDC
　zeus defense center
ZDCTBS
　zeus defense center tape and buffer system
ZDP
　zero delivery pressure
ZDR
　zeus discrimination radar
ZDT
　zero ductility transition
ZE
　zeros extended
ZEA
　zero energy assembly
ZEBRA
　zero energy breeder reactor assembly
ZEEP
　zero energy experimental pile
ZEL
　zero launch
ZENITH
　zero energy nitrogen heated thermal reactor
ZERLINA
　zero energy reactor for lattice investigations and study of new assemblies
ZES
　zero energy system
ZETA
　zero energy thermonuclear apparatus
ZETR
　zero energy thermal reactor
ZEUS
　zero energy uranium system
ZF
　zero frequency
ZFC
　zero failure criteria
ZFLIPFLOP
　zero flip-flop
ZFS
　zero field splitting
　zone field selection
ZG
　zero gravity
ZGE
　zero gravity effect

ZGS
　zero gradient synchroton
ZIE
　zimbabwe institution of engineers
ZIF
　zero insertion force
ZIFSOCKET
　zero insertion force socket
ZIP
　zinc impurity photodetector
ZIPCD
　zip code
ZIPCODE
　zoning improvement plan code
ZKW
　zero kilowatt
ZLL
　zero length launcher
ZM
　zero marker
　zero marker beacon
ZMA
　zinc meta arsenite
ZMAR
　zeus malfunction array radar
　zeus multifunction array radar
　zeus multiple array radar
ZN
　zone
ZNL
　zero memory non-linear
　zero memory non-linearity
ZNR
　zinc oxide non-linear resistance
ZO
　zone
　zone portion
ZODIAC
　zone defense integrated active capability
ZOE
　zero energy
ZOH
　zero order hold
ZOI
　zero order interpolar
ZOP
　zero order predictor
ZP
　zone punch
ZPA
　zero period acceleration
　zeus program analysis
ZPC
　zero point of charge
　zero print control

ZPEN
    zeus project engineer network
ZPI
    zone position indicator
ZPO
    zeus project office
ZPPR
    zero power plutonium reactor
ZPR
    zero power reactor
ZPRF
    zero power reactor facility
ZPT
    zero power test
ZS
    zero and substract
    zero state
    zero suppress
ZSF
    zero skip frequency
ZST
    zone standard time
ZSUP
    zero suppress
ZT
    zone time
ZTO
    zero time outage
ZURF
    zeus up range facility
ZVS
    zero voltage switching
ZWOK
    zirconium water oxidation kinetics